Annual Energy Outlook 2012

With Projections to 2035

June 2012

U.S. Energy Information Administration
Office of Integrated and International Energy Analysis
U.S. Department of Energy
Washington, DC 20585

This report was prepared by the U.S. Energy Information Administration (EIA), the statistical and analytical agency within the U.S. Department of Energy. By law, EIA's data, analyses, and forecasts are independent of approval by any other officer or employee of the United States Government. The views in this report therefore should not be construed as representing those of the Department of Energy or other Federal agencies.

Preface

The *Annual Energy Outlook 2012 (AEO2012)*, prepared by the U.S. Energy Information Administration (EIA), presents long-term projections of energy supply, demand, and prices through 2035, based on results from EIA's National Energy Modeling System (NEMS). EIA published an "early release" version of the *AEO2012* Reference case in January 2012.

The report begins with an "Executive summary" that highlights key aspects of the projections. It is followed by a "Legislation and regulations" section that discusses evolving legislative and regulatory issues, including a summary of recently enacted legislation and regulations, such as: the Mercury and Air Toxics Standards (MATS) issued by the U.S. Environmental Protection Agency (EPA) in December 2011 [1]; the Cross-State Air Pollution Rule (CSAPR) as finalized by the EPA in July 2011 [2]; the new fuel efficiency standards for medium- and heavy-duty vehicles published by the EPA and the National Highway Traffic Safety Administration (NHTSA) in September 2011 [3]; and regulations pertaining to the power sector in California Assembly Bill 32 (AB 32), the Global Warming Solutions Act of 2006 [4].

The "Issues in focus" section contains discussions of selected energy topics, including a discussion of the results in two cases that adopt different assumptions about the future course of existing policies: one case assumes the extension of a selected group of existing public policies—corporate average fuel economy (CAFE) standards, appliance standards, production tax credits, and the elimination of sunset provisions in existing energy policies; the other case assumes only the elimination of sunset provisions. Other discussions include: oil price and production trends in the *AEO2012*; potential efficiency improvements and their impacts on end-use energy demand; energy impacts of proposed CAFE standards for light-duty vehicles (LDVs), model years (MYs) 2017 to 2025; impacts of a breakthrough in battery vehicle technology; heavy-duty (HD) natural gas vehicles (NGVs); changing structure of the refining industry; changing environment for fuel use in electricity generation; nuclear power in *AEO2012*; potential impact of minimum pipeline throughput constraints on Alaska North Slope oil production; U.S. crude oil and natural gas resource uncertainty; and evolving Marcellus shale gas resource estimates.

The "Market trends" section summarizes the projections for energy markets. The analysis in *AEO2012* focuses primarily on a Reference case, Low and High Economic Growth cases, and Low and High Oil Price cases. Results from a number of other alternative cases also are presented, illustrating uncertainties associated with the Reference case projections for energy demand, supply, and prices. Complete tables for the five primary cases are provided in Appendixes A through C. Major results from many of the alternative cases are provided in Appendix D. Complete tables for all the alternative cases are available on EIA's website in a table browser at www.eia.gov/oiaf/aeo/tablebrowser.

AEO2012 projections are based generally on Federal, State, and local laws and regulations in effect as of the end of December 2011. The potential impacts of pending or proposed legislation, regulations, and standards (and sections of existing legislation that require implementing regulations or funds that have not been appropriated) are not reflected in the projections. In certain situations, however, where it is clear that a law or regulation will take effect shortly after the *AEO* is completed, it may be considered in the projection.

AEO2012 is published in accordance with Section 205c of the U.S. Department of Energy (DOE) Organization Act of 1977 (Public Law 95-91), which requires the EIA Administrator to prepare annual reports on trends and projections for energy use and supply.

Projections by EIA are not statements of what will happen but of what might happen, given the assumptions and methodologies used for any particular scenario. The Reference case projection is a business-as-usual trend estimate, given known technology and technological and demographic trends. EIA explores the impacts of alternative assumptions in other scenarios with different macroeconomic growth rates, world oil prices, and rates of technology progress. The main cases in *AEO2012* generally assume that current laws and regulations are maintained throughout the projections. Thus, the projections provide policy-neutral baselines that can be used to analyze policy initiatives.

While energy markets are complex, energy models are simplified representations of energy production and consumption, regulations, and producer and consumer behavior. Projections are highly dependent on the data, methodologies, model structures, and assumptions used in their development. Behavioral characteristics are indicative of real-world tendencies rather than representations of specific outcomes.

Energy market projections are subject to much uncertainty. Many of the events that shape energy markets are random and cannot be anticipated. In addition, future developments in technologies, demographics, and resources cannot be foreseen with certainty. Many key uncertainties in the *AEO2012* projections are addressed through alternative cases.

EIA has endeavored to make these projections as objective, reliable, and useful as possible; however, they should serve as an adjunct to, not a substitute for, a complete and focused analysis of public policy initiatives.

Updated *Annual Energy Outlook 2012* Reference case (June 2012)

The *Annual Energy Outlook 2012 (AEO2012)* Reference case included as part of this complete report, released in June 2012, was updated from the Reference case released as part of the *AEO2012* Early Release Overview in January 2012. The Reference case was updated to incorporate modeling changes and reflect new legislation or regulation that was not available when the Early Release Overview version of the Reference case was published. Major changes made in the Reference include:

- The Mercury and Air Toxics Standards (MATS) issued by the EPA in December 2011 was incorporated.
- The long-term macroeconomic projection was revised, based on the November 2011 long-term projection from IHS Global Insights, Inc.
- The Cross-State Air Pollution Rule (CSAPR), which was included in the Early Release Reference case, was kept in the final Reference case. In December 2011, a District Court delayed the rule from going into effect while in litigation.
- The California Low Carbon Fuel Standard (LCFS) was removed from the final Reference case, given the Federal court ruling in December 2011 that found some aspects of it to be unconstitutional.
- Historical data and equations for the transportation sector were revised to reflect revised data from NHTSA and FHWA.
- A new cement model was incorporated in the industrial sector.
- Photovoltaic capacity estimates for recent historical years (2009 and 2010) were updated to line up more closely with Solar Energy Industries Association (SEIA) and Interstate Renewable Energy Council (IREC) reports.
- Gulf of Mexico production data were revised downward to reflect data reported by the Bureau of Ocean Energy Management more closely.
- Data in the electricity model were revised to reflect 2009 electric utility financial data (electric utility plant in service, operations and maintenance costs, etc.) and refine the breakdown of associated costs between the generation, transmission, and distribution components.
- Higher capital costs for fabric filters were adopted in the analysis of MATS, based on EPA data.
- Reservoir-level oil data were updated to improve the API gravity and sulfur content data elements.
- The assumed volume of natural gas used at export liquefaction facilities was revised.

Future analyses using the *AEO2012* Reference case will start from the version of the Reference case released with this complete report.

Endnotes for Preface

Links current as of June 2012

1. U.S. Environmental Protection Agency, "Mercury and Air Toxics Standards," website www.epa.gov/mats.
2. U.S. Environmental Protection Agency, "Cross-State Air Pollution Rule (CSAPR)," website epa.gov/airtransport.
3. U.S. Environmental Protection Agency and National Highway Traffic Safety Administration, "Greenhouse Gas Emissions Standards and Fuel Efficiency Standards for Medium- and Heavy-Duty Engines and Vehicles; Final Rule," *Federal Register*, Vol. 76, No. 179 (September 15, 2011), pp. 57106-57513, website www.gpo.gov/fdsys/pkg/FR-2011-09-15/html/2011-20740.htm.
4. California Environmental Protection Agency, Air Resources Board, "Assembly Bill 32: Global Warming Solutions Act of 2006," website www.arb.ca.gov/cc/ab32/ab32.htm.

For further information . . .

The *Annual Energy Outlook 2012* was prepared by the U.S. Energy Information Administration (EIA), under the direction of John J. Conti (john.conti@eia.gov, 202/586-2222), Assistant Administrator of Energy Analysis; Paul D. Holtberg (paul.holtberg@eia.gov, 202/586-1284), Team Leader, Analysis Integration Team, Office of Integrated and International Energy Analysis; Joseph A. Beamon (joseph.beamon@eia.gov, 202/586-2025), Director, Office of Electricity, Coal, Nuclear, and Renewables Analysis; Sam A. Napolitano (sam.napolitano@eia.gov, 202/586-0687), Director, Office of Integrated and International Energy Analysis; A. Michael Schaal (michael.schaal@eia.gov, 202/586-5590), Director, Office of Petroleum, Natural Gas, and Biofuels Analysis; and James T. Turnure (james.turnure@eia.gov, 202/586-1762), Director, Office of Energy Consumption and Efficiency Analysis.

Complimentary copies are available to certain groups, such as public and academic libraries; Federal, State, local, and foreign governments; EIA survey respondents; and the media. For further information and answers to questions, contact:

> Office of Communications, EI-40
> Forrestal Building, Room 1E-210
> 1000 Independence Avenue, S.W.
> Washington, DC 20585

Telephone: 202/586-8800　　　　　　　　　Fax: 202/586-0727
(24-hour automated information line)　　　Website: www.eia.gov
E-mail: infoctr@eia.gov

Specific questions about the information in this report may be directed to:

Topic	Contact
General questions	Paul D. Holtberg (paul.holtberg@eia.gov, 202-586-1284)
National Energy Modeling System	Dan H. Skelly (daniel.skelly@eia.gov, 202-586-2222)
Executive summary	Paul D. Holtberg (paul.holtberg@eia.gov, 202/586-1284)
Economic activity	Kay A. Smith (kay.smith@eia.gov, 202/586-1132)
World oil prices	John L. Staub (john.staub@eia.gov, 202-586-6344)
International oil production	James P. O'Sullivan (james.osullivan@eia.gov, 202/586-2728)
International oil demand	Linda E. Doman (linda.doman@eia.gov, 202/586-1041)
Residential demand	Owen Comstock (owen.comstock@eia.gov, 202/586-4752)
Commercial demand	Erin E. Boedecker (erin.boedecker@eia.gov, 202/586-4791)
Industrial demand	Kelly A. Perl (kelly.perl@eia.gov, 202/586-1743)
Transportation demand	John D. Maples (john.maples@eia.gov, 202/586-1757)
Electricity generation, capacity	Jeff S. Jones (jeffrey.jones@eia.gov, 202/586-2038)
Electricity generation, emissions	Michael T. Leff (michael.leff@eia.gov, 202/586-1297)
Electricity prices	Lori B. Aniti (lori.aniti@eia.gov, 202/586-2867)
Nuclear energy	Laura K. Martin (laura.martin@eia.gov, 202/586-1494)
Renewable energy	Chris R. Namovicz (chris.namovicz@eia.gov, 202/586-7120)
Oil and natural gas production	Philip Budzik (philip.budzik@eia.gov, 202/586-2847)
Wholesale natural gas markets	Chetha Phang (chetha.phang@eia.gov, 202-586-4821)
Oil refining and markets	William S. Brown (william.brown@eia.gov, 202/586-8181)
Ethanol and biodiesel	Mac J. Statton (mac.statton@eia.gov, 202-586-7105)
Coal supply and prices	Michael L. Mellish (michael.mellish@eia.gov, 202/586-2136)
Carbon dioxide emissions	Perry Lindstrom (perry.lindstrom@eia.gov, 202/586-0934)

The *Annual Energy Outlook 2012* is available on the EIA website at www.eia.gov/forecasts/aeo. Assumptions underlying the projections, tables of regional results, and other detailed results will also be available, at www.eia.gov/forecasts/aeo/assumptions. Model documentation reports for the National Energy Modeling System are available at website www.eia.gov/analysis/model-documentation.cfm and will be updated for the *Annual Energy Outlook 2012* during 2012.

Other contributors to the report include Vipin Arora, Justine Barden, Joseph Benneche, Tina Bowers, Gwendolyn Bredehoeft, Phillip Budzik, Nicholas Chase, John Cochener, Michael Cole, Jim Diefenderfer, Robert Eynon, Laurie Falter, Mindi Farber-DeAnda, Adrian Geagla, Peter Gross, James Hewlett, Behjat Hojjati, Sean Hill, Kevin Jarzomski, Jim Joosten, Paul Kondis, Angelina LaRose, Thomas Lee, Tanc Lidderdale, Perry Lindstrom, Vishakh Mantri, Phyllis Martin, Elizabeth May, Carrie Milton, David Peterson, Chetha Phang, Marie Rinkowski-Spangler, Mark Schipper, Elizabeth Sendich, Joanne Shore, Robert Smith, Glen Sweetnam, Matthew Tanner, Russell Tarver, Dana Van Wagener, Diwakar Vashishat, Steven Wade, William Watson, and Peggy Wells.

Contents

Executive summary 1
Legislation and regulations 5
    Introduction 6
    1. Greenhouse gas emissions and fuel consumption standards for heavy-duty vehicles,
       model years 2014 through 2018 6
    2. Cross-State Air Pollution Rule 8
    3. Mercury and air toxics standards 9
    4. Updated State air emissions regulations 10
    5. California Assembly Bill 32: The Global Warming Solutions Act of 2006 10
    6. State renewable energy requirements and goals: Update through 2011 11
    7. California low carbon fuel standard 14

Issues in focus 17
    Introduction 18
    1. No Sunset and Extended Policies cases 18
    2. Oil price and production trends in *AEO2012* 23
    3. Potential efficiency improvements and their impacts on end-use energy demand 25
    4. Energy impacts of proposed CAFE standards for light-duty vehicles, model years 2017 to 2025 28
    5. Impacts of a breakthrough in battery vehicle technology 31
    6. Heavy-duty natural gas vehicles 36
    7. Changing structure of the refining industry 41
    8. Changing environment for fuel use in electricity generation 45
    9. Nuclear power in *AEO2012* 50
    10. Potential impact of minimum pipeline throughput constraints on Alaska North Slope oil production ... 52
    11. U.S. crude oil and natural gas resource uncertainty 56
    12. Evolving Marcellus shale gas resource estimates 63

Market trends 69
    Trends in economic activity 70
    Energy trends in the economy 71
    International energy 72
    U.S. energy demand 75
    Residential sector energy demand 77
    Commercial sector energy demand 79
    Industrial sector energy demand 81
    Transportation sector energy demand 84
    Electricity demand 86
    Electricity generation 87
    Electricity sales 88
    Electricity capacity 89
    Renewable capacity 90
    Natural gas prices 91
    Natural gas production 92
    Petroleum and other liquids consumption 94
    Petroleum and other liquids supply 95
    Coal production 98
    Coal production and prices 99
    Emissions from energy use 100

Comparison with other projections 103
    1. Economic growth 104
    2. Oil prices 104
    3. Total energy consumption 105
    4. Electricity 106
    5. Natural gas 110
    6. Liquid fuels 113
    7. Coal 113

List of acronyms 119
Notes and sources 120

Appendixes

- A. Reference case 131
- B. Economic growth case comparisons 173
- C. Price case comparisons 183
- D. Results from side cases 198
- E. NEMS overview and brief description of cases 215
- F. Regional Maps 231
- G. Conversion factors 239

Tables

Legislation and regulations

1. HD National Program vehicle regulatory categories 6
2. HD National Program standards for combination tractor greenhouse gas emissions and fuel consumption (assuming fully compliant engine) 7
3. HD National Program standards for vocational vehicle greenhouse gas emissions and fuel consumption (assuming fully compliant engine) 7
4. Renewable portfolio standards in the 30 States with current mandates 12

Issues in focus

5. Key analyses from "Issues in focus" in recent *AEOs* 18
6. Key assumptions for the residential sector in the *AEO2012* integrated demand technology cases ... 27
7. Key assumptions for the commercial sector in the *AEO2012* integrated demand technology cases ... 27
8. Estimated[a] average fuel economy and greenhouse gas emissions standards proposed for light-duty vehicles, model years 2017-2025 29
9. Vehicle types that do not rely solely on a gasoline internal combustion engine for motive and accessory power 30
10. Description of battery-powered electric vehicles 32
11. Comparison of operating and incremental costs of battery electric vehicles and conventional gasoline vehicles 33
12. Summary of key results from the Reference, High Nuclear, and Low Nuclear cases, 2010-2035 53
13. Alaska North Slope wells completed during 2010 in selected oil fields 54
14. Unproved technically recoverable resource assumptions by basin 57
15. Attributes of unproved technically recoverable resources for selected shale gas plays as of January 1, 2010 58
16. Attributes of unproved technically recoverable tight oil resources as of January 1, 2010 58
17. Estimated ultimate recovery for selected shale gas plays in three *AEOs* (billion cubic feet per well) ... 59
18. Petroleum supply, consumption, and prices in four cases, 2020 and 2035 60
19. Natural gas prices, supply, and consumption in four cases, 2020 and 2035 62
20. Marcellus unproved technically recoverable resources in *AEO2012* (as of January 1, 2010) 64
21. Marcellus unproved technically recoverable resources: *AEO2011*, USGS 2011, and *AEO2012* 64

Comparison with other projections

22. Projections of average annual economic growth, 2010-2035 104
23. Projections of oil prices, 2015-2035 (2010 dollars per barrel) 105
24. Projections of energy consumption by sector, 2010-2035 (quadrillion Btu) 106
25. Comparison of electricity projections, 2015, 2025, and 2035 (billion kilowatthours, except where noted) ... 108
26. Comparison of natural gas projections, 2015, 2025, and 2035 (trillion cubic feet, except where noted) ... 111
27. Comparison of liquids projections, 2015, 2025, and 2035 (million barrels per day, except where noted) ... 113
28. Comparison of coal projections, 2015, 2025, 2030, and 2035 (million short tons, except where noted) ... 115

Figures

Executive summary
1. Energy use per capita and per dollar of gross domestic product, 1980-2035 (index, 1980=1) 2
2. U.S. production of tight oil in four cases, 2000-2035 (million barrels per day) 2
3. Total U.S. petroleum and other liquids production, consumption, and net imports, 1970-2035 (million barrels per day) 3
4. Total U.S. natural gas production, consumption, and net imports, 1990-2035 (trillion cubic feet) 3
5. Cumulative retirements of coal-fired generating capacity, 2011-2035 (gigawatts) 4
6. U.S. energy-related carbon dioxide emissions by sector and fuel, 2005 and 2035 (million metric tons) 4

Legislation and regulations
7. HD National Program model year standards for diesel pickup and van greenhouse gas emissions and fuel consumption, 2014-2018 8
8. HD National Program model year standards for gasoline pickup and van greenhouse gas emissions and fuel consumption, 2014-2018 8
9. States covered by CSAPR limits on emissions of sulfur dioxide and nitrogen oxides 9
10. Total combined requirement for State renewable portfolio standards, 2015-2035 (billion kilowatthours) 11

Issues in focus
11. Total energy consumption in three cases, 2005-2035 (quadrillion Btu) 20
12. Consumption of petroleum and other liquids for transportation in three cases, 2005-2035 (million barrels per day) 21
13. Renewable electricity generation in three cases, 2005-2035 (billion kilowatthours) 21
14. Electricity generation from natural gas in three cases, 2005-2035 (billion kilowatthours) 22
15. Energy-related carbon dioxide emissions in three cases, 2005-2035 (million metric tons) 22
16. Natural gas wellhead prices in three cases, 2005-2035 (2010 dollars per thousand cubic feet) 23
17. Average electricity prices in three cases, 2005-2035 (2010 cents per kilowatthour) 23
18. Average annual world oil prices in three cases, 1980-2035 (2010 dollars per barrel) 24
19. World petroleum and other liquids production in the Reference case, 2000-2035 (million barrels per day) 24
20. Residential and commercial delivered energy consumption in four cases, 2010-2035 (quadrillion Btu) 25
21. Cumulative reductions in residential energy consumption relative to the 2011 Demand Technology case, 2011-2035 (quadrillion Btu) 26
22. Cumulative reductions in commercial energy consumption relative to the 2011 Demand Technology case, 2011-2035 (quadrillion Btu) 28
23. Light-duty vehicle market shares by technology type in two cases, model year 2025 (percent of all light-duty vehicle sales) 30
24. On-road fuel economy of the light-duty vehicle stock in two cases, 2005-2035 (miles per gallon) 30
25. Total transportation consumption of petroleum and other liquids in two cases, 2005-2035 (million barrels per day) 31
26. Total carbon dioxide emissions from transportation energy use in two cases, 2005-2035 (million metric tons carbon dioxide equivalent) 31
27. Cost of electric vehicle battery storage to consumers in two cases, 2012-2035 (2010 dollars per kilowatthour) 33
28. Costs of electric drivetrain nonbattery systems to consumers in two cases, 2012-2035 (2010 dollars) 33
29. Total prices to consumers for compact passenger cars in two cases, 2015 and 2035 (thousand 2010 dollars) 34
30. Total prices to consumers for small sport utility vehicles in two cases, 2015 and 2035 (thousand 2010 dollars) 34
31. Sales of new light-duty vehicles in two cases, 2015 and 2035 (thousand vehicles) 34
32. Consumption of petroleum and other liquids, electricity, and total energy by light-duty vehicles in two cases, 2000-2035 (quadrillion Btu) 35
33. Energy-related carbon dioxide emissions from light-duty vehicles in two cases, 2005-2035 (million metric tons carbon dioxide equivalent) 35
34. U.S. spot market prices for crude oil and natural gas, 1997-2012 (2010 dollars per million Btu) 36
35. Distribution of annual vehicle-miles traveled by light-medium (Class 3) and heavy (Class 7 and 8) heavy-duty vehicles, 2002 (thousand miles) 39
36. Diesel and natural gas transportation fuel prices in the HDV Reference case, 2005-2035 (2010 dollars per diesel gallon equivalent) 40
37. Annual sales of new heavy-duty natural gas vehicles in two cases, 2008-2035 (thousand vehicles) 40
38. Natural gas fuel use by heavy-duty vehicles in two cases, 2008-2035 (trillion cubic feet) 40

39. Reduction in petroleum and other liquid fuels use by heavy-duty vehicles in the HD NGV Potential case compared with the HDV Reference case, 2010-2035 (thousand barrels per day) 41
40. Diesel and natural gas transportation fuel prices in two cases, 2035 (2010 dollars per diesel gallon equivalent) 41
41. U.S. liquid fuels production industry 42
42. Mass-based overview of the U.S. liquid fuels production industry in the LFMM case, 2000, 2011, and 2035 (billion tons per year) 43
43. New regional format for EIA's Liquid Fuels Market Module (LFMM) 43
44. RFS mandated consumption of renewable fuels, 2009-2022 (billion gallons per year) 44
45. Natural gas delivered prices to the electric power sector in three cases, 2010-2035 (2010 dollars per million Btu) 47
46. U.S. electricity demand in three cases, 2010-2035 (trillion kilowatthours) 47
47. Cumulative retirements of coal-fired generating capacity by Electric Market Module region in nine cases, 2011-2035 (gigawatts) 48
48. Electricity generation by fuel in eleven cases, 2010 and 2020 (trillion kilowatthours) 49
49. Electricity generation by fuel in eleven cases, 2010 and 2035 (trillion kilowatthours) 49
50. Cumulative retrofits of generating capacity with FGD and dry sorbent injection for emissions control, 2011-2020 (gigawatts) 49
51. Nuclear power plant retirements by NERC region in the Low Nuclear case, 2010-2035 (gigawatts) 52
52. Alaska North Slope oil production in three cases, 2010-2035 (million barrels per day) 55
53. Alaska North Slope wellhead oil revenue in three cases, assuming no minimum revenue requirement, 2010-2035 (billion 2010 dollars per year) 55
54. Average production profiles for shale gas wells in major U.S. shale plays by years of operation (million cubic feet per year) 59
55. U.S. production of tight oil in four cases, 2000-2035 (million barrels per day) 61
56. U.S. production of shale gas in four cases, 2000-2035 (trillion cubic feet) 61
57. United States Geological Survey Marcellus Assessment Units 63

Market trends

58. Average annual growth rates of real GDP, labor force, and nonfarm labor productivity in three cases, 2010-2035 (percent per year) 70
59. Average annual growth rates over 5 years following troughs of U.S. recessions in 1975, 1982, 1991, and 2008 (percent per year) 70
60. Average annual growth rates for real output and its major components in three cases, 2010-2035 (percent per year) 70
61. Sectoral composition of industrial output growth rates in three cases, 2010-2035 (percent per year) 71
62. Energy end-use expenditures as a share of gross domestic product, 1970-2035 (nominal expenditures as percent of nominal GDP) 71
63. Energy end-use expenditures as a share of gross output, 1987-2035 (nominal expenditures as percent of nominal gross output) 71
64. Average annual oil prices in three cases, 1980-2035 (2010 dollars per barrel) 72
65. World petroleum and other liquids supply and demand by region in three cases, 2010 and 2035 (million barrels per day) 72
66. Total world production of nonpetroleum liquids, bitumen, and extra-heavy oil in three cases, 2010 and 2035 (million barrels per day) 73
67. North American natural gas trade, 2010-2035 (trillion cubic feet) 73
68. World energy consumption by region, 1990-2035 (quadrillion Btu) 74
69. Installed nuclear capacity in OECD and non-OECD countries, 2010 and 2035 (gigawatts) 74
70. World renewable electricity generation by source, excluding hydropower, 2005-2035 (billion kilowatthours) 75
71. Energy use per capita and per dollar of gross domestic product, 1980-2035 (index, 1980 = 1) 75
72. Primary energy use by end-use sector, 2010-2035 (quadrillion Btu) 76
73. Primary energy use by fuel, 1980-2035 (quadrillion Btu) 76
74. Residential delivered energy intensity in four cases, 2005-2035 (index, 2005 = 1) 77
75. Change in residential electricity consumption for selected end uses in the Reference case, 2010-2035 (kilowatthours per household) 77
76. Ratio of residential delivered energy consumption for selected end uses (ratio, 2035 to 2010) 78
77. Residential market penetration by renewable technologies in two cases, 2010, 2020, and 2035 (percent of households) 78
78. Commercial delivered energy intensity in four cases, 2005-2035 (index, 2005 = 1) 79
79. Energy intensity of selected commercial electric end uses, 2010 and 2035 (thousand Btu per square foot) 79
80. Efficiency gains for selected commercial equipment in three cases, 2035 (percent change from 2010 installed stock efficiency) 80

81. Additions to electricity generation capacity in the commercial sector in two cases, 2010-2035 (gigawatts) 80
82. Industrial delivered energy consumption by application, 2010-2035 (quadrillion Btu) 81
83. Industrial energy consumption by fuel, 2010, 2025 and 2035 (quadrillion Btu) 81
84. Cumulative growth in value of shipments from energy-intensive industries in three cases,
    2010-2035 (percent) 82
85. Change in delivered energy for energy-intensive industries in three cases, 2010-2035 (trillion Btu) 82
86. Cumulative growth in value of shipments from non-energy-intensive industries in three cases,
    2010-2035 (percent) 83
87. Change in delivered energy for non-energy-intensive industries in three cases, 2010-2035 (trillion Btu) 83
88. Delivered energy consumption for transportation by mode in two cases, 2010 and 2035 (quadrillion Btu) 84
89. Average fuel economy of new light-duty vehicles in two cases, 1980-2035 (miles per gallon) 84
90. Vehicle miles traveled per licensed driver, 1970-2035 (thousand miles) 85
91. Sales of light-duty vehicles using non-gasoline technologies by fuel type, 2010, 2020, and 2035
    (million vehicles sold) 85
92. Heavy-duty vehicle energy consumption, 1995-2035 (quadrillion Btu) 86
93. U.S. electricity demand growth, 1950-2035 (percent, 3-year moving average) 86
94. Electricity generation by fuel, 2010, 2020, and 2035 (billion kilowatthours) 87
95. Electricity generation capacity additions by fuel type, including combined heat and power,
    2011-2035 (gigawatts) 87
96. Additions to electricity generating capacity, 1985-2035 (gigawatts) 88
97. Electricity sales and power sector generating capacity, 1949-2035 (index, 1949 = 1.0) 88
98. Levelized electricity costs for new power plants, excluding subsidies, 2020 and 2035
    (2010 cents per kilowatthour) 89
99. Electricity generating capacity at U.S. nuclear power plants in three cases, 2010, 2025,
    and 2035 (gigawatts) 89
100. Nonhydropower renewable electricity generation capacity by energy source, including end-use capacity,
    2010-2035 (gigawatts) 90
101. Hydropower and other renewable electricity generation, including end-use generation, 2010-2035
    (billion kilowatthours) 90
102. Regional growth in nonhydropower renewable electricity generation, including end-use generation,
    2010-2035 (billion kilowatthours) 91
103. Annual average Henry Hub spot natural gas prices, 1990-2035 (2010 dollars per million Btu) 91
104. Ratio of low-sulfur light crude oil price to Henry Hub natural gas price on energy equivalent basis,
    1990-2035 91
105. Annual average Henry Hub spot natural gas prices in five cases, 1990-2035 (2010 dollars per million Btu) 92
106. Total U.S. natural gas production, consumption, and net imports, 1990-2035 (trillion cubic feet) 92
107. Natural gas production by source, 1990-2035 (trillion cubic feet) 93
108. Lower 48 onshore natural gas production by region, 2010 and 2035 (trillion cubic feet) 93
109. U.S. net imports of natural gas by source, 1990-2035 (trillion cubic feet) 94
110. Consumption of petroleum and other liquids by sector, 1990-2035 (million barrels per day) 94
111. U.S. production of petroleum and other liquids by source, 2010-2035 (million barrels per day) 95
112. Domestic crude oil production by source, 1990-2035 (million barrels per day) 95
113. Total U.S. crude oil production in six cases, 1990-2035 (million barrels per day) 96
114. Net import share of U.S. petroleum and other liquids consumption in three cases, 1990-2035 (percent) 96
115. EISA2007 RFS credits earned in selected years, 2010-2035 (billion credits) 97
116. U.S. ethanol use in blended gasoline and E85, 2000-2035 (billion gallons per year) 97
117. U.S. motor gasoline and diesel fuel consumption, 2000-2035 (million barrels per day) 98
118. Coal production by region, 1970-2035 (quadrillion Btu) 98
119. U.S. total coal production in six cases, 2010, 2020, and 2035 (quadrillion Btu) 99
120. Average annual minemouth coal prices by region, 1990-2035 (2010 dollars per million Btu) 99
121. Cumulative coal-fired generating capacity additions by sector in two cases, 2011-2035 (gigawatts) 100
122. U.S. energy-related carbon dioxide emissions by sector and fuel, 2005 and 2035 (million metric tons) 100
123. Sulfur dioxide emissions from electricity generation, 1990-2035 (million short tons) 101
124. Nitrogen oxide emissions from electricity generation, 1990-2035 (million short tons) 101

THIS PAGE INTENTIONALLY LEFT BLANK

Executive summary

Executive summary

The projections in the U.S. Energy Information Administration's (EIA's) *Annual Energy Outlook 2012 (AEO2012)* focus on the factors that shape the U.S. energy system over the long term. Under the assumption that current laws and regulations remain unchanged throughout the projections, the *AEO2012* Reference case provides the basis for examination and discussion of energy production, consumption, technology, and market trends and the direction they may take in the future. It also serves as a starting point for analysis of potential changes in energy policies. But *AEO2012* is not limited to the Reference case. It also includes 29 alternative cases (see Appendix E, Table E1), which explore important areas of uncertainty for markets, technologies, and policies in the U.S. energy economy. Many of the implications of the alternative cases are discussed in the "Issues in focus" section of this report.

Key results highlighted in *AEO2012* include continued modest growth in demand for energy over the next 25 years and increased domestic crude oil and natural gas production, largely driven by rising production from tight oil and shale resources. As a result, U.S. reliance on imported oil is reduced; domestic production of natural gas exceeds consumption, allowing for net exports; a growing share of U.S. electric power generation is met with natural gas and renewables; and energy-related carbon dioxide emissions remain below their 2005 level from 2010 to 2035, even in the absence of new Federal policies designed to mitigate greenhouse gas (GHG) emissions.

The rate of growth in energy use slows over the projection period, reflecting moderate population growth, an extended economic recovery, and increasing energy efficiency in end-use applications

Overall U.S. energy consumption grows at an average annual rate of 0.3 percent from 2010 through 2035 in the *AEO2012* Reference case. The U.S. does not return to the levels of energy demand growth experienced in the 20 years prior to the 2008-2009 recession, because of more moderate projected economic growth and population growth, coupled with increasing levels of energy efficiency. For some end uses, current Federal and State energy requirements and incentives play a continuing role in requiring more efficient technologies. Projected energy demand for transportation grows at an annual rate of 0.1 percent from 2010 through 2035 in the Reference case, and electricity demand grows by 0.7 percent per year, primarily as a result of rising energy consumption in the buildings sector. Energy consumption per capita declines by an average of 0.6 percent per year from 2010 to 2035 (Figure 1). The energy intensity of the U.S. economy, measured as primary energy use in British thermal units (Btu) per dollar of gross domestic product (GDP) in 2005 dollars, declines by an average of 2.1 percent per year from 2010 to 2035. New Federal and State policies could lead to further reductions in energy consumption. The potential impact of technology change and the proposed vehicle fuel efficiency standards on energy consumption are discussed in "Issues in focus."

Domestic crude oil production increases

Domestic crude oil production has increased over the past few years, reversing a decline that began in 1986. U.S. crude oil production increased from 5.0 million barrels per day in 2008 to 5.5 million barrels per day in 2010. Over the next 10 years, continued development of tight oil, in combination with the ongoing development of offshore resources in the Gulf of Mexico, pushes domestic crude oil production higher. Because the technology advances that have provided for recent increases in supply are still in the early stages of development, future U.S. crude oil production could vary significantly, depending on the outcomes of key uncertainties related to well placement and recovery rates. Those uncertainties are highlighted in this *Annual Energy Outlook*'s "Issues in focus" section, which includes an article examining impacts of uncertainty about current estimates of the crude oil and natural gas resources. The *AEO2012* projections considering variations in these variables show total U.S. crude oil production in 2035 ranging from 5.5 million barrels per day to 7.8 million barrels per day, and projections for U.S. tight oil production from eight selected plays in 2035 ranging from 0.7 million barrels per day to 2.8 million barrels per day (Figure 2).

Figure 1. Energy use per capita and per dollar of gross domestic product, 1980-2035 (index, 1980=1)

Figure 2. U.S. production of tight oil in four cases, 2000-2035 (million barrels per day)

Executive summary

With modest economic growth, increased efficiency, growing domestic production, and continued adoption of nonpetroleum liquids, net imports of petroleum and other liquids make up a smaller share of total U.S. energy consumption

U.S. dependence on imported petroleum and other liquids declines in the AEO2012 Reference case, primarily as a result of rising energy prices; growth in domestic crude oil production to more than 1 million barrels per day above 2010 levels in 2020; an increase of 1.2 million barrels per day crude oil equivalent from 2010 to 2035 in the use of biofuels, much of which is produced domestically; and slower growth of energy consumption in the transportation sector as a result of existing corporate average fuel economy standards. Proposed fuel economy standards covering vehicle model years (MY) 2017 through 2025 that are not included in the Reference case would further reduce projected need for liquid imports.

Although U.S. consumption of petroleum and other liquid fuels continues to grow through 2035 in the Reference case, the reliance on imports of petroleum and other liquids as a share of total consumption declines. Total U.S. consumption of petroleum and other liquids, including both fossil fuels and biofuels, rises from 19.2 million barrels per day in 2010 to 19.9 million barrels per day in 2035 in the Reference case. The net import share of domestic consumption, which reached 60 percent in 2005 and 2006 before falling to 49 percent in 2010, continues falling in the Reference case to 36 percent in 2035 (Figure 3). Proposed light-duty vehicles (LDV) fuel economy standards covering vehicle MY 2017 through 2025, which are not included in the Reference case, could further reduce demand for petroleum and other liquids and the need for imports, and increased supplies from U.S. tight oil deposits could also significantly decrease the need for imports, as discussed in more detail in "Issues in focus."

Natural gas production increases throughout the projection period, allowing the United States to transition from a net importer to a net exporter of natural gas

Much of the growth in natural gas production in the AEO2012 Reference case results from the application of recent technological advances and continued drilling in shale plays with high concentrations of natural gas liquids and crude oil, which have a higher value than dry natural gas in energy equivalent terms. Shale gas production increases in the Reference case from 5.0 trillion cubic feet per year in 2010 (23 percent of total U.S. dry gas production) to 13.6 trillion cubic feet per year in 2035 (49 percent of total U.S. dry gas production). As with tight oil, when looking forward to 2035, there are unresolved uncertainties surrounding the technological advances that have made shale gas production a reality. The potential impact of those uncertainties results in a range of outcomes for U.S. shale gas production from 9.7 to 20.5 trillion cubic feet per year when looking forward to 2035.

As a result of the projected growth in production, U.S. natural gas production exceeds consumption early in the next decade in the Reference case (Figure 4). The outlook reflects increased use of liquefied natural gas in markets outside North America, strong growth in domestic natural gas production, reduced pipeline imports and increased pipeline exports, and relatively low natural gas prices in the United States.

Power generation from renewables and natural gas continues to increase

In the Reference case, the natural gas share of electric power generation increases from 24 percent in 2010 to 28 percent in 2035, while the renewables share grows from 10 percent to 15 percent. In contrast, the share of generation from coal-fired power plants declines. The historical reliance on coal-fired power plants in the U.S. electric power sector has begun to wane in recent years.

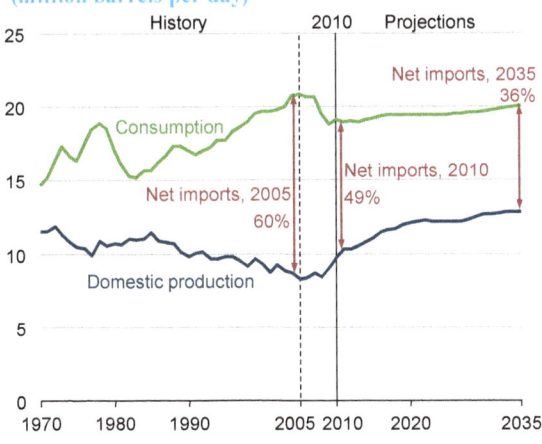

Figure 3. Total U.S. petroleum and other liquids production, consumption, and net imports, 1970-2035 (million barrels per day)

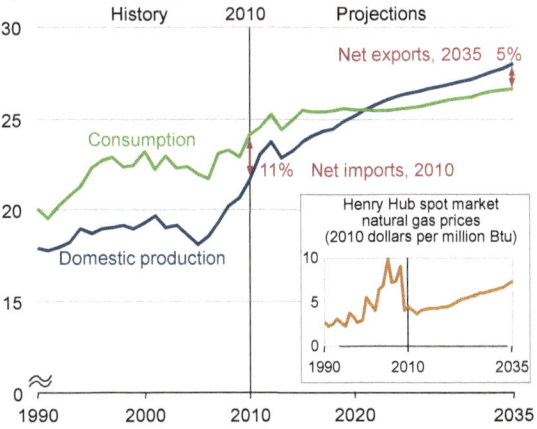

Figure 4. Total U.S. natural gas production, consumption, and net imports, 1990-2035 (trillion cubic feet)

Executive summary

Over the next 25 years, the share of electricity generation from coal falls to 38 percent, well below the 48-percent share seen as recently as 2008, due to slow growth in electricity demand, increased competition from natural gas and renewable generation, and the need to comply with new environmental regulations. Although the current trend toward increased use of natural gas and renewables appears fairly robust, there is uncertainty about the factors influencing the fuel mix for electricity generation. *AEO2012* includes several cases examining the impacts on coal-fired plant generation and retirements resulting from different paths for electricity demand growth, coal and natural gas prices, and compliance with upcoming environmental rules.

While the Reference case projects 49 gigawatts of coal-fired generation retirements over the 2011 to 2035 period, nearly all of which occurs over the next 10 years, the range for cumulative retirements of coal-fired power plants over the projection period varies considerably across the alternative cases (Figure 5), from a low of 34 gigawatts (11 percent of the coal-fired generator fleet) to a high of 70 gigawatts (22 percent of the fleet). The high end of the range is based on much lower natural gas prices than those assumed in the Reference case; the lower end of the range is based on stronger economic growth, leading to stronger growth in electricity demand and higher natural gas prices. Other alternative cases, with varying assumptions about coal prices and the length of the period over which environmental compliance costs will be recovered, but no assumption of new policies to limit GHG emissions from existing plants, also yield cumulative retirements within a range of 34 to 70 gigawatts. Retirements of coal-fired capacity exceed the high end of the range (70 gigawatts) when a significant GHG policy is assumed (for further description of the cases and results, see "Issues in focus").

Total energy-related emissions of carbon dioxide in the United States remain below their 2005 level through 2035

Energy-related carbon dioxide (CO_2) emissions grow slowly in the *AEO2012* Reference case, due to a combination of modest economic growth, growing use of renewable technologies and fuels, efficiency improvements, slow growth in electricity demand, and increased use of natural gas, which is less carbon-intensive than other fossil fuels. In the Reference case, which assumes no explicit Federal regulations to limit GHG emissions beyond vehicle GHG standards (although State programs and renewable portfolio standards are included), energy-related CO_2 emissions grow by just over 2 percent from 2010 to 2035, to a total of 5,758 million metric tons in 2035 (Figure 6). CO_2 emissions in 2020 in the Reference case are more than 9 percent below the 2005 level of 5,996 million metric tons, and they still are below the 2005 level at the end of the projection period. Emissions per capita fall by an average of 1.0 percent per year from 2005 to 2035.

Projections for CO_2 emissions are sensitive to such economic and regulatory factors due to the pervasiveness of fossil fuel use in the economy. These linkages result in a range of potential GHG emissions scenarios. In the *AEO2012* Low and High Economic Growth cases, projections for total primary energy consumption in 2035 are, respectively, 100.0 quadrillion Btu (6.4 percent below the Reference case) and 114.4 quadrillion Btu (7.0 percent above the Reference case), and projections for energy-related CO_2 emissions in 2035 are 5,356 million metric tons (7.0 percent below the Reference case) and 6,117 million metric tons (6.2 percent above the Reference case).

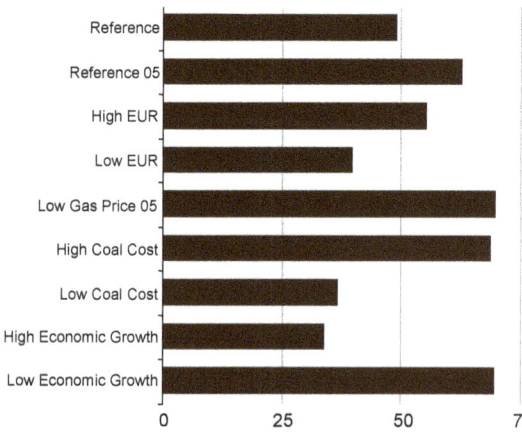

Figure 5. Cumulative retirements of coal-fired generating capacity, 2011-2035 (gigawatts)

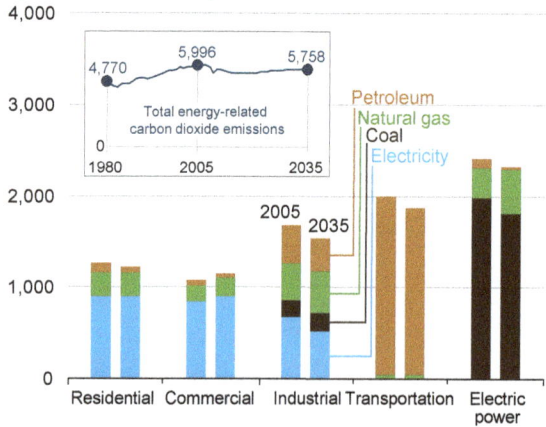

Figure 6. U.S. energy-related carbon dioxide emissions by sector and fuel, 2005 and 2035 (million metric tons)

Legislation and regulations

Legislation and regulations

Introduction

The *Annual Energy Outlook 2012 (AEO2012)* generally represents current Federal and State legislation and final implementation regulations available as of the end of December 2011. The *AEO2012* Reference case assumes that current laws and regulations affecting the energy sector are largely unchanged throughout the projection period (including the implication that laws that include sunset dates do, in fact, become ineffective at the time of those sunset dates) [5]. The potential impacts of proposed legislation, regulations, or standards—or of sections of legislation that have been enacted but require funds or implementing regulations that have not been provided or specified—are not reflected in the *AEO2012* Reference case, but some are considered in alternative cases. This section summarizes Federal and State legislation and regulations newly incorporated or updated in *AEO2012* since the completion of the *Annual Energy Outlook 2011*.

Examples of recently enacted Federal and State legislation and regulations incorporated in the *AEO2012* Reference case include:

- New greenhouse gas (GHG) emissions and fuel consumption standards for medium- and heavy-duty engines and vehicles, published by the U.S. Environmental Protection Agency (EPA) and the National Highway Transportation Safety Administration (NHTSA) in September 2011 [6]
- The Cross-State Air Pollution Rule (CSAPR), as finalized by the EPA in July 2011 [7]
- Mercury and Air Toxics Standards (MATS) rule, issued by the EPA in December 2011 [8].

There are many other pieces of legislation and regulation that appear to have some probability of being enacted in the not-too-distant future, and some laws include sunset provisions that may be extended. However, it is difficult to discern the exact forms that the final provisions of pending legislation or regulations will take, and sunset provisions may or may not be extended. Even in situations where existing legislation contains provisions to allow revision of implementing regulations, those provisions may not be exercised consistently. Many pending provisions are examined in alternative cases included in *AEO2012* or in other analyses completed by the U.S. Energy Information Administration (EIA). In addition, at the request of the Administration and Congress, EIA has regularly examined the potential implications of proposed legislation in Service Reports. Those reports can be found on the EIA website at www.eia.gov/oiaf/service_rpts.htm.

1. Greenhouse gas emissions and fuel consumption standards for heavy-duty vehicles, model years 2014 through 2018

On September 15, 2011, the EPA and NHTSA jointly announced a final rule, called the HD National Program [9], which for the first time established GHG emissions and fuel consumption standards for on-road heavy-duty trucks with a gross vehicle weight rating (GVWR) above 8,500 pounds (Classes 2b through 8) [10] and their engines. The *AEO2012* Reference case incorporates the new standards for heavy-duty vehicles (HDVs).

Due to the tremendous diversity of HDV uses, designs, and power requirements, the HD National Program separates GHG and fuel consumption standards into discrete vehicle categories within combination tractors, vocational vehicles, and heavy-duty pickups and vans (Table 1). Further, the rule recognizes that reducing GHG emissions and fuel consumption will require changes to both the engine and the body of a vehicle (to reduce the amount of work demanded by an engine). The final rule sets separate standards for the different engines used in combination tractors and vocational vehicles. *AEO2012* represents standard compliance among HDV regulatory classifications that represent the discrete vehicle categories set forth in the rule.

The HD National Program standards begin for model year (MY) 2014 vehicles and engines and are fully phased in by MY 2018. The EPA, under authority granted by the Clean Air Act, has issued GHG emissions standards that begin with MY 2014 for all engine and body categories. NHTSA, operating under regulatory timelines mandated by the Energy Independence and Security Act [11], set voluntary fuel consumption standards for MY 2014 and 2015, with the standards becoming mandatory for MY 2016 and beyond, except for diesel engine standards, which become mandatory for MY 2017 and beyond. Standards reach the most stringent levels for combination tractors and vocational vehicles in MY 2017, with subsequent standards then holding constant. Heavy-duty pickup and van standards are required to reach the highest level of stringency in MY 2018. *AEO2012* includes the HD

Table 1. HD National Program vehicle regulatory categories

Category	Description	GVWR
Combination tractors	Combination tractors are semi trucks designed to pull trailers. Standards are set separately for tractor cabs and their engines. There are no GHG or fuel consumption standards for trailers.	Class 7 and 8 (26,001 pounds and above)
Vocational vehicles	Vocational vehicles include a wide range of truck configurations, such as delivery, refuse, utility, dump, cement, fire, and tow trucks, school buses, and ambulances. The rulemaking defines vocational vehicles as all heavy-duty trucks that are not combination tractors or heavy-duty pickups or vans. Vocational vehicle standards are set separately for chassis and engines.	Class 2b through 8 (8,501 pounds and above)
Heavy-duty pickups and vans	Pickup trucks and vans are primarily 3/4-ton or 1-ton pickups used on construction sites or 12- to 15-person passenger vans.	Class 2b and 3 (8,501 to 14,000 pounds)

National Program standards beginning in MY 2014 as set by the GHG emissions portion of the rule, with standards represented by vehicle, including both the chassis and engine. AEO2012 assumes that vehicle chassis and engine manufacturers comply with the voluntary portion of the rule covering the fuel consumption standard. AEO2012 does not model the chassis and engine standards separately but allows the use of technologies to meet the HD National Program combined engine and chassis standards.

Although they are not modeled separately in AEO2012, GHG emission and fuel consumption standards for combination tractors are set for the tractor cabs and the engines used in those cabs separately in the HD National Program. Combination tractor cab standards are subdivided by GVWR (Class 7 or 8), cab type (day or sleeper), and roof type (low, mid, or high). Combination tractor engine standards are subdivided into medium heavy-duty diesel (for use in Class 7 tractors) and heavy heavy-duty diesel (for use in Class 8 tractors) (Table 2). Each tractor cab and engine combination is required to meet the GHG and fuel consumption standards for a given model year, unless they are made up by credits or other program flexibilities.

Again, although they are not modeled separately in AEO2012, GHG emission and fuel consumption standards for vocational vehicles are set separately in the HD National Program for the vehicle chassis and the engines used in the chassis. Vocational vehicle chassis standards are subdivided in the rule by GVWR (Classes 2b to 5, Classes 6 and 7, and Class 8). Vocational vehicle engine standards are subdivided into light heavy-duty diesel (for use in Classes 2b through 5), medium heavy-duty diesel (for use in Classes 6 and 7), heavy heavy-duty diesel (for use in Class 8), and spark-ignited (primarily gasoline) engines (for use in all classes) (Table 3). Each vocational vehicle chassis and engine combination is required to meet the GHG and fuel consumption standard for a given model year, unless made up by credits or other program flexibilities.

Standards for heavy-duty pickups and vans are based on the "work factor"—a weighted average of the vehicle's payload and towing capacity, adjusted for four-wheel drive capability. The standards for heavy-duty pickups and vans are different for diesel

Table 2. HD National Program standards for combination tractor greenhouse gas emissions and fuel consumption (assuming fully compliant engine)

Roof type	Day cab		Sleeper cab
	Class 7	Class 8	Class 8
2014 GHG emissions standards (grams CO_2 per ton-mile)			
Low roof	107	81	68
Mid roof	119	88	76
High roof	124	92	75
2014-2016 voluntary fuel consumption standards (gallons per 1,000 ton-miles)			
Low roof	10.5	8.0	6.7
Mid roof	11.7	8.7	7.4
High roof	12.2	9.0	7.3
2017 GHG emissions standards (grams CO_2 per ton-mile)			
Low roof	104	80	66
Mid roof	115	86	73
High roof	120	89	72
2017 fuel consumption standards (gallons per 1,000 ton-miles)			
Low roof	10.2	7.8	6.5
Mid roof	11.3	8.4	7.2
High roof	11.8	8.7	7.1

Table 3. HD National Program standards for vocational vehicle greenhouse gas emissions and fuel consumption (assuming fully compliant engine)

Standard	Light heavy-duty (Classes 2b-5)	Medium heavy-duty (Classes 6-7)	Heavy heavy-duty (Class 8)
2014 GHG emissions standard (grams CO_2 per ton-mile)	388	234	226
2016 fuel consumption standard (gallons per 1,000 ton-miles)	38.1	23.0	22.2
2017 GHG emissions standards (grams CO_2 per ton-mile)	373	225	222
2017 fuel consumption standard (gallons per 1,000 ton-miles)	36.7	22.1	21.8

and gasoline engines (Figures 7 and 8). They differ from the standards for combination tractors and vocational vehicles in that they apply to the vehicle fleet average for each manufacturer for a given model year, based on a production volume-weighted target for each model, with targets differing by work factor attribute.

The final rulemaking exempts small manufacturers of heavy-duty engines, combination tractor cabs, or vocational vehicle chassis from the GHG emissions and fuel consumption standards. Fuel consumption and GHG emissions for alternative-fuel vehicles, such as compressed natural gas vehicles, will be calculated according to their tailpipe emissions. Finally, the rulemaking contains four provisions designed to give manufacturers flexibility in meeting the GHG and fuel consumption standards. Both the EPA and NHTSA will allow for early compliance credits in MY 2013; manufacturer averaging, banking, and trading; advanced technology credits; and innovative technology credits. Those flexibility provisions are not included in the *AEO2012* Reference case.

2. Cross-State Air Pollution Rule

The CSAPR was created to regulate emissions of sulfur dioxide (SO_2) and nitrogen oxides (NO_x) from power plants greater than 25 megawatts that generate electric power from fossil fuels. CSAPR is intended to assist States in achieving their National Ambient Air Quality Standards for fine particulate matter and ground-level ozone. Limits on annual emissions of SO_2 and NO_x are designed to address fine particulate matter. The seasonal NO_x limits address ground-level ozone. Twenty-three States are subject to the annual limits, and 25 States are subject to the seasonal limits [12].

CSAPR replaces the Clean Air Interstate Rule (CAIR). CAIR is an interstate emissions cap-and-trade program for SO_2 and NO_x that would have allowed for unlimited trading among 28 eastern States. It was finalized in 2005, and requirements for emissions reductions were scheduled to begin 2009. In 2008, however, the U.S. Court of Appeals for the D.C. Circuit found that CAIR did not sufficiently meet the Clean Air Act requirements and directed the EPA to fix the flaws that it identified while CAIR remained in effect.

In July 2011, the EPA published CSAPR, with State coverage as shown in Figure 9. CSAPR consists of four individual cap-and-trade programs:

- Group 1 SO_2 covers 16 States.
- Group 2 SO_2 covers 7 States [13].
- Annual NO_x Group consists of an annual cap-and-trade program that covers all Group 1 and Group 2 SO_2 States.
- Seasonal NO_x Group covers a separate set of States, 20 of which are also in the Annual NO_x Group and 5 of which are not.

There are two SO_2 control groups, because the EPA has determined that the States in Group 1 need to meet more stringent emissions reduction requirements.

All cap-and-trade programs specified in CSAPR are included in *AEO2012*, but because the National Energy Modeling System (NEMS) does not represent electric power markets at the State level, the four group emissions caps and corresponding allowance trading could not be explicitly represented. The cap-and-trade systems for annual SO_2 and NO_x emissions are implemented for the coal demand regions by aggregating the allowance budget for each State within a region.

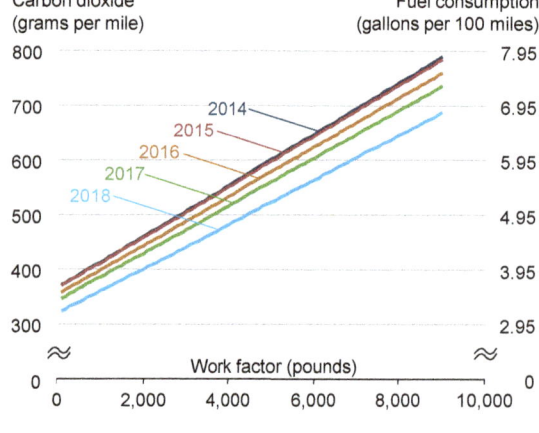

Figure 7. HD National Program model year standards for diesel pickup and van greenhouse gas emissions and fuel consumption, 2014-2018

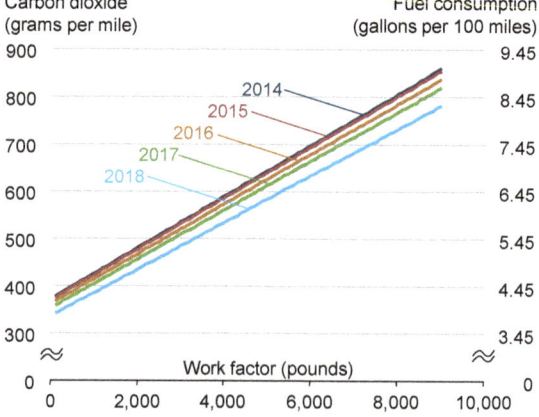

Figure 8. HD National Program model year standards for gasoline pickup and van greenhouse gas emissions and fuel consumption, 2014-2018

The EPA scheduled three annual cap-and-trade programs to commence in January 2012 and the summer season NO_x program to begin in May 2012. For three of the four programs, the initial annual cap does not change over time. For the Group 1 SO_2 program, the emissions cap across States is reduced substantially in 2014.

Emissions trading is unrestricted within a group but is not allowed across groups. Therefore, emissions allowances exist for four independent trading programs. Each State is designated an annual emissions budget, with the sum of the budgets making up the overall group emissions cap. Sources can collectively exceed State emissions budgets by close to 20 percent without any penalty. If the sources collectively exceed the State emission budget by more than the 20 percent, the sources responsible must "pay a penalty" in addition to submitting the additional allowances. The EPA set the penalties with the goal of ensuring that emissions produced by upwind States would not exceed assurance levels and contribute to air quality problems in downwind States. The emissions allowances are allocated to generating units primarily on the basis of historical energy use.

CSAPR was scheduled to begin on January 1, 2012, but the Court of Appeals issued a stay that is delaying implementation while it addresses legal challenges to the rule that have been raised by several power companies and States [14]. CSAPR is included in *AEO2012* despite the stay, because the Court of Appeals had not made a final ruling at the time *AEO2012* was completed.

3. Mercury and air toxics standards

The MATS [15] are required by Section 112 of the 1990 Clean Air Act Amendments, which requires that maximum achievable control technology be applied to power plants to control emissions of hazardous air pollutants (HAPs) [16]. The MATS rule, finalized in December 2011, regulates mercury (Hg) and other HAPs from power plants. MATS applies to Hg and hazardous acid gases, metals, and organics from coal- and oil-fired power plants with nameplate capacities greater than 25 megawatts [17]. The standards take effect in 2015.

The *AEO2012* Reference case assumes that all coal-fired generating units with capacity greater than 25 megawatts will comply with the MATS rule beginning in 2015. The MATS rule is not applied to oil-fired steam units in *AEO2012* because of their small size and limited importance. In order to comply with the MATS rule for coal, the NEMS model requires all coal-fired power plants to

Figure 9. States covered by CSAPR limits on emissions of sulfur dioxide and nitrogen oxides

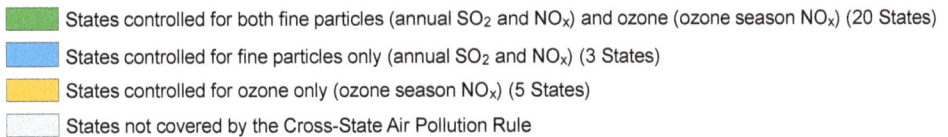

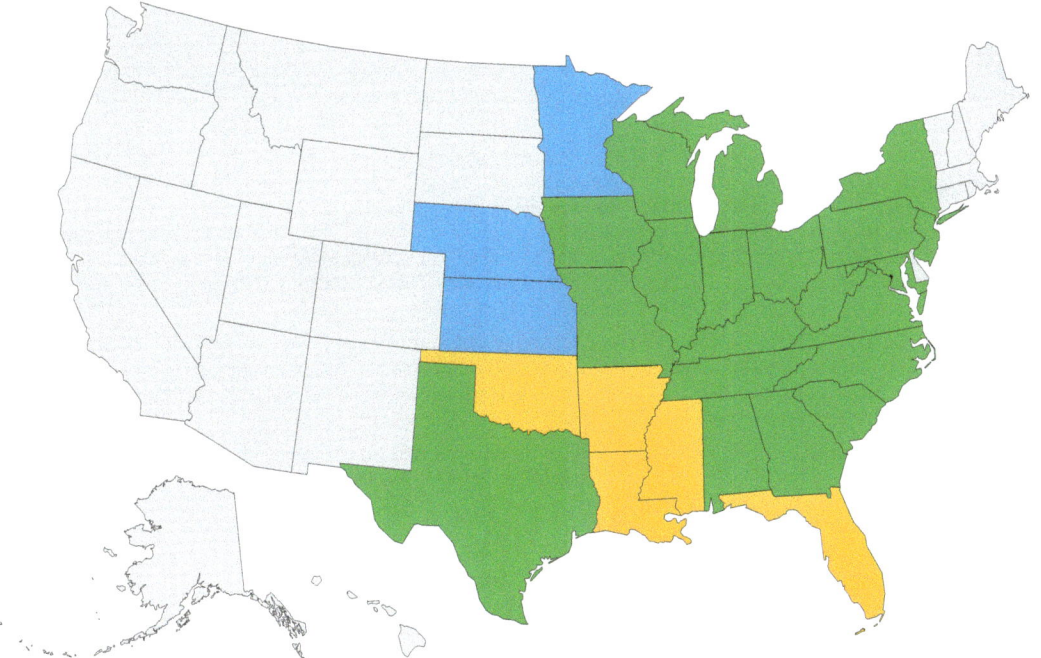

reduce Hg emissions to 90 percent below their uncontrolled emissions levels by using scrubbers and activated carbon injection controls. NEMS does not explicitly model the emissions of acid gases, toxic metals other than Hg, or organic HAPs. Therefore, in order to measure the impact of these rules, specific control technologies—either flue gas desulfurization scrubbers or dry sorbent injection systems—are assumed to be used to achieve compliance. A full fabric filter also is required to meet the limits on emissions of metals other than Hg and to improve the effectiveness of the dry sorbent injection systems. NEMS does not model the best practices associated with reductions in dioxin emissions, which also are covered by the MATS rule.

4. Updated State air emissions regulations

As its first 3-year compliance period came to a close, the Regional Greenhouse Gas Initiative (RGGI) continued to apply to fossil-fuel-fired power plants larger than 25 megawatts capacity in the northeastern United States, despite New Jersey's decision to withdraw from the program at the end of 2011. There are now nine States in the accord, which caps carbon dioxide (CO_2) emissions from covered electricity generating facilities and requires each ton of CO_2 emitted to be offset by an allowance purchased at auction. Because the program is binding, it is included in *AEO2012* as specified in the agreement.

The reduction of CO_2 emissions from the power sector in the RGGI region since 2009 is primarily a result of broader market trends. Since mid-2008, natural gas prices and electricity demand in the Northeast have fallen, while coal prices have increased. Because the RGGI baseline and projected emissions were calculated before the economic recession that began in 2008, the emissions caps are higher than actual emissions have been, leading to an excess of available allowances in recent auctions. In the past seven auctions, allowances have sold at the floor price of $1.89 per ton [18], indicating that emissions in the region are at or below the program-mandated ceiling.

As a result of the noncompetitive auctions, in which credits have not actually been traded but simply purchased at a floor price, several States have decided to retire their excess allowances permanently [19], which will result in the removal of 67 million tons of CO_2 from the RGGI emissions ceiling. Moreover, the program began a stakeholder hearing process in January 2012 that will last through the summer of 2012. The hearings, which are designed to adjust the program at the end of the first compliance period, may alter the program significantly. Because no changes have been finalized, however, modeling of the provisions in *AEO2012* is the same as in previous *Annual Energy Outlook*s.

The Western Climate Initiative is another program designed to establish a GHG emissions trading program, although the final details of the program remain undecided [20]. At the stakeholders meeting in January 2012, the commitment to emissions trading was reaffirmed. Because of the continued uncertainty over the implementation and design of the final program, it is not included in the *AEO2012* projections.

The California cap-and-trade system for GHG emissions, designed by the California Air Resources Board (CARB) in response to California Assembly Bill 32, the Global Warming Solutions Act of 2006 [21], is discussed in the following section.

5. California Assembly Bill 32: The Global Warming Solutions Act of 2006

California Assembly Bill 32 (AB 32), the Global Warming Solutions Act of 2006, authorized the CARB to set California's GHG reduction goals for 2020 and establish a comprehensive, multi-year program to reduce GHG emissions in California. As one of the major initiatives for AB 32, CARB designed a cap-and-trade program that started on January 1, 2012, with the enforceable compliance obligations beginning in 2013.

The cap-and-trade program is intended to help California achieve its goal of reducing emissions to 1990 levels by 2020. The program covers several GHGs, with the most significant being CO_2 [22]. In 2007, CARB determined that 427 million metric tons carbon dioxide equivalent ($MMTCO_2e$) was the total State-wide GHG emissions level in 1990 and, therefore, would be the 2020 emissions target. All electric power plants, large industrial facilities, suppliers of transportation fuel, and suppliers of natural gas in California are required to submit emissions allowances for each ton of CO_2 or CO_2-equivalent emissions they produce, in order to comply with the final rule [23]. Emissions resulting from electricity generated outside California but consumed in the State also are subject to the cap.

The cap-and-trade program applies to multiple economic sectors throughout the State's economy, but for *AEO2012*, due to modeling limitations, it is assumed to be implemented only in the electric power sector. *AEO2012* places limits on emissions from electric power plants and cogeneration facilities in California, as well as power plants in other States that sell power to California. The cap is set to begin in 2013 and to decline linearly to 85 percent of the 2013 value by 2020.

The enforceable cap goes into effect in 2013, and there are three compliance periods—multi-year periods for which the compliance obligation is calculated for covered entities. The first compliance period lasts for 2 years, and the second and third periods last for 3 years each, as follows:

- Compliance Period 1: 2013-2014
- Compliance Period 2: 2015-2017
- Compliance Period 3: 2018-2020.

The electricity and industrial sectors are required to comply with the cap starting in 2013. Suppliers of natural gas and transportation fuels are required to comply starting in 2015, when the second compliance period begins. For the first compliance period, covered entities are required to submit allowances for up to 30 percent of their annual emissions in each year; however, at the end of 2014 they are required to account for all the emissions for which they were responsible during the 2-year period.

Annual GHG allowance budgets for the State (i.e., emissions caps) are set by the final rule [24] as follows: for 2013, 162.8 MMTCO$_2$e; for 2014, 159.7 MMTCO$_2$e; for 2015, 394.5 MMTCO$_2$e; for 2016, 382.4 MMTCO$_2$e; for 2017, 370.4 MMTCO$_2$e; for 2018, 358.3 MMTCO$_2$e; for 2019, 346.3 MMTCO$_2$e; and for 2020, 334.2 MMTCO$_2$e.

A majority of the allowances (51 percent) [25] allocated over the initial 8 years of the program will be distributed through auctions, which will be held quarterly when the program commences. Auctions are set to begin in 2012, and the program caps will take effect in 2013. Revenue gained from the auctions is intended to be used for purposes related to AB 32, as determined by the Governor and the State Legislature.

Twenty-five percent of the allowances are allocated directly to electric utilities that sell electricity to consumers in the State. The utilities are then required to put their allowances up for auction and use the revenue generated from the auction to credit ratepayers. An exception is made for public power agencies, which will be able to keep allowances for compliance.

Seventeen percent of the allowances are allocated directly to industrial facilities covered by the rule, in order to mitigate the economic impact of the cap on the industrial sector. Over the 2013-2020 period, the number of allowances allocated annually to the industrial sector declines linearly, by a total of 50 percent.

The remaining 7 percent of the allowances issued in a given year go into a cost containment reserve and forward reserve auction. The cost containment reserve is intended to be called on only if allowance prices rise above a set amount. Each entity can also use offsets to meet up to 8 percent of its compliance obligation. Offsets used as part of the program must be approved by the CARB.

6. State renewable energy requirements and goals: Update through 2011

To the extent possible, *AEO2012* incorporates the impacts of State laws requiring the addition of renewable generation or capacity by utilities doing business in the States. Currently, 30 States and the District of Columbia have an enforceable renewable portfolio standard (RPS) or similar laws (Table 4). Under such standards, each State determines its own levels of renewable generation, eligible technologies [26], and noncompliance penalties. *AEO2012* includes the impacts of all laws in effect at the end of 2011 (with the exception of Alaska and Hawaii, because NEMS provides electricity market projections for the contiguous lower 48 States only). However, the projections do not include policies with either voluntary goals or targets that can be substantially satisfied with nonrenewable resources. In addition, the model is not able to treat fuel-specific provisions—such as those for solar and offshore wind energy—as distinct targets. Where applicable, these distinct targets (sometimes referred to as "tiers," "set-asides," or "carve-outs") may be subsumed into the broader targets, or are not modeled because they may be met with existing capacity and/or projected growth based on modeled economic and policy factors.

In the *AEO2012* Reference case, States generally are assumed to meet their ultimate RPS targets. The RPS compliance constraint in most regions is approximated, because NEMS is not a State-level model, and each State generally represents only a portion of one of the NEMS electricity regions. Compliance costs in each region are tracked, and the projection for total renewable generation is checked for consistency with any State-level cost-control provisions, such as caps on renewable credit prices, limits on State compliance funding, or impacts on consumer electricity prices. In general, EIA has confirmed the States' requirements through original documentation, although the Database of State Incentives for Renewables & Efficiency was also used to support those efforts [27].

Figure 10. Total combined requirement for State renewable portfolio standards, 2015-2035 (billion kilowatthours)

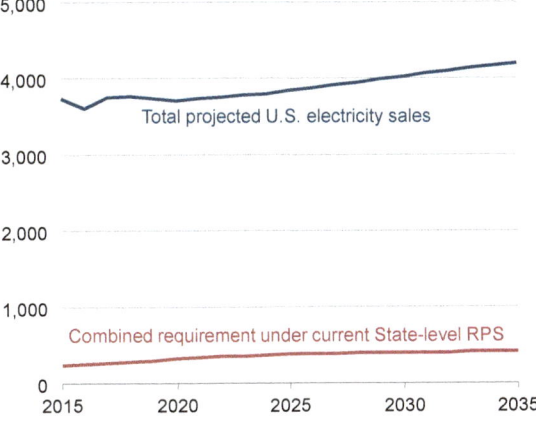

No new RPS programs were enacted over the past year; however, some States with existing RPS programs made modifications in 2011. The aggregate RPS requirement for the various State programs, as modeled in *AEO2012*, is shown in Figure 10. By 2025, these targets account for about 10 percent of U.S. sales. The requirement is derived from the legal targets and projected sales, and does not account for any discretionary or nondiscretionary waivers or limits on compliance found in most State RPS programs. State RPS policies are not the only driver of growth in renewable generation, and a more complete discussion of those factors can be found in "Market trends." The following sections detail the significant changes made by the States. In addition, Table 4 provides a summary of all State RPS laws.

Legislation and regulations

Table 4. Renewable portfolio standards in the 30 States with current mandates

State	Program mandate
AZ	Arizona Corporate Commission Decision No. 69127 requires 15 percent of electricity sales to be renewable by 2025, with interim goals increasing annually. A specific percentage of the target must be from distributed generation. Multiple credits may be provided to solar generation and systems manufactured in-State.
CA	SBX1-2, enacted in 2011, requires that 33 percent of electricity sales be met by renewable sources by 2020. The legislation codifies the 33 percent requirement in Executive Order S-21-09, which served as a continuation of California's first RPS, in which investor-owned utilities (IOUs) were required to deliver 20 percent of sales from renewable sources. Under SBX1-2, both IOUs and publicly owned municipal utilities are subject to the RPS.
CO	Enacted in March of 2010, House Bill (HB) 1001 strengthens the State's existing RPS program by requiring that 20 percent of electricity generated by IOUs in 2015 be renewable, increasing to 30 percent in 2020. There is also a distributed generation requirement. In-State generation receives a 25-percent credit premium.
CT	Public Act 07-242 mandates a 27-percent renewable sales requirement by 2020, including a 4-percent mandate for higher efficiency or combined heat and power systems. Of the overall total, 3 percent may be met by waste-to-energy and conventional biomass facilities.
DE	Senate Substitute 1 amended Senate Bill (SB) 119 to extend the increasing RPS targets to 2025; 25 percent of generation is now required to come from renewable sources in 2025. There is a separate requirement for solar generation (3.5 percent of the total in 2025), and there are penalty payments for compliance failure. Offshore wind generation receives 3.5 times the credit amount, and solar technologies receive 3 times the credit amount.
HI	HB 1464 sets the renewable mandate at 40 percent by 2030. All existing renewable facilities are eligible to meet the target, which has two interim milestones. (Not included in NEMS.)
IL	Public Act 095-0481 created an agency responsible for overseeing the mandate of 25-percent renewable sales by 2025, with escalating annual targets. In addition, 75 percent of the required sales must be generated from wind, 6 percent from solar, and 1 percent from distributed generation. The plan also includes a cap on the incremental costs resulting from the penetration of renewable generation. In 2009, the rule was modified to cover sales outside a utility's home territory.
IA	In 1983, a capacity mandate of 105 megawatts of renewable energy capacity was adopted. By the end of 2010, Iowa had well over 3,000 megawatts of wind-powered capacity alone.
KS	In 2009, HB 2369 established a requirement that 20 percent of installed capacity must use renewable resources by 2020.
ME	In 2007, Public Law 403 was added to the State's RPS requirements. The law requires that 10 percent of sales come from new renewable capacity by 2017, and that level must be maintained in subsequent years. The years leading up to 2017 also have new generation milestones. Generation from eligible community-owned facilities receives a 10-percent credit premium.
MD	In April 2008, HB 375 revised the preceding RPS to contain a 20-percent target by 2022, including a 2-percent solar target. HB 375 also raised penalty payments for "Tier 1" compliance shortfalls to 4 cents per kilowatthour. SB 277, while preserving the 2-percent by 2022 solar target, made the interim solar requirements and penalty payments slightly less stringent. In 2011, SB 717 extended the eligibility of the solar target to include solar water heating systems.
MA	The State RPS has a goal of a 15-percent renewable share of total sales by 2020 and includes necessary payments for compliance shortfalls. Eligible biomass is restricted to low-carbon life cycle emission sources. A Solar Carve-Out Program was also added, which seeks to establish 400 megawatts of solar generating capacity.
MI	Public Act 295, enacted in 2008, established an RPS that will require 10 percent of all electricity sales to be generated from renewable sources by 2015. Double credits are given to solar energy. In addition, the State's large utilities are required to procure an additional combined total of 1,100 megawatts of renewable capacity by 2015, although generation from those facilities may be counted toward the generation-based RPS.
MN	SF 4 created a 30-percent renewable requirement by 2020 for Xcel, the State's largest supplier, and a 25-percent requirement by 2025 for other suppliers. The 30-percent requirement for Xcel consists of 24 percent that must be from wind, 1 percent that can be from wind or solar, and 5 percent that can be from other resources.
MO	In November 2008, Missouri voters approved Proposition C, which mandates a 2-percent renewable energy requirement in 2011, increasing incrementally to 15 percent of generation in 2021. Bonus credits are given to renewable generation within the State.
MT	HB 681, approved in April 2007, expanded the State RPS provisions to all suppliers. Initially the law covered only regulated utilities. A 15-percent share of sales must be renewable by 2015. The State operates a renewable energy credit market.
NV	The State has an escalating renewable target, established in 1997 and most recently revised in 2009 by SB 358, which mandates a 25-percent renewable generation share of sales by 2025. Up to one-quarter of the 25-percent share may be met through efficiency measures. There is also a minimum requirement for photovoltaic systems, which receive bonus credits.

(continued on next page)

Table 4. Renewable portfolio standards in the 30 States with current mandates (continued)

State	Program mandate
NH	HB 873, passed in May 2007, legislated that 23.8 percent of electricity sales must be met by renewables in 2025. Compliance penalties vary by generation type.
NJ	In 2006, the New Jersey Board of Public Utilities revised the State RPS to increase the renewable generation target to 22.5 percent of sales by 2021, with interim targets. Assembly Bill (AB) 3520, enacted in 2010, further refines the mandate to include 5,300 gigawatthours of solar generation by 2026, with the percentage-based RPS component to reach 20.38 percent by 2021, not including the required solar generation. SB 2036 has a specific provision for offshore wind, with a goal to develop 1,100 megawatts of capacity.
NM	SB 418, passed in March 2007, directs investor-owned utilities to derive 20 percent of their sales from renewable generation by 2020. The renewable portfolio must consist of diversified technologies, with wind and solar each accounting for 20 percent of the target. There is a separate standard of 10 percent by 2020 for cooperatives.
NY	The Public Service Commission issued updated RPS rules in January 2010 that expand the program to a 30-percent requirement by 2015. There is also a separate end-use standard. The program is administered and funded by the State.
NC	In 2007, SB 3 created an RPS of 12.5 percent by 2021 for investor-owned utilities. There is also a 10-percent requirement by 2018 for cooperatives and municipals. Through 2018, 25 percent of the target may be met through efficiency standards, increasing to 40 percent in later years. Verifiable electricity demand reduction can also satisfy the RPS, with no upper limit.
OH	SB 221, passed in May 2008, requires 25 percent of electricity sales to be produced from alternative energy resources by 2025, including low-carbon and renewable technologies. One-half of the target must come from renewable sources. Municipals and cooperatives are exempt.
OR	SB 838, signed into law in June 2007, requires that renewable generation account for 25 percent of sales by 2025 for large utilities, and 5 to 10 percent of sales by 2025 for smaller utilities. Renewable electricity on line after 1995 is considered eligible.
PA	The Alternative Energy Portfolio Standard, signed into law in November 2004, has an 18-percent requirement by 2020. Most of the qualifying generation must be renewable, but there is also a provision that allows waste coal resources to receive credits.
RI	The Renewable Energy Standard was signed into law in 2004. The program requires that 16 percent of total sales be renewable by 2019. The interim program targets escalate more rapidly in later years. If the target is not met, a generator must pay an alternative compliance penalty. State utilities also must procure 90 megawatts of new renewable capacity, including 3 megawatts of solar, by 2014.
TX	SB 20, passed in August 2005, strengthened the State RPS by mandating 5,880 megawatts of renewable capacity by 2015. There is also a target of 500 megawatts of renewable capacity other than wind.
WA	In November 2006, Washington voters approved Initiative 937, which specifies that 15 percent of sales from the State's largest generators must come from renewable sources by 2020. There is an administrative penalty of 5 cents per kilowatthour for noncompliance. Generation from any otherwise qualified facility that came on line after 1999 is eligible.
WV	HB 103, passed in June 2009, established a requirement that 25 percent of electricity sales must come from alternative energy resources by 2025. Alternative energy was defined to include various renewables, along with several different fossil energy technologies.
WI	SB 459, passed in March 2006, strengthened the State RPS with a requirement that, by 2015, 10 percent of electricity sales must be generated from renewable resources, and that the renewable share of total generation must be at least 6 percentage points above the average renewable share from 2001 to 2003.

California

The State codified its RPS of 33 percent by 2020 through the passage of SBX1-2, the California Renewable Energy Resources Act [28]. The California Public Utilities Commission and California Energy Commission are the primary implementing authorities for SBX1-2, which builds on California's prior RPS mandate for 20 percent of electricity sales by 2010 [29]. SBX1-2 extends the application of the RPS to local publicly owned utilities, which had greater flexibility under the State's previous RPS mandate. SBX1-2 supersedes the 2009 Executive Order that charged the CARB with implementing the 33-percent RPS; however, CARB does retain an enforcement role over publicly owned local utilities. Because implementing regulations were not available at the time the *AEO2012* projections were being developed, the 2009 Executive Order was modeled. Although the targets specified in the two programs are similar, enforcement mechanisms may differ significantly.

Connecticut

Public Act 11-80 adds a solar-specific component to the existing RPS target, which requires that renewables should account for 27 percent of sales by 2020 [30]. The State's Clean Energy Finance and Investment Authority is tasked with creating an investment program that will result in the procurement of 30 megawatts of residential solar installations that can be counted toward the general RPS requirement.

Delaware

Delaware enacted SB 124, which extends the list of sources eligible to meet the State's RPS to include fuel cells under certain conditions [31]. Fuel cell projects that can be fueled by renewable sources and that are owned or operated by qualified providers can apply to earn renewable energy credits and, on a limited basis, solar renewable energy credits.

Illinois

With the enactment of SB 1652, the State augmented its existing RPS to include a distributed generation requirement [32]. SB 1652 requires that 1 percent of the renewable target (25 percent of sales from renewable sources by 2025 for large utilities) be fulfilled by distributed generation by mid-2015, with incremental targets beginning to take effect in 2013.

Maryland

The State enacted two pieces of legislation that allow for additional flexibility in meeting the existing RPS target of 20 percent of sales from renewable generation by 2022. SB 690 extends the designation of waste-to-energy facilities as qualifying to meet the 20-percent target beyond 2022, rather than sunsetting [33]. In addition, SB 717 specifies that solar water heating systems may also fulfill the solar set-aside requirement, which requires that solar sources account for 2 percent of electricity sales by 2022 [34].

North Carolina

North Carolina enacted SB 75, which allows reductions in electricity demand to qualify toward meeting the State's existing renewable energy and energy efficiency portfolio standard. The legislation defines electricity demand reduction as a "measureable reduction in the electricity demand of a retail electric customer that is voluntary, under the real-time control of both the electric power supplier and the retail electric customer, and measured in real time, using two-way communications devices that communicate on the basis of standards" [35]. There is no upper limit on the portion of the RPS requirement that can be met by electricity demand reduction.

7. California low carbon fuel standard

The Low Carbon Fuel Standard (LCFS), administered by the CARB [36], was signed into law in January 2010. Regulated parties under the legislation generally are the fuel producers and importers who sell motor gasoline or diesel fuel in California. The LCFS legislation is designed to reduce the carbon intensity (CI) of motor gasoline and diesel fuels sold in California by 10 percent between 2012 and 2020 through the increased sale of alternative "low-carbon" fuels. Each alternative low-carbon fuel has its own CI, based on life-cycle analyses conducted under the guidance of CARB for a number of approved fuel pathways. The CIs are calculated on an energy-equivalent basis, measured in grams of CO_2 equivalent emissions per megajoule.

In December 2011, the U.S. District Court for the Eastern Division of California ruled in favor of several trade groups that claimed the LCFS violated the interstate commerce clause of the U.S. Constitution by seeking to regulate farming and ethanol production practices in other States, and granted an injunction blocking enforcement by CARB [37]. The future of the LCFS program remains uncertain. After the initial ruling, a request for a stay of the injunction was quickly filed by CARB, which would have allowed the LCFS to remain in place during the appeal process; however, that request was denied by the same judge who initially blocked enforcement of the LCFS [38]. A new request for a stay of injunction while CARB appeals the original ruling was filed with the U.S. Ninth District Court of Appeals and was granted as of April 23, 2012 [39]. A decision on the appeal filed by CARB is yet to be made. As a result of the initial ruling's timing, along with EIA's prior completion of modeling efforts, the LCFS is not included in the *AEO2012* Reference case [40].

Endnotes for Legislation and regulations

Links current as of June 2012

5. A complete list of the laws and regulations included in *AEO2012* is provided in *Assumptions to the Annual Energy Outlook 2012*, Appendix A, website www.eia.gov/forecasts/aeo/assumptions/pdf/0554(2012).pdf.

6. U.S. Environmental Protection Agency and National Highway Traffic Safety Administration, "Greenhouse Gas Emissions Standards and Fuel Efficiency Standards for Medium- and Heavy-Duty Engines and Vehicles; Final Rule," *Federal Register*, Vol. 76, No. 179 (Washington, DC: September 15, 2011), pp. 57106-57513, website www.gpo.gov/fdsys/pkg/FR-2011-09-15/html/2011-20740.htm.

7. U.S. Environmental Protection Agency, "Cross-State Air Pollution Rule (CSAPR)," website epa.gov/airtransport.

8. U.S. Environmental Protection Agency, "Mercury and Air Toxics Standards," website www.epa.gov/mats.

9. U.S. Environmental Protection Agency and National Highway Traffic Safety Administration, "Greenhouse Gas Emissions Standards and Fuel Efficiency Standards for Medium- and Heavy-Duty Engines and Vehicles; Final Rule," *Federal Register*, Vol. 76, No. 179 (Washington, DC: September 15, 2011), website www.gpo.gov/fdsys/pkg/FR-2011-09-15/html/2011-20740.htm.

10. For purposes of this final rulemaking, heavy-duty trucks are those with a gross vehicle weight rating of at least 8,501 pounds, except those Class 2 b vehicles of 8,501 to 10,000 pounds that are currently covered under light-duty vehicle fuel economy and greenhouse gas emissions standards.

11. Congressional Research Service, *Energy Independence and Security Act of 2007: A Summary of Major Provisions*, Order Code RL34294 (Washington, DC: December 2007), website www.seco.noaa.gov/Energy/2007_Dec_21_Summary_Security_Act_2007.pdf.

12. U.S. Environmental Protection Agency, *Cross-State Air Pollution Rule: Reducing Air Pollution, Protecting Public Health* (Washington, DC: December 15, 2011), website www.epa.gov/airtransport/pdfs/CSAPRPresentation.pdf.

13. U.S. Environmental Protection Agency, *Cross-State Air Pollution Rule: Reducing Air Pollution, Protecting Public Health* (Washington, DC: December 15, 2011), Slide 3, website www.epa.gov/airtransport/pdfs/CSAPRPresentation.pdf.

14. T. Schoenberg, B. Wingfield, and J. Johnsson, "EPA Cross-State Emissions Rule Put on Hold by Court," *Bloomberg Businessweek* (January 4, 2012), website www.businessweek.com/news/2012-01-04/epa-cross-state-emissions-rule-put-on-hold-by-court.html.

15. The *AEO2012* Early Release Reference case was prepared before the final MATS rule was issued and, therefore, did not include MATS.

16. U.S. Environmental Protection Agency, "National Emission Standards for Hazardous Air Pollutants From Coal- and Oil-Fired Electric Utility Steam Generating Units and Standards of Performance for Fossil-Fuel-Fired Electric Utility, Industrial-Commercial-Institutional, and Small Industrial-Commercial-Institutional Steam Generating Units," *Federal Register*, Vol. 77, No. 32 (Washington, DC: February 16, 2012), pp. 9304-9513, website www.gpo.gov/fdsys/pkg/FR-2012-02-16/pdf/2012-806.pdf.

17. The Clean Air Act, Section 112(a)(8), defines an electric generating unit.

18. Regional Greenhouse Gas Initiative, "CO_2 Auctions, Tracking & Offsets," website www.rggi.org/market.

19. M. Navarro, "Regional Cap-and-Trade Effort Seeks Greater Impact by Cutting Carbon Allowances," *The New York Times* (January 26, 2012), website www.nytimes.com/2012/01/27/nyregion/in-greenhouse-gas-initiative-many-unsold-allowances.html?_r=2.

20. Western Climate Initiative, *WCI Emissions Trading Program Update* (San Francisco, CA: January 12, 2012), website www.westernclimateinitiative.org/document-archives/Partner-Meeting-Materials/Jan-12-Stakeholder-Update-Presentation/%20.

21. California Code of Regulations, Subchapter 10 Climate Change, Article 5, Sections 95800 to 96023, Title 17, "California Cap on Greenhouse Gas Emissions and Market-Based Compliance Mechanisms" (Sacramento, CA: July 2011), website www.arb.ca.gov/regact/2010/capandtrade10/candtmodreg.pdf.

22. California Code of Regulations, Subchapter 10 Climate Change, Article 5, Sections 95800 to 96023, Title 17, "California Cap on Greenhouse Gas Emissions and Market-Based Compliance Mechanisms" (Sacramento, CA: July 2011), website www.arb.ca.gov/regact/2010/capandtrade10/candtmodreg.pdf.

23. California Code of Regulations, Subchapter 10 Climate Change, Article 5, Section 95810, "Covered Gases" (Sacramento, CA: July 2011), website www.arb.ca.gov/regact/2010/capandtrade10/candtmodreg.pdf.

24. California Code of Regulations, Subchapter 10 Climate Change, Article 5, Section 95841, "Annual Allowance Budgets for Calendar Years 2013-2020" (Sacramento, CA: July 2011), website www.arb.ca.gov/regact/2010/capandtrade10/candtmodreg.pdf.

Legislation and regulations

25. California Air Resources Board, *Proposed Regulation to Implement the California Cap-and-Trade Program*, Appendix J, "Allowance Allocation" (Sacramento, CA: October 2010), p. 12, website www.arb.ca.gov/regact/2010/capandtrade10/capv4appj.pdf.
26. The eligible technology, and even the definition of the technology or fuel category, will vary by State. For example, one State's definition of renewables may include hydroelectric power generation, while another's definition may not. Table 4 provides more detail on how the technology or fuel category is defined by each State.
27. More information about the Database of State Incentives for Renewables & Efficiency can be found at website www.dsireusa.org/about.
28. State of California, Senate Bill 2, "California Renewable Energy Resources Act" (Sacramento, CA: April 2011), website www.leginfo.ca.gov/pub/11-12/bill/sen/sb_0001-0050/sbx1_2_bill_20110412_chaptered.html.
29. State of California, Public Utilities Code, Sections 399.11 to 399.31, website www.leginfo.ca.gov/cgi-bin/displaycode?section=puc&group=00001-01000&file=399.11-399.31.
30. State of Connecticut, Public Act 11-80, "An Act Concerning the Establishment of the Department of Energy and Environmental Protection and Planning for Connecticut's Energy Future" (Hartford, CT: July 1, 2011), website www.cga.ct.gov/2011/ACT/PA/2011PA-00080-R00SB-01243-PA.htm.
31. State of Delaware, Senate Bill 124, "An Act To Amend Title 26 Of The Delaware Code Relating To Delaware's Renewable Energy Portfolio Standards And Delaware-Manufactured Fuel Cells" (Dover, DE: July 7, 2011), website www.legis.delaware.gov/LIS/lis146.nsf/vwLegislation/SB+124/$file/legis.html?open.
32. State of Illinois, Senate Bill 1652, "An Act Concerning Public Utilities" (Springfield, IL: October 26, 2011), website www.ilga.gov/legislation/97/SB/PDF/09700SB1652lv.pdf.
33. State of Maryland, Senate Bill 690, "An Act Concerning Renewable Energy Portfolio – Waste-to-Energy and Refuse-Derived Fuel" (Annapolis, MD: May 29, 2011), website mlis.state.md.us/2011rs/bills/sb/sb0690e.pdf.
34. State of Maryland, Senate Bill 717, "An Act Concerning Renewable Energy Portfolio Standard – Renewable Energy Credits – Solar Water Heating Systems" (Annapolis, MD: May 29, 2011), website http://mlis.state.md.us/2011rs/bills/sb/sb0717e.pdf.
35. General Assembly of North Carolina, Senate Bill 75, "An Act to Promote the Use of Electricity Demand Reduction to Satisfy Renewable Energy Portfolio Standards" (Raleigh, NC: April 28, 2011), website www.ncleg.net/Sessions/2011/Bills/Senate/PDF/S75v4.pdf.
36. California Code of Regulations, Subchapter 10 Climate Change, Article 4, Sections 95480 to 95490, Title 17, Subarticle 7, "Low Carbon Fuel Standard," (Sacramento, CA: July 2011), website www.arb.ca.gov/regact/2009/lcfs09/finalfro.pdf.
37. State of California, "Low Carbon Fuel Standard (LCFS) Supplemental Regulatory Advisory 10-04B" (Sacramento, CA: December 2011), website www.arb.ca.gov/fuels/lcfs/123111lcfs-rep-adv.pdf.
38. Renewable Fuels Association, "Judge Denies California Attempt to Reimplement LCFS" (January 23, 2012), website www.ethanolrfa.org/news/entry/judge-denies-california-attempt-to-reimplement-lcfs.
39. State of California, "LCFS Enforcement Injunction is Lifted" (Sacramento, CA: April 24, 2012), website www.arb.ca.gov/fuels/lcfs/LCFS_Stay_Granted.pdf.
40. The LCFS was included in the *AEO2012* Early Release Reference case, which was completed before the ruling by the Court.

Issues in focus

Issues in focus

Introduction

The "Issues in focus" section of the *Annual Energy Outlook (AEO)* provides an in-depth discussion on topics of special interest, including significant changes in assumptions and recent developments in technologies for energy production and consumption. Detailed quantitative results are available in Appendix D. The first topic updates a discussion included in the *Annual Energy Outlook 2011 (AEO2011)* that compared the results of two cases with different assumptions about the future course of existing energy policies. One case assumes the elimination of sunset provisions in existing energy policies; that is, the policies are assumed not to sunset as they would under current law. The other case assumes the extension or expansion of a selected group of existing policies—corporate average fuel economy (CAFE) standards, appliance standards, and production tax credits (PTCs)—in addition to the elimination of sunset provisions.

Other topics discussed in this section as identified by subsection number include (2) oil price and production trends in the *Annual Energy Outlook 2012 (AEO2012)*; (3) potential efficiency improvements and their impacts on end-use energy demand; (4) energy impacts of proposed CAFE standards for light-duty vehicles (LDVs), model years (MYs) 2017 to 2025; (5) impacts of a breakthrough in battery vehicle technology; (6) heavy-duty (HD) natural gas vehicles (NGVs); (7) changing structure of the refining industry; (8) changing environment for fuel use in electricity generation; (9) nuclear power in *AEO2012*; (10) potential impact of minimum pipeline throughput constraints on Alaska North Slope oil production; (11) U.S. crude oil and natural gas resource uncertainty; and (12) evolving Marcellus shale gas resource estimates.

The topics explored in this section represent current and emerging issues in energy markets; but many of the topics discussed in *AEO*s published in recent years also remain relevant today. Table 5 provides a list of titles from the 2011, 2010, and 2009 *AEO*s that are likely to be of interest to today's readers—excluding topics that are updated in *AEO2012*. The articles listed in Table 5 can be found on the U.S. Energy Information Administration (EIA) website at www.eia.gov/analysis/reports.cfm?t=128.

1. No Sunset and Extended Policies cases

Background

The *AEO2012* Reference case is best described as a "current laws and regulations" case, because it generally assumes that existing laws and regulations will remain unchanged throughout the projection period, unless the legislation establishing them sets a sunset date or specifies how they will change. The Reference case often serves as a starting point for the analysis of proposed legislative or regulatory changes. While the definition of the Reference case is relatively straightforward, there may be considerable interest in a variety of alternative cases that reflect the updating or extension of current laws and regulations. In that regard, areas of particular interest include:

- Laws or regulations that have a history of being extended beyond their legislated sunset dates. Examples include the various tax credits for renewable fuels and technologies, which have been extended with or without modifications several times since their initial implementation.

Table 5. Key analyses from "Issues in focus" in recent *AEO*s

AEO2011	AEO2010	AEO2009
Increasing light-duty vehicle greenhouse gas and fuel economy standards for model years 2017 to 2025	Energy intensity trends in *AEO2010*	Economics of plug-in hybrid electric vehicles
Fuel consumption and greenhouse gas emissions standards for heavy-duty vehicles	Natural gas as a fuel for heavy trucks: Issues and incentives	Impact of limitations on access to oil and natural gas resources in the Federal Outer Continental Shelf
Potential efficiency improvements in alternative cases for appliance standards and building codes	Factors affecting the relationship between crude oil and natural gas prices	Expectations for oil shale production
Potential of offshore crude oil and natural gas resources	Importance of low permeability natural gas reservoirs	Bringing Alaska North Slope natural gas to market
Prospects for shale gas	U.S. nuclear power plants: Continued life or replacement after 60?	Natural gas and crude oil prices in *AEO2009*
Cost uncertainties for new electric power plants	Accounting for carbon dioxide emissions from biomass energy combustion	Greenhouse gas concerns and power sector planning
Carbon capture and storage: Economics and issues		Tax credits and renewable generation
Power sector environmental regulations on the horizon		

- Laws or regulations that call for the periodic updating of initial specifications. Examples include appliance efficiency standards issued by the U.S. Department of Energy (DOE), and CAFE and greenhouse gas (GHG) emissions standards for vehicles issued by the National Highway Traffic Safety Administration (NHTSA) and the U.S. Environmental Protection Agency (EPA).
- Laws or regulations that allow or require the appropriate regulatory agency to issue new or revised regulations under certain conditions. Examples include the numerous provisions of the Clean Air Act that require the EPA to issue or revise regulations if it finds that an environmental quality target is not being met.

To provide some insight into the sensitivity of results to scenarios in which existing tax credits do not sunset, two alternative cases are discussed in this section. No attempt is made to cover the full range of possible uncertainties in these areas, and readers should not view the cases discussed as EIA projections of how laws or regulations might or should be changed.

Analysis cases

The two cases prepared—the No Sunset and Extended Policies cases—incorporate all the assumptions from the *AEO2012* Reference case, except as identified below. Changes from the Reference case assumptions in these cases include the following.

No Sunset case

- Extension through 2035 of the PTC for cellulosic biofuels of up to $1.01 per gallon (set to expire at the end of 2012).
- Extension of tax credits for renewable energy sources in the utility, industrial, and buildings sectors or for energy-efficient equipment in the buildings sector, including:
  - The PTC of 2.2 cents per kilowatthour or the 30-percent investment tax credit (ITC) available for wind, geothermal, biomass, hydroelectric, and landfill gas resources, currently set to expire at the end of 2012 for wind and 2013 for the other eligible resources, are assumed to be extended indefinitely.
  - For solar power investment, a 30-percent ITC that is scheduled to revert to a 10-percent credit in 2016 is, instead, assumed to be extended indefinitely at 30 percent.
  - In the buildings sector, tax credits for the purchase of energy-efficient equipment, including photovoltaics (PV) in new houses, are assumed to be extended indefinitely, as opposed to ending in 2011 or 2016 as prescribed by current law. The business ITCs for commercial-sector generation technologies and geothermal heat pumps are assumed to be extended indefinitely, as opposed to expiring in 2016; and the business ITC for solar systems is assumed to remain at 30 percent instead of reverting to 10 percent.
  - In the industrial sector, the ITC for combined heat and power (CHP) that ends in 2016 in the *AEO2012* Reference case is assumed to be preserved through 2035, the end of the projection period.

Extended Policies case

The Extended Policies case includes additional updates in Federal equipment efficiency standards that were not considered in the Reference case or No Sunset case. Residential end-use technologies subject to updated standards are not eligible for tax credits in addition to the standards. Also, the PTC for cellulosic biofuels beyond 2012 is not included because the renewable fuel standard (RFS) program that is already included in the *AEO2012* Reference case tends to be the binding driver of cellulosic biofuels use. Other than these exceptions, the Extended Policies case adopts the same assumptions as the No Sunset case, plus the following:

- Federal equipment efficiency standards are updated at periodic intervals, consistent with the provisions in the existing law, with the levels based on ENERGY STAR specifications, or Federal Energy Management Program (FEMP) purchasing guidelines for Federal agencies. Standards are also introduced for products that are not currently subject to Federal efficiency standards.
- Updated Federal residential and commercial building energy codes reach 30-percent improvement in 2020 relative to the 2006 International Energy Conservation Code in the residential sector and the American Society of Heating, Refrigerating and Air-Conditioning Engineers Building Energy Code 90.1-2004 in the commercial sector. Two subsequent rounds in 2023 and 2026 each add an assumed 5-percent incremental improvement to building energy codes.

  The equipment standards and building codes assumed for the Extended Policies case are meant to illustrate the potential effects of these policies on energy consumption for buildings. No cost-benefit analysis or evaluation of impacts on consumer welfare was completed in developing the assumptions. Likewise, no technical feasibility analysis was conducted, although standards were not allowed to exceed "maximum technologically feasible" levels described in DOE's technical support documents.

- The *AEO2012* Reference, No Sunset, and Extended Policies cases include both the attribute-based CAFE standards for LDVs for MY 2011 and the joint attribute-based CAFE and vehicle GHG emissions standards for MY 2012 to MY 2016. However, the Reference and No Sunset cases assume that LDV CAFE standards increase to 35 miles per gallon (mpg) by MY 2020, as called for in the Energy Independence and Security Act of 2007 (EISA2007), and that the CAFE standards are then held constant in subsequent model years, although the fuel economy of new LDVs continues to rise modestly over time.

  The Extended Policies case modifies the assumption in the Reference and No Sunset cases by assuming the incorporation of the proposed CAFE standards recently announced by the EPA and NHTSA for MY 2017 through MY 2025, which call for an

Issues in focus

annual average increase in fuel economy for new LDVs of 3.9 percent. After 2025, CAFE standards are assumed to increase at an average annual rate of 1.5 percent through 2035.

- In the industrial sector, the ITC for CHP is extended to cover all system sizes (limited to only capacities between 25 and 50 megawatts in the Reference case), which may include multiple units. Also, the ITC is modified to increase the eligible CHP unit cap from 15 megawatts to 25 megawatts. These extensions are consistent with previously proposed or pending legislation.

Analysis results

The changes made to Reference case assumptions in the No Sunset and Extended Policies cases generally lead to lower estimates for overall energy consumption, increased use of renewable fuels, particularly for electricity generation, and reduced energy-related emissions of carbon dioxide (CO_2). Because the Extended Policies case includes most of the assumptions in the No Sunset case but adds others, the impacts in the Extended Policies case tend to be greater than those in the No Sunset case. Although these cases show lower energy prices—because the tax credits and end-use efficiency standards lead to lower energy demand and reduce the cost of renewable fuels—consumers spend more on appliances that are more efficient in order to comply with the tighter appliance standards, and the Government receives lower tax revenues as consumers and businesses take advantage of the tax credits.

Energy consumption

Total energy consumption in the No Sunset case is close to the level in the Reference case (Figure 11). Improvements in energy efficiency lead to reduced consumption in this case, but somewhat lower energy prices lead to higher relative consumption, offsetting some of the impact of the improved efficiency.

Total energy consumption growth in the Extended Policies case is markedly below the Reference case projection. In 2035, total energy consumption in the Extended Policies case is nearly 6 percent below its projected level in the Reference case.

Buildings energy consumption

The No Sunset case extends tax credits for residential and commercial renewable energy systems and for the purchase of energy-efficient residential equipment. The Extended Policies case builds on the No Sunset case by assuming updated Federal equipment efficiency standards and new standards for some products that are not currently subject to standards. For residential end-use technologies subject to standards, updated standards are assumed to replace any extension of incentives from the No Sunset case. Federal residential and commercial building energy codes are also improved as described above. Renewable distributed generation (DG) technologies (PV systems and wind turbines) provide much of the buildings-related energy savings in the No Sunset case. Extended tax credits in the No Sunset case spur increased adoption of renewable DG systems, leading to 110 billion kilowatthours of onsite electricity generation in 2035—more than four times the amount of onsite electricity generated in 2035 in the Reference case. Similar adoption of renewable DG takes place in the Extended Policies case. With the additional efficiency gains from assumed future standards and more stringent building codes, delivered energy consumption for buildings in 2035 is 6.8 percent (1.5 quadrillion Btu) lower in the Extended Policies case than in the Reference case, a reduction nearly five times as large as the 1.4-percent (0.3 quadrillion Btu) reduction in the No Sunset case.

Electricity use shows the largest reduction relative to the Reference case, with buildings electricity consumption 2.4 percent and 8.2 percent lower, respectively, in the No Sunset and Extended Policies cases in 2035. Space heating and cooling are affected by both assumed standards and building codes, leading to significant savings in energy consumption for heating and cooling in the Extended Policies case. In 2035, energy use for space heating in buildings is 6.9 percent lower, and energy use for space cooling is 17.3 percent lower, in the Extended Policies case than in the Reference case. In addition to improved standards and codes, extended tax credits for PV prompt increased adoption, offsetting some of the purchased electricity for cooling. New standards for televisions and for personal computers (PCs) and related equipment in the Extended Policies case lead to savings of 20.6 percent and 18.2 percent, respectively, in residential electricity use by this equipment in 2035 relative to the Reference case. Residential and commercial natural gas use declines from 8.3 quadrillion Btu in 2010 to 7.9 quadrillion Btu in 2035 in the Extended Policies case, representing a 6.2-percent reduction from the Reference case in 2035.

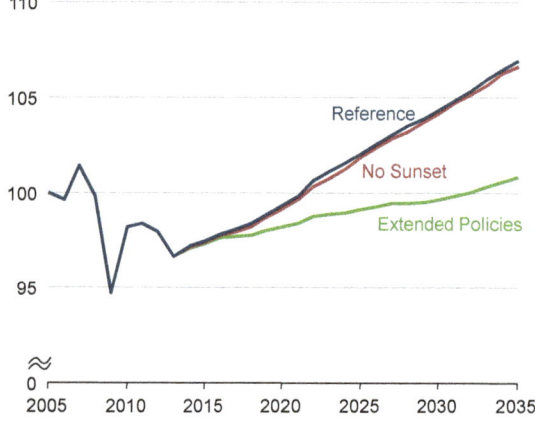

Figure 11. Total energy consumption in three cases, 2005-2035 (quadrillion Btu)

Industrial energy consumption

The Extended Policies case modifies the Reference case by extending the existing industrial CHP ITC through the end of the projection period, expanding it to include all industrial CHP system sizes, and raising the maximum credit that can be claimed from 15 megawatts of installed capacity to 25 megawatts. These assumptions are based on the current proposals in H.R. 2750 and H.R. 2784 of the 112th Congress. The changes result in 2.7 gigawatts of additional industrial CHP capacity over the Reference case level in 2035. Natural gas consumption in the industrial sector (excluding refining) increases from 7.3 quadrillion Btu in the Reference case to 7.4 quadrillion Btu in the Extended Policies case, a 1.6-percent rise. Electricity purchases are nearly unchanged in the Extended Policies case, as additional demand for electricity relative to the Reference case is fulfilled almost exclusively by increased generation from CHP.

Transportation energy consumption

The Extended Policies case modifies the Reference case and No Sunset case by assuming the incorporation of the CAFE standards recently proposed by the EPA and NHTSA for MY 2017 through 2025, which call for a 3.9-percent annual average increase in fuel economy for new LDVs, with CAFE standards applicable after 2025 assumed to increase at an average annual rate of 1.5 percent through 2035. Sales of vehicles that do not rely solely on a gasoline internal combustion engine for both motive and accessory power (including those that use diesel, alternative fuels, and/or hybrid electric systems) play a substantial role in meeting the higher fuel economy standards, growing to almost 80 percent of new LDV sales in 2035, compared with about 35 percent in the Reference case.

LDV energy consumption declines in the Extended Policies case, from 16.6 quadrillion Btu (8.9 million barrels per day) in 2010 to 12.9 quadrillion Btu (7.3 million barrels per day) in 2035, about a 20-percent reduction from the Reference case in 2035. Petroleum and other liquids fuels consumption in the transportation sector declines in the Extended Policies case, from 13.8 million barrels per day in 2010 to 12.7 million barrels per day in 2035, compared to an increase in the Reference case to 14.4 million barrels per day (Figure 12).

Renewable electricity generation

The extension of tax credits for renewables through 2035 would, over the long run, lead to more rapid growth in renewable generation than in the Reference case. When the renewable tax credits are extended without extending energy efficiency standards, as is assumed in the No Sunset case, there is a significant increase in renewable generation in 2035 relative to the Reference case (Figure 13). Extending both renewable tax credits and energy efficiency standards (Extended Policies case) results in more modest growth in renewable generation, because renewable generation in the near term is a significant source of new generation to meet load growth, and enhanced energy efficiency standards tend to reduce overall electricity consumption and the need for new generation resources.

In the No Sunset and Extended Policies cases, renewable generation more than doubles from 2010 to 2035, as compared with a 77-percent increase in the Reference case. In 2035, the share of total electricity generation accounted for by renewables is between 19 and 20 percent in both the No Sunset and Extended Policies cases, as compared with 15 percent in the Reference case.

In all three cases, the most rapid growth in renewable capacity occurs in the very near term, largely as the result of projects already under construction or planned. After that, the growth slows through 2020 before picking up again. Some of the current surge of renewable capacity additions is occurring in anticipation of the expiration of Federal incentives within the next year (for wind) or two (for other renewable fuels except solar). Results from the No Sunset and Extended Policies cases indicate that, given sufficient

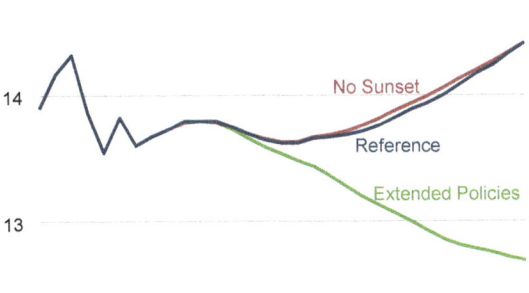

Figure 12. Consumption of petroleum and other liquids for transportation in three cases, 2005-2035 (million barrels per day)

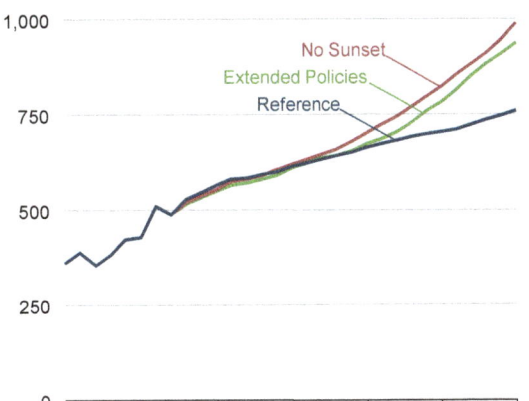

Figure 13. Renewable electricity generation in three cases, 2005-2035 (billion kilowatthours)

Issues in focus

lead time, a long-term extension of these expiring provisions could result in the postponement of some near-term activity to better match projected patterns of load growth. With slow growth in electricity demand and the addition of capacity stimulated by renewable incentives, little new capacity is needed between 2015 and 2020. In addition, in some regions, attractive low-cost renewable resources already have been developed, leaving only less favorable sites that may require significant investment in transmission as well as other additional infrastructure costs. Starting around 2020, significant new sources of renewable generation also appear on the market as a result of cogeneration at biorefineries built primarily to produce renewable liquid fuels to meet the Federal RFS, where combustion of waste products to produce electricity is an economically attractive option.

Between 2020 and 2025, renewable generation in the No Sunset and Extended Policies cases starts to increase more rapidly than in the Reference case, and, as a result, generation from nuclear and fossil fuels is reduced from the levels in the Reference case. Natural gas represents the largest source of displaced generation. In 2035, electricity generation from natural gas is 11 percent lower in the No Sunset case and 15 percent lower in the Extended Policies case than in the Reference case (Figure 14).

Energy-related CO_2 emissions

In the No Sunset and Extended Policies cases, lower overall energy demand leads to lower levels of energy-related CO_2 emissions than in the Reference case. The Extended Policies case shows much larger emissions reductions than the No Sunset and Reference cases, due in part to the inclusion of tighter LDV fuel economy standards for MY 2017 through MY 2035. From 2010 to 2035, energy-related CO_2 emissions are reduced by a cumulative total of 4.3 billion metric tons (a 3.0-percent reduction over the period) in the Extended Policies case from the Reference case projection, as compared with 0.9 billion metric tons (a 0.6-percent reduction over the period) in the No Sunset case (Figure 15). The increase in fuel economy standards assumed for new LDVs in the Extended Policies case is responsible for more than 40 percent of the total reduction in CO_2 emissions in 2035 in comparison with the Reference case. The balance of the reduction in CO_2 emissions is a result of greater improvement in appliance efficiencies and increased penetration of renewable electricity generation.

The majority of the emissions reductions in the No Sunset case result from increases in renewable electricity generation. Consistent with current EIA conventions and EPA practice, emissions associated with the combustion of biomass for electricity generation are not counted, because they are assumed to be balanced by carbon uptake when the feedstock is grown. A small reduction in transportation sector emissions in the No Sunset case is counterbalanced by an increase in emissions from refineries during the production of synthetic fuels that receive tax credits. Relatively small incremental reductions in emissions are attributable to renewables in the Extended Policies case, mainly because electricity demand is lower than in the Reference case, reducing the consumption of all fuels used for generation, including biomass.

In the residential sector, in both the No Sunset and Extended Policies cases, water heating, space cooling, and space heating together account for most of the emissions reductions from Reference case levels. In the commercial sector, only the Extended Policies case projects substantial reductions of emissions in those categories. In the industrial sector, the Extended Policies case projects reduced emissions as a result of decreases in electricity purchases and petroleum use that are partially offset by increased reliance on natural gas—for example, increased use of natural gas fired industrial CHP.

Energy prices and tax credit payments

With lower levels of overall energy use and more consumption of renewable fuels in the No Sunset and Extended Policies cases, energy prices are lower than in the Reference case. In 2035, natural gas wellhead prices are $0.44 per thousand cubic feet (6.6

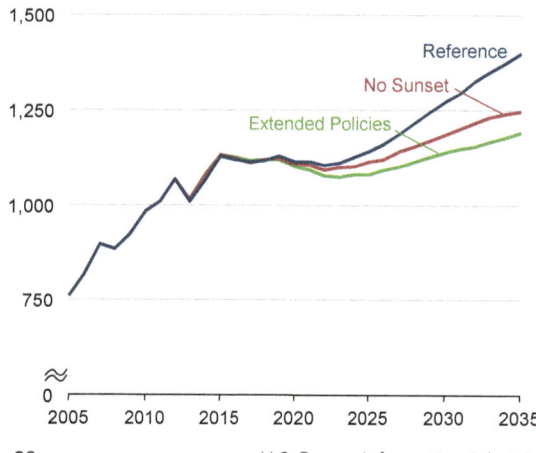

Figure 14. Electricity generation from natural gas in three cases, 2005-2035 (billion kilowatthours)

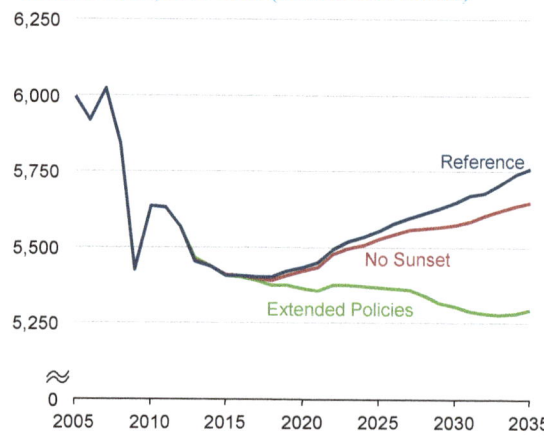

Figure 15. Energy-related carbon dioxide emissions in three cases, 2005-2035 (million metric tons)

percent) and $0.82 per thousand cubic feet (12.3 percent) lower in the No Sunset and Extended Policies cases, respectively, than in the Reference case (Figure 16), and electricity prices are about 2 percent and 5 percent lower than in the Reference case (Figure 17).

The reductions in energy consumption and CO_2 emissions in the Extended Policies case are accompanied by higher equipment costs for consumers and revenue reductions for the U.S. Government. From 2012 to 2035, residential and commercial consumers spend, on average, an additional $19 billion per year (in 2010 dollars) for newly purchased end-use equipment, distributed generation systems, and residential building shell improvements in the Extended Policies case as compared with the Reference case. On the other hand, they save an average of $22 billion per year on energy purchases.

Tax credits paid to consumers in the buildings sector (or, from the Government's perspective, reduced revenue) in the No Sunset case average $5 billion (real 2010 dollars) more per year than in the Reference case, which assumes that existing tax credits expire as currently scheduled, mostly by 2016.

The largest response to Federal tax incentives for new renewable generation is seen in the No Sunset case, with extension of the PTC and the 30-percent ITC resulting in annual average reductions in Government tax revenues of approximately $2.5 billion from 2011 to 2035, as compared with $520 million per year in the Reference case. Additional reductions in Government tax revenue in the No Sunset case result from extensions of the cellulosic biofuels PTC. These reductions increase rapidly from $52 million in 2013 to $7.2 billion (2010 dollars) in 2035 (a cumulative total of $75.1 billion) in comparison with the Reference case.

2. Oil price and production trends in *AEO2012*

The oil price in *AEO2012* is defined as the average price of light, low-sulfur crude oil delivered in Cushing, Oklahoma, which is similar to the price for light, sweet crude oil, West Texas Intermediate (WTI), traded on the New York Mercantile Exchange. *AEO2012* also includes a projection of the U.S. annual average refiners' acquisition cost of imported crude oil, which is more representative of the average cost of all crude oils used by domestic refiners. Currently there is a price differential between WTI and similar-quality marker crude oils delivered to international ports via tanker (e.g., Brent and Louisiana Light Sweet crudes). The *AEO2012* Reference case assumes that the large discrepancy will fade over time, as construction of more adequate pipeline capacity between Cushing and the Gulf of Mexico eases transportation of crude oil supplies to and from U.S. refineries.

Oil prices are influenced by a number of factors, including some that have mainly short-term impacts. Other factors, such as the Organization of the Petroleum Exporting Countries (OPEC) production decisions and expectations about future world demand for petroleum and other liquids, affect prices in the longer term. Supply and demand in the world oil market are balanced through responses to price movements, and the factors underlying supply and demand expectations are both numerous and complex. The key factors determining long-term supply, demand, and prices for petroleum and other liquids can be summarized in four broad categories: the economics of non-OPEC supply, OPEC investment and production decisions, the economics of other liquids supply, and world demand for petroleum and other liquids.

AEO2012 includes projections of future supply and demand for "petroleum and other liquids." The term "petroleum" refers to crude oil (including tight oil from shale [also referred to as shale oil], chalk, and other low-permeability formations), lease condensate, natural gas plant liquids, and refinery gain. The term "other liquids" refers to biofuels, bitumen (oil sands), coal-to-liquids (CTL), biomass-to-liquids (BTL), gas-to-liquids (GTL), extra-heavy oils (technically petroleum but grouped in "other liquids" in this report), and oil shale [41].

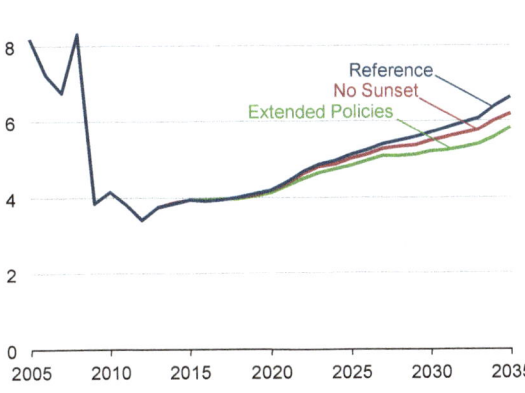

Figure 16. Natural gas wellhead prices in three cases, 2005-2035 (2010 dollars per thousand cubic feet)

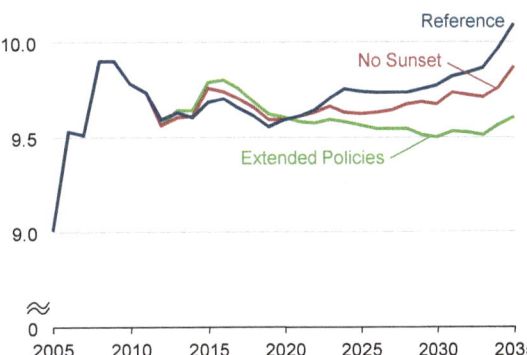

Figure 17. Average electricity prices in three cases, 2005-2035 (2010 cents per kilowatthour)

Reference case

The global oil market projections in the *AEO2012* Reference case are based on the assumption that current practices, politics, and levels of access will continue in the near to mid-term. The Reference case assumes that continued robust economic growth in the non-Organization for Economic Cooperative Development (OECD) nations, including China and India, will more than offset slower growth projected for many OECD nations. In the Reference case, non-OECD petroleum and other liquids consumption is about 21 million barrels per day higher in 2035 than it was in 2010, but OECD consumption grows by less than 2 million barrels per day over the same period. Total world consumption of petroleum and other liquids grows to 106 million barrels per day in 2030 and 110 million barrels per day in 2035.

The Reference case also assumes that limitations on access to resources in many areas restrain the growth of non-OPEC petroleum liquids production over the projection period, and that OPEC production maintains a relatively constant share of total world petroleum and other liquids supply—between 40 and 42 percent. With those constraining factors, satisfying the growing world demand for petroleum and other liquids in coming decades requires production from higher-cost resources, particularly for non-OPEC producers with technically challenging supply projects. In the Reference case, the increased cost of non-OPEC supplies, a constant OPEC market share, and easing of Cushing WTI infrastructure constraints combine to support average increases in real oil prices of about 5 percent per year from 2010 to 2020 and about 1 percent per year from 2020 to 2035. In 2035, the average real price of crude oil in the Reference case is $145 per barrel in 2010 dollars (Figure 18). The rapid increase in the near term is based on the assumption that the WTI price will return to parity with Brent by 2016 as current constraints on pipeline capacity between Cushing and the Gulf of Mexico are eliminated.

Increases in non-OPEC production of petroleum and other liquids in the Reference case come primarily from high-cost petroleum liquids projects in areas with inconsistent or unreliable fiscal or political regimes and from increasingly expensive other liquids projects that are made economical by rising oil prices and advances in production technology (Figure 19). Bitumen production in Canada and biofuels production mostly from the United States and Brazil are the most important components of the world's incremental supply of other liquids from 2010 to 2035 in the Reference case.

Low Oil Price case

In the Low Oil Price case, non-OECD economic growth is lower than in the Reference case, leading to slower growth in demand for petroleum and other liquids. Lower demand, combined with greater access to and production of petroleum liquids resources, results in sustained lower oil prices. In particular, the Low Oil Price case focuses on demand in non-OECD countries, where uncertainty about future growth is much higher than in the mature economies of the OECD. The Low Oil Price case assumes that oil prices fall steadily after 2011 to about $58 per barrel in 2017, then rise slowly to $62 per barrel in 2035. Growth in world demand for petroleum and other liquids is slowed by lower gross domestic product (GDP) growth in the non-OECD countries than is projected in the Reference case. Average annual GDP growth in the non-OECD nations is assumed to be 1.5 percentage points lower than in the Reference case, increasing by only 3.5 percent per year from 2010 to 2035. As a result, non-OECD demand for petroleum and other liquids in 2035 is 7 million barrels per day lower than in the Reference case, and total world consumption in 2035 is 2 million barrels per day lower, at 107 million barrels per day.

In the Low Oil Price case, the market power of OPEC producers is weakened, and they lose the ability to control prices and limit production. As a result, the OPEC market share of world petroleum and other liquids production is 46 percent in 2035, as

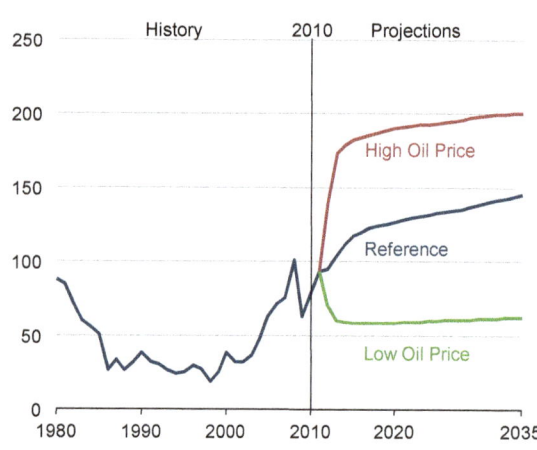

Figure 18. Average annual world oil prices in three cases, 1980-2035 (2010 dollars per barrel)

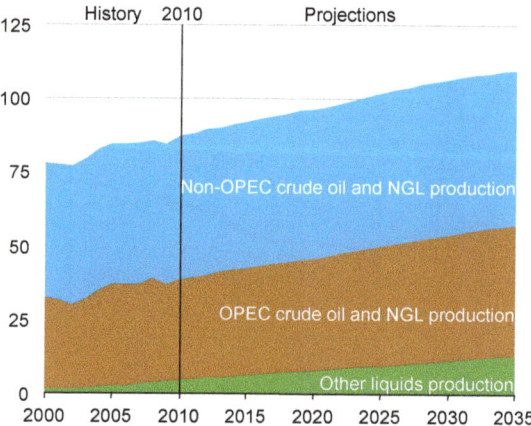

Figure 19. World petroleum and other liquids production in the Reference case, 2000-2035 (million barrels per day)

compared with 40 to 42 percent in the Reference case. Despite lower prices, non-OPEC levels of petroleum liquids production are maintained until about 2020, as projects currently underway or planned are completed and begin production. After 2020, non-OPEC petroleum liquids production declines as existing fields are depleted and not fully replaced by production from new fields and higher cost enhanced recovery technologies.

The Low Oil Price case assumes that technologies for producing biofuels, bitumen, CTL, BTL, GTL and extra-heavy oils achieve much lower costs than in the Reference case. As a result, production of those liquids increases to 16 million barrels per day in 2035 despite significantly lower oil prices.

High Oil Price case

In the High Oil Price case, the assumption of high demand for petroleum and other liquids in the non-OECD nations, combined with more constrained supply availability, results in higher oil prices than in the Reference case. Oil prices ramp up quickly to $186 per barrel (2010 dollars) in 2017 and continue rising slowly thereafter, to about $200 per barrel in 2035. The higher prices result from higher demand for petroleum and other liquid fuels in the non-OECD nations, resulting from the assumption of higher economic growth than in the Reference case. Specifically, GDP growth rates for China and India in 2012 are 1.0 percentage point higher than in the Reference case, and 0.3 percentage point higher in 2035. For most other non-OECD regions, GDP growth rates average about 0.5 percentage point above the Reference case in 2012. For the OECD regions, where prices rather than a higher economic growth rate are the main factor affecting demand, consumption of petroleum and other liquids remains fairly flat over the projection.

On the supply side, OPEC countries are assumed to reduce their market share somewhat, to less than 41 percent through 2035. Non-OPEC petroleum liquids resources outside the United States are assumed to be less accessible and/or more costly to produce than in the Reference case, and higher prices make other liquids supply more attractive. In 2035, other liquids production totals 17 million barrels per day in the High Oil Price case, about 4 million barrels per day above the Reference case level, and other liquids account for 15 percent of the total supply of petroleum and other liquids.

3. Potential efficiency improvements and their impacts on end-use energy demand

In 2010, the residential and commercial buildings sectors used 20.4 quadrillion Btu of delivered energy, or 28 percent of total U.S. energy consumption. The residential sector accounted for 57 percent of that energy use and the commercial sector 43 percent. In the *AEO2012* Reference case, delivered energy for buildings increases by a total of 9 percent, to 22.2 quadrillion Btu in 2035, which is modest relative to the rate of increase in the number of buildings and their occupants. In contrast, the U.S. population increases by 25 percent, commercial floorspace increases by 27 percent, and the number of households increases by 28 percent. Accordingly, energy use in the buildings sector on a per-capita basis declines in the projection. The decline of buildings energy use per capita in past years has been attributable in part to improvements in the efficiencies of appliances and building shells, and efficiency improvements continue to play a key role in projections of buildings energy consumption.

Existing policies, such as Federal appliance standards, along with evolving State policies, and market forces, are drivers of energy efficiency in the United States. A number of recent changes in the broader context of the U.S. energy system that affect energy prices, such as advances in shale gas extraction and the economic slowdown, also have the potential to affect the dynamics of energy efficiency improvement in the U.S. buildings sector. Although these influences are important, technology improvement remains a critical factor for energy use in the buildings sector. The emphasis for this analysis is on fundamental factors, particularly technology factors, that affect energy efficiency, rather than on potential policy or regulatory options.

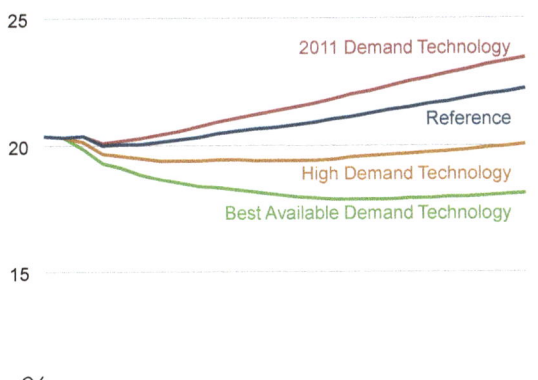

Figure 20. Residential and commercial delivered energy consumption in four cases, 2010-2035 (quadrillion Btu)

Three alternative cases in *AEO2012* illustrate the impacts of different assumptions for rates of technology improvement on delivered energy use in the residential and commercial sectors (Figure 20). These cases are in addition to the Extended Policies and No Sunset cases discussed earlier, and they are intended to provide a broader perspective on changes in demand-side technologies. In the High Demand Technology case, high-efficiency technologies are assumed to penetrate end-use markets at lower consumer hurdle rates, with related assumptions in the transportation and industrial sectors. In the Best Available Demand Technology case, new equipment purchases are limited to the most efficient versions of technologies available in the residential and commercial buildings sectors regardless of cost. In the

2011 Demand Technology case, future equipment purchases are limited to the options available in 2011 ("frozen technology"), and 2011 building codes remain unchanged through 2035. Like the High Demand and Best Available Demand Technology cases, the 2011 Demand Technology case includes all current Federal standards.

Without the benefits of technology improvement, buildings energy use in the 2011 Demand Technology case grows to 23.4 quadrillion Btu in 2035, as compared with 22.2 quadrillion Btu in the Reference case. In the High Demand Technology case, energy delivered to the buildings sectors only reaches about 20 quadrillion Btu for any year in the projection period, and in the Buildings Best Available Demand Technology case it declines to 17.9 quadrillion Btu in 2026 before rising slightly to 18.1 quadrillion Btu in 2035.

Background

The residential and commercial sectors together are referred to as the "buildings sector." The cases discussed here are not policy-driven scenarios but rather "what-if" cases used to illustrate the impacts of alternative technology penetration trajectories on buildings sector energy use. In a general sense, this approach can be understood as reflecting uncertainty about technological progress itself, or uncertainty about consumer behavior, in that the market response to a new technology is uncertain. This type of uncertainty is being studied through market research, behavioral economics, and related disciplines that examine how purchasers perceive options, differentiate products, and react to information over time. By varying technology progress across the full range of end uses, the integrated demand cases provide estimates of potential changes in energy savings that, in reality, are likely to be less uniform and more specific to certain end uses, technologies, and consumer groups. Specific assumptions for each of the cases are summarized in Tables 6 and 7.

Results for the residential sector

To emphasize that efficiency is persistent and its effects accumulate over time, energy use is discussed in terms of cumulative reductions (2011-2035) relative to a case with no future advances in technology after 2011. An extensive range of residential equipment is covered by Federal efficiency standards, and the continuing effects of those standards contribute to the cumulative reduction in delivered energy use of 12.3 quadrillion Btu through 2035 in the Reference case relative to the 2011 Demand Technology case. Electricity and natural gas account for more than 85 percent of the difference, each showing a cumulative reduction greater than 5 quadrillion Btu over the period. Energy use for space heating shows the most improvement in the Reference case, affected by improvements in building shells and heating equipment (Figure 21). Televisions and PCs and related equipment use 1.9 quadrillion Btu less energy over the projection period, as devices with energy-saving features continue to penetrate the market, and laptops continue to gain market share over desktop PCs.

Cumulative savings in residential energy use from 2011 to 2035 total 31.6 quadrillion Btu in the High Demand Technology case and 56.2 quadrillion Btu in the Best Available Demand Technology case in comparison with the 2011 Demand Technology case. Electricity accounts for the largest share of the reductions in the High Demand Technology case (49 percent) and the Best Available Demand Technology case (51 percent). In addition to adopting more optimistic assumptions in the High Demand Technology and Best Available Demand Technology cases for end-use equipment, residential PV and wind technologies are assumed to have greater cost declines than in the Reference case, contributing to reductions in purchased electricity. In 2035, residential PV and wind systems produce 23 billion kilowatthours more electricity in the Best Available Demand Technology case than in the 2011 Demand Technology case.

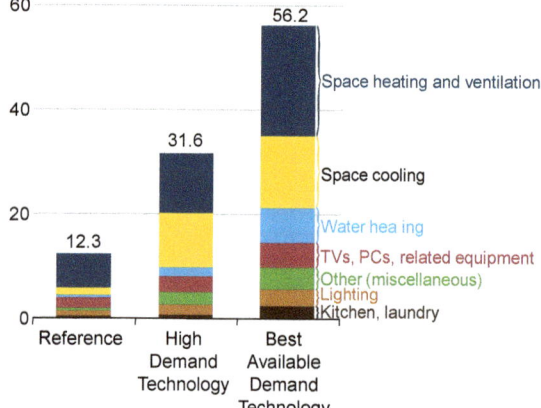

Figure 21. Cumulative reductions in residential energy consumption relative to the 2011 Demand Technology case, 2011-2035 (quadrillion Btu)

In the High Demand Technology and Best Available Demand Technology cases, energy use for residential space heating again shows the most improvement relative to the 2011 Demand Technology case. Large kitchen and laundry appliances claim a small share of the reductions, as Federal standards limit increases in energy consumption for those uses even in the 2011 Demand Technology case. Light-emitting diodes (LED) lighting provide the potential for further savings in the High and Best Available Demand Technology cases beyond the reductions realized as a result of the EISA2007 (Public Law 110-140) lighting standards.

Results for the commercial sector

Like the residential sector, analysis results for the commercial sector are discussed here in terms of cumulative reductions relative to the 2011 Demand Technology case, in order to illustrate the effect of efficiency improvements over the period from 2011 to 2035. Buildings in the commercial sector are less homogeneous than those in the residential sector, in terms of both form and function. Although many commercial products

Table 6. Key assumptions for the residential sector in the *AEO2012* integrated demand technology cases

Assumptions	Integrated 2011 Demand Technology	Integrated High Demand Technology[a]	Integrated Buildings Best Available Demand Technology[a]
End-use equipment	Limited to technology menu available in 2011. Promulgated standards still take effect.	Earlier availability, lower cost, and/or higher efficiencies for advanced equipment.	Purchases limited to highest available efficiency for each technology class, regardless of cost.
Hurdle rates	Same as Reference case distribution; varies by end-use technology.	All energy efficiency investments evaluated at 7-percent real interest rate.	All energy efficiency investments evaluated at 7-percent real interest rate.
Building shells	Fixed at 2011 levels.	New buildings meet ENERGY STAR specifications after 2016. Efficiency improvement for existing buildings is 50 percent greater than in the Reference case.	New buildings meet most efficient specifications. Efficiency improvement for existing buildings is 100 percent greater than in the Reference case.
Distributed and combined heat and power generation	No improvement in technology cost or performance after 2011. Learning rates same as in the Reference case.	PV and wind costs based on Advanced Case in EIA Technology reports.[b] Learning rates adjusted for all technologies.	PV and wind costs reduced by twice the difference between the Reference and High Technology costs. Learning rates adjusted for all technologies.
Personal computers	ENERGY STAR sales and enabling rates; LCD and laptop shares fixed at 2011 values.	ENERGY STAR sales and enabling rates. LCD and laptop shares higher than in the Reference case.	ENERGY STAR sales and enabling rates. LCD share approaches 100 percent. Laptop share higher than in the Reference case.
TVs, cable boxes, and satellite systems	Fixed at 2011 values.	Unit energy consumption (UEC) values are average of Reference and Best Available Demand Technology cases.	Per-unit consumption levels reduced to ENERGY STAR specifications.
Miscellaneous electricity end uses	Unit energy consumption (UEC) values fixed at 2011 values.	Most efficient equipment selected after 2014.	Most efficient equipment selected in all years.

[a]All changes from the Reference case start in 2012 unless otherwise stated.
[b]U.S. Energy Information Administration, *Photovoltaic (PV) Costs and Performance Characteristics for Residential and Commercial Applications, Final Report* (August 2010), and *The Cost and Performance of Distributed Wind Turbines, 2010-2035, Final Report* (August 2010).

Table 7. Key assumptions for the commercial sector in the *AEO2012* integrated demand technology cases

Assumptions	Integrated 2011 Demand Technology	Integrated High Demand Technology[a]	Integrated Buildings Best Available Demand Technology[a]
End-use equipment	Limited to technology menu available in 2011. Promulgated standards still take effect.	Earlier availability, lower cost, and/or higher efficiencies for advanced equipment.	Purchases limited to highest available efficiency for each technology class, regardless of cost.
Hurdle rates	Same as Reference case distribution.	All energy efficiency investments evaluated at 7-percent real interest rate.	All energy efficiency investments evaluated at 7-percent real interest rate.
Building shells	Fixed at 2011 levels.	25 percent more improvement than in the Reference case by 2035.	50 percent more improvement than in the Reference case by 2035.
Distributed and combined heat and power generation	No improvement in technology cost or performance after 2011. Learning same as in the Reference case.	PV and wind costs, CHP cost and performance based on Advanced Case in EIA Technology reports.[b] Learning rates adjusted for advanced technologies.	PV and wind costs reduced by twice the difference between the Reference and High Technology costs. CHP based on Advanced Case in EIA Technology reports.[b] Learning rates adjusted for advanced technologies.
PC-related office equipment	ENERGY STAR sales and enabling rates; LCD and laptop shares fixed at 2011 values.	ENERGY STAR sales and enabling rates. LCD and laptop shares higher than in the Reference case.	ENERGY STAR sales and enabling rates. LCD share approaches 100 percent. Laptop share higher than in the Reference case.
Non-PC Office Equipment	Same as Reference case except for elimination of data center efficiency improvements.	Partial adoption of network power management for copiers, etc. Use of higher-efficiency power supplies for servers.	Greater adoption of network power management for copiers, etc. Use of higher-efficiency power supplies and continuous power management for servers.
Miscellaneous electricity	Less efficiency improvement than in the Reference case for uninterruptible power supplies (UPSs), network equipment, elevators, and water services.	Savings from high-efficiency UPSs and network equipment.	Greater savings from high-efficiency UPSs and network equipment.

[a]All changes from the Reference case start in 2012 unless otherwise stated.
[b]U.S. Energy Information Administration, *Photovoltaic (PV) Costs and Performance Characteristics for Residential and Commercial Applications, Final Report* (August 2010), *The Cost and Performance of Distributed Wind Turbines, 2010-2035, Final Report* (August 2010), and *Commercial and Industrial CHP Technology Costs and Performance Data* (June 2010).

are subject to Federal efficiency standards, FEMP guidelines, and ENERGY STAR specifications, coverage is not as comprehensive as in the residential sector. Still, those initiatives and the ensuing efficiency improvements contribute to a cumulative reduction in commercial delivered energy use of 4.1 quadrillion Btu in the Reference case relative to the 2011 Demand Technology case (Figure 22). Virtually all of the reduction is in purchased electricity. Increased adoption of DG and CHP accounts for 0.4 quadrillion Btu (115 billion kilowatthours) of the cumulative reduction in purchased electricity in the Reference case. Commercial natural gas use is actually slightly higher in the Reference case because of the increased penetration of CHP. Office-related computer equipment sees the most significant end-use energy savings relative to the 2011 Demand Technology case, primarily because laptop computers gain market share from desktop computers.

Commercial heating, ventilation and cooling account for almost 50 percent of the 17.1 quadrillion Btu in cumulative energy savings in the High Demand Technology case relative to the 2011 Demand Technology case. The more optimistic assumptions for end-use equipment in the High Demand Technology case offset the additional energy consumed as a result of greater adoption of CHP, resulting in a cumulative reduction in natural gas consumption of 0.9 quadrillion Btu. The increase in distributed and CHP generation contributes 0.8 quadrillion Btu (231 billion kilowatthours) to the cumulative reduction in purchased electricity use.

Technologies such as LED lighting result in almost as much improvement as space heating and ventilation in the Best Available Demand Technology case relative to the 2011 Demand Technology case. Significant reductions are seen for all end-use services, with a cumulative reduction in energy consumption of 24.6 quadrillion Btu. Even when consumers choose the most efficient type of each end-use technology, the more optimistic assumptions regarding technology learning for advanced CHP technologies result in more natural gas use in the Best Available Demand Technology case relative to the 2011 Demand Technology case.

In comparison to a case that restricts future equipment to the efficiencies available in 2011, the alternative cases show the potential for reductions in energy consumption from the adoption of more energy-efficient technologies. In the Reference case, technology improvement reduces residential energy consumption by 12.3 quadrillion Btu—equivalent to 4.1 percent of total residential energy use—from 2011 to 2035 in comparison with the 2011 Demand Technology case. In the commercial sector, energy consumption is reduced by 4.1 quadrillion Btu—equivalent to 1.7 percent of total commercial energy use—over the same period. With greater technology improvement in the High Demand Technology case, cumulative energy savings from 2011 to 2035 rise by an additional 6.4 percent and 5.5 percent in the residential and commercial sectors, respectively. In the Best Available Demand Technology case, the cumulative reductions in energy consumption grow by an additional 8.2 percent and 3.1 percent in the residential and commercial sectors, respectively. In the Reference case, a cumulative total of 16.4 quadrillion Btu of energy consumption is avoided over the projection period relative to the 2011 Demand Technology case. That reduction is roughly equivalent to 80 percent of the energy that the buildings sectors consumed in 2010. In the Best Available Demand Technology case, cumulative energy consumption is reduced by an additional 64.3 quadrillion Btu from 2011 to 2035.

4. Energy impacts of proposed CAFE standards for light-duty vehicles, model years 2017 to 2025

In response to environmental, economic, and energy security concerns, EPA and NHTSA in December 2011 jointly issued a proposed rule covering GHG emissions and CAFE standards for passenger cars and light-duty trucks in MY 2017 through MY 2025 [42]. EPA and NHTSA expect to announce a final rule in the second half of 2012. In this section, EIA uses the National Energy Modeling System (NEMS), which has been updated since last year but, due to the timing of the modeling process, does not incorporate all information from the pending rulemaking process, to assess potential energy impacts of the regulatory proposal.

EPA is proposing GHG emissions standards that will reach a fleetwide LDV average of 163 grams CO_2 per mile (54.5 mpg equivalent) in MY 2025, or 49.6 mpg for the CAFE-only portion (Table 8). Passenger car standards are made more stringent by reducing the average annual CO_2 emissions allowed by 5 percent per year from MY 2016 through MY 2025. Average annual CO_2 emissions from light-duty trucks are reduced by 3.5 percent per year from MY 2016 through MY 2021, with larger average reductions for smaller light-duty trucks and smaller average reductions for larger light-duty trucks. For MY 2021 through MY 2025, light-duty trucks would be required to achieve a 5-percent average annual reduction rate. In this section, EIA assumes that the reductions in GHG emissions required under EPA standards exceed the reductions required under the NHTSA CAFE standards and are achieved through changes other than those that would provide further improvement in fuel economy as tested for compliance with the NHTSA standards.

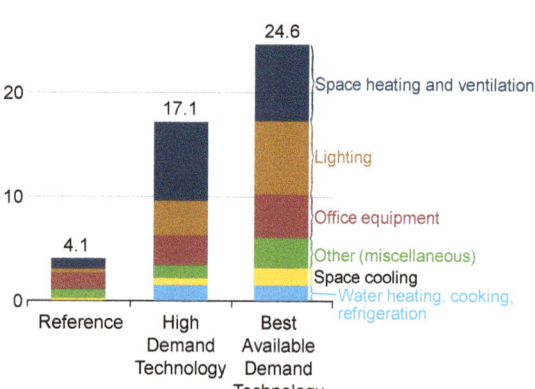

Figure 22. Cumulative reductions in commercial energy consumption relative to the 2011 Demand Technology case, 2011-2035 (quadrillion Btu)

NHTSA has proposed CAFE standards for LDVs that will reach a fleetwide average of 49.6 mpg in MY 2025, based on the projected inclusion of reductions in GHG emissions that are achieved by means other than improvements in fuel economy. CAFE standards are proposed for MY 2017 through MY 2021, and conditionally for MY 2022 through MY 2025. The proposed standards for passenger cars increase by 4.1 percent per year for MY 2017 through MY 2021 and 4.3 percent for MY 2022 through MY 2025. For light-duty trucks, the CAFE standards would increase by 2.9 percent per year for MY 2017 through MY 2021, with greater improvement required for smaller light-duty trucks and somewhat smaller improvement required for larger light-duty trucks. For MY 2022 through MY 2025, CAFE standards for all light-duty trucks would increase by 4.7 percent per year. Although there are complex dynamics in play among the CAFE standards and other policies, including those related to biofuels [43] and other gasoline alternatives, CAFE standards are the single most powerful regulatory mechanism affecting energy use in the U.S. transportation sector.

AEO2012 includes a CAFE Standards case that incorporates the proposed NHTSA fuel economy standards for MY 2017 through MY 2025. Fuel economy and GHG emissions standards for MY 2011 through MY 2016 have been promulgated already as final rules and are represented in the *AEO2012* Reference case. Further, the Reference case assumes that CAFE standards rise slightly to meet the requirement that LDVs reach 35 mpg by 2020 mandated in EISA2007.

As modeled by EIA, compliance with the more stringent fuel economy standards in the CAFE Standards case leads to a change in the vehicle sales mix. Vehicles that use electric power stored in batteries, or use a combination of a liquid fuel (including gasoline) and electric power stored in batteries for motive and/or accessory power—such as hybrid electric vehicles (HEVs) or plug-in hybrid electric vehicles (PHEVs)—or that use liquid fuels other than gasoline, such as diesel or E85, play a larger role than in the Reference case. The CAFE Standards case also projects a significant improvement in the fuel economy of traditional vehicles with gasoline internal combustion engines with and without micro hybrid technologies. In the analysis, vehicles that combine gasoline internal combustion engines with micro hybrid systems are projected to have the largest increase in sales relative to the Reference case (Figure 23 and Table 9).

Gasoline-only vehicles retain the single largest share of new vehicle sales in 2025. In order to meet increased fuel economy requirements, the average fuel economy of gasoline vehicles, including micro hybrids, is raised by the introduction of new fuel-efficient technologies and improved vehicle designs. The fuel economy of gasoline-only passenger cars, including micro hybrids, increases from 32 mpg in 2010 to 51 mpg in 2025 in the CAFE Standards case, compared with 38 mpg in 2025 in the Reference case. The fuel economy of gasoline-powered light-duty trucks, including micro hybrids, rises similarly, from 24 mpg in 2010 to 37 mpg in 2025 in the CAFE Standards case, compared with 31 mpg in 2025 in the Reference case.

As vehicle attributes, such as horsepower and weight, change in response to the more stringent fuel economy standards, some consumers switch from passenger cars to light trucks. Light-duty trucks account for 39 percent of new LDV sales in 2025 in the CAFE Standards case, higher than their 37 percent share in 2025 in the Reference case but still much lower than their 2005 share of more than 50 percent. In 2025, new passenger cars average 56 mpg and light-duty trucks average 40 mpg in the CAFE Standards case, compared with 41 mpg and 31 mpg, respectively, in the Reference case. Although more stringent standards stimulate sales of vehicles with higher fuel economy, it takes time for new vehicles to penetrate the vehicle fleet in numbers that are sufficiently large to affect the average fuel economy of the entire U.S. LDV stock. Currently there are about 230 million LDVs on the road in the United States, projected to increase to 276 million in 2035. As a consequence of the gradual scrapping of older vehicles and the introduction of new, more fuel-efficient models, the average on-road fuel economy of the LDV stock,

Table 8. Estimated[a] average fuel economy and greenhouse gas emissions standards proposed for light-duty vehicles, model years 2017-2025

	2016 (base)	2017	2018	2019	2020	2021	2022	2023	2024	2025
Fuel economy only (miles per gallon)										
Passenger cars	37.8	40.0	41.4	43.0	44.7	46.6	48.8	51.0	53.5	56.0
Light-duty trucks	28.8	29.4	30.0	30.6	31.2	33.3	34.9	36.6	38.5	40.3
All light-duty vehicles	34.1	35.3	36.4	37.5	38.8	40.9	42.9	45.0	47.3	49.6
Carbon dioxide emissions (grams per mile)										
Passenger cars	225	213	202	192	182	173	165	158	151	144
Light-duty trucks	298	295	285	277	270	250	237	225	214	203
All light-duty vehicles	250	243	232	223	213	200	190	181	172	163

[a]Based on projected mix of LDV sales.

Issues in focus

representing the fuel economy realized by all vehicles in use, increases from around 20 mpg in 2010 to 22 mpg in 2016, 27.5 mpg in 2025, and 34.5 mpg in 2035, as compared with 28 mpg in 2035 in the Reference case (Figure 24).

More stringent fuel economy standards lead to reductions in total energy consumption. Total cumulative delivered energy consumption by LDVs from 2017 to 2035 is 8 percent lower in the CAFE Standards case than in the Reference case. LDV delivered energy consumption is 6 percent lower in 2025 in the CAFE Standards case than in the Reference case and 17 percent lower in 2035. Total consumption of petroleum and other liquids in the transportation sector is 0.5 million barrels per day lower in 2025 and 1.4 million barrels per day lower in 2035 in the CAFE Standards case than in the Reference case (Figure 25). The existing standards are modestly exceeded in the Reference case. If the standards are just met, the reduction in liquids consumption is 0.5 million barrels per day in 2025 and 1.6 million barrels per day in 2035 in the CAFE Standards case relative to the Reference case. The reductions in total delivered energy use and liquid fuel consumption become more pronounced later in the projection, as more of the total vehicle stock consists of vehicles with higher fuel economy.

The more stringent regulatory standards in the CAFE Standards case change the composition of the vehicle fleet by fuel type and shift the mix of fuels consumed. Nevertheless, motor gasoline, including gasoline blended with up to 15 percent ethanol (used in vehicles manufactured in MY 2001 and after), remains the predominant fuel by far for LDVs in the CAFE Standards case, accounting for 84 percent of LDV delivered energy consumption in 2035—only slightly less than its 86-percent share in 2035 in the Reference case.

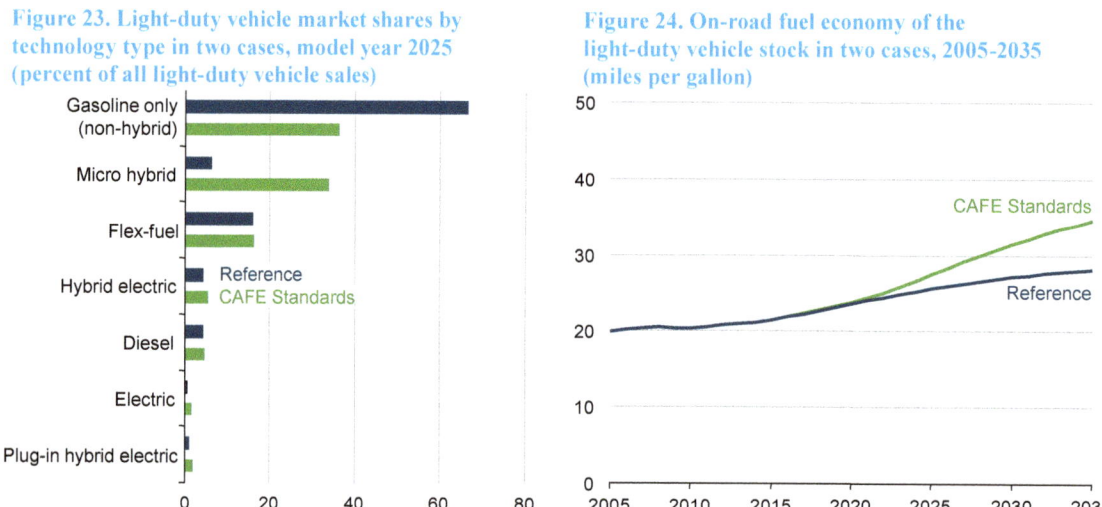

Figure 23. Light-duty vehicle market shares by technology type in two cases, model year 2025 (percent of all light-duty vehicle sales)

Figure 24. On-road fuel economy of the light-duty vehicle stock in two cases, 2005-2035 (miles per gallon)

Table 9. Vehicle types that do not rely solely on a gasoline internal combustion engine for motive and accessory power

Vehicle type	Description
Micro hybrid	Vehicles with gasoline engines, larger batteries, and electrically powered auxiliary systems that allow the engine to be turned off when the vehicle is coasting or idling and then quickly restarted. Regenerative braking recharges the batteries but does not provide power to the wheels for traction.
Hybrid electric (gasoline or diesel)	Vehicles that combine internal combustion and electric propulsion engines but have limited all-electric range and batteries that cannot be recharged with grid power.
Diesel	Vehicles that use diesel fuel in a compression-ignition internal combustion engine.
Plug-in hybrid electric	Vehicles that use battery power for driving some distance, until a minimum level of battery power is reached, at which point they operate on a mixture of battery and internal combustion power. Plug-in hybrids also can be engineered to run in a "blended mode," where an onboard computer determines the most efficient use of battery and internal combustion power. The batteries can be recharged from the grid by plugging a power cord into an electrical outlet.
Electric	Vehicles that operate by electric propulsion from batteries that are recharged exclusively by electricity from the grid or through regenerative braking.
Flex-fuel	Vehicles that can run on gasoline or any gasoline-ethanol blend up to 85 percent ethanol.

Total motor gasoline demand for LDVs is 19 percent lower in the CAFE Standards case in 2035 than in the Reference case, and lower demand for motor gasoline reduces the amount of ethanol used in E10 and E15 gasoline blends. As a consequence, more E85 fuel is sold to meet the RFS. E85 accounts for 10 percent of delivered energy consumption by LDVs in 2035, compared with 8 percent in the Reference case. Diesel fuel accounts for 5 percent of LDV delivered energy consumption in 2035, similar to its share in the Reference case. Electricity use by LDVs grows in the CAFE Standards case but still makes up less than 1 percent of LDV delivered energy demand in 2035.

Reductions in LDV delivered energy consumption reduce GHG emissions from the transportation sector. From 2017 and 2035, cumulative CO_2 emissions from transportation are 357 million metric tons (mmt) lower in the CAFE Standards case compared to the Reference case, a reduction of 5 percent. Transportation GHG emissions decline from 1,876 mmt in 2010 to 1,759 mmt in 2025 and to 1,690 mmt in 2035, reductions of 4 percent and 10 percent from the Reference case, respectively (Figure 26).

5. Impacts of a breakthrough in battery vehicle technology

The transportation sector's dependence on petroleum-based fuels has prompted significant efforts to develop technology and alternative fuel options that address associated economic, environmental, and energy security concerns. Electric drivetrain vehicles, including HEVs, PHEVs, and plug-in electric vehicles (EVs), are particularly well suited to meet those objectives, because they reduce petroleum consumption by improving vehicle fuel economy and, in the case of PHEVs and EVs, substitute electric power for gasoline use (see Table 10 for a descriptive list of electric drivetrain technologies).

AEO2012 includes a High Technology Battery case that examines the potential impacts of significant breakthroughs in battery electric vehicle technology on vehicle sales, energy demand, and CO_2 emissions. Breakthroughs may include a dramatic reduction in the cost of battery and nonbattery systems, success in addressing overheating and life-cycle concerns, as well as the introduction of battery-powered electric vehicles in several additional vehicle size classes. A brief summary of the results of the High Technology Battery case follows a discussion of the current market for battery electric vehicles.

Sales of light-duty HEVs, introduced in the United States more than a decade ago, peaked at about 350,000 new sales in 2007 and have maintained a roughly 3-percent share of total LDV sales through 2011. PHEVs were introduced in the United States at the end of 2010 with the production of the Chevy Volt, a PHEV-40 (PHEV with a 40-mile range). Although manufacturer plans call for increased production of PHEVs, sales in the first full year were under 10,000 units [44]. EVs were first introduced in the early 1900s, and manufacturers again made EVs available in the 1990s but with a focus on niche markets. The Nissan Leaf, an EV-100 (EV with a 100-mile range) introduced around the same time as the Chevy Volt, has sparked interest in the wider commercial prospects for EVs; however, sales in 2011 remained below 10,000 units.

The individual decision to purchase a vehicle is influenced by many factors, including style, performance, comfort, environmental values, expected use, refueling capability, and expectations of future fuel prices. In general, one of the single most important factors consumers consider when deciding to purchase a vehicle is cost. Specifically, they generally are more willing to purchase new vehicle technologies, such as battery electric systems, instead of conventional gasoline internal combustion engines (ICEs) if the economic benefit over a period of ownership is greater than the initial price of the vehicle. Additional costs and benefits—such as refueling time or difficulty of refueling, increased or decreased maintenance, and resale value—also may enter into vehicle choice decisions. Further, consumers may be unwilling to spend more to purchase a vehicle, even if it accrues fuel cost savings beyond the initial cost over a relatively short period, because they are unfamiliar with the new technology or alternative fuel.

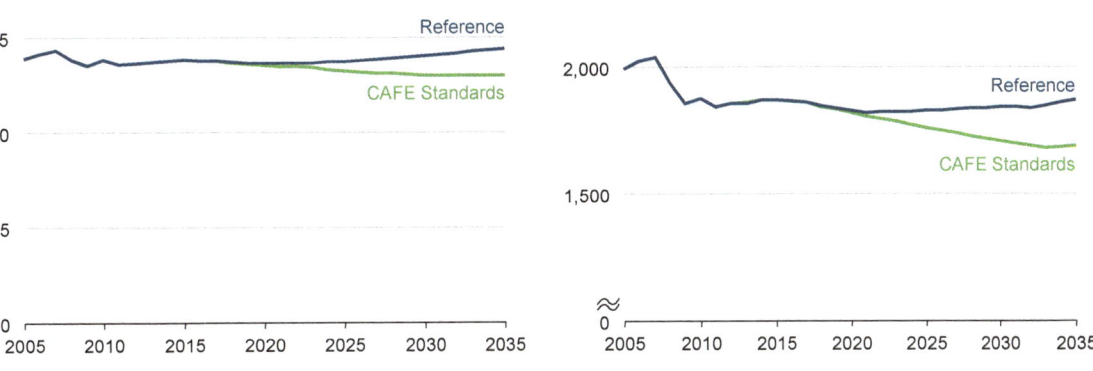

Figure 25. Total transportation consumption of petroleum and other liquids in two cases, 2005-2035 (million barrels per day)

Figure 26. Total carbon dioxide emissions from transportation energy use in two cases, 2005-2035 (million metric tons carbon dioxide equivalent)

Issues in focus

Battery electric vehicles offer an economic benefit to consumers over conventional gasoline ICEs in terms of significant fuel cost savings from both increased fuel economy for HEVs and PHEVs and the displacement of gasoline with electricity for PHEVs and EVs. Currently available battery electric vehicles such as the Toyota Prius (HEV), Chevy Volt (PHEV), and Nissan Leaf (EV) achieve much higher fuel economy (mpg) and, with the higher efficiency of electric motors, higher gasoline-equivalent mpg in electric mode, providing consumers with lower fueling costs. The Toyota Prius achieves an EPA-estimated 39 to 53 mpg, depending on trim and driving test cycle. The Chevy Volt achieves 35 to 40 mpg in charge-sustaining mode [45] and 93 to 95 mpg equivalent in charge-depleting mode. The Nissan Leaf achieves 99 mpg equivalent. In comparison, the Toyota Corolla, a passenger car generally similar to the Prius, achieves 26 to 34 mpg; the Chevy Cruze, a passenger car in the compact car size class similar to the Volt, achieves 25 to 42 mpg; and the Nissan Versa, a subcompact passenger car similar to the Leaf [46], achieves 24 to 34 mpg.

The inclusion of advanced battery technology that increases fuel economy and, in the case of PHEVs and EVs, displaces gasoline with electricity increases the initial cost of the vehicle to the consumer. The Toyota Prius has a manufacturer's suggested retail price (MSRP) between $24,000 and $29,500 (compared with $16,130 to $17,990 for the Toyota Corolla); the Chevy Volt has an MSRP between $39,145 and $42,085 (compared with $16,800 to $23,190 for the Chevy Cruze); and the Nissan Leaf has an MSRP between $35,200 and $37,250 (compared with $14,480 to $18,490 for the Nissan Versa) [47]. Based on these MSRPs, the current incremental consumer purchase cost of a battery electric vehicle relative to a comparable conventional gasoline vehicle is around $7,000 for an HEV and $20,000 for a PHEV or EV, before accounting for Federal and State tax incentives.

Although consumers may value high-cost battery electric vehicles for a variety of reasons, it is unlikely that they can achieve wide-scale market penetration while their additional purchase costs remain significantly higher than the present value of future fuel savings. Currently, the discounted fuel savings achieved, assuming five years of ownership with future fuel savings discounted at 7 percent, are significantly less than the incremental purchase cost of the vehicles (Table 11). This result is true even if gasoline is $6.00 per gallon. This calculation does not take into account any difference in maintenance cost or refueling infrastructure.

Recognizing the potential of HEVs, PHEVs, and EVs to reduce U.S. petroleum consumption and save consumers refueling costs, efforts are underway at both the public and private levels to address several of the barriers to wide-scale adoption of battery electric vehicle technology. Paramount among the barriers are reducing the cost of battery electric vehicles by lowering battery and nonbattery system costs and solving battery life-cycle and overheating limitations that will allow battery storage to downsize while maintaining a given driving range. For example, battery and nonbattery systems costs could be reduced by improving the manufacturing process, changing battery chemistry, or improving the electric motor. Solving battery life-cycle and overheating

Table 10. Description of battery-powered electric vehicles

Vehicle type	Description
Micro or "mild" hybrid	Vehicles with ICEs, larger batteries, and electrically powered auxiliary systems that allow the engine to be turned off when the vehicle is coasting or idle and then be quickly restarted. Regenerative braking recharges the batteries but does not provide power to the wheels for traction. Micro and mild hybrids are not connected to the electrical grid for recharging and are not considered as HEVs in this analysis.
Full hybrid electric (HEV)	Vehicles that combine an internal combustion engine with electric propulsion from an electric motor and battery. The vehicle battery is recharged by capturing some of the energy lost during braking. Stored energy is used to eliminate engine operation during idle, operate the vehicle at slow speeds for limited distances, and assist the ICE drivetrain throughout its drive cycle. Full HEV systems are configured in parallel, series, or power split systems, depending on how power is delivered to the drivetrain. HEVs are not connected to the electric grid for recharging.
Plug-in hybrid electric (PHEV)	Vehicles with larger batteries to provide power to drive the vehicle for some distance in charge-depleting mode, until a minimum level of battery power is reached (a "minimum state of charge"), at which point they operate on a mixture of battery and internal combustion power ("charge-sustaining mode"). The minimum state of charge is engineered to about 25 percent of full charge to ensure that the battery's life cycle matches the expected life of the vehicle. PHEVs also can be engineered to run in a "blended mode," using an onboard computer to determine the most efficient use of battery and internal combustion power. The battery can be recharged either from the grid by plugging a power cord into an electrical outlet or by the internal combustion engine. Current PHEV batteries are designed to recharge to about 75 percent of capacity for safety reasons related to battery overheating, leaving a depth of discharge of around 50 percent of total battery capacity. Typically, the distance a fully charged PHEV can travel in charge-depleting mode is indicated by its designation. For example, a PHEV-40 is engineered to travel around 40 miles on battery power alone before switching to charge-sustaining operation.
Plug-in electric (EV)	Vehicles that operate solely on an electric drivetrain with a large battery and electric motor and do not have an ICE to provide motive power. EVs are recharged primarily from the electrical grid by plugging into an electrical outlet, with some additional energy captured through regenerative braking. EV batteries also have a working depth of discharge capacity that is limited to both lower and upper levels due to life-cycle and safety concerns. EVs are designated by the distance a fully charged vehicle can travel in all-electric mode. For example, an EV-100 is designed to travel around 100 miles on battery power. EVs lack the "range extender" capability of PHEVs, which can switch instantly to an ICE when the battery reaches a minimum state of charge.

concerns would allow battery capacity to be downsized, which would improve the depth of discharge and make the battery less expensive. In addition, public and private efforts to address other obstacles to wider adoption of plug-in battery vehicles are underway, including the development of public charging infrastructure.

The *AEO2012* High Technology Battery case examines the potential impacts of battery technology breakthroughs by assuming the attainment of program goals established by DOE's Office of Energy Efficiency and Renewable Energy (EERE) for high-energy battery storage cost, maximum depth of discharge, and cost of a nonbattery traction drive system for 2015 and 2030 (Figures 27 and 28) [48]. EERE's program goals represent significant breakthroughs in battery and nonbattery systems, in terms of costs and life-cycle and safety concerns, in comparison with current electric vehicle technologies. Further, with breakthroughs in battery electric vehicle technology, more vehicle size classes are assumed to be available for passenger cars and light-duty trucks.

Reduced costs for battery and nonbattery systems in the High Technology Battery case lead to significantly lower HEV, PHEV, and EV costs to the consumer (Figures 29 and 30). The Reference case already projects a much lower real price to consumers for battery electric vehicles in 2035 relative to 2010 as a result of cost reductions for battery and nonbattery systems. Those declines are furthered in the High Technology Battery case. The prices of HEVs and PHEVs with a 10-mile range decline by an additional $1,500, or 5 percent, in 2035 in the High Technology Battery case relative to the Reference case. For PHEVs with a 40-mile range the relative decline is $3,500, or 11 percent, in 2035. For EVs with 100-mile (EV100) and 200-mile (EV200) ranges the relative declines are $3,600 and $13,300, or 13 percent and 30 percent, respectively, in 2035 relative to the Reference case.

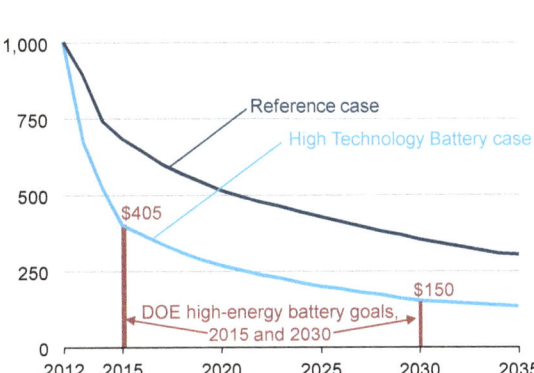

Figure 27. Cost of electric vehicle battery storage to consumers in two cases, 2012-2035 (2010 dollars per kilowatthour)

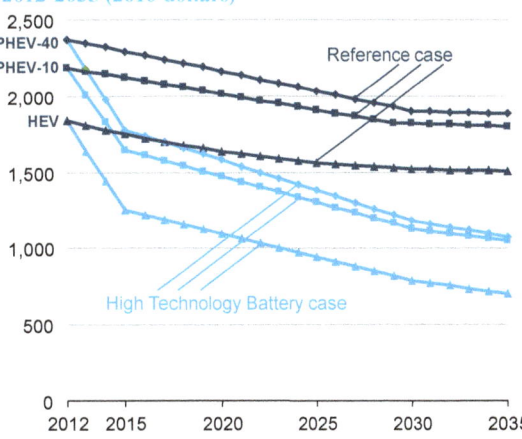

Figure 28. Costs of electric drivetrain nonbattery systems to consumers in two cases, 2012-2035 (2010 dollars)

Table 11. Comparison of operating and incremental costs of battery electric vehicles and conventional gasoline vehicles

Characteristics	Hybrid electric vehicle (Prius)	Plug-in hybrid electric vehicle (Volt)	Plug-in electric vehicle (Leaf)
Fuel efficiency (mpg equivalent)	45	38 (charge-sustaining mode) 94 (charge-depleting mode)	99 (charge-depleting mode)
Annual vehicle miles traveled			12,500
Percent vehicle miles traveled electric only	0	58	100
Fuel savings vs. conventional gasoline ICE vehicle (at $3.50 per gallon)[a]	$1,169	$2,036	$3,314
Fuel savings vs. conventional gasoline ICE vehicle (at $6.00 per gallon)[a]	$2,004	$4,340	$7,071
Incremental vehicle cost (2010 dollars) relative to cost of 35-mpg conventional gasoline ICE vehicle[b]	$7,000	$20,000	$20,000

[a]5-year net present value of fuel savings, assuming 35 mpg for ICE, 7% discount rate, and $0.10 per kilowatthour electricity price.
[b]Does not include Federal, State, or local tax credits.

Lower vehicle prices lead to greater penetration of battery electric vehicle sales in the High Technology Battery case than projected in the Reference case. Battery electric vehicles, excluding mild hybrids, grow from 3 percent of new LDV sales in 2013 to 24 percent in 2035, compared with 8 percent in 2035 in the Reference case (Figure 31). Due to the still prohibitive incremental cost, EV200 vehicles do not achieve noticeable market penetration.

Plug-in vehicles, including both PHEVs and EVs, show the largest growth in sales in the High Technology Battery case, resulting from the relatively larger incremental reduction in vehicle costs. Plug-in vehicle sales grow to just over 13 percent of new vehicle sales in 2035, compared with 3 percent in 2035 in the Reference case, with EV sales growing to 8 percent of new LDV sales in 2035, compared with 2 percent in 2035 in the Reference case. Virtually all sales of plug-in vehicles are EVs with a 100-mile range, given the prohibitive cost, even in 2035, of batteries for EVs with a 200-mile range. PHEVs grow to just under 6 percent of total sales, compared with 2 percent in 2035 in the Reference case. Most PHEV sales are vehicles with a 10-mile all-electric range.

Although plug-in vehicle sales increase substantially in the High Technology Battery case, that growth is tempered by the lack of widespread high-speed recharging infrastructure. In the absence of such public infrastructure, consumers must rely almost entirely on recharging at home. According to data from the 2009 Residential Energy Consumption Survey, 49 percent of households that own vehicles park within 20 feet of an electrical outlet [49]. A widespread publicly available infrastructure was not considered as part of the High Technology Battery case, which limits the maximum market potential of PHEVs and EVs.

HEV sales, including an ICE powered by either diesel fuel or gasoline, increase in the High Technology Battery case from 3 percent of sales in 2013 to 11 percent in 2035, compared with 5 percent in 2035 in the Reference case. Although the cost declines for HEVs are modest relative to those for other battery electric vehicle types, HEVs benefit from being unconstrained by the lack of recharging infrastructure.

Increased sales of battery electric vehicles in the High Technology Battery case lead to their gradual penetration throughout the LDV fleet. In 2035, HEVs represent 9 percent of the 276 million LDV stock, as compared with 4 percent in the Reference case. EVs and PHEVs each account for about 5 percent of the LDV stock in the High Technology Battery case in 2035, compared with 1 percent each in the Reference case.

The penetration of battery electric vehicles with relatively higher fuel economy and efficient electric motors reduces total energy use by LDVs from 15.6 quadrillion Btu in 2013 to 14.8 quadrillion Btu in 2035 in the High Technology Battery case, compared with 15.5 quadrillion Btu in 2035 in the Reference case (Figure 32). LDV liquid fuel use declines to

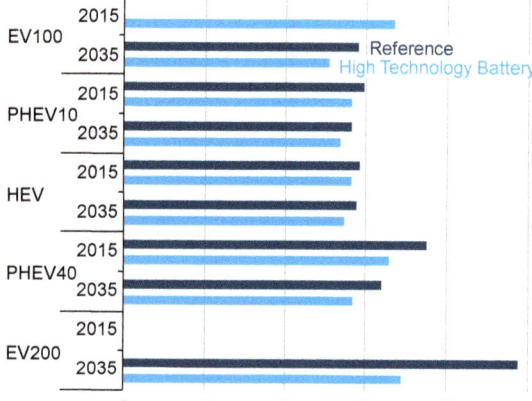

Figure 29. Total prices to consumers for compact passenger cars in two cases, 2015 and 2035 (thousand 2010 dollars)

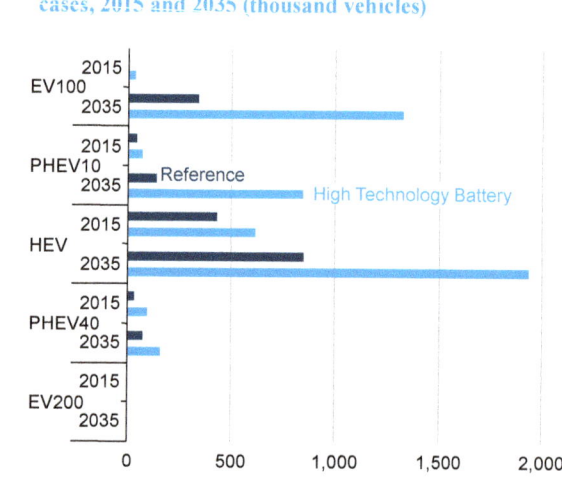

Figure 30. Total prices to consumers for small sport utility vehicles in two cases, 2015 and 2035 (thousand 2010 dollars)

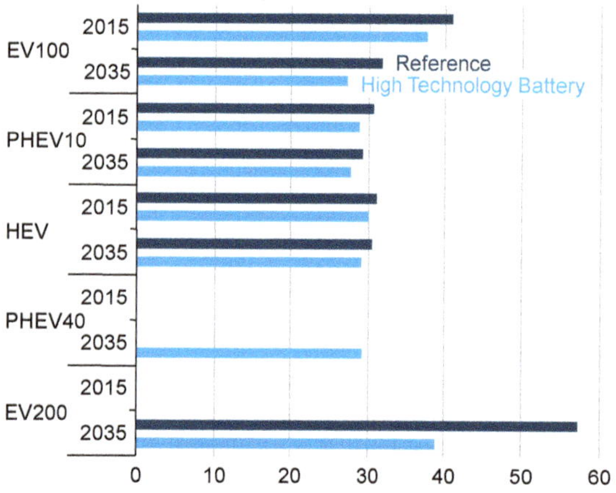

Figure 31. Sales of new light-duty vehicles in two cases, 2015 and 2035 (thousand vehicles)

14.6 quadrillion Btu in 2035 in the High Technology Battery case, and their electricity use increases to 0.2 quadrillion Btu—as compared with 15.4 quadrillion Btu of liquid fuel consumption and essentially no electricity consumption in 2035 in the Reference case. The reduction in liquid fuel consumption in the High Technology Battery case lowers U.S. net imports of petroleum from 8.5 million barrels per day in 2013 to 6.9 million barrels per day in 2035, compared with 7.2 million barrels per day in 2035 in the Reference case.

The reduction in total energy consumption by LDVs and displacement of petroleum and other liquid fuels with electricity decreases LDV energy-related CO_2-equivalent emissions from 1,030 million metric tons in 2013 to 935 million metric tons in 2035 in the High Technology Battery case, which represents a 2-percent decrease from 958 million metric tons in 2035 in the Reference case (Figure 33). CO_2 and other GHG emissions from the electric power consumed by PHEVs and EVs is treated as representative of the national electricity grid and not regionalized. Ultimately, the CO_2 and other GHG emissions of plug-in vehicles will depend on the fuel used in generating electricity.

The High Technology Battery case assumes a breakthrough in the costs of batteries and nonbattery systems for battery electric vehicles. Yet, despite the assumed dramatic decline in battery and nonbattery system costs, battery electric vehicles still face obstacles to wide-scale market penetration.

First, prices for battery electric vehicles remain above those for conventional gasoline counterparts, even with the assumption of technology breakthroughs throughout the projection period. The decline in sales prices relative to those for conventional vehicles may be enough to justify purchases by consumers who drive more frequently, consider relatively longer payback periods, or would purchase a more expensive but environmentally cleaner vehicle for a moderate additional cost. However, relatively more expensive battery electric vehicles may not pay back the higher purchase cost over the ownership period for a significant population of consumers.

In addition, EVs face the added constraint of plug-in infrastructure availability. Currently, there are about 8,000 public locations in the United States with at least one outlet for vehicle recharging, about 2,000 of which are in California [50]. In comparison, there are some 150,000 gasoline refueling stations available for public use. Without the construction of a much larger recharging network, consumers will have to rely on residential recharging, which is available for only around 40 percent of U.S. dwellings.

Further, recharging times differ dramatically depending on the voltage of the outlet. Typical 120-volt outlets can take up to 20 hours for a full EV battery to recharge; a 240-volt outlet can reduce the recharging time to about 7 hours [51]. Quick-recharging 480-volt outlets are under consideration for 30-minute "ultra-quick" recharges, but they may raise concerns related to safety and residential or commercial building codes. Even with ultra-quick recharging, EVs still would require substantially longer times for refueling than are required for ICE vehicles using liquid fuels. Given the concerns about availability and duration of recharging, the obstacle of severe range limitation, which does not affect PHEVs or HEVs, may inhibit the adoption of EVs by consumers.

Finally, another obstacle to wide-scale adoption of battery electric vehicles and other types of alternative-fuel vehicles is the increase in fuel economy for conventional gasoline vehicles and other types of AFVs resulting from higher fuel economy standards for LDVs. Final standards for LDV fuel economy currently are in place through MY 2016, and new CAFE standards proposed for MY 2017 through MY 2025 would increase combined LDV fuel economy to 49.6 mpg (56.0 mpg for passenger cars and 40.3 mpg for light-duty trucks) [52]. While the standards themselves may promote the adoption of battery electric vehicles, they also could considerably change the economic payback of electric drivetrain vehicles by decreasing consumer refueling costs for

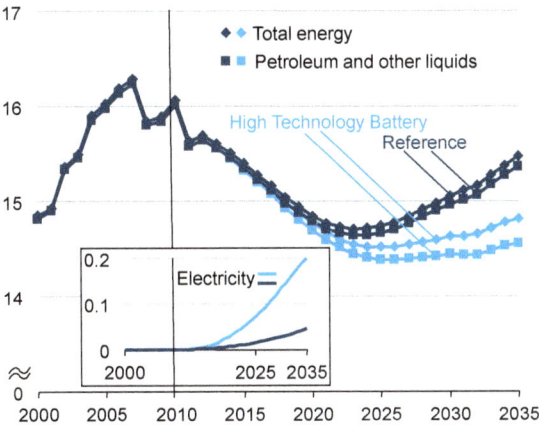

Figure 32. Consumption of petroleum and other liquids, electricity, and total energy by light-duty vehicles in two cases, 2000-2035 (quadrillion Btu)

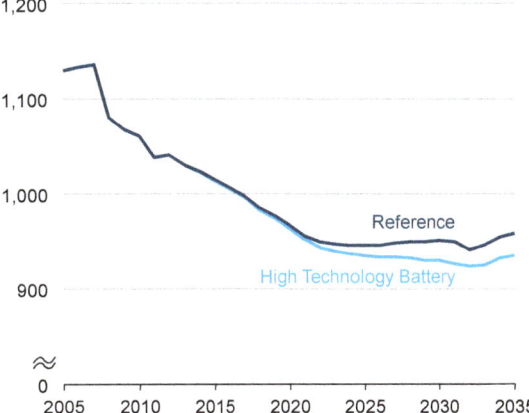

Figure 33. Energy-related carbon dioxide emissions from light-duty vehicles in two cases, 2005-2035 (million metric tons carbon dioxide equivalent)

6. Heavy-duty natural gas vehicles

Environmental and energy security concerns, together with recent optimism about natural gas supply and recent lower natural gas prices, have led to significant interest in the potential for fueling heavy-duty vehicles (HDVs) with natural gas produced domestically. Key market uncertainties with regard to natural gas as a fuel for HDVs include fuel and infrastructure issues (such as the build-out process for refueling stations and whether there will be sufficient demand for refueling to cover the required capital outlays, and retail pricing and taxes for liquefied natural gas [LNG] and compressed natural gas [CNG] fuels); and vehicle issues (including incremental costs for HDVs fueled by natural gas, availability of fueling infrastructure, cost-effectiveness in view of average vehicle usage, vehicle residual value, vehicle weight, and vehicle refueling time).

Current state of the market

At present, HDVs in the United States are fueled almost exclusively by petroleum-based diesel fuel [53]. In 2010, use of petroleum-based diesel fuel by HDVs accounted for 17 percent (2.2 million barrels per day) of total petroleum consumption in the transportation sector (12.8 million barrels per day) and 12 percent of the U.S. total for all sectors (18.3 million barrels per day). Consumption of petroleum-based diesel fuel by HDVs increases to 2.3 million barrels per day in 2035 in the *AEO2012* Reference case, accounting for 19 percent of total petroleum consumption in the transportation sector (12.1 million barrels per day) and 14 percent of the U.S. total for all sectors (17.2 million barrels per day).

Historically, natural gas has played a negligible role as a highway transportation fuel in the United States. In 2010, there were fewer than 40,000 total natural gas HDVs on the road, or 0.4 percent of the total HDV stock of nearly 9 million vehicles. Sales of new HDVs fueled by natural gas peaked at about 8,000 in 2003, and fewer than 1,000 were sold in 2010 out of a total of more 360,000 HDVs sold. With relatively few vehicles on the road, natural gas accounted for 0.3 percent of total energy used by HDVs in 2010.

As of May 2012, there were 1,047 CNG fueling stations and 53 LNG fueling stations in the United States, with 53 percent of the CNG stations and 57 percent of the LNG stations being privately owned and not open to the public [54]. Further, the stations were not evenly distributed across the United States, with 22 percent (227) of the CNG stations and 68 percent (36) of the LNG stations located in California. In comparison, nationwide, there were more than 157,000 stations selling motor gasoline in 2010 [55].

Developments in natural gas and petroleum markets in recent years have led to significant price disparities between the two fuels and sparked renewed interest in natural gas as a transportation fuel. Led by technological breakthroughs in the production of natural gas from shale formations, domestic production of dry natural gas increased by about 14 percent from 2008 to 2011. In the *AEO2012* Reference case, U.S. natural gas production (including supplemental gas) increases from 21.6 trillion cubic feet in 2010 to 28.0 trillion cubic feet in 2035. Further, although the world market for oil and petroleum products is highly integrated, with prices set in the global marketplace, natural gas markets are less integrated, with significant price differences across regions of the world. With the recent growth in U.S. natural gas production, domestic natural gas prices in 2012 are significantly lower than crude oil prices on an energy-equivalent basis (Figure 34).

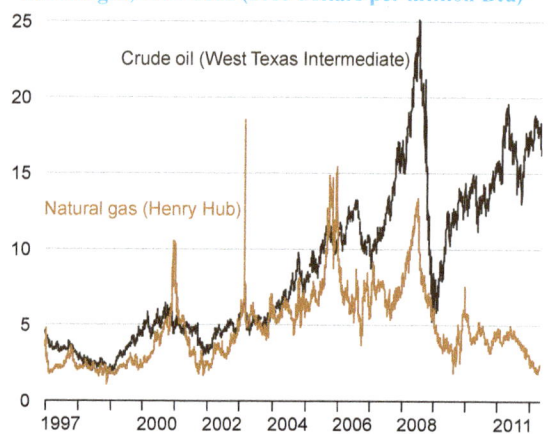

Figure 34. U.S. spot market prices for crude oil and natural gas, 1997-2012 (2010 dollars per million Btu)

Fuel and infrastructure issues

Even when it appears that an emerging technology can be profitable with significant market penetration, achieving significant penetration can be difficult and, potentially, unattainable. Refueling stations for NGVs are unlikely to be built without some assurance that there will be sufficient numbers of NGVs to be refueled, soon enough to allow for recovery of the capital investment within a reasonable period of time. In terms of estimating the prices that will be charged for NGV fuels beyond the cost of the dry natural gas itself, and the issue of expected utilization rates, there are additional uncertainties related to capital and operating costs, taxes, and the potential of prices being set on the basis of the prices of competing fuels.

Basic fuel issues

Diesel fuel falls into the category of distillate fuels, which have constituted more than 25 percent of U.S. refinery output in recent years. The cost of diesel fuel is linked closely to the value of crude oil inputs for the refining process. In 2011, the spot price of Gulf Coast ultra-low sulfur diesel fuel averaged $2.97 per gallon. The wholesale diesel price reflects crude oil costs, as well as the difference between the wholesale price at the refinery gate and the cost of crude oil input, commonly referred to as the "crack spread," which reflects the costs and profits of refineries. Beyond the wholesale price, the pump price of diesel fuel reflects distribution costs, Federal, State, and local fuel taxes, retailing costs, and profits. For diesel fuel, with an average energy content of 138,690 Btu per gallon, the 2011 national average retail price of $3.84 per gallon is equivalent to about $27.80 per million Btu.

Although early models of NGVs sometimes were less fuel-efficient than comparable diesel-fueled vehicles, current technologies allow for natural gas to be used as efficiently as diesel in HDV applications. Therefore, comparisons between natural gas and diesel fueling costs can be based on the price of energy-equivalent volumes of fuel. For this analysis, the cost and price of natural gas fuels are expressed in terms of diesel gallon equivalent (dge). For example, with an energy content of approximately 84,820 Btu per gallon, 1 gallon of LNG is equivalent in energy terms to 0.612 gallons of diesel fuel.

Fuel costs for LNG and CNG vehicles depend on the cost of natural gas used to produce the fuels, the cost of the liquefaction or compression process (including profits), the cost of moving fuel from production to refueling sites (if applicable), taxes, and retailing costs. Costs can vary with the scale of operations, but the significant disparity between current natural gas and crude oil prices suggests that the cost of CNG and LNG fuels in dge terms could be significantly below the price of diesel fuel.

There are different wholesale natural gas prices and capital costs associated with CNG and LNG stations. CNG retail stations, which typically have connections to the pipeline distribution network and thus require compression equipment and special refueling pumps, are likely to pay prices for natural gas that are similar to those paid by commercial facilities. For LNG stations, insulated LNG storage tanks and special refueling pumps are needed. LNG typically would be delivered from a liquefaction facility that, depending on its scale, would pay a natural gas price similar to the prices paid by electric power plants. The costs of liquefying and transporting the fuel to the retail station would ultimately be included in the retail price.

In a competitive market, retail fuel prices should reflect costs, including input, processing, distribution, and retailing costs, normal profit margins for processors, distributors, and retailers, and taxes. For example, the market for diesel fuel, which is produced by a large number of foreign and domestic refiners and is sold through numerous distributors and retail outlets, generally is considered to be a competitive market, in which retail prices follow costs.

CNG and LNG markets, at least in their initial stages, may not be as competitive as diesel fuel markets. For example, at public refueling stations, LNG and CNG currently sell at prices significantly higher than would be suggested by a long-term analysis of cost-based pricing. According to DOE's April 2012 "Clean Cities Alternative Fuel Price Report," the average nationwide nominal retail price for LNG was $3.05 per dge, and the average for CNG was $2.32 per dge [56].

If the use of LNG and/or CNG to fuel HDVs starts to grow, it is likely to take some time before fuel production and refueling infrastructure become sufficiently widespread for competition among fuel providers alone to assure that fuel prices are more closely linked to cost-based levels. However, even without many fuel providers, operators of an LNG and/or CNG vehicle fleet may be in a position to negotiate cost-based fuel prices with refueling station operators seeking to lock in demand for their initial investments in refueling infrastructure. Such arrangements provide an alternative to reliance on centrally fueled fleets as a means of circumventing the problem of how to introduce NGVs and natural gas refueling infrastructures concurrently.

Build-out process for refueling stations

It is not clear how NGVs and an expanded natural gas refueling infrastructure ultimately will evolve. One view is that a "hub-and-spoke" model for refueling infrastructure will expand sufficiently in multiple areas for a point-to-point system to take hold eventually. The "hubs" in the model would include the local refueling infrastructure, currently in place primarily to support local fleets. The "spokes" would ensure that refueling infrastructure is in place on the main transportation corridors connecting the hubs.

Several regional efforts are in place to encourage such "hub-and-spoke" growth for NGV refueling facilities. They include the Texas Clean Transportation Triangle [57], a strategic plan for CNG and LNG refueling stations between Dallas, San Antonio, and Houston; and the Interstate Clean Transportation Corridor [58], which aims to provide LNG fueling stations between such major western cities as Los Angeles, Las Vegas, Phoenix, Reno, Salt Lake City, and San Francisco. There also is a plan for a Pennsylvania Clean Transportation Corridor [59], which would provide CNG and LNG fueling stations between Pittsburgh, Harrisburg, Scranton, and Philadelphia.

In several corridors, Federal and State incentives are subsidizing both the construction of refueling stations and the production of heavy-duty LNG vehicles [60], in an effort to ensure that both demand and supply will be in place concurrently. A major question is whether gaps between isolated targeted markets can be bridged to provide a nationwide refueling structure that will allow heavy-duty NGVs to travel almost anywhere.

Sufficiency of demand for refueling to cover capital outlay

The cost of providing refueling services for NGVs depends on a number of factors and is distinctly different for CNG and LNG vehicles. Investment decisions are likely to be based on levels of demand. NGV refueling capability can be added at an existing facility or at a separate dedicated facility (which would require an additional investment). The costs depend in part on the number of fueling hoses added. LNG stations in particular benefit from higher volumes, but they also require significant additional land to accommodate storage tank(s), and they must satisfy special safety requirements—both of which add costs that can vary significantly from place to place. One added cost in operating an LNG station is the need for safety suits and specialized training for station attendants who dispense the fuel.

LNG typically is delivered to refueling stations via tanker truck from a separate liquefaction facility, the proximity of which is a major factor in the cost and frequency of deliveries. Any significant expansion of LNG refueling capacity also will require expanded liquefaction capacity, which currently is not sufficiently dispersed throughout the country to support a nationwide LNG refueling infrastructure. Although there are several dedicated large-scale natural gas liquefaction facilities in the United States, primarily in the West, there are smaller liquefaction plants and LNG storage tanks currently in use for meeting peak-shaving needs of utilities and pipelines during times of high demand. There are more than 100 such facilities in the United States, with a combined liquefaction capacity of more than 6 billion cubic feet per day. The majority are concentrated in the Northeast and Southeast [61].

Retail prices and taxes for LNG and CNG fuels

Even if the costs are fully known, retail prices for CNG and LNG transportation fuels remain uncertain, given questions about whether dispensers would charge higher prices in order to recover costs more rapidly if the facility were underutilized or would set prices to be competitive with the price of diesel. Prices charged at private stations for fleet vehicles presumably would be based on cost. With the number of refueling stations limited, competition between retailers is likely to be limited, at least initially. However, NGV refueling stations presumably would want to provide sufficient economic incentive in terms of the competitiveness of fuel prices to encourage more purchases of NGVs.

NGV fuel is taxed at State and Federal levels. Currently, on a Federal level, CNG is taxed at the same rate as gasoline on an energy-equivalent basis ($0.18 per gasoline gallon equivalent, or $0.21 per dge). However, LNG is taxed at a higher effective rate than diesel fuel, because it is taxed volumetrically at $0.24 per LNG gallon equivalent ($0.40 per dge) rather than on the basis of energy content [62]. State taxes vary, averaging $0.15 per dge for CNG and $0.24 per dge for LNG.

Vehicle Issues

Incremental vehicle cost

NGVs have significant incremental costs relative to their diesel-powered counterparts because of the need for pressurization and insulation of CNG or LNG tanks and the lower energy content of natural gas as a fuel. Total incremental costs relative to diesel HDVs range from about $9,750 to $36,000 for Class 3 trucks (GVWR 10,001 to 14,000 pounds), $34,150 to $69,250 for Class 4 to 6 trucks (GVWR 14,001 to 26,000 pounds), and $49,000 to $86,125 for Class 7 and 8 trucks (GVWR greater than 26,001 pounds). The incremental costs of heavy-duty NGVs depend in large part on the volume of the vehicle's CNG or LNG storage tank, which can be sized to match its typical daily driving range. Non-storage-tank incremental costs average about $2,000 for Class 3 vehicles, $20,000 for Class 4 to 6 vehicles, and $30,000 for Class 7 to 8 vehicles [63]. Fuel storage costs are about $350 per gallon diesel equivalent for CNG, with the incremental cost for Class 3 CNG vehicle storage tanks ranging between about $8,000 and $30,000; and about $475 per gallon diesel equivalent for LNG, with the incremental cost for Class 4 to 8 LNG vehicle storage tanks ranging between about $14,000 and $52,000. Natural gas fuel storage technology is relatively mature, leaving only modest opportunity for cost reductions.

Availability of fueling infrastructure

The absence of widespread public refueling infrastructure can impose a serious constraint on heavy-duty NGV purchases. Owners who typically refuel vehicles at a private central location do not face an absolute constraint based on infrastructure, however, and heavy-duty NGVs currently in operation have tended to be purchased by fleet operators who refuel consistently at a specific central location or in areas where their vehicles routinely operate on dedicated routes.

Cost-effectiveness with average vehicle usage

In order to take advantage of potential fuel cost savings from switching to NGVs, owners must operate the vehicles enough to pay back the higher incremental cost in a reasonable period of time. The payback period varies with miles driven and is shorter for trucks that are used more intensively. Payback periods for the upfront incremental costs of NGVs are greater than 5 years for Class 3 vehicles unless they are driven at least 20,000 to 40,000 miles per year, and for Class 7 and 8 vehicles unless they are driven at least 60,000 to 80,000 miles per year. Shorter payback periods, 3 years or less, may reflect typical owner expectations more accurately [64], but they require much more intensive use: around 60,000 to 80,000 miles annually for Class 3 vehicles and more than 100,000 miles annually for Class 7 and 8 vehicles. For example, for a Class 7 or 8 compression ignition NGV with average fuel economy of 6 miles per gallon (which has a similar fuel economy compared to a diesel counterpart) and an incremental cost of $80,000, the payback period would be just over 3 years if the vehicle were driven 100,000 miles per year, assuming a diesel fuel price of $4.00 per gallon and an LNG fuel price of $2.50 per gallon. If the same Class 7 or 8 vehicle were driven 40,000 miles per year, the payback period would be about 8 years. Further, without a widely available infrastructure, heavy-duty NGVs tend to be considered by centrally refueled fleets, which may have less mileage-intensive vehicle use.

According to the Department of Transportation's Vehicle Inventory and Use Survey [65], last completed in 2002, a large segment of the HDV market simply does not drive enough to justify the purchase of an NGV (Figure 35). Around 30 percent of Class 3 vehicles and 75 percent of Class 7 and 8 vehicles are not driven enough to reach the 5-year payback threshold mentioned above. This is a significant portion of the market that would require either more favorable fuel economics or lower vehicle costs before the purchase of an NGV could be justified.

Other market uncertainties

Other factors may also affect market acceptance of heavy-duty NGVs. First, the purchase decision could be affected by the considerable additional weight of CNG or LNG tanks. For owners who typically "weight-out" a vehicle (driving with a full payload), adding heavy CNG or LNG tanks necessitates a reduction in freight payload. The EPA and NHTSA have estimated that about one-third of Class 8 sleeper tractors routinely are "weighted-out" [66].

A diesel tractor with 200 gallons of tank capacity and a fuel economy of 6 miles per gallon can drive 1,200 miles on a single refueling. The same tractor would need up to 110 dge of LNG tank capacity, at a considerable weight penalty and an incremental cost of more than $80,000, to allow for a range of about 650 miles on a single refueling. Because owner/operators typically stop several times per day, the reduction in unrefueled maximum range would not require additional breaks for vehicles with large CNG or LNG tanks. However, CNG and LNG vehicles that do not opt for large tanks because of either weight or incremental cost considerations might have to refuel more frequently.

Finally, the owner perception of the balance of risk and reward for large capital investment is an uncertainty. Higher upfront capital costs can prove economically prohibitive for some potential owners. Even if the payback period for an investment in natural gas vehicles seemed acceptable, financing constraints or returns available on competing investment options could preclude the purchase. Additionally, the residual value of natural gas HDVs could, in theory, affect market uptake. With little natural gas refueling infrastructure in existence, the potential resale market is constrained to owners of centrally operated fleets. However, lease terms tend to limit the importance of this factor.

The complex set of factors influencing the potential for natural gas as a fuel for HDVs includes several areas for which policy mechanisms have been discussed. Most policy debates to date have considered the possibility of subsidies to reduce the incremental cost of natural gas vehicles (for example, in Senate and House versions of the New Alternative Transportation to Give Americans Solutions Act [67]) and Federal grant-based or other financial support for fueling station infrastructure. In addition, market hurdles related to consumer acceptance or payback periods might also be addressed through loan guarantees or related financial support policies, both for the vehicles and for the refueling infrastructure.

HD NGV Potential case results

The *AEO2012* HD NGV Potential case examines issues associated with expanded use of heavy-duty NGVs, under an assumption that the refueling infrastructure exists to support such an expansion. The HD NGV Potential case differs from an earlier sensitivity case completed as part of the *Annual Energy Outlook 2010*, which focused on possible subsidies to expand the market potential for heavy-duty NGVs and limited its attention to vehicles operating within 200 miles of a central CNG refueling facility.

The *AEO2012* HD NGV Potential case permits expansion of the HDV market to allow a gradual increase in the share of HDV owners who would consider purchasing an NGV if justified by the fuel economics over a payback distribution with a weighted average of 3 years. The gradual increase in the maximum natural gas market share reflects the fact that a national natural gas refueling program would require time to build out. The natural gas refueling infrastructure is expanded in the HD NGV Potential case simply by assumption; it is not clear how (or whether) specific barriers to natural gas refueling infrastructure investment can be overcome.

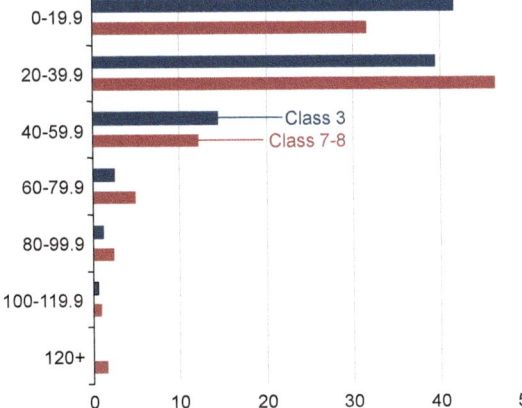

Figure 35. Distribution of annual vehicle-miles traveled by light-medium (Class 3) and heavy (Class 7 and 8) heavy-duty vehicles, 2002 (thousand miles)

Incremental costs for NGVs in the HD NGV Potential case differ from those in the Reference case. In the HD NGV Potential case, incremental costs are determined by assuming a set cost for CNG or LNG engines plus a CNG or LNG tank cost based on the average amount of daily travel and vehicle size class. The HD NGV Potential case includes separate delivered CNG and LNG fuel prices for fleet and nonfleet operators. Added per-unit charges to recover infrastructure are set and held constant in real terms throughout the projection period, based on the assumptions that refueling stations would be utilized at a sufficiently high rate to warrant the capital investment, and that the prices charged for the fuel would be cost-based (i.e., station operators would not set prices on the basis of prices for competing fuels). Motor fuels taxes are assumed to remain at their current levels in nominal terms, maintaining the higher energy-equivalent tax on LNG relative to diesel fuel.

In defining CNG and LNG prices for the HD NGV Potential case, EIA examined current motor fuel taxes and any charges added to the commodity price of dry natural gas sold at private central refueling stations (fleets) and at retail stations where actual data were available. Accordingly, an HDV Reference case was developed from the *AEO2012* Reference case, by including the updated fleet and retail CNG and LNG prices, to provide a consistent basis for comparison with the HD NGV Potential case (Figure 36). The HDV Reference case assumes that Class 3 through 6 vehicles use CNG, obtained from either fleet operators (using fleet prices) or nonfleet operators (using retail prices), and that Class 7 and 8 vehicles, both fleet and nonfleet, use LNG.

Sales of heavy-duty NGVs rise dramatically in the HD NGV Potential case, based on the national availability of refueling infrastructure and expanded market potential (Figure 37). Sales of new heavy-duty NGVs increase from 860 in 2010 (0.2 percent of total new HDV sales) to about 275,000 in 2035 (34 percent of total new vehicle sales), as compared with 26,000 in the HDV Reference case (3 percent of total new HDV sales). New heavy-duty NGVs gradually claim a more significant share of the vehicle stock, from 0.4 percent in 2010 to 21.8 percent (2,750,000 vehicles) in 2035, as compared with 2.4 percent (300,000 vehicles) in 2035 in the HDV Reference case.

As a result of the large projected increase in sales of new heavy-duty NGVs, natural gas demand in the HDV sector rises from about 0.01 trillion cubic feet in 2010 to 1.8 trillion cubic feet in 2035 in the HD NGV Potential case, as compared with 0.1 trillion cubic feet in the HDV Reference case (Figure 38). The natural gas share of total energy use by HDVs grows from 0.2 percent in 2010 to 32 percent in 2035 in the HD NGV Potential case, compared with 1.6 percent in the HDV Reference case.

Figure 36. Diesel and natural gas transportation fuel prices in the HDV Reference case, 2005-2035 (2010 dollars per diesel gallon equivalent)

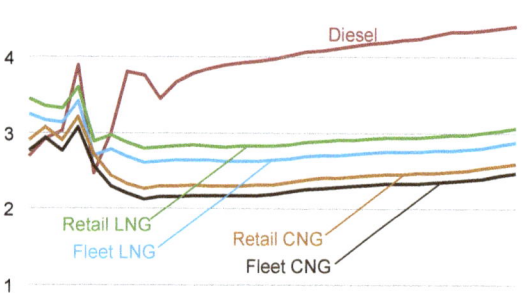

Roughly speaking, about 1 trillion cubic feet of natural gas consumed per year replaces 0.5 million barrels per day of petroleum and other liquids. Thus, natural gas consumption by HDVs in the HD NGV Potential case displaces about 850,000 barrels per day of petroleum and other liquids consumption in 2035 (Figure 39). Without a major impact on world oil prices, which is not expected to result from the gradual but significant adoption of natural gas as a fuel for U.S. HDVs, nearly all the reduction in petroleum and other liquids use by U.S. HDVs would be reflected by a decline in imports.

In the HD NGV Potential case, projected total U.S. natural gas consumption in 2035 is 1.4 trillion cubic feet (5 percent) higher than in the Reference case, as the increase in natural gas use by vehicles is partially offset by lower consumption in other sectors, in response to higher natural gas prices (Figure 40). The electric power and industrial sectors account for the

Figure 37. Annual sales of new heavy-duty natural gas vehicles in two cases, 2008-2035 (thousand vehicles)

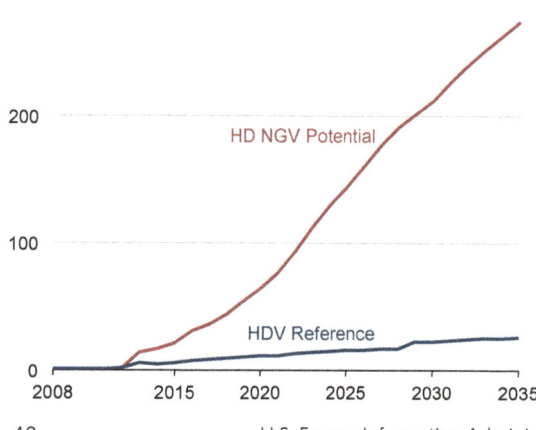

Figure 38. Natural gas fuel use by heavy-duty vehicles in two cases, 2008-2035 (trillion cubic feet)

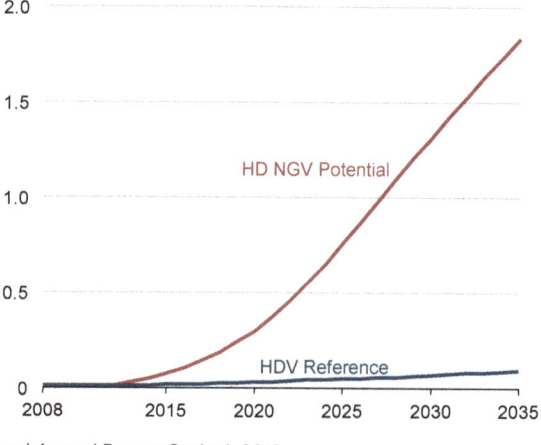

bulk of the consumption offsets, as their 2035 natural gas use is, respectively, 0.3 trillion cubic feet (3.1 percent) and 0.2 trillion cubic feet (2.7 percent) lower than in the Reference case.

In 2035, U.S. domestic natural gas production in the HD NGV Potential case is 1.1 trillion cubic feet (3.9 percent) higher than in the HDV Reference case. The higher level of natural gas production needed to support the growth in HDV fuel use results in a 10-percent increase in natural gas prices—$0.76 per million Btu (2010 dollars)—at the Henry Hub in 2035 in comparison with the HDV Reference case. Percentage increases in delivered natural gas prices to other sectors, which include transmission and distribution costs that are not affected by higher prices to producers, are smaller, with delivered natural gas prices increasing by 4.9 percent in the residential sector, 5.9 percent in the commercial sector, 8.9 percent in the industrial sector, and 7.9 percent in the electricity generation sector in comparison with the HDV Reference case in 2035.

7. Changing structure of the refining industry

Petroleum-based liquid fuels represent the largest source of U.S. energy consumption, accounting for about 37 percent of total energy consumption in 2010. The mix and composition of liquids, however, have changed in recent years in response to changes in regulations and other factors, and the structure of the liquid fuels production industry has changed in response [68]. The changes in the industry require that analytical tools used for market analysis of the liquid fuels produced by the industry also be reevaluated.

In recognition of the fundamental changes in the liquid fuels production industry, EIA is developing a new Liquid Fuels Market Module (LFMM), which it intends to use in place of the existing Petroleum Market Module (PMM) to produce the *Annual Energy Outlook 2013*. The LFMM will allow EIA to address more adequately the current and anticipated domestic and international market environments, to analyze the implications of emerging technologies and fuel alternatives, and to evaluate the impact of complex emerging energy-related policy, legislative, and regulatory issues. Some results from an early simulation of the LFMM, the LFMM case, are provided here.

The landscape for both production and consumption of liquid fuels in the United States continues to evolve, leading to changes in the mix of liquid fuel feedstocks, with greater emphasis on renewable fuels. The liquid fuels markets are not homogeneous; regional differences have become more pronounced. Furthermore, U.S. policymakers are paying more attention to evolving markets for liquid fuels and the potential for improving the efficiency of liquid fuels consumption, reducing GHG emissions associated with the production and consumption of liquid fuels, and improving the Nation's energy security by reducing reliance on imports. Major industry changes and their implications are discussed below.

New feedstocks and technologies

Over the past 25 years, the U.S. liquid fuels production industry has changed from being based primarily on domestic petroleum to using a variety of feedstocks and finished products from sources around the world. Regulatory and policy changes have resulted in the use of feedstocks other than crude oil, such as natural gas and renewable biomass, and could lead to the use of other feedstocks (such as coal) in the coming years. These changes have resulted in a transition from a relatively straightforward supply chain relying on crude oil and finished products to an increasingly complex system, which must be reflected in models to produce valid projections.

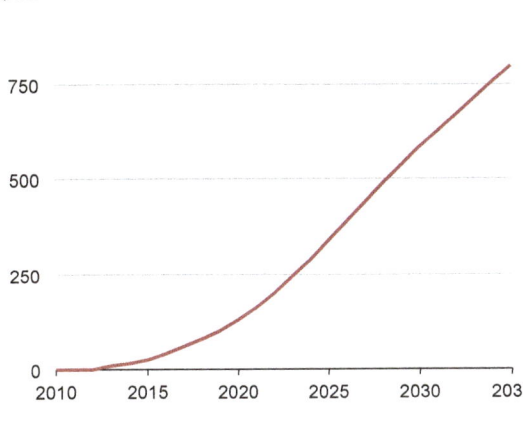

Figure 39. Reduction in petroleum and other liquid fuels use by heavy-duty vehicles in the HD NGV Potential case compared with the HDV Reference case, 2010-2035 (thousand barrels per day)

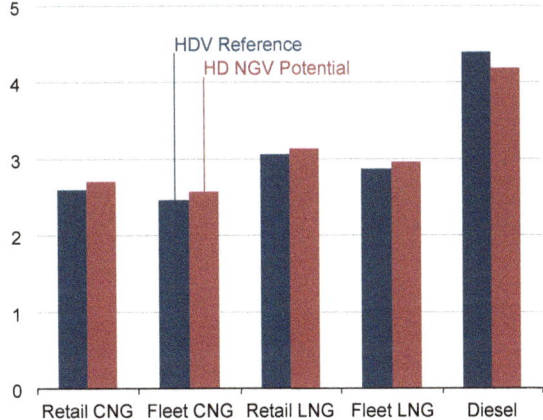

Figure 40. Diesel and natural gas transportation fuel prices in two cases, 2035 (2010 dollars per diesel gallon equivalent)

Issues in focus

The term "liquid fuels production industry" refers to all the participants in the production and delivery of liquid fuels, from production of feedstocks to delivery of both liquid and non-liquid end-use products to customers. It includes participants in the more traditional petroleum refining sector, relying on crude oil as a primary feedstock; in the nonpetroleum fossil fuel sector, using natural gas and coal to produce liquid fuels; and in the biofuel sector, using biomass to produce biofuels such as ethanol and biodiesel. The complexity of the industry supply chain is inadequately described by nomenclature predicated on specific feedstocks (e.g., crude oil), processes (e.g. refinery hydrotreating), or end-use products (e.g., diesel fuel and gasoline), which fail to capture the significant economic implications of non-liquid-fuel products for the industry.

The components of the U.S. liquid fuels production industry—including petroleum, nonpetroleum fossil fuel, and biofuel sectors—are shown in Figure 41, along with examples illustrating processes and products. Figure 41 also highlights the differences between the new expanded "liquid fuels production industry," which the entire figure represents, and the less extensive "petroleum and other liquids industry," the components of which are highlighted in red.

Nonpetroleum feedstocks are used in many new and emerging technologies, such as fermentation, enzymatic conversion, GTL, CTL, biomass-to-liquids, and algae-based biofuels. The new technologies provide valuable non-liquid-fuel co-products—such as chemical feedstocks, distiller's grains, and vegetable oils—that significantly affect the economics of liquid fuels production. The emergence of renewable biofuels has led to the introduction of midstream components such as ethanol and biodiesel, which are blended with petroleum products such as gasoline and diesel fuel during the final stages of the supply chain at refineries, blending sites, or retail pumps. The increase in biofuel production has led to new distribution channels and infrastructure investments and recognition of new production regions, such as the high concentration of ethanol producers in the Midwest. The new LFMM will include the entire liquid fuels production industry, providing greater flexibility for integrating new technologies and their associated products into the liquid fuels supply chain, better reflecting the industry's evolution.

In *AEO2012*, the "petroleum and other liquids" category includes the petroleum sector and those non-petroleum-based liquid products shaded in red in Figure 41, such as ethanol and biodiesel, which are blended with petroleum products to make end-use liquid fuels. Because this approach treats nonpetroleum products as exogenously produced feedstocks, the petroleum and other liquids concept used in *AEO2012* does not explicitly link the industrial processes that yield nonpetroleum liquid fuels (nor their feedstocks, nonpetroleum fossil fuels and biomass) with liquids production. The more inclusive definition of the liquid fuels production industry illustrated in Figure 41 is necessary to capture and model the full range of product flows and economic drivers of decisionmaking by firms involved in this complex industry.

Nonpetroleum feedstocks do not exist in traditional liquid form, and they require a different analytical approach for analysis of their conversion to liquid fuels. Traditional volumetric measures, such as process gain, are not applicable to an analysis of the liquids produced from nonpetroleum feedstocks. It is more appropriate to use the fundamental principles of mass and energy balance to evaluate process performance, market penetration, and supply/demand dynamics when the uses of nonpetroleum feedstocks are being examined. This approach allows for comparison among the different sectors of the liquid fuels production industry. Figure 42 provides an overview of the liquid fuels production industry on a mass basis.

The variety and changing dynamics of nonpetroleum feedstocks and the resulting end-use products also are illustrated in Figure 42. In recent history, biomass has taken significant market share from petroleum feedstocks, correlated with shifts in product yields—a trend that is expected to continue in the future, along with further diversification into nonpetroleum fossil feedstocks. In 2000, nearly all liquid fuels were derived from petroleum. Since then, however, the share of petroleum has dropped while the shares of biomass and other fossil fuels have increased. In 2011, the combined biomass and other fossil fuels share of feedstocks was almost 18 percent, measured on a mass basis. In the LFMM case, the biomass share of feedstock consumption increases to

Figure 41. U.S. liquid fuels production industry

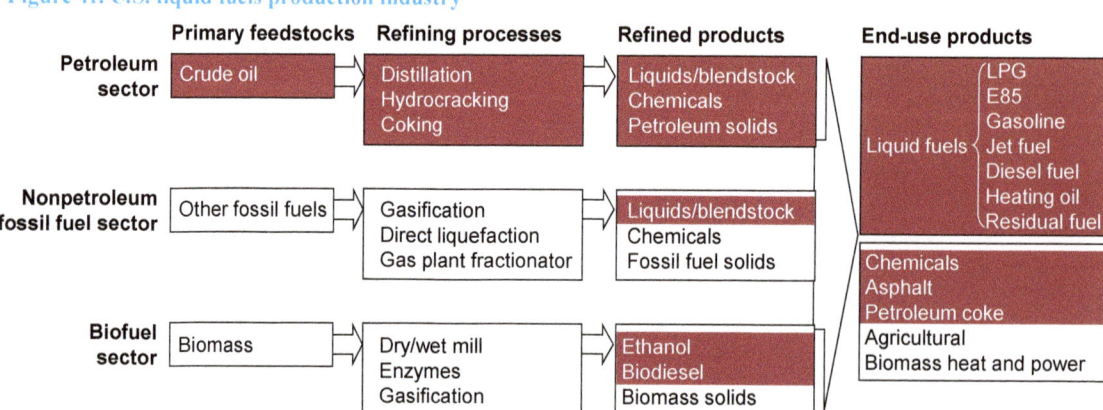

30 percent in 2035, and the petroleum share falls to about 57 percent. The biomass share of end-use products increases only to 10 percent in 2035, reflecting differences in conversion efficiencies between petroleum and nonpetroleum feedstocks, as highlighted by the growing but still small nonpetroleum content of gasoline and distillates.

Changes in crude oil types

Economic growth in the developing countries over the past decade has increased global demand for crude oil. Over the same period, new technologies for recovering crude oil, changes in the yields of existing crude oil fields, and a global increase in exploration have expanded the number and variety of crude oil types. The United States currently imports more than 100 different types of crude oil from around the world, including a growing number from Canada and Mexico, with a wide range of API gravities (between 10.4 and 64.6) and sulfur content (between 0.02 and 5.5 percent). Consequently, it is difficult to group them according to the categories used in the existing NEMS PMM. A new and more comprehensive representation of the numerous crude types is required, as well as flexibility to add new sources.

The United States increasingly is using crude oil extracted from oil sands and oil shale, as well as other nontraditional petroleum sources that require additional processing. The new sources have led to shifts in crude oil flows and changes in the distribution network. The increased variety and regional availability of certain crude types has created new market dynamics and pricing relationships that are difficult to capture using existing methods, especially considering the rapid emergence of "tight oil" production, which, to date, has been substantially different in quality from the crude oil previously expected to be available to U.S. refineries. For example, light sweet crude oil sourced from the Bakken shale formation in North Dakota has been sold to refiners on the Gulf Coast in recent years at a substantial discount relative to heavier imported crudes, because of limitations in the delivery infrastructure.

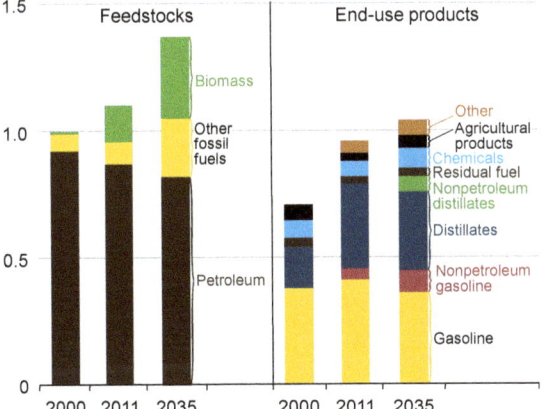

Figure 42. Mass-based overview of the U.S. liquid fuels production industry in the LFMM case, 2000, 2011, and 2035 (billion tons per year)

The growing number of sources, changes in characteristics of crudes, and shifting price relationships in crude oil markets require an updated representation of different crude types in NEMS. The model also needs an updated and more dynamic representation of the crude oil distribution network in order to provide better estimates of changes in crude oil flows and potential new regional sources in the future.

Regional updates

The Petroleum Administration for Defense Districts (PADD), which were developed by the Department of Defense during World War II, have been traditionally used as the regional framework for analyzing liquid fuels production. Because the topology and configuration of the liquid fuels market have changed significantly, and new feedstocks have emerged from regions that are subsets of PADDs, the regional definitions for processing liquid fuels need to be redefined. Toward this end, EIA has redefined the refining regions on the basis of market potential and availability of feedstocks. The redefined regions will be further divided as market conditions change. The new regional configuration of the NEMS LFMM will use eight domestic regions and adds a new international region (Figure 43).

Each new refining region has unique characteristics. PADD 1 has been left unchanged in the new configuration, but can be further divided based on recent and possible future refinery closures and shifts in imports from Europe. PADD 2 was subdivided into the Great Lakes and Inland regions due to the concentrated

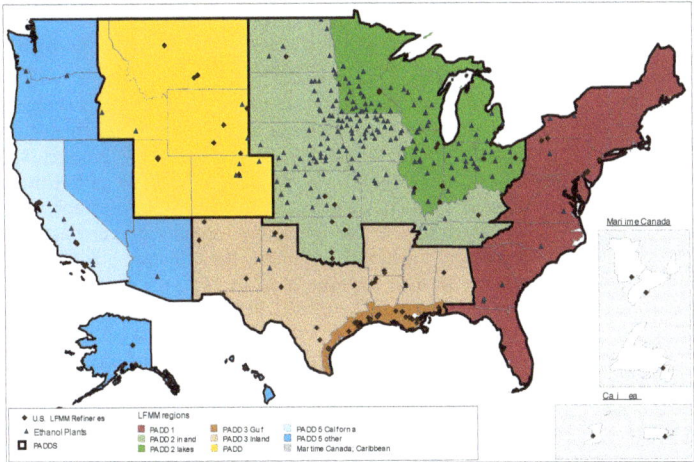

Figure 43. New regional format for EIA's Liquid Fuels Market Module (LFMM)

Issues in focus

production of biofuels and access to Canadian crudes. PADD 3 was divided into the Gulf Coast and Inland regions due to the inability of the interior refineries to handle heavy sour crude. PADD 4 was left unchanged. California was separated from the rest of PADD 5 due to the State's unique gasoline and diesel specifications and regulatory policies. A new international region was added comprising Maritime Canada and the Caribbean.

The modified regional refinery format will allow EIA's analyses to more accurately capture regional refinery trends and potential regional regulatory policies that affect the liquid fuels market. For example, California often enacts its own regulatory policies earlier than the rest of its PADD region, and its individual actions could not be represented accurately in the PADD framework. As a further example, recent refinery closures and other developments on the East Coast evidence the need for a dynamic and flexible representation of the refinery regions that supply the U.S. market.

Changing product markets

Crude oil is still the most important and valuable feedstock for the liquid fuels production industry. More than 650 refineries, located in more than 116 countries, have the capacity to refine 86 million barrels of crude oil per day. In the past, most of the complex refineries that could transform a wide variety of crudes into numerous different products to meet demand were located in the United States. Now, however, complex refineries are becoming more common in Europe and the developing countries of Asia and Latin America, and the products from export-focused merchant refineries in those countries have the potential to compete with U.S. products. An example is the regular export of surplus gasoline from refiners in Europe to the Northeast United States.

Traditional measures of profitability, such as the 3-2-1 crack spread, require modification in NEMS in view of the changing market for liquid fuels. The calculation of margins requires consideration of multiple feedstocks and multiple products produced in refineries, biorefineries, and production facilities for nonpetroleum fuels. Operators in the liquid fuels production industry are faced with a choice of investing in facilities and modifying their configurations to meet changing market demand, or exchanging domestic feedstocks and products with merchant refineries in a global market. For example, increased U.S. efficiency standards for LDVs have reduced demand for gasoline and increased demand for diesel fuel, which has led to more gasoline exports and more investment to increase diesel output from domestic refineries.

EIA's new LFMM representation of the liquid fuels production industry will need to account for global competition for both crude oil and end-use products. As refineries around the world become larger and more complex, smaller refineries may not be able to compete with imports produced at low margins. Therefore, it is necessary to have a more robust and dynamic representation of the liquid fuel producers, as well as additional flexibility to adjust inputs, refinery configurations, and crude and product demands as the industry evolves.

Regulations and policies

It is important for EIA's models to represent existing laws and regulations accurately, in addition to being flexible enough to model proposed laws and regulations. One of the most important regulations currently affecting the U.S. liquid fuels industry is the RFS, which not only has increased production and use of renewable fuels, but also has changed how fuels are distributed and consumed both here and abroad. The RFS mandates the use of biofuels that are consumed primarily as blends with traditional petroleum products, such as gasoline and diesel fuel (Figure 44). Because of their chemical properties, ethanol, biodiesel, and other first-generation biofuels generally require their own distribution networks or investments in new infrastructure. In addition, because they are produced outside traditional petroleum refineries, the new products are added at different points in the supply chain, either at blending terminals or at retail sites via blender pumps. Modeling those changes requires an update to the traditional PADD regional format used to represent the liquid fuels market, as well as an update to the transportation network that distributes the fuels.

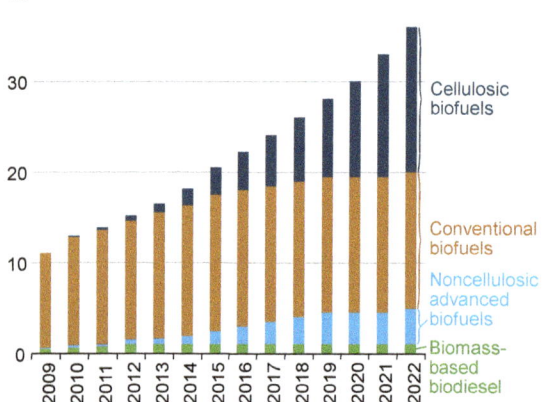

Figure 44. RFS mandated consumption of renewable fuels, 2009-2022 (billion gallons per year)

The RFS also requires consideration of many new technologies and increases the complexity of decisionmaking in the liquid fuels production industry. Fuel volumes by product are mandated by the RFS. For each year, regulated parties must make the decision to either buy the available renewable fuels in proportion to their RFS requirements or purchase the necessary credits. For example, the cellulosic biofuel credit price is set as the greater of $0.25 cents per gallon or $3.00 per gallon minus the wholesale gasoline price, both based on 2008 real dollars. The RFS also contains a general waiver based on technical, economic, or environmental feasibility that the EPA Administrator has discretionary authority to act on to reduce the mandates for advanced and total biofuels.

In addition, use of biofuels has broader implications for the global market, in terms of both feedstocks and the fuels themselves. A good example is ethanol. Its primary feedstocks are corn and sugar, both of which are global commodities in high demand as food sources as well as biofuel feedstocks. U.S. ethanol producers compete globally in other countries, such as Brazil, that have their own renewable fuels mandates.

Finally, coproducts from biofuels production have a significant influence on their economics. For example, the value of the dried distillers grains coproduct from corn ethanol production, which can be sold to the agricultural sector, can offset up to one-third of the purchase cost for the corn feedstock. Thus, the economics of biofuels production are complex, and they require a model that accounts for numerous investment decisions, feedstock markets, and global interactions. The RFS adds to the liquids fuels market a number of fuel technologies, midstream products and coproducts, evolving regional production and distribution networks, and complex domestic and global market interactions.

The U.S. liquid fuels market has evolved substantially over the past 20 years in terms of available fuel types, production regions, global market dynamics, and regulations and policies. The transition has resulted in a liquid fuels market that uses both petroleum- and nonpetroleum-based inputs, distributes them around the country by a variety of methods, and makes investment decisions based on both economic and regulatory factors. The changes are significant enough to make the framework and metrics used in traditional refinery models no longer adaptable or robust enough for proper modeling of the transformed liquid fuels market. EIA currently is in the process of updating its framework to allow better representation of the transformed industry.

8. Changing environment for fuel use in electricity generation

Introduction
The *AEO2012* Reference case shows considerable change in the mix of generating technologies over the next 25 years. Coal remains the dominant source of electricity generation in the Reference case, with a 38-percent share of total generation in 2035, but that is down from shares of 45 percent in 2010 and nearly 50 percent in 2005. The decrease in coal's share of total generation is offset primarily by increases in the shares of natural gas and renewables. Key factors contributing to the shift away from coal are sustained low natural gas prices, higher coal prices, slow growth in electricity demand, and the implementation of Mercury and Air Toxics Standards (MATS) [69] and Cross-State Air Pollution Rule (CSAPR) [70]. These factors influence how existing plants are used, which plants are retired, and what types of new plants are built.

Fuel prices and dispatch of power plants
The price of fuel is a major component of a power plant's variable operating costs [71]. The fuel-related variable cost of generating electricity is a function of the fuel price and the efficiency of the plant's conversion of the fuel into electricity, also referred to as the heat rate. Although natural gas prices declined dramatically in the second half of 2011 and the first half of 2012, coal-fired power plants have generally had the advantage of lower fuel prices and the disadvantage of higher heat rates in comparison to combined-cycle plants fueled by natural gas.

Power plants are dispatched primarily on the basis of their variable costs of operation. Plants with the lowest operating costs generally operate continuously. Plants with higher variable costs are brought on line sequentially as demand for generation increases. Because fuel prices influence variable costs, changes in fuel prices can affect the choice of plants dispatched. For instance, if the price of natural gas decreases, the variable costs for combined-cycle plants may fall below those for competing coal-fired plants, and, as a result, the combined-cycle plant may be dispatched before the coal-fired plant. Coal and natural gas plants can vary their outputs on the basis of fuel prices, but there are some cases in which plants may cycle off completely until they can be operated economically. In order to examine the overall impacts of changes in projected fuel price trends on the electric power sector, *AEO2012* includes alternative cases that assume higher and lower prices for natural gas and coal.

Demand for electricity
Electricity demand determines how much generating capacity is needed. When demand increases, plants with higher operating costs are brought into service, increasing average operating costs and, as a result, average electricity prices. Higher prices, in turn, provide economic incentives for the construction of new capacity. Conversely, when demand declines, plants with higher operating costs are taken off line or run at lower intensities, and the economic incentives for new plant construction are reduced. If a plant is not profitable, the owner may decide to retire it.

Mercury and Air Toxics Standards and Cross-State Air Pollution Rule
Both MATS and CSAPR are included in the *AEO2012* Reference case [72]. Both rules have significant implications for the U.S. generating fleet, especially coal-fired power plants. MATS requires all U.S. coal- and oil-fired power plants with capacities greater than 25 megawatts to meet emission limits consistent with the average performance of the top 12 percent of existing units—known as the maximum achievable control technology. MATS applies to three pollutants: mercury, hydrogen chloride (HCl), and fine particulate matter ($PM_{2.5}$). HCl and $PM_{2.5}$ are intended to serve as surrogate pollutants for acid gases and nonmercury metals, respectively. CSAPR is a cap-and-trade program that sets caps on sulfur dioxide (SO_2) and nitrogen oxide (NO_x) emissions from all fossil-fueled plants greater than 25 megawatts in 28 States in most of the eastern half of the United States. CSAPR is scheduled

to begin in 2012, although implementation was delayed by a court-issued stay at the time this article was completed [73]. See also "Cross-State Air Pollution Rule" in the "Legislation and regulations" section of this report.

Although the two rules differ in their makeup and the pollutants covered, the technologies that can be used to meet their requirements are not mutually exclusive. For instance, in order to meet the MATS acid gas standard, it is assumed that coal-fired plants without appropriate existing controls will need to install either flue-gas desulfurization (FGD) or dry sorbent injection (DSI) systems, which also reduce SO_2 emissions. Therefore, by complying with the MATS standards for acid gases, plants will lower overall SO_2 emissions, facilitating compliance with CSAPR.

AEO2012 assumes that all coal-fired power plants will be required to reduce mercury emissions to 90 percent below their pre-control levels in order to comply with MATS. The *AEO2012* NEMS explicitly models mercury emissions from power plants. Reductions in mercury emissions can be achieved with a combination of FGDs and selective catalytic reduction, which is primarily used to reduce SO_2 and NO_x emissions, or by installing activated carbon injection (ACI) systems. FGD systems may be effective in reducing mercury emissions from bituminous coal (due to its chemical makeup), but ACI systems may be necessary to remove mercury emissions from plants burning subbituminous and lignite coal.

NEMS does not explicitly model emissions of acid gases or toxic metals other than mercury. In order to represent the MATS limits for those emissions, *AEO2012* assumes that plants must install either FGD or DSI systems to meet the acid gas standard and, in the absence of a scrubber, a full fabric filter to meet the MATS standard for nonmercury metals. *AEO2012* assumes that the appropriate control technologies will be installed by 2015 in order to meet the MATS requirements.

DSI and wet and dry FGD systems are technologies that will allow plants to meet the MATS standards for acid gases. As of 2010, 43 percent of U.S. generating capacity already had FGDs installed [74]. For a number of the remaining, uncontrolled plants, operators will need to assess the effectiveness of installing FGD or DSI systems to comply with MATS. There are economic and engineering tradeoffs between the two technologies. FGD systems require significant upfront investment but have relatively low operating costs. DSI systems generally do not require significant capital expenses but may use significant quantities of sorbent to operate effectively, which increases their operating costs. Waste disposal for DSI also may be a significant variable cost, whereas the waste products from FGD systems can be sold as feedstock for industrial processes.

The EPA set an April 2015 compliance deadline for MATS, but the rule allows State environmental permitting agencies to extend the deadline by a year. Beyond 2016, the EPA stated that it will handle noncompliant units that need to operate for reliability purposes on a case-by-case basis [75]. *AEO2012* assumes that all plants will comply with MATS by the beginning of 2015.

Economics of plant retirements

The decision to retire a power plant is an economic one. Plant owners must determine whether a plant's future operations will be profitable. Environmental regulations, low natural gas prices, higher coal prices, and future demand for electricity all are key factors in the decision. Coal plants without FGD systems and with high heat rates, high delivered coal costs, and strong competition from neighboring natural gas plants in regions with slow growth in electricity demand may be especially prone to retirement.

Greenhouse gas policy in *AEO2012*

Uncertainty about possible future regulation of GHG emissions will continue to influence investment decisions in the power sector. Despite a lack of Congressional action, many utilities include simulations with a future CO_2 emissions price when evaluating long-term investment decisions. A carbon price would increase the cost of generation for all fossil fuel plants, but the largest impact would be on coal-fired plants. Thus, plant owners could be reluctant to retrofit existing coal plants to control for non-GHG pollutants, given the possibility that GHG regulations might be enacted in the near future. This uncertainty may influence the assumptions plant owners make about the economic lives of particular facilities.

In the Reference case, the costs of environmental retrofits are assumed to be recovered over a 20-year period. Two alternative cases assume that the costs would be recovered over 5 years, reflecting concern that future laws or regulations aimed at limiting GHG emissions will have significant negative effects on the economics of investing in existing coal plants.

AEO2012 also includes two alternative cases that assume enactment of an explicit GHG control policy. In each case, a CO_2 price is applied across all sectors starting in 2013 and increased at a 5-percent annual real rate through 2035. The price starts at $25 per metric ton in the GHG25 case and $15 per metric ton in the GHG15 case. The CO_2 price is applied across sectors and has a significant impact on the cost of generating electricity from fossil fuels, particularly coal.

Alternative cases

In order to illustrate the impacts of the various influences on the electric power sector, *AEO2012* includes several alternative cases that include varying assumptions about fuel prices, electricity demand, and the cost recovery period for environmental control equipment investments:

- The Reference 05 case assumes that the cost recovery period for investments in new environmental controls is reduced from 20 years to 5 years.

- The Low Estimated Ultimate Recovery (EUR) case assumes that the EUR per tight oil or shale gas well is 50 percent lower than in the Reference case, increasing the per-unit cost of developing the resource and, ultimately, the price of natural gas used at power plants (Figure 45).
- The High EUR case assumes that the EUR per tight oil or shale gas well is 50 percent higher than in the Reference case, decreasing the per-unit cost of developing the resource and the price of natural gas for power plants.
- The Low Gas Price 05 case combines the more optimistic assumptions about future volumes of shale gas production from the High EUR case with a 5-year recovery period for investments in new environmental controls.
- The High Coal Cost case assumes lower mining productivity and higher costs for labor, mine equipment, and coal transportation, which ultimately result in higher coal prices for electric power plants.
- The Low Coal Cost case assumes higher mining productivity and lower costs for labor, mine equipment, and coal transportation, which ultimately result in lower coal prices for electric power plants.
- The Low Economic Growth case assumes lower growth rates for population and labor productivity, higher interest rates, and lower growth in industrial output, which ultimately reduce demand for electricity (Figure 46), which is reflected in electricity sales, relative to the Reference case.
- The High Economic Growth case assumes higher growth rates for population and labor productivity. With higher productivity gains and employment growth, inflation and interest rates are lower than in the Reference case, and, consequently, economic output grows at a higher rate, ultimately increasing demand for electricity, which is reflected in electricity sales, relative to the Reference case.
- In the GHG15 case, the CO_2 price is set at $15 per metric ton in 2013 and increases at a real annual rate of 5 percent per year over the projection period. Price is set to target the same reduction in CO_2 emissions as in the *AEO2011* GHG Price Economywide case.
- In the GHG25 case, the CO_2 price is set at $25 per metric ton in 2013 and increases at a real annual rate of 5 percent per year over the projection period. Price is set to target the same dollar amount as in the *AEO2011* GHG Price Economywide case.

Analysis results

Coal-fired plant retirements

Significant amounts of coal-fired generating capacity are retired in all the alternative cases considered (Figure 47). (For a map of the electricity regions projected, see Appendix F.) In the Reference 05 case, 63 gigawatts of coal-fired capacity is retired through 2035, 28 percent higher than in the Reference case. In the High EUR case, 55 gigawatts of coal-fired capacity is retired, as lower wholesale electricity prices and competition from natural gas combined-cycle units makes the operation of some coal plants uneconomical. In the Low Economic Growth case, 69 gigawatts of coal-fired capacity is retired, because lower demand for electricity reduces the need for new capacity and makes investments in older plants unattractive.

The High Economic Growth case results in fewer retirements, as existing coal-fired capacity is needed to meet growing electricity demand, and higher economic growth pushes up natural gas prices. In the Low Coal Cost case, the lower relative coal prices increase the profit margins for coal-fired power plants, making it more likely that investments in retrofit equipment will be recouped over the life of the plants.

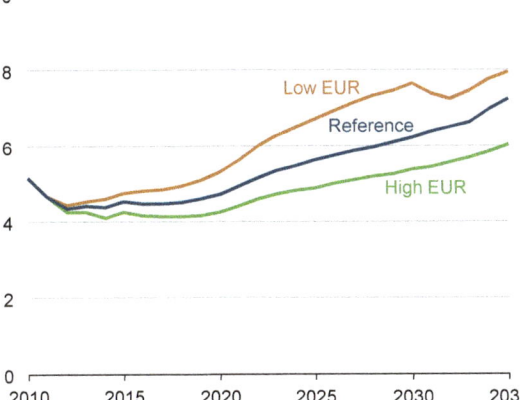

Figure 45. Natural gas delivered prices to the electric power sector in three cases, 2010-2035
(2010 dollars per million Btu)

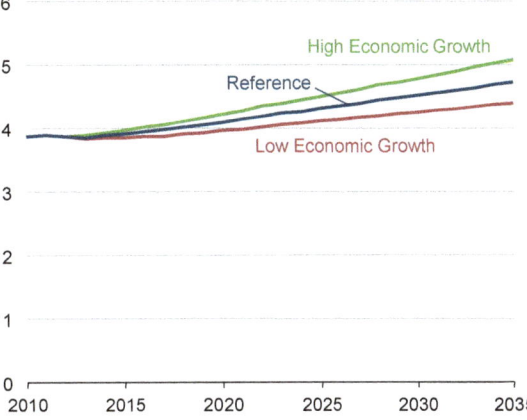

Figure 46. U.S. electricity demand in three cases, 2010-2035 (trillion kilowatthours)

Issues in focus

Coal-fired capacity retirements are concentrated in two North American Electric Reliability Corporation (NERC) regions: the SERC Reliability Corporation (SERC) region, which covers the Southeast region, and the Reliability First Corporation (RFC), which includes most of the Mid-Atlantic and Ohio Valley region [76]. Many coal-fired plants in those regions are sensitive to the factors that influence retirement decisions, as discussed above. In the SERC and RFC regions, which in 2010 accounted for 65 percent of U.S. coal-fired generating capacity, 43 percent of the coal-fired plants do not have FGD units installed. Coal plants in the RFC and SERC regions are fueled primarily by bituminous coal, generally the coal with the highest cost. Projected demand for electricity in the early years of the Reference case is low nationwide and, especially, in the RFC region, where demand in 2015 is slightly lower than in 2010. In both the GHG15 and GHG25 cases, even larger amounts of coal-fired capacity are retired by 2035 than in the non-GHG policy cases.

Generation by fuel

Coal

In all cases, generation from coal is lower in 2020 than in 2010. Higher coal prices, relatively low natural gas prices, retirements of coal-fired capacity, and slow growth in electricity demand are responsible for the decrease. Generation from coal is lower than in the Reference case in the Reference 05, High EUR, Low Gas Price 05, High Coal Cost, and Low Economic Growth cases as a result of additional retirements of coal-fired capacity, lower natural gas prices, higher coal prices, or lower electricity demand. In cases where the opposite assumptions are incorporated, coal-fired generation is higher.

Generation from coal begins to recover after 2020, as electricity demand and natural gas prices start to rise. The strongest increases in coal-fired electricity generation occur in the Low EUR, Low Coal Cost, and High Economic Growth cases. When lower natural gas prices, lower economic growth, and/or higher coal prices are assumed, coal-fired generation still increases after 2020 but at a slower rate. In all cases, utilization of existing coal-fired power plants increases, because there is no significant growth in new coal-fired capacity. In the most optimistic case, the High Economic Growth case, only 3.3 gigawatts of new coal-fired capacity is added from 2017 to 2035 [77].

Despite a declining share of the generation mix, coal still has the highest share of total electricity generation in 2035 in all non-GHG or High TRR cases. However, it never again reaches the 2010 share of 45 percent, even in the Low EUR case (where it reaches 40 percent in 2035). Conversely, the coal share of total generation in 2035 is 34 percent in the Low Gas Price 05 case. The lower coal share is offset by increased generation from natural gas, which grows significantly in all the cases. The natural gas share of total generation almost equals that of coal in the Low Gas Price 05 case. In the GHG15 and GHG25 cases, coal-fired generation drops to 16 percent and 4 percent, respectively, of the total generation mix in 2035, and in both cases generation from coal declines significantly as the explicit price on CO_2 emissions increases costs. In the GHG15 and GHG25 cases, decreases in coal-fired generation are offset by a mix of natural gas, nuclear, and renewable generation.

Natural gas

In the *AEO2012* Reference case, electricity generation from natural gas in 2020 is 13 percent above the 2010 level, despite an increase of only 5 percent in overall electricity generation. Low natural gas prices result in greater utilization of existing combined-cycle plants as well as the addition of 16 gigawatts of natural gas combined-cycle capacity from 2010 to 2020. The same trends are amplified in cases with lower natural gas prices and more coal-fired capacity retirements and muted in cases with higher natural gas prices and fewer coal-fired capacity retirements. Generation from combustion turbines does not change significantly across the cases, demonstrating that changes in the relative economics of coal and natural gas affect primarily the dispatch of combined-cycle plants to meet base and intermediate load requirements, not combustion turbines to meet peak load requirements.

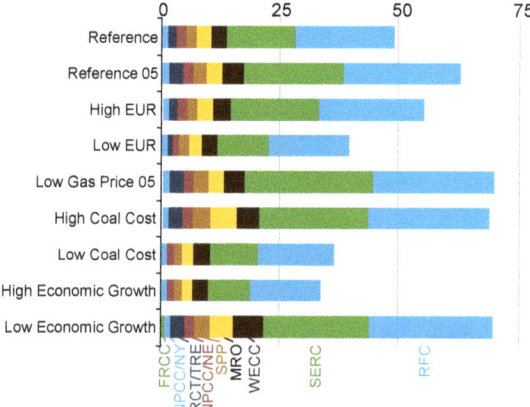

Figure 47. Cumulative retirements of coal-fired generating capacity by Electric Market Module region in nine cases, 2011-2035 (gigawatts)

In the Reference case, 58 gigawatts of natural gas combined-cycle capacity is added from 2020 to 2035, causing an increase in generation from natural gas during the period (Figures 48 and 49). In the Low EUR and Low Coal Cost cases, growth in natural gas combined-cycle capacity is slower. Although generation from natural gas increases overall with the addition of new capacity, utilization of existing combined-cycle plants drops slightly as higher natural gas prices reduce the frequency at which combined-cycle plants are dispatched.

In the GHG15 and GHG25 cases, electricity generation from natural gas exceeds generation from coal in 2020. Natural gas has one-half the CO_2 emissions of coal, and at relatively low CO_2 prices, natural gas generation is seen as an attractive alternative to coal. However, as CO_2 prices rise over the projection period, the increasing cost of generating electricity with natural gas causes the growth in natural gas generation to slow. In the GHG25 case, natural gas combined-cycle plants with CCS play a role in CO_2 mitigation, with 34 gigawatts of natural gas combined-cycle capacity added between 2022 and 2035.

Nuclear

Generation from nuclear power plants does not change significantly from Reference case levels in any of the non-GHG cases, due to the high cost of new nuclear plant construction relative to natural gas and renewables. In the GHG15 and GHG25 cases, nuclear power plants become more competitive with fossil plants, because they do not emit CO_2 and are needed to replace coal-fired capacity that is retired due to the cost of CO_2 emissions. In the GHG15 and GHG25 cases, generation from nuclear power is 57 percent and 121 percent higher, respectively, in 2035 than in 2010.

Renewables

Generation from renewable energy sources grows by 77 percent from 2010 to 2035 in the Reference case. Most of the growth in renewable electricity generation is a result of State RPS requirements, Federal tax credits, and—in the case of biomass—the availability of low-cost feedstocks. The change in renewable generation over the 2010-2035 period varies from a 102-percent increase in the High Economic Growth case to a 62-percent increase in the Low Economic Growth case. The largest growth in renewable generation is projected in the GHG15 and GHG25 cases, where renewable generation increases by about 150 percent from 2010 and 2035 in both cases. A price on CO_2 emissions makes generation from renewables more competitive with fossil plants without CCS.

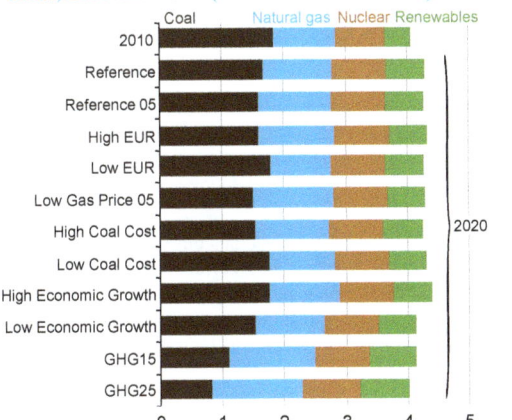

Figure 48. Electricity generation by fuel in eleven cases, 2010 and 2020 (trillion kilowatthours)

Installations of retrofit equipment

As discussed above, it is assumed that all coal-fired plants must have either FGD or DSI systems installed by 2015 to comply with environmental regulations. Because retirement is the only other option, cases with more retirements have fewer retrofits and vice versa (Figure 50). In the Reference 05 and Low Gas Price 05 cases, the relative cost of FGD units is higher because of the short payback period, making DSI a relatively more attractive option.

Emissions

SO_2 emissions are significantly below 2010 levels in 2015 in all cases, as a result of coal-fired capacity retirements and the installation of pollution control equipment to comply with MATS. *AEO2012* assumes that a DSI system, combined with a fabric filter, will remove 70 percent of a coal plant's SO_2 emissions, and an FGD unit 95 percent. As a result of the requirement for FGD or DSI systems, all coal plants larger than 25 megawatts that did not have FGD units installed in 2010 significantly reduce their SO_2 emissions after 2015 by

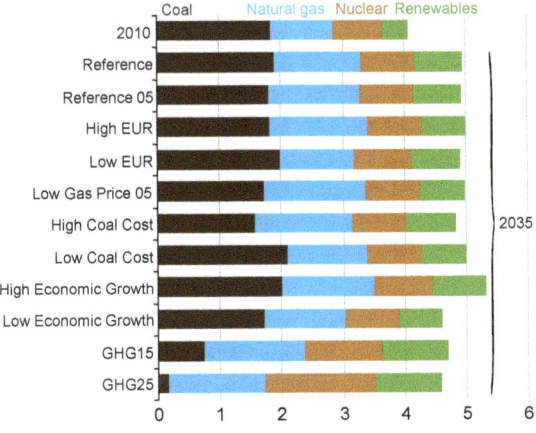

Figure 49. Electricity generation by fuel in eleven cases, 2010 and 2035 (trillion kilowatthours)

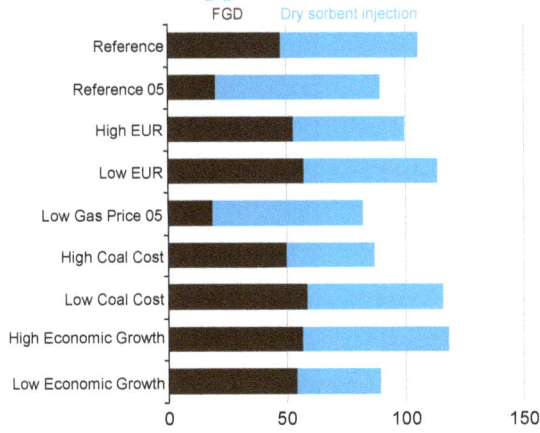

Figure 50. Cumulative retrofits of generating capacity with FGD and dry sorbent injection for emissions control, 2011-2020 (gigawatts)

installing control equipment. In all cases, coal-fired generation is down overall, which also contributes to the decline in emissions. SO_2 emissions increase after 2020 in all non-GHG cases, as coal-fired generation increases with rising natural gas prices. Because DSI and FGD retrofits do not remove all the SO_2 from coal-fired power plant emissions, increases in coal-fired generation result in higher SO_2 emissions, although they are still much lower than comparable 2010 levels. Also, the level of SO_2 reduction is proportional to the amount of coal-fired generation, and therefore the cases with the highest projected levels of coal-fired generation also project the highest levels of SO_2 emissions.

The projections for mercury emissions are similar. After a sharp drop in 2015, mercury emissions begin to rise slowly as coal-fired generation increases in all non-GHG cases. However, mercury emissions in 2035 still are significantly below 2010 levels, as the requirement for a 90-percent reduction in uncontrolled emissions of mercury remains binding throughout the projection.

NO_x emissions are not directly affected by MATS, but both annual and seasonal cap-and-trade programs are included in CSAPR. Emissions reductions relative to 2010 levels are small throughout the projection period in most cases, mainly because compliance with CSAPR NO_x regulations is required in only 26 States, and 2010 emissions levels already were close to the cap.

CO_2 emissions from the electric power sector fall slightly in cases that project declines in coal use, but the largest reductions occur in the GHG15 and GHG25 cases. In the GHG15 case, CO_2 emissions from the electric power sector are 46 percent below 2010 levels in 2035, and in the GHG25 case they are 76 percent below 2010 levels.

Electricity prices

Real electricity prices in 2035 are 3 percent above the 2010 level in the Reference case. The increase is relatively modest because natural gas prices increase slowly, and several alternatives for complying with the environmental regulations are available. When lower natural gas prices are assumed, real electricity prices decline relative to the Reference case. Both the GHG15 and GHG25 cases assume that costs for CO_2 emission allowances are passed through directly to customers. Therefore, average electricity prices in the GHG15 and GHG25 cases in 2035 are 25 percent and 33 percent higher, respectively, than in the Reference case. The GHG15 and GHG25 cases do not include any of the rebates to electricity consumers included in some other GHG policy proposals, which would reduce the impact on electricity prices.

9. Nuclear power in *AEO2012*

In the *AEO2012* Reference case, electricity generation from nuclear power in 2035 is 10 percent above the 2010 total. The nuclear share of overall generation, however, declines from 20 percent in 2010 to 18 percent in 2035, reflecting increased shares for natural gas and renewables.

In the Reference case, 15.8 gigawatts of new nuclear capacity is added from 2010 through 2035, including both new builds (a total of 8.5 gigawatts) and power uprates at operating nuclear power plants (7.3 gigawatts). A total of 6.1 gigawatts of nuclear capacity is retired in the Reference case, with most of the retirements coming after 2030. However, given the current uncertainty about likely lifetimes of nuclear plants now in operation and the potential for new builds, *AEO2012* includes several alternative cases to examine the impacts of different assumptions about future nuclear power plant uprates and operating lifetimes.

Uprates

Power plant uprates involve projects that are intended to increase the licensed capacity of existing nuclear power plants and permit those plants to generate more electricity. The U.S. Nuclear Regulatory Commission (NRC) must approve all uprate projects before they are undertaken and verify that the reactors will be able to operate safely at higher levels of output. Power plant uprates can increase plant capacity by 1 to 20 percent, depending on the size and type of the uprate project. Capital expenditures may be small (e.g., installing a more accurate sensor) or significant (e.g., replacing key plant components, such as turbines).

In developing projections for nuclear power, EIA relies on both reported data and estimates. Reported data come from Form EIA-860 [78], which requires all nuclear power plant owners to report any plans for building new plants or making major modifications to existing plants (such as uprates) over the next 10 years. In 2010, operators reported that they intended to complete uprate projects sometime during the next 10 years, which together would add a total of 0.8 gigawatts of new capacity. In addition to the reported plans for capacity uprates, EIA assumed that additional power uprates over the period from 2011 to 2035 would add another 6.5 gigawatts of capacity, based on interactions with EIA stakeholders with significant experience in implementing power plant uprates.

New builds

Building a new nuclear power plant is a tremendously complex project that can take many years to complete. Specialized high-wage workers, expensive materials and components, and engineering and construction expertise are required, and only a select group of firms worldwide can provide them. In the current economic environment of low natural gas prices and flat demand for electricity, the overall market conditions for new nuclear power plants are challenging.

Nuclear power plants are among the most expensive options for new generating capacity available today [79]. In the *AEO2012* Reference case, the overnight capital costs associated with building a nuclear power plant planned in 2012 are assumed to be $5,335 per kilowatt of capacity, which translates to $11.7 billion for a dual-unit 2,200-megawatt power plant. The overnight costs do not include additional costs such as financing, interest carried forward, and peripheral infrastructure updates [80]. Despite the cost, however, deployment of new nuclear capacity supports the long-term resource plans of many utilities, by allowing fuel diversification and providing a hedge in the future against potential GHG emissions regulations or natural gas prices that are higher than expected.

Incentive programs exist to encourage the construction of new reactors in the United States. At the Federal level, the Energy Policy Act of 2005 (EPACT05) established a loan guarantee program for new nuclear plants completed and in operation by 2020 [81]. A total of $18.5 billion is available, of which $8.3 billion has been conditionally committed to the construction of Southern Company's Vogtle Units 3 and 4 [82]. EPACT05 also provides a PTC of $18 per megawatthour for electricity produced during the first 8 years of operation for a new nuclear plant [83]. New nuclear plants must be operational by 2021 to be eligible for the PTC, and the credit is limited to the first 6 gigawatts of new nuclear plant capacity. In addition to Federal incentives, several States provide favorable regulatory environments for new nuclear plants by allowing plant owners to recover their investments through retail electricity rates.

Several utilities are moving forward with plans to deploy new nuclear power plants in the United States. The Reference case reflects those plans by including 6.8 gigawatts of new nuclear capacity over the projection period. As reported on Form EIA-860, 5.5 gigawatts of new capacity (Vogtle Units 3 and 4, Summer Units 2 and 3, and Watts Bar Unit 2) are expected to be operational by 2020 [84]. The Reference case also includes 1.3 gigawatts associated with the construction of Bellefonte Unit 1, which the Tennessee Valley Authority reflects in its Integrated Resource Plan [85].

In addition to reported plans for new nuclear power plants, 1.8 gigawatts of unplanned capacity is built in the later years of the Reference case. Higher natural gas prices, recovering demand for electricity, and the need to make up for the loss of a limited amount of nuclear capacity all play a role in the additional builds.

Long-term operation of the existing nuclear power fleet

The NRC has the authority to issue initial operating licenses for commercial nuclear power plants for a period of 40 years. As of December 31, 2011, there were 7 reactors that received their initial full power operating licenses over 40 years ago. Among this set of reactors, Oyster Creek Unit 1 was the first reactor to operate for over 40 years, after receiving its initial full power operating license in August 1969. Oyster Creek Unit 1 was followed by Dresden Units 2 and 3, H.B. Robinson Unit 2, Monticello, Point Beach 1, and R.E. Ginna. The decision to apply for an operating license renewal is made by nuclear power plant owners, typically based on economics and the ability to meet NRC requirements. As of January 2012, the NRC had granted license renewals to 71 of the 104 operating reactors in the United States, allowing them to operate for a total of 60 years [86]. Currently, the NRC is reviewing license renewal applications for 15 reactors and expects to receive applications from another 14 reactors between 2012 and 2016 [87].

NRC regulations do not limit the number of license renewals a nuclear power plant may be granted. The nuclear power industry is preparing applications for license renewals that would allow continued operation beyond 60 years. The first application seeking approval to operate for 80 years is tentatively scheduled to be submitted by 2013. Some aging nuclear plants may, however, pose a variety of issues that could lead to decisions not to apply for a second license renewal, such as high operation and maintenance costs or the need for large capital expenditures to meet NRC requirements. Industry research on long-term reactor operations and aging management is focused on identifying challenges that aging facilities might encounter and formulating potential approaches to meet those challenges [88]. Typical challenges involve materials degradation, safety margins, and assessing the integrity of concrete structures. In the Reference case, 6.1 gigawatts of nuclear power plant capacity is retired by 2035, based on uncertainty related to issues associated with long-term operations and aging management [89].

It should be noted that although the Oyster Creek Generating Station in Lacey Township, New Jersey, received a license renewal and could operate until 2029, the plant's owner has reported to EIA that it will be retired in 2019, after 50 years of operation. The *AEO2012* Reference case includes this reported early retirement. Also, given the evolving nature of the NRC's regulatory response to the accident at Japan's Fukushima Daiichi nuclear power plant in March 2011, the Reference case does not include retirements directly related to the accident (for example, retirements prompted by potential new NRC regulatory requirements for safety retrofits).

Sensitivity cases

The *AEO2012* Low Nuclear case assumes that only the planned nuclear plant uprates already reported to EIA will be completed. Uprates that are currently under review or expected to be submitted to the NRC are not included. The Low Nuclear case also assumes that all nuclear power plants will be retired after 60 years of operation, resulting in a 30.9-gigawatt reduction in U.S. nuclear power capacity from 2010 to 2035. Figure 51 shows nuclear capacity retirements in the Low Nuclear case by NERC region. It should be noted that after the retirement of Oyster Creek in 2019, the next nuclear plant retirement occurs in 2029 in the Low Nuclear case. No new nuclear plants are built in the Low Nuclear case beyond the 6.8 gigawatts already planned.

In the High Nuclear case, in addition to plants already under construction, plants with active license applications at the NRC are constructed, provided that they have a tentatively scheduled mandatory hearing before the NRC or Atomic Safety and Licensing Board and deploy a currently certified design for the nuclear steam supply system, such as the AP1000. With this assumption, an additional 6.2 gigawatts of new nuclear capacity is added relative to the Reference case. The High Nuclear case also assumes that all existing nuclear power plants will receive their second license renewals and will operate through 2035. Uprates in the

Issues in focus

High Nuclear case are consistent with those in the Reference case. The only retirement included in the High Nuclear case is the announced early retirement of Oyster Creek in 2019.

Results

In the Reference case, 8.5 gigawatts of new nuclear power plant capacity is added from 2010 to 2035, including the 6.8 gigawatts reported to EIA (referred to as "planned") and 1.8 gigawatts built endogenously in NEMS (referred to as "unplanned"). Unplanned capacity is added starting in 2030 in response to rising natural gas prices, which make new nuclear power plants a more competitive option for new electric capacity. In the High Nuclear case, planned capacity additions are almost double those in the Reference case, but unplanned additions are lower. The price of natural gas delivered to the power sector in the High Nuclear case is lower than in the Reference case, making the economics of nuclear power plants slightly less attractive. The additional planned capacity in the High Nuclear case also reduces the need for new unplanned capacity. No unplanned capacity is added in the Low Nuclear case.

Nuclear power generation in 2035 reflects the differences in capacity that occur in the nuclear cases. In the High Nuclear case, nuclear generation in 2035 is 10 percent higher than in the Reference case, and the nuclear share of total generation is 20 percent, as compared with 18 percent in the Reference case. The increase in nuclear capacity in the High Nuclear case contributes to an increase in total electricity generation, in spite of lower levels of generation from natural gas (4 percent lower than in the Reference case in 2035) and coal and renewables (less than 1 percent lower for each fuel).

In the Low Nuclear case, generation from nuclear power in 2035 is 30 percent lower than in the Reference case, due to the loss of 30.9 gigawatts of nuclear capacity that is retired after 60 years of operation. As a result, the nuclear share of total generation is reduced to 13 percent. The loss of generation is made up primarily by increased generation from natural gas (12 percent higher than in the Reference case in 2035), coal (1 percent higher), and renewables (3 percent higher).

Real average electricity prices in 2035 are 1 percent lower in the High Nuclear case than in the Reference case, as slightly less natural gas capacity is dispatched, lowering the marginal price of electricity. In the Low Nuclear case, average electricity prices in 2035 are 5 percent higher than in the Reference case as a result of the retirement of a significant amount of nuclear capacity, which has relatively low operating costs, and its replacement with natural gas capacity, which has higher fuel costs that are passed through to consumers in retail electricity prices. With all nuclear power plants being retired after 60 years of operation in the Low Nuclear case, an additional 12 gigawatts of nuclear capacity would be shut down between 2035 and 2040.

The impacts of nuclear plant retirements on retail electricity prices in the Low Nuclear case are more apparent in regions with relatively large amounts of nuclear capacity. For example, electricity prices in the Low Nuclear case are 7 percent higher than in the Reference case for the NERC MRO Region, and 6 percent higher in the Northeast, Mid-Atlantic, and Southeast regions. Even in regions where no nuclear capacity is retired, there are small increases in electricity prices relative to the Reference case, because higher demand for natural gas in regions with nuclear plant retirements affect prices nationwide.

The Reference case projections for CO_2 emissions also are affected by changes in assumptions about nuclear plant lifetimes. In the Low Nuclear case, CO_2 emissions from the electric power sector in 2035 are 3 percent higher than in the Reference case as a result of switching from nuclear generation to natural gas and coal, both which produce more CO_2 emissions. In the High Nuclear case, CO_2 emissions from the power sector are slightly (1 percent) lower than in the Reference case. Table 12 summarizes key results from the *AEO2012* Reference, High Nuclear, and Low Nuclear cases.

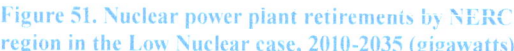

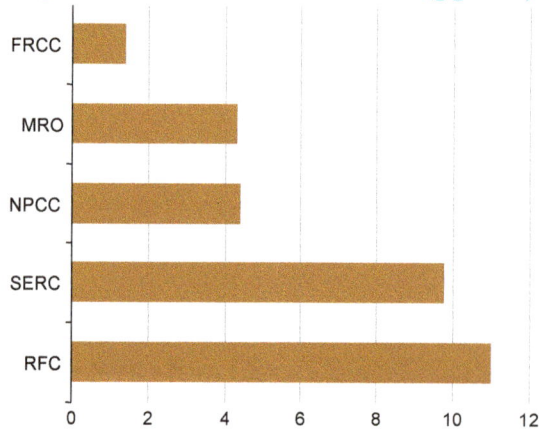

Figure 51. Nuclear power plant retirements by NERC region in the Low Nuclear case, 2010-2035 (gigawatts)

10. Potential impact of minimum pipeline throughput constraints on Alaska North Slope oil production

Introduction

Alaska's North Slope oil production has been declining since 1988, when average annual production peaked at 2.0 million barrels per day. In 2010, about 600,000 barrels per day of oil was produced on the North Slope. Although new North Slope oil fields have started production since 1988, the decline of North Slope production has resulted largely from depletion of the North Slope's two largest fields, Prudhoe Bay and Kuparuk River. Recently, Alyeska Pipeline Service Company (Alyeska), the operator of the Trans-Alaska Pipeline System (TAPS), stated that oil pipeline transportation problems could begin when throughput falls below 550,000 barrels per day and become increasingly severe with further declines [90].

Alyeska estimates that TAPS operational problems could become considerable when throughput falls below 350,000 barrels per day. The decline of both North Slope oil production and TAPS throughput raises the possibility that North Slope oil production might be shut down, with the existing oil fields plugged and abandoned sometime before 2035. That possibility is discussed here, as well as alternatives that could prolong the life of North Slope oil fields and TAPS beyond 2035.

Background

Declining TAPS throughput

TAPS is an 800-mile crude oil pipeline that transports North Slope oil production south to the Alyeska marine terminal in Valdez, Alaska. The crude oil is then transported by tankers to West Coast refineries. TAPS currently is the only means for transporting North Slope crude oil to refineries and the petroleum consumption markets they serve.

From 2004 through 2006, Alyeska reconfigured and refurbished TAPS, spending about $400 million to $500 million [91] both to reduce operating expenses and to permit TAPS to operate at lower flow rates, with a potential minimum mechanical throughput rate thought to be about 200,000 barrels per day at that time [92]. As North Slope oil production has declined, however, concern about TAPS operation under low flow conditions has grown [93]. In August 2008, Alyeska initiated its Low Flow Impact Study, which was released on June 15, 2011 [94].

The Alyeska study identified the following potential problems that might occur as TAPS throughput declines from the current production levels:

- Water dropout from the crude oil, which could cause pipeline corrosion
- Ice formation in the pipe if the oil temperature drops below freezing
- Wax precipitation and deposition
- Soil heaving.

Other potential operational issues at low flow rates include sludge dropout, reduced ability to remove wax, reduction in pipeline leak detection efficiency, pipeline shutdown and restart, and the running of pipeline pigs that both clean the pipeline and check its integrity.

Although TAPS low flow problems could begin at volumes around 550,000 barrels per day in the absence of any mitigation, their severity is expected to increase as throughput declines further. As the types and severity of problems multiply, the investment required to mitigate these is expected to increase significantly. Because of the many and diverse operational problems expected to occur at throughput volumes below 350,000 barrels per day, considerable investment could be required to keep the pipeline operational below that threshold. The Alyeska study does not provide any estimates of what it might cost to keep the pipeline operational below either 550,000 or 350,000 barrels per day. Currently, Alyeska is conducting tests and analyses to determine the likely efficacy and costs of different remedies.

Mitigating the decline of North Slope oil production

Although much of the public focus has been on the operational capability of TAPS at low flow rates, the more fundamental issue is declining oil production. The TAPS low flow issue would be alleviated most readily by discovery and production of large new sources of oil on the North Slope. Potential sources of significant North Slope oil production are located offshore in the Chukchi and Beaufort Seas and onshore in shale and heavy oil deposits. The Arctic National Wildlife Refuge (ANWR) is also estimated to hold approximately 10.4 billion barrels of technically recoverable oil resources, but Federal oil and gas leasing in ANWR currently is prohibited [95]. Another potential source of new TAPS volumes would be the conversion of North Slope natural gas resources to either methanol or Fischer-Tropsch petroleum products that could be transported to market via TAPS. Finally, in the absence of new North Slope petroleum supplies, alternative crude oil transportation facilities could be developed, such as a new small-diameter pipeline running parallel to the TAPS route [96] or a new offshore oil terminal for North Slope production.

Table 12. Summary of key results from the Reference, High Nuclear, and Low Nuclear cases, 2010-2035

Projection	Reference	High Nuclear	Low Nuclear
Nuclear plant cumulative retirements (gigawatts)	6.1	0.6	30.9
Generating capacity cumulative additions (gigawatts)			
Coal	16.6	16.1	18.9
Natural gas	141.6	126.2	147.6
Nuclear capacity uprates	7.3	7.3	0.8
Planned nuclear capacity additions	6.8	13.5	6.8
Unplanned nuclear capacity additions	1.8	1.3	--
Renewables	67.4	64.5	73.4
Average delivered electricity price, 2035 (2010 cents per kilowatthour)	10.1	10.0	10.6
Average delivered natural gas price for electric power, 2035 (2010 dollars per million Btu)	7.21	7.00	8.03
CO_2 emissions from electric power generation, 2035 (million metric tons)	2,330	2,301	2,404

Which of these potential low-flow solutions (or combination thereof) may ultimately come to fruition is impossible to determine at this time. Moreover, each solution comes with its own unique set of costs, risks, and lead times. Not only does each solution entail its own set of risks, there is also a significant risk that production from existing North Slope fields might decline much faster than anticipated and/or that the cost of operating those fields might escalate much faster than expected. Under those circumstances, there is a risk that any solution(s) could be both too little and too late, because the North Slope oil fields would be shut down before a TAPS solution could be implemented.

How quickly TAPS flows will decline, the types of low flow problems that might develop, and the degree of mitigation required depend on the success or failure of current offshore and onshore oil exploration and development programs and the quality of the oil produced. For example, low-viscosity oil is less problematic to TAPS operations than heavy, viscous oil. Because the future success of North Slope oil exploration and development is unknown, it is prudent to consider the circumstances under which North Slope oil production might cease altogether, causing a shutdown of the TAPS pipeline.

Aside from the question of what it might cost to keep TAPS operating at lower flow rates, an additional question is what it might cost to keep the existing North Slope oil fields producing. Even if the continued operation of TAPS were not in question, each North Slope oil field's production will eventually decline to a point at which it is no longer economical to keep the field operating. Oil and gas fields typically are shut down and abandoned when operating and maintenance costs exceed production revenues. At that point, wells are plugged and abandoned, surface equipment is removed, and the land is remediated to meet State and Federal requirements.

Although the cost structure of North Slope field production as production declines is unknown, production generally can be sustained profitably at lower production rates when oil prices are higher. Similarly, the economic feasibility of mitigating the problems arising from TAPS low flow rates improves when oil prices are higher. Consequently, revenues generated by North Slope oil production will play a pivotal role in determining the continued economic viability of existing North Slope oil fields, the development of new fields, the continued operation of TAPS at lower flow rates, and the potential development of new transportation facilities.

Several basic strategies have been employed to mitigate declining oil production and revenues from existing oil fields. First, the field operator can drill in-fill wells into those portions of the reservoir where oil cannot flow to existing production wells. Second, the operator can use enhanced oil recovery (EOR) that involves injecting steam or gases (along with water) to reduce viscosity and increase oil volumes as an aid to moving oil to the production wells. Currently, methane and natural gas liquids are being reinjected with water into many North Slope oil fields to achieve this outcome, which is referred to as "miscible hydrocarbon" EOR [97].

Drilling in-fill and EOR injection wells requires investments that are paid for through "maintenance" capital expenditures [98]. Both activities provide diminishing returns over time, as less oil typically is recovered with each new in-fill or EOR well, causing the cost per barrel of oil recovered to rise over time. Table 13 shows the number of in-fill and gas/water injection wells completed in 2010 at the three largest North Slope oil fields.

The diminishing returns from new in-fill and EOR wells is demonstrated in recent remarks by a ConocoPhillips official who noted that approximately $630 million was to be spent on maintenance capital expenditures in 2011, compared with about $240 million in 2001 [99]. In 2001 and 2010, ConocoPhillips provided 37.4 percent and 39.1 percent, respectively, of total North Slope oil production [100]. Using those percentages to scale up ConocoPhillips maintenance capital expenditures so that they represent total capital expenditures for North Slope maintenance, then total North Slope maintenance costs can be estimated at about $640 million in 2001 and $1.6 billion in 2011—a 150-percent increase over a period in which total North Slope oil production declined from 931,000 barrels per day to 562,000 barrels per day. If maintenance capital expenditures increased at the same rate (150 percent) over the next 10 years, they could be as high as $4 billion in 2021.

Another method for extending oil production is to produce increasing amounts of water relative to oil [101]. As oil is produced from a reservoir, water typically enters the formation, causing the water-to-oil ratio to increase exponentially over time as oil production volumes decline [102]. Because the cost per barrel for handling and reinjecting reservoir water typically is relatively constant, the operating cost per barrel of oil produced increases exponentially over time.

Shutdown and abandonment assumptions

According to the Alyeska study, a TAPS throughput of about 350,000 barrels per day appears to be the threshold at which significant investment would be required to permit lower TAPS throughput. *AEO2012* adopts the 350,000 barrel per day figure as

Table 13. Alaska North Slope wells completed during 2010 in selected oil fields

Production unit	Miscible hydrocarbon EOR	In-fill development wells	Gas/water injection wells	Total wells
Colville River	Yes	8	6	14
Kuparuk River	Yes	25	26	51
Prudhoe Bay	Yes	68	8	76
Subtotal		101	40	141
Total North Slope				168

the threshold for either making significant investments in TAPS or the alternatives, or shutting down and decommissioning TAPS and the North Slope oil fields [103].

In the *AEO2012* analysis, the shutdown and decommissioning of TAPS and the North Slope oil fields are also conditional on whether North Slope wellhead oil production revenues fall below a specific level. The appropriate revenue threshold is uncertain, because there is little or no information available to the public on operating and maintenance costs for existing oil fields, how those costs have grown historically as production has declined, or how they might grow in the future. Similarly, there are no public data available on what it might cost to keep TAPS operating as throughput declines [104]. Given the lack of public information, this analysis endeavors to determine both future North Slope production revenues in alternative oil price cases and an order-of-magnitude estimate of wellhead production costs.

AEO2012 assumes that, in order for the North Slope fields to be shut down, plugged, and abandoned, two conditions would need to be met simultaneously: TAPS throughput at or below 350,000 barrels per day and total North Slope oil production revenues at or below $5 billion per year. It is also assumed that if those two conditions were met, TAPS would be decommissioned and dismantled, and North Slope oil exploration and production activities would cease [105].

The $5 billion threshold for North Slope oil production revenue used in *AEO2012* is not intended to be conclusive regarding the conditions under which the North Slope oil fields and TAPS would remain in operation. As noted earlier, in-fill and EOR well drilling requirements could escalate to about $4 billion per year by 2021 [106]. Moreover, with the State of Alaska royalty rate currently at about 18.5 percent [107], a $5 billion revenue level would equate to almost $1 billion in royalties.

Also, an order of magnitude estimate of operating costs can be made by examining what oil companies report for their annual production expenses. For example, ExxonMobil reported a range of regional production costs per barrel of oil equivalent (excluding taxes) of $6.17 to $20.07 per barrel in 2010, with the U.S. average production cost being $10.67 per barrel [108]. At 350,000 barrels per day, a North Slope operating expense of $10 to $20 per barrel would equate to $1.28 to $2.56 billion per year in annual operating expenses. Of course, production costs could well exceed $20 per barrel as North Slope oil production declines.

Although the $5 billion North Slope revenue figure is not conclusive with regard to the actual annual costs faced by North Slope field operators in the future, it is a reasonable estimate in light of the sum of current maintenance capital expenditures ($1.6 billion), estimated operating expenses at 350,000 barrels per day ($1.28 to $2.56 billion), and a royalty cost of about $1 billion. As discussed below, the oil production revenue threshold serves to either advance or delay the date when TAPS and North Slope oil production would be shut down.

The final assumption is that a complete shutdown of North Slope oil production would occur in the year in which both the throughput and revenue criteria are satisfied. In reality, the actual shutdown of North Slope oil production might be extended over a number of years and could begin either before or after the year in which the criteria employed by North Slope producers are met.

Projections

A shutdown of North Slope oil production before 2035 is projected only in the Low Oil Price case, which shows both TAPS throughput and North Slope oil revenues falling below the 350,000 barrels per day and $5 billion per year thresholds, respectively, in 2026 (Figures 52 and 53). In both the Reference and High Oil Price cases, oil prices are sufficiently high both to stimulate the

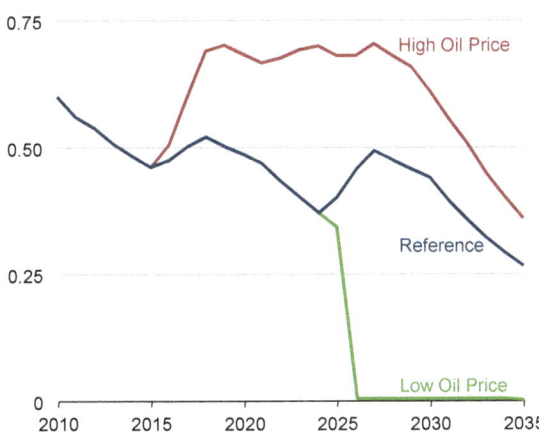

Figure 52. Alaska North Slope oil production in three cases, 2010-2035 (million barrels per day)

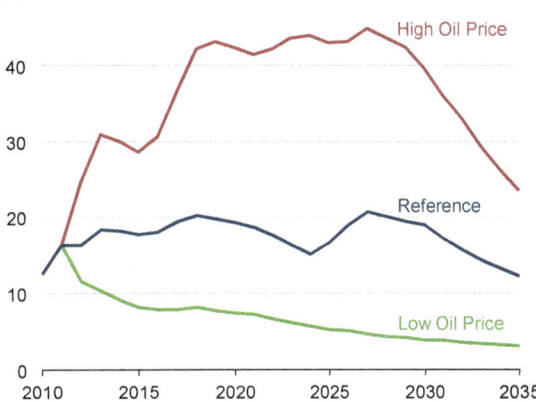

Figure 53. Alaska North Slope wellhead oil revenue in three cases, assuming no minimum revenue requirement, 2010-2035 (billion 2010 dollars per year)

development of new North Slope oil fields, especially offshore, and to provide sufficient oil production revenues to keep the North Slope producing oil through 2035.

Figure 53 shows the projected North Slope oil production revenue stream over time in the three price cases, with North Slope oil production continuing even after production volume and revenue requirements are no longer met in the Low Oil Price case. Thus, if the minimum North Slope revenue requirement were $7.5 billion, a shutdown of North Slope production could occur as soon as 2020, but only in the Low Oil Price case.

There is considerable uncertainty about the long-term viability of North Slope oil production and continued operation of TAPS through 2035. The two most important determinants of their future viability are the wellhead oil price that North Slope producers receive and the availability and cost of developing new North Slope oil resources. Those two factors will determine whether new oil fields are developed, whether existing oil fields remain sufficiently profitable to continue operating, and whether the investments required to keep TAPS operating at flow rates below 350,000 barrels per day are economically feasible.

The *AEO2012* Low and High Oil Price cases suggest that North Slope oil production will remain viable across a wide range of oil prices. Only in the Low Oil Price case are North Slope wellhead oil revenues sufficiently low to cause a shutdown of North Slope oil production. If the Low Oil Price case represents a low-probability outer boundary for future oil prices, then the likely future outcome is that North Slope oil production will continue until at least 2035, if not longer.

11. U.S. crude oil and natural gas resource uncertainty

A common measure of the long-term viability of U.S. domestic crude oil and natural gas as an energy source is the remaining technically recoverable resource (TRR). Estimates of TRR are highly uncertain, however, particularly in emerging plays where few wells have been drilled. Early estimates tend to vary and shift significantly over time as new geological information is gained through additional drilling, as long-term productivity is clarified for existing wells, and as the productivity of new wells increases with technology improvements and better management practices. TRR estimates used by EIA for each *AEO* are based on the latest available well production data and on information from other Federal and State governmental agencies, industry, and academia.

The remaining TRR consist of "proved reserves" and "unproved resources." *Proved reserves* of crude oil and natural gas are the estimated volumes expected to be produced, with reasonable certainty, under existing economic and operating conditions [*109*]. Proved reserves are also company financial assets reported to investors, as determined by U.S. Securities and Exchange Commission regulations. *Unproved resources* are additional volumes estimated to be technically recoverable without consideration of economics or operating conditions, based on the application of current technology [*110*]. As wells are drilled and field equipment is installed, unproved resources become proved reserves and, ultimately, production.

AEO estimates of TRR for shale gas and tight oil [*111*] have changed significantly in recent years (Table 14) [*112*]. In particular, the estimates of shale gas TRRs have changed significantly since the *AEO2011* was published, based on new well performance data and United States Geological Survey (USGS) resource assessments. For example, in the past year the USGS has released resource assessments for five basins: Appalachian (Marcellus only), Arkoma, Texas-Louisiana-Mississippi Salt, Western Gulf, and Anadarko [*113*]. The shale gas and tight oil formations in those five basins were the primary focus of EIA's resource revisions for *AEO2012*. In 2002, the USGS estimated Marcellus TRR at 1.9 trillion cubic feet; in 2011, the updated USGS estimate for Marcellus was 84 trillion cubic feet (see the following article for more discussion). For the four other basins, shale gas and tight oil TRR had not been assessed previously. The USGS has not published an assessment of the Utica play in the Appalachian Basin.

The remainder of this discussion describes how estimates of remaining U.S. unproved technically recoverable resources of shale gas and tight oil are developed for *AEO*, and how uncertainty in those estimates could affect U.S. crude oil and natural gas markets in the future.

Estimating technically recoverable resources of shale gas and tight oil

The remaining unproved TRR for a continuous-type shale gas or tight oil area is the product of (1) land area, (2) well spacing (wells per square mile), (3) percentage of area untested, (4) percentage of area with potential, and (5) EUR per well [*114*]. The USGS periodically publishes shale gas resource assessments that are used as a guide for selection of key parameters in the calculation of the TRR used in the *AEO*. The USGS seeks to assess the recoverability of shale gas and tight oil based on the wells drilled and technologies deployed at the time of the assessment.

The *AEO* TRRs incorporate current drilling, completion, and recovery techniques, requiring adjustments to the USGS estimates, as well as the inclusion of shale gas and tight oil resources not yet assessed by USGS. When USGS assessments and underlying data become publicly available, the USGS assumptions for land area, well spacing, and percentage of area with potential typically are used by EIA to develop the *AEO* TRR estimates. EIA may revise the well spacing assumptions in future *AEOs* to reflect evolving drilling practices. If well production data are available, EIA analyzes the decline curve of producing wells to calculate the expected EUR per well from future drilling.

Of the five basins recently assessed by the USGS, underlying details have been published only for the Marcellus shale play in the Appalachian basin. *AEO2012* assumptions for the other shale plays are based on geologic surveys provided from State agencies (if available), analysis of available production data, and analogs from current producing plays with similar geologic properties (Table 15). For *AEO2012*, only eight plays are included in the tight oil category (Table 16). Additional tight oil resources are expected to be included in the tight oil category in future *AEOs* as more work is completed in identifying currently producing reservoirs that may be categorized as tight formations, and as new tight oil plays are identified and incorporated.

A key assumption in evaluating the expected profitability of drilling a well is the EUR of the well. EURs vary widely not only across plays but also within a single play. To capture the economics of developing each play, the unproved resources for each play within each basin are divided into subplays—first across States (if applicable), and then into three productivity categories: best, average, and below average. Although the average EUR per well for a play may not change by much from one *AEO* to the next, the range of well performance encompassed by representative EURs can change substantially (Table 17).

For every *AEO*, the EUR for each subplay is determined by fitting a hyperbolic decline curve to the latest production history, so that changes in average well performance can be captured. Annual reevaluations are particularly important for shale gas and tight oil formations that have undergone rapid development. For example, because there has been a dramatic change from drilling vertical wells to drilling horizontal wells in most tight oil and shale gas plays since 2003, EURs for those plays based on vertical well performance are less useful for estimating production from future drilling, given that most new wells are expected to be primarily horizontal.

In addition, the shape of the annual well production profiles associated with the EUR varies substantially across the plays (Figure 54). For example, in the Marcellus, Fayetteville, and Woodford shale gas plays, nearly 65 percent of the well EUR is produced in the first 4 years. In contrast, in the Haynesville and Eagle Ford plays, 95 percent and 82 percent, respectively, of the well EUR is produced in the first four years. For a given EUR level, increased "front loading" of the production profile improves well economics, but it also implies an increased need for additional drilling to maintain production levels.

At the beginning of a shale play's development, high initial well production rates result in significant production growth as drilling activity in the play increases. The length of time over which the rapid growth can be sustained depends on the size of the

Table 14. Unproved technically recoverable resource assumptions by basin

Basin	AEO2006 (as of 1/1/2004)	AEO2007 (as of 1/1/2005)	AEO2008 (as of 1/1/2006)	AEO2009 (as of 1/1/2007)	AEO2010 (as of 1/1/2008)	AEO2011 (as of 1/1/2009)	AEO2012 (as of 1/1/2010)
Shale gas (trillion cubic feet)							
Appalachian	15	15	14	51	59	441	187
Fort Worth	40	39	38	60	60	20	19
Michigan	11	11	11	10	10	21	18
San Juan	10	10	10	10	10	12	10
Illinois	3	3	3	4	4	11	11
Williston	4	4	4	4	4	7	3
Arkoma	--	42	42	49	45	54	27
Anadarko	--	3	3	7	6	3	13
TX-LA-MS Salt	--	--	--	72	72	80	66
Western Gulf	--	--	--	--	18	21	59
Columbia	--	--	--	--	51	41	12
Uinta	--	--	--	--	7	21	11
Permian	--	--	--	--	--	67	27
Greater Green River	--	--	--	--	--	18	13
Black Warrior	--	--	--	--	--	4	5
Shale gas total	**83**	**126**	**125**	**267**	**347**	**827**	**482**
Tight oil (billion barrels)							
Williston	--	3.7	3.7	3.7	3.6	3.6	5.4
San Joaquin/Los Angeles	--	--	--	--	15.4	15.4	13.7
Rocky Mountain basins	--	--	--	--	5.1	5.1	6.5
Western Gulf	--	--	--	--	5.6	5.6	5.7
Permian	--	--	--	--	--	1.6	1.6
Anadarko	--	--	--	--	--	0.2	0.3
Tight oil total	**--**	**3.7**	**3.7**	**3.7**	**29.7**	**31.5**	**33.2**

Issues in focus

technically recoverable resource in each play, the rate at which drilling activity increases, and the extent of the play's "sweet spot" area [115]. In the longer term, production growth tapers off as high initial production rates of new wells in "sweet spots" are offset by declining rates of existing wells, and as drilling activity moves into less-productive areas. As a result, in the later stages of a play's resource development, maintaining a stable production rate requires a significant increase in drilling.

Table 15. Attributes of unproved technically recoverable resources for selected shale gas plays as of January 1, 2010

Basin/Play	Area (square miles)	Average well spacing (wells per square mile)	Percent of area untested	Percent of area with potential	Average EUR (billion cubic feet per well)	Number of potential wells	TRR (billion cubic feet)
Appalachian							
Marcellus	104,067	5	99	18	1.56	90,216	140,565
Utica	16,590	4	100	21	1.13	13,936	15,712
Arkoma							
Woodford	3,000	8	98	23	1.97	5,428	10,678
Fayetteville	5,853	8	93	23	1.30	10,181	13,240
Chattanooga	696	8	100	29	0.99	1,633	1,617
Caney	2,890	4	100	29	0.34	3,369	1,135
TX-LA-MS Salt							
Haynesville/Bossier	9,320	8	98	34	2.67	24,627	65,860
Western Gulf							
Eagle Ford	7,600	6	99	47	2.36	21,285	50,219
Pearsall	1,420	6	100	85	1.22	7,242	8,817
Anadarko							
Woodford	3,350	4	99	29	2.89	3,796	10,981
Total, selected shale gas plays						181,714	318,825
Total, all U.S. shale gas plays						410,722	481,783

Table 16. Attributes of unproved technically recoverable tight oil resources as of January 1, 2010

Basin/Play	Area (square miles)	Average well spacing (wells per square mile)	Percent of area untested	Percent of area with potential	Average EUR (million barrels per well)	Number of potential wells	TRR (million barrels)
Western Gulf							
Austin Chalk	16,078	3	72	61	0.13	21,165	2,688
Eagle Ford	3,200	5	100	54	0.28	8,665	2,461
Anadarko							
Woodford	3,120	6	100	88	0.02	16,375	393
Permian							
Avalon/Bone Springs	1,313	4	100	78	0.39	4,085	1,593
Spraberry	1,085	6	99	72	0.11	4,636	510
Rocky Mountain basins							
Niobrara	20,385	8	97	80	0.05	127,451	6,500
Williston Bakken[a]	6,522	2	77	97	0.55	9,767	5,372
San Joaquin/Los Angeles							
Monterey/Santos	2,520	12	98	93	0.50	27,584	13,709
Total tight oil						219,729	33,226

[a] Includes Sanish-Three Forks formation.

Issues in focus

The amount of drilling that occurs each year depends on company budgets and finances and the economics of drilling, completing, and operating a well—determined largely by wellhead prices for oil and natural gas in the area. For example, current high crude oil prices and low natural gas prices are directing drilling toward those plays or portions of plays with a high concentration of liquids (crude oil, condensates, and natural gas plant liquids). Clearly, not all the wells that would be needed to develop each play fully can be drilled in one year—for example, more than 630,000 new wells would be needed to bring total U.S. shale gas and tight oil resources into production. In 2010, roughly 37,500 total oil and natural gas wells were drilled in the United States. It takes time and money to evaluate, develop, and produce hydrocarbon resources.

Although changes in the overall TRR estimates are important, the economics of developing the TRR and the timing of the development determine the projections for production of domestic crude oil and natural gas. TRR adjustments that affect resources which are not economical to develop during the projection period do not affect the AEO projections. Thus, significant variation in the overall TRR does not always result in significant changes in projected production.

EUR sensitivity cases and results

Estimated ultimate recovery per well is a key component in estimates of both technically recoverable resources and economically recoverable resources of tight oil and shale gas. The EUR for future wells is highly uncertain, depending on the application of new and/or improved technologies as well as the geology of the formation where the wells will be drilled. EUR assumptions typically have more impact on projected production than do any of the other parameters used to develop TRR estimates. For AEO2012, two cases were created to examine the impacts of higher and lower TRR for tight oil and shale gas by varying the assumed EUR per well.

These High and Low EUR cases are not intended to represent a confidence interval for the resource base, but rather to illustrate how different EUR assumptions can affect projections of domestic production, prices, and consumption. To emphasize this point, an additional case was developed that combines a change in the assumed well spacing for all shale gas and tight oil plays with the EUR assumptions in the High EUR case. Well spacing is also highly uncertain, depending on the application of new and/or improved technologies as well as the geology of the formation where the well is being drilled. In the AEO2012 Reference case, the well spacing for shale gas and tight oil drilling ranges from 2 to 12 wells per square mile.

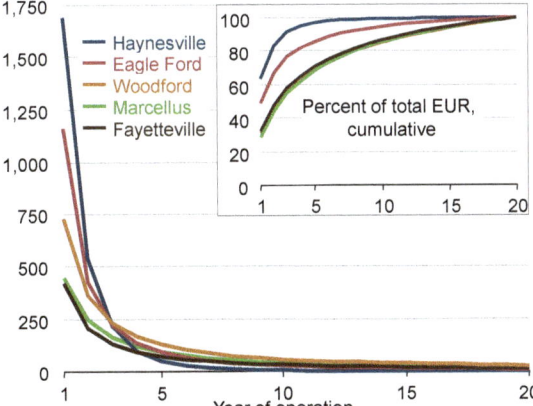

Figure 54. Average production profiles for shale gas wells in major U.S. shale plays by years of operation (million cubic feet per year)

Table 17. Estimated ultimate recovery for selected shale gas plays in three AEOs (billion cubic feet per well)

Basin/Play	AEO2010 Range	AEO2010 Average	AEO2011 Range	AEO2011 Average	AEO2012 Range	AEO2012 Average
Appalachian						
Marcellus	0.25–0.74	0.49	0.86–4.66	1.62	0.02–7.80	1.56
Utica	--	--	--	--	0.10–2.75	1.13
Arkoma						
Woodford	1.43–4.28	2.85	3.00–5.32	4.06	0.40–4.22	1.97
Fayetteville	0.91–2.73	1.82	0.86–2.99	2.03	0.19–3.22	1.30
Chattanooga	--	--	--	--	0.14–1.94	0.99
Caney	--	--	--	--	0.05–0.66	0.34
TX-LA-MS Salt						
Haynesville/Boosier	2.30–6.89	4.59	1.13–8.65	3.58	0.08–5.76	2.67
Western Gulf						
Eagle Ford	1.10–3.29	2.19	1.73–7.32	2.63	0.41–4.93	2.36
Pearsall	--	--	--	--	0.12–2.91	1.22
Anadarko						
Woodford	--	--	2.65–4.54	3.42	0.68–5.37	2.89

Issues in focus

Low EUR case. In the Low EUR case, the EUR per tight oil or shale gas well is assumed to be 50 percent lower than in the Reference case, increasing the per-unit cost of developing the resource. The total unproved tight oil TRR is decreased to 17 billion barrels, and the shale gas TRR is decreased to 241 trillion cubic feet, as compared with 33 billion barrels of tight oil and 482 trillion cubic feet of shale gas in the Reference case.

High EUR case. In the HIGH EUR case, the EUR per tight oil or shale gas well is assumed to be 50 percent higher than in the Reference case, decreasing the per-unit cost of developing the resource. The total unproved tight oil TRR is increased to 50 billion barrels and the shale gas TRR is increased to 723 trillion cubic feet.

High TRR case. In the High TRR case, the well spacing for all tight oil and shale gas plays is assumed to be 8 wells per square mile (i.e., each well has an average drainage area of 80 acres), and the EUR per tight oil or shale gas well is assumed to be 50 percent higher than in the Reference case. In addition, the total unproved tight oil TRR is increased to 89 billion barrels and the shale gas TRR is increased to 1,091 trillion cubic feet, more than twice the TRRs for tight oil and shale gas wells in the Reference case.

The effects of the changes in assumptions in the three cases on supply, demand, and prices for oil and for natural gas are significantly different in magnitude, because the domestic oil and natural gas markets are distinctly different markets. Consequently, the following discussion focuses first on how the U.S. oil market is affected in the three sensitivity cases, followed by a separate discussion of how the U.S. natural gas market is affected in the three cases.

Crude oil and natural gas liquid impacts

The primary impact of the Low EUR, High EUR, and High TRR cases with respect to oil production is a change in production of tight oil and natural gas plant liquids (NGPL) (Table 18). NGPL production is discussed in conjunction with tight oil production, because significant volumes of NGPL are produced from tight oil and shale gas formations. Thus, changing the EURs directly affects NGPL production. Relative to the Reference case, tight oil production increases more slowly in the Low EUR case and more rapidly in the High EUR and High TRR cases. On average, tight oil production from 2020 to 2035 is approximately 450,000 barrels per day lower in the Low EUR case, 410,000 barrels per day higher in the High EUR case, and 1.3 million barrels per day higher in the High TRR case than in the Reference case (Figure 55). NGPL production in 2035 is more than 350,000 barrels per day lower in the Low EUR case than in the Reference case, nearly 320,000 barrels per day higher in the High EUR case, and 1.0 million barrels per day higher in the High TRR case.

Tight oil production is highest in the High TRR case, which assumes both higher EUR per well and generally lower drainage area per well than in the Reference case. In the High TRR case, tight oil production increases from roughly 400,000 barrels per day in 2010 to nearly 2.8 million barrels per day in 2035, with the Bakken formation accounting for most of the increase. The TRR estimate for the Bakken is more than 7 times higher in the High TRR case than in the Reference case—39.3 billion barrels compared to 5.4 billion barrels—which supports a continued dramatic production increase through 2015 and a longer plateau at a much higher production level through 2035 than in the Reference case. Bakken crude oil production (excluding NGPLs) increases from roughly 270,000 barrels per day in 2010 to nearly 800,000 barrels per day in 2015 before reaching over 1 million barrels per day in 2021 and remaining at that level through 2035 in the High TRR case, compared with peak tight oil production of roughly 530,000 barrels per day in the Reference case. Cumulative crude oil production from the Bakken from 2010 to 2035 is roughly 8.5 billion barrels in the High TRR case, compared with 4.3 billion barrels in the Reference case.

Table 18. Petroleum supply, consumption, and prices in four cases, 2020 and 2035

Projection	2010	2020 Reference	2020 Low EUR	2020 High EUR	2020 High TRR	2035 Reference	2035 Low EUR	2035 High EUR	2035 High TRR
Low-sulfur light crude oil price (2010 dollars per barrel)	79	127	128	125	122	145	147	143	140
Total U.S. production of crude oil and natural gas plant liquids (million barrels per day)	7.5	9.6	8.8	10.3	11.6	9.0	8.1	10.0	11.8
Tight oil	0.4	1.2	0.9	1.5	2.2	1.2	0.7	1.7	2.8
Natural gas plant liquids	2.1	2.9	2.6	3.1	3.6	3.0	2.7	3.3	4.0
Other U.S. crude oil	5.1	5.5	5.3	5.6	5.7	4.8	4.8	4.9	5.0
Tight oil share of total U.S. crude oil and NGPL production (percent)	5	12	10	15	19	14	9	17	23
U.S. net import share of petroleum product supplied (percent)	50	37	41	34	27	36	41	32	24

Every incremental barrel of domestic crude oil production displaces approximately one barrel of imports, because U.S. consumption of liquid fuels varies little across the cases. Consequently, the projected share of net petroleum imports in total U.S. liquid fuel consumption in 2035 varies considerably across the EUR and TRR cases, from 41 percent in the Low EUR case to 24 percent in the High TRR case, as compared with 36 percent in the Reference case. However, additional downstream infrastructure may be required to process the high levels of NGPL production in the High EUR and High TRR cases.

Changes in domestic oil production have only a modest impact on domestic crude oil and petroleum product prices, because any change in domestic oil production is diluted by the much larger world oil market. The United States produced 5.5 million barrels per day, or 7 percent of total world crude oil production of 73.9 million barrels per day in 2010 and is projected generally to maintain that share of world crude oil production through 2035 in the Reference case.

Natural gas impacts

The EUR and TRR cases show more significant impacts on U.S. natural gas supply, consumption, and prices than that projected for crude oil and petroleum products for two reasons (Table 19). First, the U.S. natural gas market constitutes the largest regional submarket within the relatively self-contained North American natural gas market. Second, in the Reference case, shale gas production accounts for 49 percent of total U.S. natural gas production in 2035, while tight oil production accounts for only 14 percent of total U.S. crude oil and NGPL production and 1 percent of world crude oil production. As a result, changes in shale gas production have a commensurately larger impact on North American natural gas prices than tight oil production has on world oil prices.

The projections for domestic shale gas production are highly sensitive to the assumed EUR per well. In 2035, total shale gas production varies from 9.7 trillion cubic feet in the Low EUR case to 16.0 trillion cubic feet in the High EUR case and 20.5 trillion cubic feet in the High TRR case, as compared with 13.6 trillion cubic feet in the Reference case (Figure 56). Because shale gas production accounts for such a large proportion of total natural gas production in 2035, the large changes in shale gas production result in commensurately large swings in total U.S. natural gas production. In 2035, total U.S. natural gas production ranges from 26.1 trillion cubic feet in the Low EUR case to 34.1 trillion cubic feet in the High TRR case, a difference of 8.0 trillion cubic feet production between the two cases.

In comparison with the Reference case, per-unit production costs are nearly double in the Low EUR case and about one-half in the High EUR case. In the Low EUR case, the Henry Hub natural gas price of $8.26 per million Btu in 2035 (2010 dollars) is $0.89 per million Btu higher than the Reference case price of $7.37 per million Btu. In the High EUR case, the 2035 Henry Hub natural gas price of $5.99 per million Btu is $1.38 per million Btu lower than the Reference case price. In the High TRR case, the 2035 Henry Hub natural gas price of $4.25 per million Btu is $3.12 per million Btu less than the Reference case price.

The natural gas prices projected in the Low EUR case are sufficiently high to enable completion of an Alaska gas pipeline, with operations beginning in 2031. Because an Alaska gas pipeline would make up for some of the reduction in Lower 48 shale gas production, differences between the Reference and Low EUR case projections for natural gas production, prices, and consumption in 2035 are somewhat less than would otherwise be expected.

The 2035 price spread of $4.01 per million Btu across the cases is reflected in the projected levels of U.S. natural gas consumption. Higher natural gas prices in the Low EUR case reduce total natural gas consumption to 25.0 trillion cubic feet in 2035, compared with 26.6 trillion cubic feet in the Reference case; and lower natural gas prices in the High EUR and High TRR cases increase consumption in 2035 to 28.4 trillion cubic feet and 31.9 trillion cubic feet, respectively.

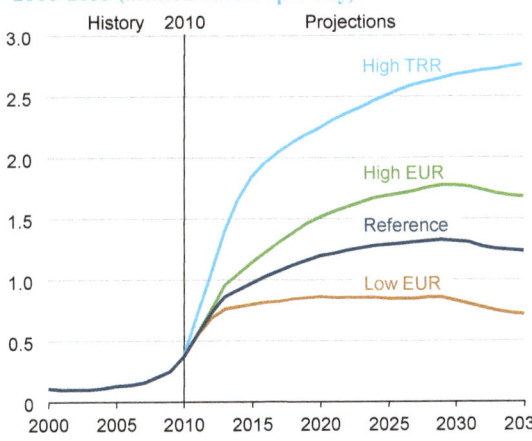

Figure 55. U.S. production of tight oil in four cases, 2000-2035 (million barrels per day)

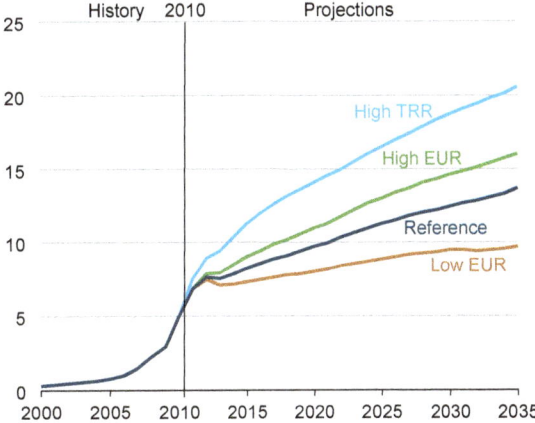

Figure 56. U.S. production of shale gas in four cases, 2000-2035 (trillion cubic feet)

Issues in focus

The variation in total U.S. natural gas consumption between the High EUR and High TRR cases is reflected to some degree in each end-use category. The electric power sector shows the greatest sensitivity to natural gas prices, with natural gas use for electricity generation being more responsive to changes in fuel prices than is consumption in the other sectors, because much of the electric power sector's fuel consumption is determined by the dispatching of existing generation units based on the operating cost of each unit, which in turn is determined largely by the costs of competing fuels—especially coal and natural gas. Natural gas consumption in the electric power sector in 2035 totals 7.7 trillion cubic feet in the Low EUR case, compared with 9.0 trillion cubic feet in the Reference case, 10.1 trillion cubic feet in the High EUR case, and 12.6 trillion cubic feet in the High TRR case.

In the end-use consumption sectors, opportunities to switch fuels generally are limited to when a new facility is built or when a facility's existing equipment is retired and replaced. Collectively, for all the end-use sectors, natural gas consumption in 2035 varies by only about 1.9 trillion cubic feet across the cases, from 17.3 trillion cubic feet in the Low EUR case to 19.2 trillion cubic feet in the High TRR case, as compared with 17.7 trillion cubic feet in the Reference case.

In 2035, the United States is projected to be a net exporter of natural gas in all the cases. The projected volumes of net exports vary, with lower natural gas prices resulting in higher net exports. However, the High TRR, High EUR, and Low EUR cases assume that U.S. gross exports of LNG remain constant at 0.9 trillion cubic feet from 2020 through 2035, because of the inherent complexities and uncertainties of projecting foreign natural gas production, consumption, and trade. It is likely, however, that actual levels of net LNG exports would be affected by changes in U.S. prices, which in turn, would dampen the extent of the price difference across the resource cases.

The variation in levels of net U.S. natural gas exports shown in Table 20 reflects the impact of domestic natural gas prices on natural gas pipeline imports and exports. Generally, lower natural gas prices, as in the High TRR case, result in lower natural gas imports from Canada and higher natural gas exports to Mexico. In 2035, net natural gas exports from the United States vary from 1.2 trillion cubic feet in the Low EUR case to 2.4 trillion cubic feet in the High TRR case, as compared with 1.4 trillion cubic feet in the Reference case.

The sensitivity cases in this discussion are not intended to provide a confidence interval for estimates of recoverable resources of domestic tight oil and shale gas but rather to illustrate the significance of key assumptions underlying the tight oil and shale

Table 19. Natural gas prices, supply, and consumption in four cases, 2020 and 2035

Projection	2010	2020				2035			
		Reference	Low EUR	High EUR	High TRR	Reference	Low EUR	High EUR	High TRR
Henry Hub natural gas spot price (2010 dollars per million Btu)	4.39	4.58	5.31	4.04	3.02	7.37	8.26	5.99	4.25
Total U.S. natural gas production (trillion cubic feet)	21.6	25.1	23.6	26.3	29.1	27.9	26.1	30.1	34.1
Onshore lower 48	18.7	22.5	21.0	23.6	26.6	25.0	21.2	27.2	31.7
Shale gas	5.0	9.7	8.0	10.9	14.0	13.6	9.7	16.0	20.5
Other natural gas	13.7	12.8	12.9	12.7	12.6	11.3	11.4	11.2	11.1
Offshore lower 48	2.6	2.3	2.4	2.3	2.2	2.7	3.1	2.6	2.3
Alaska	0.4	0.3	0.3	0.3	0.3	0.2	1.8	0.2	0.2
Shale gas production as percent of total U.S. natural gas production	23	39	34	42	48	49	37	53	60
Total net U.S. imports of natural gas (trillion cubic feet)	2.6	0.3	0.5	0.2	-0.2	-1.4	-1.2	-1.7	-2.4
Total U.S. consumption of natural gas (trillion cubic feet)	24.1	25.5	24.2	26.5	28.9	26.6	25.0	28.4	31.9
Electric Power	7.4	7.9	6.8	8.7	10.5	9.0	7.7	10.1	12.6
Residential	4.9	4.8	4.8	4.9	4.9	4.6	4.6	4.7	4.8
Commercial	3.2	3.4	3.4	3.5	3.6	3.6	3.5	3.7	4.0
Industrial	6.6	7.1	7.0	7.1	7.4	7.0	6.9	7.2	7.6
Other	2.0	2.3	2.2	2.3	2.5	2.4	2.4	2.6	2.8

gas TRRs used in *AEO2012*. TRR estimates are highly uncertain and can be expected to change in subsequent *AEOs* as additional information is gained through continued exploration, development, and production.

12. Evolving Marcellus shale gas resource estimates

As discussed in the preceding article, estimates of crude oil and natural gas TRR are uncertain. Estimates of the Marcellus shale TRR, which have received considerable attention over the past year, are no exception. TRR estimates are likely to continue evolving as drilling continues and more information becomes publicly available. The Marcellus shale gas play covers more than 100,000 square miles in parts of eight States, but most of the drilling to date has been in two areas of northeast Pennsylvania and southwest Pennsylvania/northern West Virginia. Until 2010, the State of Pennsylvania had maintained a 5-year embargo on the release of well-level production data, which severely limited the publicly available information about Marcellus well production. Now Pennsylvania provides well production data on a cumulative basis—annually for the years before 2010 and semi-annually starting in the second half of 2010. Even with more data available, however, it is still a challenge to estimate TRR for the Marcellus play.

In 2002, the USGS estimated that 0.8 trillion cubic feet to 3.7 trillion cubic feet of technically recoverable shale gas resources existed in the Marcellus, with a mean estimate of 1.9 trillion cubic feet [116]. At that time, most of the well production data available were for vertical wells drilled in West Virginia. Since 2003, technological improvements have led to more-productive and less-costly wells. The newer horizontal wells have higher EURs [117] than the older vertical wells. In 2011, the USGS released an updated assessment for the Marcellus resource, with a mean estimate of 84 trillion cubic feet of undiscovered TRR (ranging from 43 trillion cubic feet to 144 trillion cubic feet) [118]. For its 2011 assessment, the USGS evaluated well production data from Pennsylvania and West Virginia that were available in early 2011 and determined that the data were "not sufficient for the construction of individual well Estimated Ultimate Recovery distributions" [119]. Instead, the USGS chose analogs from other U.S. shale gas plays to determine the EUR distributions for its three Marcellus assessment units—Foldbelt, Interior, and Western Margin (Figure 57).

Estimates of the TRR for U.S. shale gas are updated each year for the *AEO*. For *AEO2011*, an independent consultant was hired to estimate the Marcellus TRR as the available USGS TRR estimate issued in 2003 was clearly too low, since cumulative production from the Marcellus shale was on a path to exceed it within a year or two. For *AEO2012*, EIA adopted the 2011 USGS estimates of the Marcellus assessment areas, well spacing, and percent of area with potential. However, EIA examines available well production data each year to estimate shale EURs for use in the *AEO* (Table 20).

The revised Marcellus EUR for *AEO2012* is close to the EUR used in *AEO2011* but nearly 70 percent higher than the EUR used in the 2011 USGS assessment. The Interior Assessment Unit EURs developed by EIA reflects the current practice of horizontal drilling and well production data through June 2011 for Pennsylvania and West Virginia [120]. Because there has been very little, if any, drilling in the Western Margin and Foldbelt Assessment Units, the USGS EURs were used for the States in those areas. The resulting *AEO2012* estimate for the Marcellus TRR is 67 percent lower than the *AEO2011* estimate, primarily as a result of increased well spacing (132 acres per well vs. 80 acres per well) and a lower percentage of area with potential (18 percent vs. 34 percent) (Table 21).

Figure 57. United States Geological Survey Marcellus Assessment Units

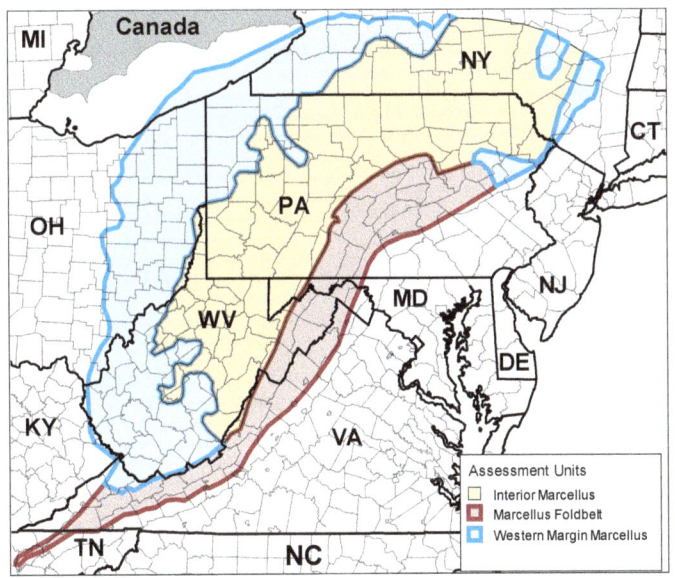

The estimation of Marcellus shale gas resources is highly uncertain, given both the short production history of current producing wells and the concentration of most producing wells in two small areas, Northeast Pennsylvania and Southwest Pennsylvania/Northern West Virginia. The Marcellus EURs are expected to change as additional data are released and the methodology for developing EURs is refined. Also, as more wells are drilled over a broader area, and as operators optimize well spacing to account for evolving drilling practices, the assumption for average well spacing may be revised. Although the Marcellus shale resource estimate will be updated for every *AEO*, revisions will not necessarily have a significant impact on projected natural gas production, consumption, and prices.

Issues in focus

Table 20. Marcellus unproved technically recoverable resources in *AEO2012* (as of January 1, 2010)

Assessment Unit/State	Area (square miles)	Well spacing (wells per square mile)	Percent of area untested	Percent of area with potential	EUR (billion cubic feet per well)				TRR (billion cubic feet)
					High	Mid	Low	Average	
Foldbelt	19,063	4	100	5	0.50	0.18	0.03	0.21	757
Maryland	435	4	100	5	0.50	0.18	0.03	0.21	17
Pennsylvania	7,951	4	100	5	0.50	0.18	0.03	0.21	316
Tennessee	353	4	100	5	0.50	0.18	0.03	0.21	14
Virginia	7,492	4	100	5	0.50	0.18	0.03	0.21	298
West Virginia	2,833	4	100	5	0.50	0.18	0.03	0.21	113
Interior	45,161	4	99	37	6.33	1.41	0.06	1.95	137,677
Maryland	763	4	100	37	2.02	0.30	0.02	0.52	629
New York	10,381	4	100	37	7.80	1.79	0.07	2.43	40,124
Ohio	361	4	99	37	2.02	0.30	0.02	0.52	296
Pennsylvania	23,346	4	98	37	7.80	1.79	0.07	2.43	88,182
Virginia	321	4	100	37	2.02	0.30	0.02	0.52	264
West Virginia	9,989	4	99	37	2.02	0.30	0.02	0.52	8,182
Western	39,844	5	100	7	0.35	0.11	0.03	0.13	2,107
Kentucky	207	5	100	7	0.35	0.11	0.03	0.13	11
New York	7,985	5	100	7	0.35	0.11	0.03	0.13	424
Ohio	13,515	5	100	7	0.35	0.11	0.03	0.13	718
Pennsylvania	6,582	5	100	7	0.35	0.11	0.03	0.13	350
Virginia	653	5	100	7	0.35	0.11	0.03	0.13	35
West Virginia	10,901	5	98	7	0.35	0.11	0.03	0.13	569
Total Marcellus	104,067	5	99	18	5.05	1.13	0.05	1.56	140,541

Table 21. Marcellus unproved technically recoverable resources: *AEO2011*, USGS 2011, and *AEO2012*

Estimate	Area (square miles)	Well spacing		Percent of area untested	Percent of area with potential	Average EUR (billion cubic feet per well)	TRR (billion cubic feet)
		Acres	Wells per square mile				
AEO2011 (as of 1/1/2009)							
Marcellus	94,893	80	8	99%	34%	1.62	410,374
USGS (2011 assessment)							
Marcellus	104,067	132	4.9	99%	18%	0.93	84,198
Foldbelt	19,063	149	4.3	100%	5%	0.21	765
Interior	45,156	149	4.3	99%	37%	1.15	81,374
Western	39,844	117	5.5	99%	7%	0.13	2,059
AEO2012 (as of 1/1/2010)							
Marcellus	104,067	132	4.9	99%	18%	1.56	140,541
Foldbelt	19,063	149	4.3	100%	5%	0.21	757
Interior	45,161	149	4.3	99%	37%	1.95	137,677
Western	39,844	117	5.5	100%	7%	0.13	2,107

Endnotes for Issues in focus

Links current as of June 2012

41. Oil shale liquids, derived from heating kerogen, are distinct from shale oil and also from tight oil, which is classified by EIA as crude oil. Oil shale is not expected to be produced in significant quantities in the United States before 2035.
42. U.S. Environmental Protection Agency and National Highway Transportation Safety Administration, "2017 and Later Model Year Light-Duty Vehicle Greenhouse Gas Emissions and Corporate Average Fuel Economy Standards; Proposed Rule," Federal Register, Vol. 76, No. 231 (Washington, DC: December 1, 2011), website www.nhtsa.gov/staticfiles/rulemaking/pdf/cafe/2017-25_CAFE_NPRM.pdf.
43. The EISA 2007 RFS requirement for increasing volumes of biofuels results in a significant number of FFVs in both the Reference case and the CAFE case.
44. S. Bianco, "Chevy Volt Has Best Month Ever, But Nissan Leaf Still Wins 2011 Plug-in Sales Contest," autobloggreen, website green.autoblog.com/2012/01/04/chevy-volt-has-best-month-ever-but-nissan-leaf-still-wins-2011.
45. Battery electric vehicle charge-depleting mode occurs when the vehicle relies on battery power for operation. Charge-sustaining mode occurs when battery electric power is coupled with power provided by the internal combustion engine. Vehicles can be designed to operate on a blended mode that uses both charge-depleting and charge-sustaining modes while in operation, depending on the drive cycle.
46. Toyota, "Toyota Cars, Trucks, SUVs, and Accessories," website www.toyota.com; Nissan USA, "Nissan Cars, Trucks, Crossovers, & SUVs," website www.nissanusa.com; and Chevrolet, "2012 Cars, SUVs, Trucks, Crossovers & Vans," website www.chevy.com. **Note**: Miles per gallon equivalent, as listed by automotive manufacturers, is derived by the U.S. Environmental Protection Agency, www.fueleconomy.gov.
47. Toyota, "Toyota Cars, Trucks, SUVs, and Accessories," website www.toyota.com; Nissan USA, "Nissan Cars, Trucks, Crossovers, & SUVs," website www.nissanusa.com; and Chevrolet, "2012 Cars, SUVs, Trucks, Crossovers & Vans," website www.chevy.com. **Note**: Miles per gallon equivalent, as listed by automotive manufacturers, is derived by the U.S. Environmental Protection Agency, www.fueleconomy.gov.
48. U.S. Department of Energy, Office of Energy Efficiency and Renewable Energy, "Vehicle Technologies Program," website www.eere.energy.gov/vehiclesandfuels/technologies/systems/index.html.
49. U.S. Energy Information Administration, "Residential Energy Consumption Survey (RECS), 2009 RECS Survey Data," website 205.254.135.7/consumption/residential/data/2009.
50. U.S. Department of Energy, Office of Energy Efficiency and Renewable Energy, "Alternative Fuels & Advanced Vehicles Data Center," website www.afdc.energy.gov.
51. Indiana University, School of Public and Environmental Affairs, "Plug-in Electric Vehicles: A Practical Plan for Progress," website www.indiana.edu/~spea/pubs/TEP_combined.pdf.
52. U.S. Environmental Protection Agency and National Highway Transportation Safety Administration, "2017 and Later Model Year Light-Duty Vehicle Greenhouse Gas Emissions and Corporate Average Fuel Economy Standards; Proposed Rule," Federal Register, Vol. 76, No. 231 (Washington, DC: December 1, 2011), website www.nhtsa.gov/staticfiles/rulemaking/pdf/cafe/2017-25_CAFE_NPRM.pdf.
53. For this analysis, heavy-duty vehicles include trucks with a Gross Vehicle Weight Rating of 10,001 pounds and higher, corresponding to Gross Vehicle Weight Rating classes 3 through 8 vehicles.
54. U.S. Department of Energy, Office of Energy Efficiency and Renewable Energy, "Alternative Fueling Station Database Custom Query" (Washington, DC: June 3, 2010), website www.afdc.energy.gov/afdc/fuels/stations_query.html. Accessed June 30, 2012.
55. National Petroleum News, Market Facts 2011.
56. U.S. Department of Energy, Office of Energy Efficiency and Renewable Energy, *Clean Cities Alternative Fuel Price Report* (Washington, DC: April, 2012), website www.afdc.energy.gov/afdc/pdfs/afpr_apr_12.pdf.
57. The Texas Clean Transportation Triangle is supported by Texas State Senate Bill 20, which provides vehicle rebates and fueling grants. See West, Williams, House Research Organization, "Bill Analysis: SB 20" (Austin, TX: May 21, 2011), website www.hro.house.state.tx.us/pdf/ba82r/sb0020.pdf.
58. The Interstate Clean Transportation Corridor was developed in 1996. The corridor is now partially established with LNG truck refueling infrastructure in California and to Reno, Las Vegas, and Phoenix. See Gladstein, Neandross & Associates, "Interstate Clean Transportation Corridor" (Santa Monica, CA: February 2, 2012), website ictc.gladstein.org.

Issues in focus

59. The Pennsylvania Clean Transportation Corridor was proposed in a report, "A Road Map to a Natural Gas Vehicle Future" (Canonsburg, PA: April 5, 2011), sponsored by the Marcellus Shale Coalition, website marcelluscoalition.org/wp-content/uploads/2011/04/MSC_NGV_Study.pdf.

60. The American Recovery and Reinvestment Act has provided more than $300 million toward cost-sharing projects related to alternative fuels. U.S. Department of Energy, Office of Energy Efficiency and Renewable Energy, "American Recovery and Reinvestment Act Project Awards" (Washington, DC: September 7, 2011) website www1.eere.energy.gov/cleancities/projects.html.

61. For a map of U.S. LNG peak shaving, see U.S. Energy Information Administration, "U.S. LNG Peaking Shaving and Import Facilities, 2008" (Washington, DC: December, 2008), website www.eia.gov/pub/oil_gas/natural_gas/analysis_publications/ngpipeline/lngpeakshaving_map.html.

62. The LNG Excise Tax Equalization Act of 2012, proposed in the U.S. House of Representatives, would require the tax treatment of LNG and diesel fuel to be equivalent on the basis of heat content. See Civic Impulse, LLC, "H.R. 3832: LNG Excise Tax Equalization Act of 2012" (Washington, DC: May 29, 2012), website legacy.govtrack.us/congress/bill.xpd?bill=h112-3832.

63. Developed from e-mail correspondence with Graham Williams, 4/11/12.

64. U.S. Environmental Protection Agency and National Highway Transportation Safety Administration, "Greenhouse Gas Emissions Standards and Fuel Efficiency Standards for Medium- and Heavy-Duty Engines and Vehicles," Federal Register Vol. 76, No. 179 (Washington, DC: September 15, 2011), website www.federalregister.gov/articles/2011/09/15/2011-20740/greenhouse-gas-emissions-standards-and-fuel-efficiency-standards-for-medium--and-heavy-duty-engines#p-3.

65. U.S. Census Bureau, "Vehicle Inventory and Use Survey (VIUS) (discontinued after 2002)" (Washington, DC: May 29, 2012), website www.census.gov/econ/overview/se0501.html.

66. U.S. Environmental Protection Agency and National Highway Transportation Safety Administration, "Greenhouse Gas Emissions Standards and Fuel Efficiency Standards for Medium- and Heavy-Duty Engines and Vehicles," Federal Register Vol. 76, No. 179 (Washington, DC: September 15, 2011), website www.federalregister.gov/articles/2011/09/15/2011-20740/greenhouse-gas-emissions-standards-and-fuel-efficiency-standards-for-medium--and-heavy-duty-engines#p-3.

67. For information on the New Alternative Transportation to Give Americans Solutions Act of 2012, see Civic Impulse, LLC, "H.R. 1380: New Alternative Transportation to Give Americans Solutions Act of 2011" (Washington, DC: May 29, 2012), website legacy.govtrack.us/congress/bill.xpd?bill=h112-1380.

68. The liquid fuels production industry includes all participants involved in the production of liquid fuels: producers of feedstocks, petroleum- and nonpetroleum-based refined products and blendstocks, and liquid and non-liquid end-use products.

69. U.S. Environmental Protection Agency, "Mercury and Air Toxics Standards" (Washington, DC: March 27, 2012), website www.epa.gov/mats.

70. U.S. Environmental Protection Agency, "Cross-State Air Pollution Rule (CSAPR)" (May 25, 2012), website www.epa.gov/airtransport.

71. Other components of variable cost include emissions control technology, waste disposal, and emissions allowance credits.

72. The *AEO2012* Early Release Reference case was prepared before the final MATS rule was issued and, therefore, did not include MATS.

73. United States Court of Appeals for the District of Columbia Circuit, "EME Homer City Generation, L.P., v. Environmental Protection Agency" (Washington, DC: December 30, 2011), website www.epa.gov/airtransport/pdfs/CourtDecision.pdf.

74. U.S. Energy Information Administration, *Electric Power Annual 2010* (Washington, DC, November 2011), Table 3.10, "Number and Capacity of Existing Fossil-Fuel Steam-Electric Generators with Environmental Equipment, 1991 through 2010," website www.eia.gov/electricity/annual/html/table3.10.cfm.

75. U.S. Environmental Protection Agency, Office of Enforcement and Compliance Assurance, "The Environmental Protection Agency's Enforcement Response Policy for Use of Clean Air Act Section 113(a) Administrative Orders in Relation to Electric Reliability and the Mercury and Air Toxics Standard" (Washington, DC: December 16, 2011), website www.epa.gov/compliance/resources/policies/civil/erp/mats-erp.pdf.

76. See Appendix F for a map of the EMM regions.

77. The EPA is proposing that new fossil-fuel-fired power plants begin meeting an output-based standard of 1,000 pounds CO_2 per megawatthour. See U.S. Environmental Protection Agency, "Carbon Pollution Standard for New Power Plants" (Washington, DC: May 23, 2012), website www.epa.gov/carbonpollutionstandard/actions.html. Existing coal plants without CCS will not be able to meet that standard, and the proposed rule does not apply to plants already under construction. The EPA proposal is not included in *AEO2012*.

78. U.S. Energy Information Administration, Form EIA-860, "Annual Electric Generator Report" (Washington, DC: November 30, 2011), website www.eia.gov/cneaf/electricity/page/eia860.html.
79. U.S. Energy Information Administration, "Levelized Cost of New Generation Resources in the Annual Energy Outlook 2012" (Washington, DC: March 2012), website www.eia.gov/forecasts/aeo/electricity_generation.cfm.
80. U.S. Energy Information Administration, "Assumptions to *AEO2012*" (Washington, DC: June 2012), website www.eia.gov/forecasts/aeo/assumptions.
81. U.S. Government Printing Office, "Energy Policy Act of 2005, Public Law 109-58, Title XVII—Incentives for Innovative Technologies" (Washington, DC: August 8, 2005), website www.gpo.gov/fdsys/pkg/PLAW-109publ58/html/PLAW-109publ58.htm.
82. U.S. Department of Energy, Loan Programs Office, "Loan Guarantee Program: Georgia Power Company" (Washington, DC: June 4, 2012), website lpo.energy.gov/?projects=georgia-power-company.
83. U.S. Government Printing Office, "Energy Policy Act of 2005, Public Law 109-58, Title XVII—Incentives for Innovative Technologies, paras. 638, 988, and 1306" (Washington, DC, August 2005), website www.gpo.gov/fdsys/pkg/PLAW-109publ58/html/PLAW-109publ58.htm.
84. U.S. Energy Information Administration, Form EIA-860, "Annual Electric Generator Report," website www.eia.gov/cneaf/electricity/page/eia860.html.
85. Tennessee Valley Authority, "Integrated Resource Plan" (Knoxville, TN: March 2011), website www.tva.com/environment/reports/irp/index.htm.
86. U.S. Nuclear Regulatory Commission, "Status of License Renewal Applications and Industry Activities: Completed Applications" (Washington, DC: May 22, 2012), website www.nrc.gov/reactors/operating/licensing/renewal/applications.html#completed.
87. U.S. Nuclear Regulatory Commission, "Status of License Renewal Applications and Industry Activities: Completed Applications" (Washington, DC: May 22, 2012), website www.nrc.gov/reactors/operating/licensing/renewal/applications.html#completed.
88. Electric Power Research Institute, "Long-Term Operations (QA)" (Palo Alto, CA: June 4, 2012), website portfolio.epri.com/ProgramTab.aspx?sId=NUC&rId=210&pId=6177.
89. International Forum for Reactor Aging Management (IFRAM), "Inaugural Meeting of the International Forum for Reactor Aging Management (IFRAM)" (Colorado Springs, CO: August 5, 2011), website ifram.pnnl.gov.
90. Alyeska Pipeline Service Company, *Low Flow Impact Study, Final Report* (Anchorage, AL: June 15, 2011), at www.alyeska-pipe.com/Inthenews/LowFlow/LoFIS_Summary_Report_P6%2027_FullReport.pdf.
91. Tim Bradner, "Alyeska Invests in New Methods to Extend Pipeline Life," *Alaska Journal of Commerce* (June 1, 2009), website www.alaskajournal.com/Alaska-Journal-of-Commerce/May-2009/Alyeska-invests-in-new-methods-to-extend-pipeline-life/.
92. U.S. Department of Energy, National Energy Technology Laboratory, *Alaska North Slope Oil and Gas – A Promising Future or an Area in Decline? (Addendum Report)*, DOE/NETL-2009/1385 (Washington, DC: April 8, 2009), website www.netl.doe.gov/technologies/oil-gas/publications/AEO/ANS_Potential.pdf, pp. 1-4 and 1-5.
93. Alan Bailey, "TAPS transitioning to a low flow future," *Petroleum News*, Vol. 14, No. 29 (Anchorage, AK: July 19, 2009), website www.petroleumnews.com/pntruncate/5456274.shtml (subscription site).
94. Alyeska Pipeline Service Company, *Low Flow Impact Study, Final Report* (Anchorage, AL: June 15, 2011), at www.alyeska-pipe.com/Inthenews/LowFlow/LoFIS_Summary_Report_P6%2027_FullReport.pdf.
95. U.S. Department of the Interior, U.S. Geological Survey, *The Oil and Gas Resource Potential of the Arctic National Wildlife Refuge 1002 Area, Alaska,* Open File Report 98-34 (Washington, DC: May 1998), website pubs.usgs.gov/of/1998/ofr-98-0034/ANWR1002.pdf; U.S. Geological Survey, *Arctic National Wildlife Refuge, 1002 Area, Petroleum Assessment, 1998, Including Economic Analysis,* USGS Fact Sheet FS-028-01 (Washington, DC: April 2001), website pubs.usgs.gov/fs/fs-0028-01/fs-0028-01.pdf; and David W. Houseknecht and Kenneth J. Bird, *Oil and Gas Resources of the Arctic Alaska Petroleum Province,* U.S. Geological Survey Professional Paper 1732-A (Washington, DC: October 31, 2006), website pubs.usgs.gov/pp/pp1732/pp1732a/pp1732a.pdf.
96. In 2004, BP commissioned a study that examined the possibility of building a 20-inch pipeline to Fairbanks and using the Alaska railroad to transport the oil to Valdez, at an estimated cost of about $3 billion. Source: Alan Bailey, "A TAPS bottom line," *Petroleum News*, Volume 17, Number 3 (Anchorage, AK: January 15, 2012), website www.petroleumnews.com/pntruncate/225019711.shtml.

Issues in focus

97. The most common miscible gas EOR technique is to alternate the injection of gas and water, referred to as water-alternating-gas or WAG. Source: Oil and Gas Journal, Special Report: EOR/Heavy Oil Survey: 2010 worldwide EOR survey, Volume 108, Issue 14, published April 19, 2010.

98. Capital expenditures can be split into two categories—maintenance and development—with development expenditures allocated to the development of new fields that have not yet reached peak production.

99. Source for 2011 CP capital expenditures—*Petroleum News*, "Eagle Ford Could Nudge Alaska for COP" (May 8, 2011); source for 2001 CP capital expenditures—*Petroleum News*, "Sunrise or Sunset for ConocoPhillips in Alaska?" (October 27, 2002); source for 2001 and 2011 CP split in capital expenditures—*Petroleum News*, "Johansen: Urgency Lacking on Throughput" (October 16, 2011).

100. These figures were derived from the CP ownership shares of the Colville River, Kuparuk River, and Prudhoe Bay field units and from the oil production reports of the Alaska Department of Natural Resources—Oil and Gas Division.

101. The volume of water produced relative to the volume of oil produced is referred to as the "water cut."

102. U.S. Geological Survey, *Economics of Undiscovered Oil in Federal Lands on the National Petroleum Reserve—Alaska*, by Emil Attanasi, Open-File Report 03-44 (January 2003), Figures A-2 (Alpine Field) and A-3 (Kuparuk Field).

103. In fact, these decisions would have to be made some time before the 350,000-barrel-per-day threshold is reached so they would be ready for implementation either prior to reaching the threshold or when that threshold is reached.

104. The owners of TAPS and operators of the North Slope fields might not know either at this junction what these future costs might be for both operating TAPS and the North Slope fields as volumes decline; at best they have estimates that might or might not turn out to be true.

105. The assumption that all North Slope exploration activity would cease with the decommissioning of TAPS might not be entirely realistic because some offshore oil fields might be economic to develop using floating production, storage, and offloading facilities (FPSO). This would be especially true in the Chukchi Sea, which has much less of an ice pack problem during the winter than the Beaufort Sea.

106. Maintenance capital expenditures could also decline if the field operators determined that drilling more wells was unprofitable.

107. *Petroleum News*, "Who Produces Crude Oil in Alaska?" Vol. 16, No. 43 (October 23, 2011).

108. ExxonMobil, 2010 Financial & Operating Review, Table entitled: "Oil and Gas Exploration and Production Earnings," p. 70.

109. See also EIA, "U.S. Crude Oil, Natural Gas, and Natural Gas Liquids Reserves," November 30, 2010, website www.eia.gov/oil_gas/natural_gas/data_publications/crude_oil_natural_gas_reserves/cr.html.

110. The further delineation of unproved resources into inferred reserves and undiscovered resources is not applicable to continuous resources since the extent of the formation is geologically known. For continuous resources, the USGS undiscovered technically recoverable resources are comparable to the EIA unproved resources. The USGS methodology for assessing continuous petroleum resources is at pubs.usgs.gov/ds/547/downloads/DS547.pdf.

111. "Tight oil" refers to crude oil and condensates produced from low-permeability sandstone, carbonate, and shale formations.

112. See shale gas map at www.eia.gov/oil_gas/rpd/shale_gas.pdf for basin locations.

113. Appalachian: pubs.usgs.gov/of/2011/1298/; Arkoma: pubs.usgs.gov/fs/2010/3043/; TX-LA-MS Salt and Western Gulf: pubs.usgs.gov/fs/2011/3020/; Anadarko: pubs.usgs.gov/fs/2011/3003/.

114. A well's estimated ultimate recovery (EUR) equals the cumulative production of that well over a 30-year productive life, using current technology without consideration of economic or operating conditions.

115. "Sweet spot" is an industry term for those select and limited areas within a shale or tight play where the well EURs are significantly greater than the rest of the play, sometimes as much as ten times greater than the lower production areas within a play.

116. USGS Fact Sheet FS-009-03. pubs.usgs.gov/fs/fs-009-03/FS-009-03-508.pdf.

117. A well's EUR equals the cumulative production of that well over a 30-year productive life, using current technology without consideration of economic or operating conditions.

118. USGS Fact Sheet 2011-3092, pubs.usgs.gov/fs/2011/3092/pdf/fs2011-3092.pdf.

119. USGS Open-File Report 2011-1298, pubs.usgs.gov/of/2011/1298/OF11-1298.pdf, page 2.

120. Well-level production from Pennsylvania is provided in two time intervals (annual and semi-annual). To estimate production on a comparable basis, well-level production is converted to an average daily rate by dividing gas quantity by gas production days. Because wells drilled before 2008 are vertical wells and do not reflect the technology currently being deployed, only wells drilled after 2007 are considered in the EUR evaluation. Well-level production for wells drilled in West Virginia is provided on a monthly basis.

Market trends

Projections by the U.S. Energy Information Administration (EIA) are not statements of what will happen but of what might happen, given the assumptions and methodologies used for any particular case. The Reference case projection is a business-as-usual estimate, given known technology, as well as market, demographic, and technological trends. Most cases in the *Annual Energy Outlook 2012 (AEO2012)* generally assume that current laws and regulations are maintained throughout the projections. Such projections provide a baseline starting point that can be used to analyze policy initiatives. EIA explores the impacts of alternative assumptions in other cases with different macroeconomic growth rates, world oil prices, rates of technology progress, and policy changes.

While energy markets are complex, energy models are simplified representations of energy production and consumption, regulations, and producer and consumer behavior. Projections are highly dependent on the data, methodologies, model structures, and assumptions used in their development. Behavioral characteristics are indicative of real-world tendencies rather than representations of specific outcomes.

Energy market projections are subject to much uncertainty. Many of the events that shape energy markets are random and cannot be anticipated. In addition, future developments in technologies, demographics, and resources cannot be foreseen with certainty. Many key uncertainties in the *AEO2012* projections are addressed through alternative cases.

EIA has endeavored to make these projections as objective, reliable, and useful as possible; however, they should serve as an adjunct to, not as a substitute for, a complete and focused analysis of public policy initiatives.

Trends in economic activity

Recovery in real gross domestic product growth continues at a modest rate

Figure 58. Average annual growth rates of real GDP, labor force, and nonfarm labor productivity in three cases, 2010-2035 (percent per year)

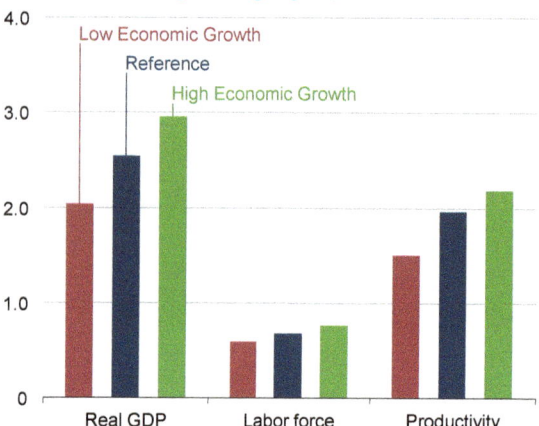

Slow consumption growth, fast investment growth, and an ever-improving trade surplus

Figure 60. Average annual growth rates for real output and its major components in three cases, 2010-2035 (percent per year)

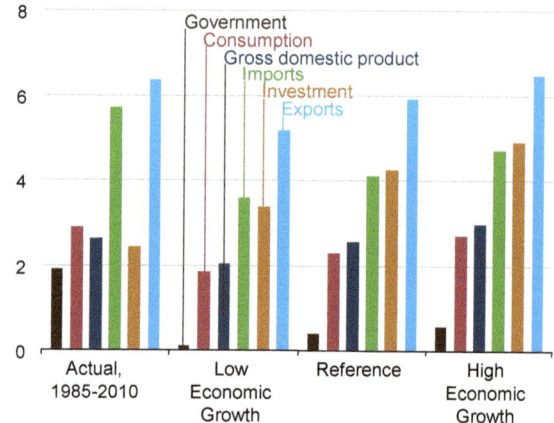

AEO2012 presents three views of U.S. economic growth (Figure 58). In 2011, the world economy experienced shocks that included turmoil in the Middle East and North Africa, a Greek debt crisis with financial impacts spreading to other Eurozone countries, and an earthquake in Japan, all leading to slower economic growth. U.S. growth projections in part reflect those world events.

U.S. recovery from the 2007-2008 recession has been slower than past recoveries (Figure 59). A feature of economic recoveries since 1975 has been slowing employment gains, and, following the most recent recession, growth in nonfarm employment has been slower than in any other post-1960 recovery [121]. The average rates of growth are strong starting from the trough of the recessions.

Figure 59. Average annual growth rates over 5 years following troughs of U.S. recessions in 1975, 1982, 1991, and 2008 (percent per year)

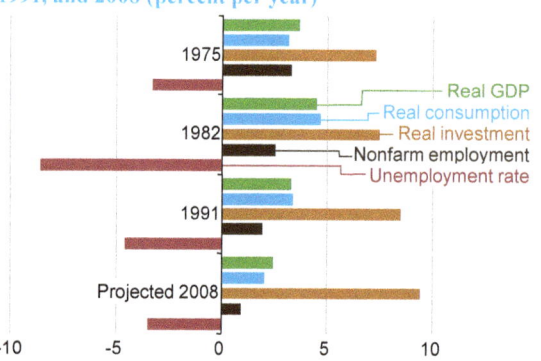

AEO2012 presents three economic growth cases: Reference, High, and Low. The High Economic Growth case assumes high growth and low inflation; the Low Economic Growth case assumes low growth and high inflation. Figure 60 compares the average annual growth rates for output and its major components in each of the three cases.

The short-term outlook (5 years) in each case represents current thinking about economic activity in the United States and the rest of the world; about the impacts of domestic fiscal and monetary policies; and about potential risks to economic activity. The long-term outlook projects smooth economic growth, assuming no shocks to the economy.

Differences among the Reference case and the High and Low Economic Growth cases reflect different expectations for growth in population (specifically, net immigration), labor force, capital stock, and productivity, which are above trend in the High Economic Growth case and below trend in the Low Economic Growth case. The average annual growth rate for real gross domestic product (GDP) from 2010 to 2035 in the Reference case is 2.5 percent, as compared with about 3.0 percent in the High Economic Growth case and about 2.0 percent in the Low Economic Growth case.

Compared with the 1985-2010 period, investment growth from 2010 to 2035 is faster in all three cases, whereas consumption, government expenditures, and imports grow more slowly in all three cases. Opportunities for trade are assumed to expand in each of the three cases, resulting in real trade surpluses by 2018 that continue through 2035.

Energy trends in the economy

Output growth for energy-intensive industries remains slow

Figure 61. Sectoral composition of industrial output growth rates in three cases, 2010-2035 (percent per year)

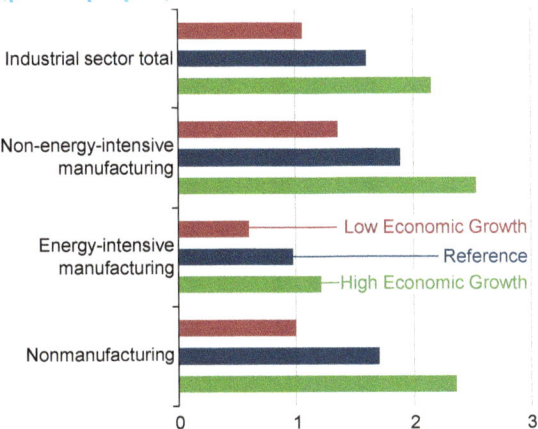

Industrial sector output has grown more slowly than the overall economy in recent decades, with imports meeting a growing share of demand for industrial goods, whereas the service sector has grown more rapidly [122]. In the *AEO2012* Reference case, real GDP grows at an average annual rate of 2.5 percent from 2010 to 2035, while both the industrial sector as a whole and its manufacturing component grow by 1.6 percent per year (Figure 61). As the economy recovers from the 2008-2009 recession, growth in U.S. manufacturing output in the Reference case accelerates from 2010 through 2020. After 2020, growth in manufacturing output slows due to increased foreign competition, slower expansion of domestic production capacity, and higher energy prices. These factors weigh heavily on the energy-intensive manufacturing sectors, which taken together grow at a slower rate of about 1.0 percent per year from 2010 to 2035, with variation by industry ranging from 0.8-percent annual growth for bulk chemicals to 1.5-percent annual growth for food processing.

A decline in U.S. dollar exchange rates, combined with modest growth in unit labor costs, stimulates U.S. exports, eventually improving the U.S. current account balance. From 2010 to 2035, real exports of goods and services grow by an average of 5.9 percent per year, and real imports of goods and services grow by an average of 4.1 percent per year. Strong growth in exports is an important component of projected growth in the transportation equipment, electronics, and machinery industries.

Energy expenditures decline relative to gross domestic product and gross output

Figure 62. Energy end-use expenditures as a share of gross domestic product, 1970-2035 (nominal expenditures as percent of nominal GDP)

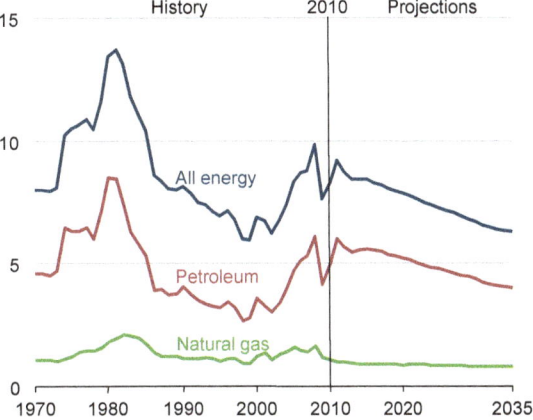

Total U.S. energy expenditures decline relative to GDP in the *AEO2012* Reference case (Figure 62) [123]. The projected share of energy expenditures falls from 2011 through 2035, averaging 7.5 percent from 2010 to 2035, which is below the historical average of 8.8 percent from 1970 to 2010.

Gross output corresponds roughly to sales in the U.S. economy. Figure 63 provides an approximation of total energy expenditures relative to total sales. Energy expenditures as a share of gross output show roughly the same pattern as do energy expenditures as a share of GDP. The projected average shares of gross output relative to expenditures for total energy, petroleum, and natural gas are close to their historical averages, at 4.1 percent, 2.1 percent, and 0.5 percent, respectively.

Figure 63. Energy end-use expenditures as a share of gross output, 1987-2035 (nominal expenditures as percent of nominal gross output)

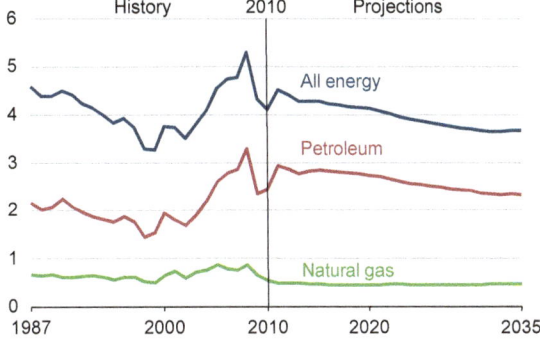

International energy

Oil price cases depict uncertainty in world oil markets

Figure 64. Average annual oil prices in three cases, 1980-2035 (2010 dollars per barrel)

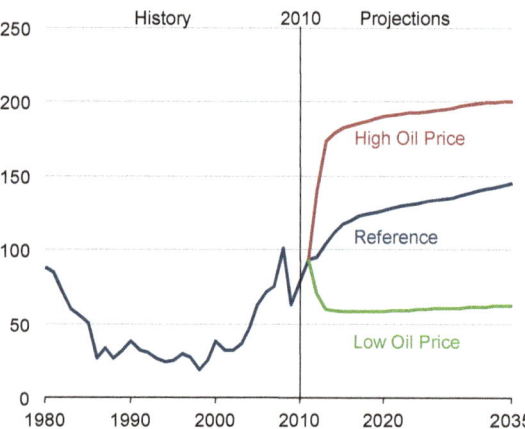

Oil prices in AEO2012, defined in terms of the average price of low-sulfur, light crude oil (West Texas Intermediate [WTI]) delivered to Cushing, Oklahoma, span a broad range that reflects the inherent volatility and uncertainty of oil prices (Figure 64). The AEO2012 price paths are not intended to reflect absolute bounds for future oil prices but rather to provide a basis for analysis of the implications of world oil market conditions that differ from those assumed in the AEO2012 Reference case. The Reference case assumes that the current price discount for WTI relative to similar "marker" crude oils (such as Brent and Louisiana Light Sweet) will fade when adequate pipeline capacity is built between Cushing and the Gulf of Mexico.

In the Low Oil Price case, GDP growth in countries outside the Organization of the Petroleum Exporting Countries (non-OPEC) is slower than in the Reference case, resulting in lower demand for petroleum and other liquids, and producing countries develop stable fiscal policies and investment regimes that encourage resource development. OPEC nations increase production, achieving approximately a 46-percent market share of total petroleum and other liquids production in 2035.

The High Oil Price case depicts a world oil market in which total GDP growth in countries outside the Organization for Economic Cooperation and Development (non-OECD) is faster than in the Reference case, driving up demand for petroleum and other liquids. Production of crude oil and natural gas liquids (NGL) is restricted by political decisions and limits on access to resources (such as the use of quotas and fiscal regimes) compared with the Reference case. Petroleum and other liquids production in the major producing countries is reduced (for example, the OPEC share averages 40 percent), and the consuming countries turn to more expensive production from other liquids sources to meet demand.

Trends in petroleum and other liquids markets are defined largely by the developing nations

Figure 65. World petroleum and other liquids supply and demand by region in three cases, 2010 and 2035 (million barrels per day)

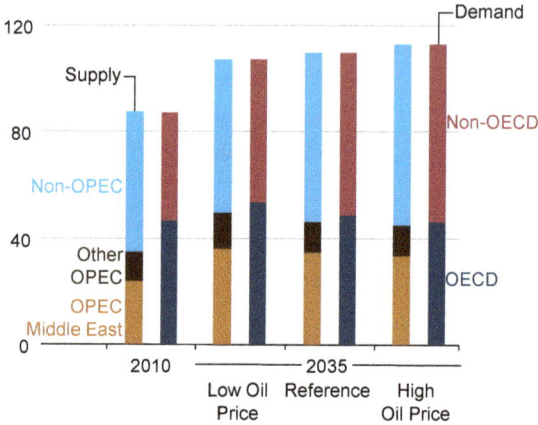

Total use of petroleum and other liquids in the AEO2012 Reference, High Oil Price, and Low Oil Price cases in 2035 ranges from 107 to 113 million barrels per day (Figure 65). The alternative oil price cases reflect shifts in both supply and demand, with the result that total consumption and production levels do not vary widely. Although demand in the OECD countries is influenced primarily by price, demand in non-OECD regions—where future economic uncertainty is greatest—drives the price projections. That is, non-OECD petroleum and other liquids consumption is lower in the Low Oil Price case and higher in the High Oil Price case than it is in the Reference case.

OECD petroleum and other liquids use grows in the Reference case to 48 million barrels per day in 2035, while non-OECD use grows to 61 million barrels per day. In the Low Oil Price case, OECD petroleum and other liquids use in 2035 is higher than in the Reference case, at 53 million barrels per day, but demand in the slow-growing non-OECD economies in the Low Price case rises to only 54 million barrels per day. In the High Oil Price case the opposite occurs, with OECD consumption falling to 46 million barrels per day in 2035 and fast-growing non-OECD use—driven by higher GDP growth—increasing to 67 million barrels per day in 2035.

The supply response also varies across the price cases. In the Low Oil Price case, OPEC's ability to constrain market share is weakened, and low prices have a negative impact on non-OPEC crude oil supplies relative to the Reference case. Because non-crude oil technologies achieve much lower costs in the Low Price case, supplies of other liquids are more plentiful than in the Reference case. In the High Oil Price case, OPEC restricts production, non-OPEC resources become more economic, and high prices make other liquids more attractive.

International energy

Production from resources other than crude oil and natural gas liquids increases

Figure 66. Total world production of nonpetroleum liquids, bitumen, and extra-heavy oil in three cases, 2010 and 2035 (million barrels per day)

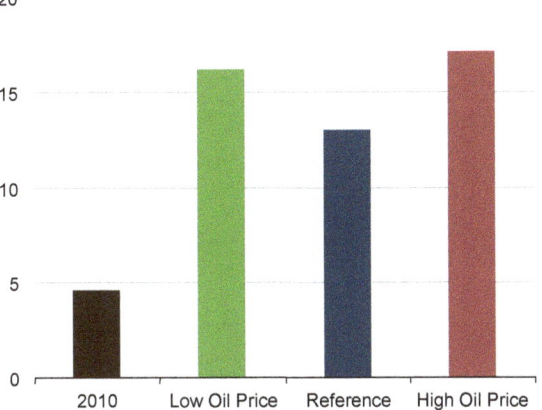

In 2010, world production of liquid fuels from resources other than crude oil and NGL totaled 4.6 million barrels per day, or about 5 percent of all petroleum and other liquids production. Production from those other sources grows to 13.0 million barrels per day (about 12 percent of total global production of petroleum and other liquids) in 2035 in the AEO2012 Reference case, 16.2 million barrels per day (15 percent of the total) in the Low Oil Price case, and 17.1 million barrels per day (15 percent of the total) in the High Oil Price case (Figure 66). The higher levels of production from other resources result from declining technology costs in the Low Oil Price case and from higher oil prices in the High Oil Price case.

Assumptions about the development of other liquids resources differ across the three cases. In the Reference case, increasingly expensive projects become more economically competitive as a result of rising oil prices and advances in production technology. Bitumen in Canada and biofuels in the United States and Brazil are the most important components of production from sources other than crude oil and NGL. Excluding crude oil and NGL, U.S. and Brazilian biofuels and Canadian bitumen account for more than 70 percent of the total world increase in petroleum and other liquids production from 2010 to 2035 in the Reference case.

In the High Oil Price case, rising prices support increased development of nonpetroleum liquids, bitumen, and extra-heavy oil. A smaller increase is projected in the Low Oil Price case, which assumes significant declines in technology costs, particularly for extra-heavy oil production. Bitumen and biofuels continue to be the most important contributors to this supply category through 2035.

U.S. reliance on imported natural gas from Canada declines as exports grow

Figure 67. North American natural gas trade, 2010-2035 (trillion cubic feet)

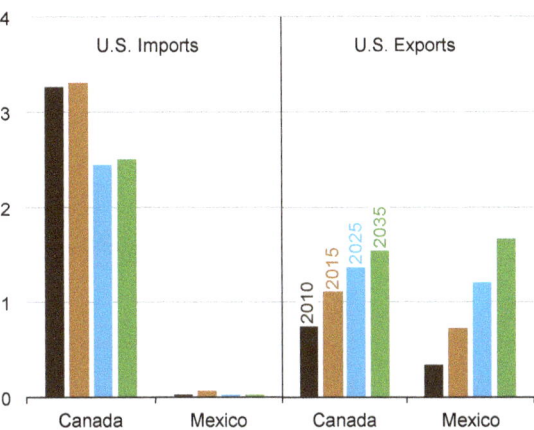

The energy markets of the three North American nations (United States, Canada, and Mexico) are well integrated, with extensive infrastructure that allows cross-border trade between the United States and both Canada and Mexico. The United States, which is by far the region's largest energy consumer, currently relies on Canada and Mexico for supplies of petroleum and other liquid fuels. Canada and Mexico were the largest suppliers of U.S. petroleum and other liquids imports in 2010, providing 2.5 and 1.3 million barrels per day, respectively. In addition, Canada supplies the United States with substantial natural gas supplies, exporting 3.3 trillion cubic feet to U.S. markets in 2010 (Figure 67).

In the AEO2012 Reference case, energy trade between the United States and the two other North American countries continues. In 2035, the United States still imports 3.4 million barrels per day of petroleum and other liquid fuels from Canada in the Reference case, but imports from Mexico fall to 0.8 million barrels per day. With prospects for domestic U.S. natural gas production continuing to improve, the need for imported natural gas declines. U.S. imports of natural gas from Canada fall to 2.4 trillion cubic feet in 2025 in the Reference case and remain relatively flat through the end of the projection. On the other hand, U.S. natural gas exports to both Canada and Mexico increase. Canada's imports of U.S. natural gas grow from 0.7 trillion cubic feet in 2010 to 1.5 trillion cubic feet in 2035, and Mexico's imports grow from 0.3 trillion cubic feet in 2010 to 1.7 trillion cubic feet in 2035 in the AEO2012 Reference case.

International energy

China and India account for half the growth in world energy use

Figure 68. World energy consumption by region, 1990-2035 (quadrillion Btu)

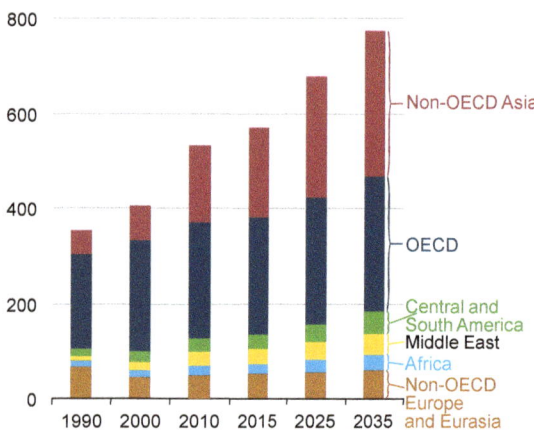

After Fukushima, prospects for nuclear power dim in Japan and Europe but not elsewhere

Figure 69. Installed nuclear capacity in OECD and non-OECD countries, 2010 and 2035 (gigawatts)

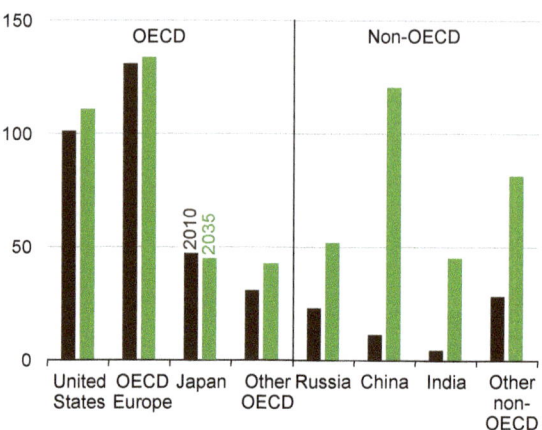

World energy consumption increases by 47 percent from 2010 through 2035 in the AEO2012 Reference case (Figure 68). Most of the growth is projected for emerging economies outside the OECD, where robust economic growth is accompanied by increased demand for energy. Total non-OECD energy use grows by 72 percent, compared with an 18-percent increase in OECD energy use.

Energy consumption in non-OECD Asia, led by China and India, shows the most robust growth among the non-OECD regions, rising by 91 percent from 2010 to 2035. However, strong growth also occurs in much of the rest of the non-OECD regions: 69 percent in Central and South America, 65 percent in Africa, and 62 percent in the Middle East. The slowest growth among the non-OECD regions is projected for non-OECD Europe and Eurasia (including Russia), where substantial gains in energy efficiency are achieved through replacement of inefficient Soviet-era capital equipment.

Worldwide, the use of energy from all sources increases in the projection. Given expectations that oil prices will remain relatively high, petroleum and other liquids are the world's slowest-growing energy sources. High energy prices and concerns about the environmental consequences of greenhouse gas (GHG) emissions lead a number of national governments to provide incentives in support of the development of alternative energy sources, making renewables the world's fastest-growing source of energy in the outlook.

The earthquake and tsunami that hit northeastern Japan in March 2011 caused extensive loss of life and infrastructure damage, including severe damage to several reactors at the Fukushima Daiichi nuclear power plant. In the aftermath, governments in several countries that previously had planned to expand nuclear capacity—including Japan, Germany, Switzerland, and Italy—reversed course. Even China announced a temporary suspension of its approval process for new reactors pending a thorough safety review.

Before the Fukushima event, EIA had projected that all regions of the world with existing nuclear programs would expand their nuclear power capacity. Now, however, Japan's nuclear capacity is expected to contract by about 3 gigawatts from 2010 to 2035 (Figure 69). In OECD Europe, Germany's outlook has been revised to reflect a phaseout of all nuclear power by 2025. As a result, the projected net increase in OECD Europe's nuclear capacity in the AEO2012 Reference case is only 3 gigawatts from 2010 to 2035.

Significant expansion of nuclear power is projected to continue in the non-OECD region as a whole, with total nuclear capacity more than quadrupling. From 2010 to 2035, nuclear power capacity increases by a net 109 gigawatts in China, 41 gigawatts in India, and 28 gigawatts in Russia, as strong growth in demand for electric power and concerns about security of energy supplies and the environmental impacts of fossil fuel use encourage further development of nuclear power in non-OECD countries.

U.S. energy demand

Wind power leads rise in world renewable generation, solar power also grows rapidly

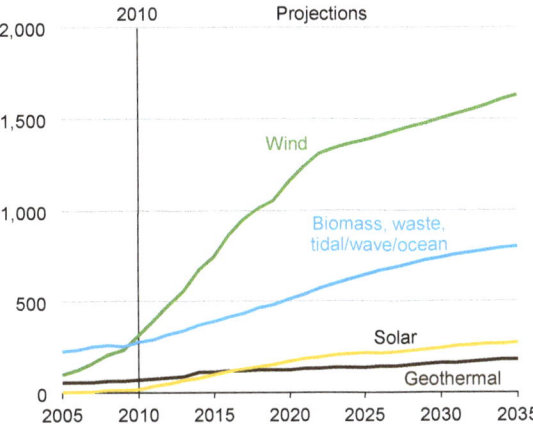

Figure 70. World renewable electricity generation by source, excluding hydropower, 2005-2035 (billion kilowatthours)

Renewable energy is the world's fastest-growing source of marketed energy in the AEO2012 Reference case, increasing by an average of 3.0 percent per year from 2010 to 2035, compared to an average of 1.6 percent per year for total world energy consumption. In many parts of the world, concerns about the security of energy supplies and the environmental consequences of GHG emissions have spurred government policies that support rapid growth in renewable energy installations.

Hydropower is well-established worldwide, accounting for 83 percent of total renewable electricity generation in 2010. Growth in hydroelectric generation accounts for about one-half of the world increase in renewable generation in the Reference case. In Brazil and the developing nations of Asia, significant builds of mid- and large-scale hydropower plants are expected, and the two regions together account for two-thirds of the total world increase in hydroelectric generation from 2010 to 2035.

Solar power is the fastest-growing source of renewable energy in the outlook, with annual growth averaging 11.7 percent. However, because it currently accounts for only 0.4 percent of total renewable generation, solar remains a minor part of the renewable mix even in 2035, when its share reaches 3 percent. Wind generation accounts for the largest increment in nonhydro-power renewable generation—60 percent of the total increase, as compared with solar's 12 percent (Figure 70). The rate of wind generation slows markedly after 2020 because most government wind goals are achieved and wind must then compete on the basis of economics with fossil fuels. Wind-powered generating capacity has grown swiftly over the past decade, from 18 gigawatts of installed capacity in 2000 to an estimated 179 gigawatts in 2010.

In the United States, average energy use per person declines from 2010 to 2035

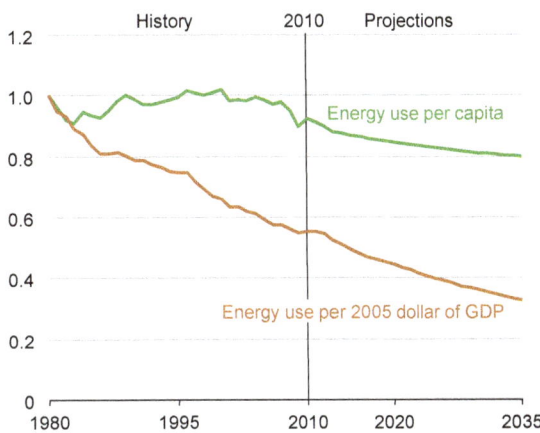

Figure 71. Energy use per capita and per dollar of gross domestic product, 1980-2035 (index, 1980 = 1)

Growth in energy use is linked to population growth through increases in housing, commercial floorspace, transportation, and goods and services. These changes affect not only the level of energy use but also the mix of fuels consumed.

Changes in the structure of the economy and in the efficiency of the equipment deployed throughout the economy also have an impact on energy use per capita. The shift in the industrial sector away from energy-intensive manufacturing toward services is one reason for the projected decline in industrial energy intensity (energy use per dollar of GDP), but its impact on energy consumption per capita is less direct (Figure 71). From 1990 to 2007, the service sectors increased from a 69-percent share of total industrial output to a 75-percent share, but energy use per capita remained fairly constant, between 330 and 350 million British thermal units (Btu) per person, while energy use per dollar of GDP dropped from about 10,500 to 7,700 Btu. Increases in the efficiency of freight vehicles and the shift toward output from the service sectors are projected to continue through 2035, lowering energy use in relation to GDP. Energy use per dollar of GDP is projected to be about 4,400 Btu in 2035, or about one-third of the 1980 level.

Efficiency gains in household appliances and personal vehicles have a direct, downward impact on energy use per capita, as do efficiency gains in the electric power sector, as older, inefficient coal and other fossil steam electricity generating plants are retired in anticipation of lower electricity demand growth, changes in fuel prices, and new environmental regulations. As a result, U.S. energy use per capita declines to 274 million Btu in 2035.

U.S. energy demand

Industrial and commercial sectors lead U.S. growth in primary energy use

Figure 72. Primary energy use by end-use sector, 2010-2035 (quadrillion Btu)

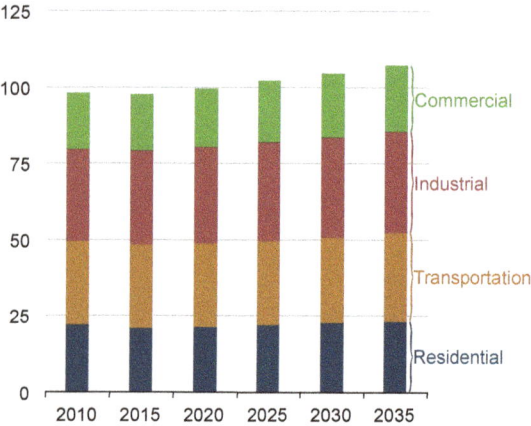

Total primary energy consumption, including fuels used for electricity generation, grows by 0.3 percent per year from 2010 to 2035, to 106.9 quadrillion Btu in 2035 in the *AEO2012* Reference case (Figure 72). The largest growth, 3.3 quadrillion Btu from 2010 to 2035, is in the commercial sector, which currently accounts for the smallest share of end-use energy demand. Even as standards for building shells and energy efficiency are being tightened in the commercial sector, the growth rate for commercial energy use, at 0.7 percent per year, is the highest among the end-use sectors, propelled by 1.0 percent average annual growth in commercial floorspace.

The industrial sector, which was more severely affected than the other end-use sectors by the 2008-2009 economic downturn, shows the second-largest increase in total primary energy use, at 3.1 quadrillion Btu from 2010 to 2035. The total increase in industrial energy consumption is 2.1 quadrillion Btu from 2008 to 2035, attributable to increased production of biofuels to meet the Energy Independence and Security Act of 2007 (EISA2007) renewable fuels standard (RFS) as well as increased use of natural gas in some industries, such as food and paper, to generate their own electricity.

Primary energy use in both the residential and transportation sectors grows by 0.2 percent per year, or by just over 1 quadrillion Btu each from 2010 to 2035. In the residential sector, increased efficiency reduces energy use for space heating, lighting, and clothes washers and dryers. In the transportation sector, light-duty vehicle (LDV) energy consumption declines after 2012 to 14.7 quadrillion Btu in 2023 (the lowest point since 1998) before increasing through 2035, when it is still 4 percent below the 2010 level.

Renewable energy sources lead rise in primary energy consumption

Figure 73. Primary energy use by fuel, 1980-2035 (quadrillion Btu)

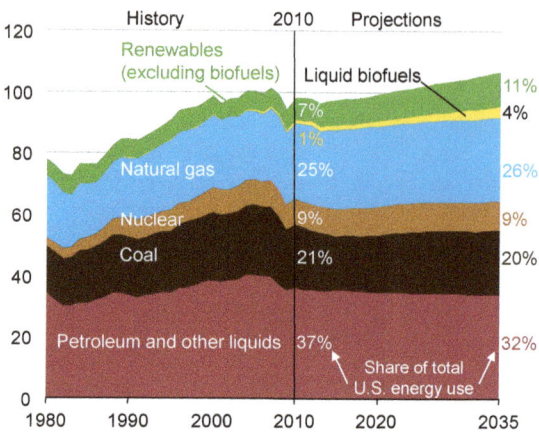

With the exception of petroleum and other liquids, which falls through 2032 before increasing slightly in the last 3 years of the projection, consumption of all fuels increases in the *AEO2012* Reference case. In addition, coal consumption increases at a relatively weak average rate of less than 0.1 percent per year from 2010 to 2035, remaining below 2010 levels until after 2031. As a result, the aggregate fossil fuel share of total energy use falls from 83 percent in 2010 to 77 percent in 2035, while renewable fuel use grows rapidly (Figure 73). The renewable share of total energy use (including biofuels) increases from 8 percent in 2010 to 14 percent in 2035 in response to the Federal RFS, availability of Federal tax credits for renewable electricity generation and capacity, and State renewable portfolio standard (RPS) programs.

The petroleum and other liquids share of fuel use declines as consumption of other liquids increases. Almost all consumption of liquid biofuels is in the transportation sector. Biofuels, including biodiesel blended into diesel, E85, and ethanol blended into motor gasoline (up to 15 percent), account for 10 percent of all petroleum and other liquids consumption in 2035.

Natural gas consumption grows by about 0.4 percent per year from 2010 to 2035, led by the use of natural gas in electricity generation. Growing production from tight shale keeps natural gas prices below their 2005-2008 levels through 2035.

By the end of 2012, a total of 9.3 gigawatts of coal-fired power plant capacity currently under construction is expected to come online, and another 1.7 gigawatts is added after 2017 in the Reference case, including 0.9 gigawatts with carbon sequestration capability. Additional coal is consumed in the coal-to-liquids (CTL) process to produce heat and power, including electricity generation at CTL plants.

Residential sector energy demand

Residential energy use per household declines for a range of technology assumptions

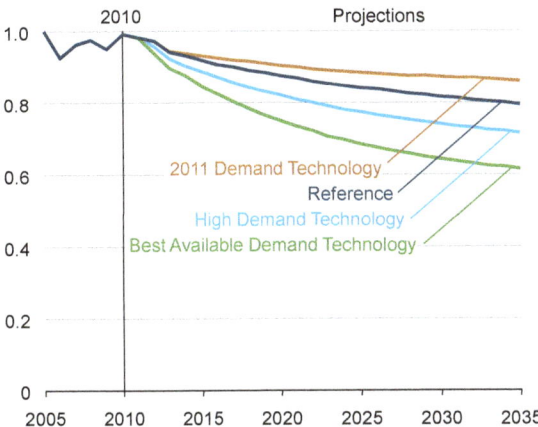

Figure 74. Residential delivered energy intensity in four cases, 2005-2035 (index, 2005 = 1)

Electricity use increases with number of households despite efficiency improvement

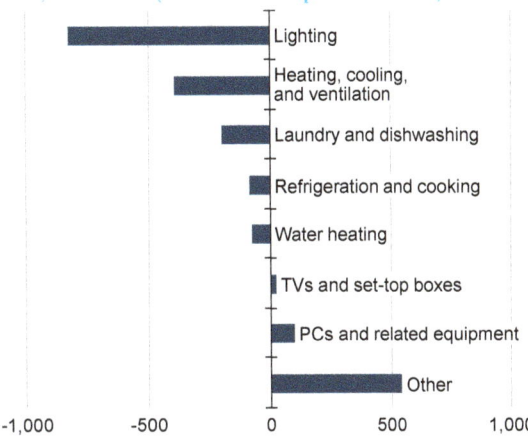

Figure 75. Change in residential electricity consumption for selected end uses in the Reference case, 2010-2035 (kilowatthours per household)

In the *AEO2012* Reference case, residential sector energy intensity, defined as average energy use per household per year, declines by 19.8 percent, to 81.9 million Btu per year in 2035 (Figure 74). Total delivered energy use in the residential sector remains relatively constant from 2010 to 2035, but a 27.5-percent growth in the number of households reduces the average energy intensity of each household. Most residential end-use services become less energy-intensive, with space heating accounting for more than one-half of the decrease. Population shifts to warmer and drier climates also contribute to a reduction in demand for space heating.

Three alternative cases show how different technology assumptions affect residential energy intensity. The 2011 Demand Technology case assumes no improvement in efficiency for end-use equipment or building shells beyond those available in 2011. The High Demand Technology case assumes higher efficiency, earlier availability, lower cost, and more frequent energy-efficient purchases for some advanced equipment. The Best Available Demand Technology case limits customers who purchase new and replacement equipment to the most efficient model available in the year of purchase—regardless of cost—and assumes that new homes are constructed to the most energy-efficient specifications.

From 2010 to 2035, household energy intensity declines by 27.7 percent in the High Demand Technology case and by 37.9 percent in the Best Available Demand Technology case. In the 2011 Demand Technology case, household energy intensity also falls as older appliances are replaced with 2011 vintage equipment. Without further gains in efficiency for residential equipment and building shells, the total decline from 2010 to 2035 is only 13.2 percent.

Despite a decrease in electricity consumption per household, total delivered electricity use in the residential sector grows at an average rate of 0.7 percent per year in the *AEO2012* Reference case, while natural gas use and petroleum and other liquids use fall by 0.2 percent and 1.3 percent per year, respectively, from 2010 to 2035. The increase in efficiency, driven by new standards and improved technology, is not high enough to offset the growth in the number of households and electricity consumption in "other" uses.

Portions of the Federal lighting standards outlined in EISA 2007 went into effect on January 1, 2012. Over the next two years, general-service lamps that provide 310 to 2,600 lumens of light are required to consume about 30 percent less energy than typical incandescent bulbs. High-performance incandescent, compact fluorescent, and light-emitting diode (LED) lamps continue to replace low-efficacy incandescent lamps. In 2035, delivered energy for lighting per household in the Reference case is 827 kilowatthours per household lower, or 47 percent below the 2010 level (Figure 75).

Electricity consumption for three groups of electricity end uses increases on a per-household basis in the Reference case. Electricity use for televisions and set-top boxes grows by an average of 1.1 percent per year, accounting for 7.3 percent of total delivered electricity consumption in 2035. Personal computers (PCs) and related equipment account for 4.6 percent of residential electricity consumption in 2035, averaging 1.8-percent annual growth from their 2010 level. Electricity use by other household electrical devices, for which market penetration increases with little coverage by efficiency standards, increases by 1.8 percent annually and accounts for nearly one-fourth of total residential electricity consumption in 2035.

Residential sector energy demand

Residential consumption varies depending on efficiency assumptions

Figure 76. Ratio of residential delivered energy consumption for selected end uses
(ratio, 2035 to 2010)

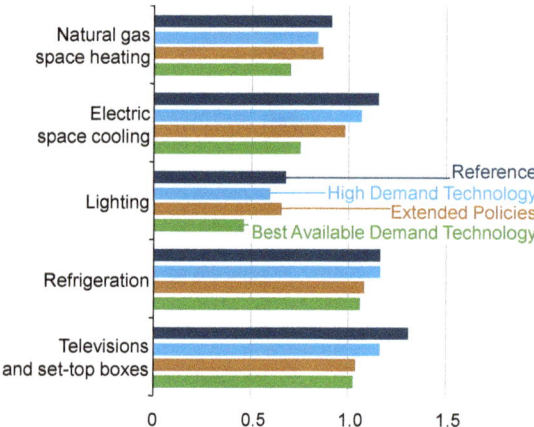

Tax credits could spur growth in renewable energy equipment in the residential sector

Figure 77. Residential market penetration by renewable technologies in two cases, 2010, 2020, and 2035 (percent of households)

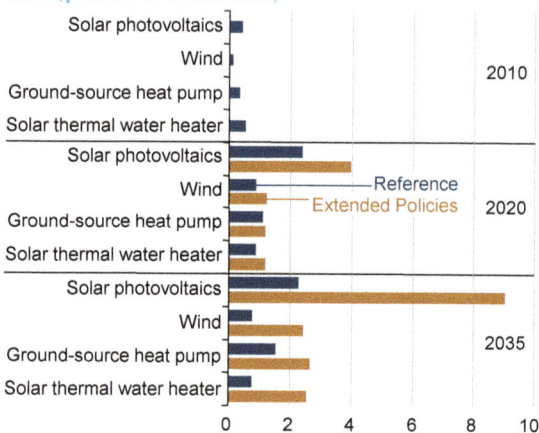

The *AEO2012* Reference case and three alternative cases demonstrate opportunities for improved energy efficiency to reduce energy consumption in the residential sector. The Reference, High Demand Technology, and Best Available Demand Technology cases include different levels of efficiency improvement without anticipating the enactment of new appliance standards. The Extended Policies case assumes the enactment of new rounds of standards, generally based on improvements seen in current ENERGY STAR equipment.

Despite continued growth in the number of households and number of appliances, energy consumption for some end uses is lower in 2035 than in 2010, implying that improved energy efficiency offsets the growth in service demand. In the case of natural gas space heating, population shifts towards warmer and drier climates also reduce consumption; the opposite is true for electric space cooling.

In the Extended Policies case, the enactment of new standards is based on the U.S. Department of Energy's multi-year schedule. For lighting, which already has an EISA2007-based standard that is scheduled to go into effect in 2020, future standards are not assumed until 2026. Among electric end uses, lighting has the largest percentage decline in energy use (more than 50 percent) in the Best Available Demand Technology case from 2010 to 2035 (Figure 76).

Televisions and set-top boxes, which are not currently covered by Federal standards, are assumed to have new standards in 2016 and 2018, respectively, in the Extended Policies case. The enactment of these new standards holds energy use for televisions and set-top boxes at or near their 2010 levels through 2035.

Consistent with current law, existing investment tax credits (ITCs) expire at the end of 2016 in the *AEO2012* Reference case. The current credits can offset 30 percent of installed costs for a variety of distributed generation (DG) technologies, fostering their adoption. Installations slow dramatically after the ITCs expire, and in several cases their overall market penetration falls because growth in households exceeds the rise in new renewable installations (Figure 77). In the *AEO2012* Extended Policies case, the ITCs are extended through 2035, and penetration rates for all renewable technologies continue to rise.

In the Reference case, photovoltaic (PV) and wind capacities grow by average rates of 10.8 percent and 9.2 percent per year, respectively, from 2010 to 2035. In the Extended Policies case, residential PV capacity increases to 54.6 gigawatts in 2035, with annual growth averaging 18.1 percent, and wind capacity grows to 11.0 gigawatts in 2035, averaging 15.9 percent per year.

The ITCs also affect the penetration of renewable space-conditioning and water-heating equipment. Ground-source heat pumps reach a 2.6-percent market share in 2035 in the Extended Policies case, after adding nearly 3.5 million units. In the Reference case, without the ITC extension, their market penetration is only 1.5 percent in 2035, with 1.6 million fewer installations than in the Extended Policies case.

Market penetration of solar water heaters in the Extended Policies case is 2.5 percent in 2035, more than triple the Reference case share. In the Reference case, installations increase by 2.5 percent annually from 2010 to 2035, compared with 7.5 percent annually in the Extended Policies case.

Commercial sector energy demand

For commercial buildings, pace of decline in energy intensity depends on technology

Figure 78. Commercial delivered energy intensity in four cases, 2005-2035 (index, 2005 = 1)

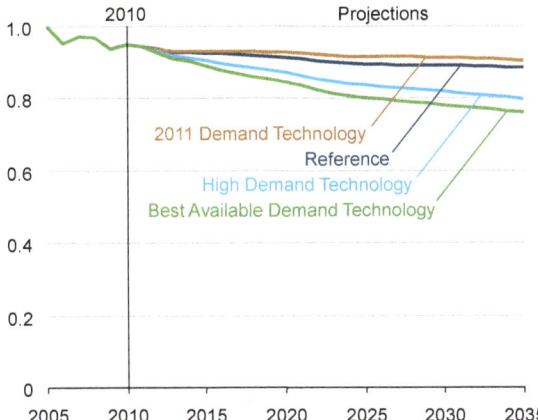

In the *AEO2012* Reference case, average delivered energy use per square foot of commercial floorspace declines by 7.0 percent from 2010 to 2035 (Figure 78). Growth in commercial floorspace (26.9 percent) leads to an increase in delivered energy use (18.1 percent), but efficiency improvements in equipment and building shells reduce energy intensity in commercial buildings. Space heating, space cooling, and lighting contribute most to the decrease in intensity, with space heating accounting for significantly more than cooling and lighting combined.

Three alternative cases show the potential impact of energy-efficient technologies on energy intensity in commercial buildings. The 2011 Demand Technology case limits equipment and building shell technologies in later years to the options available in 2011. The High Demand Technology case assumes higher efficiencies for equipment and building shells, lower costs, earlier availability of some advanced equipment, and decisions by commercial customers that place greater importance on future energy savings. The Best Available Technology case assumes more efficient buildings shells for new and existing buildings than in the High Demand Technology case and also requires commercial customers to choose among the most efficient models for each technology when replacing old or purchasing new equipment.

From 2010 to 2035, the intensity of commercial energy use in the 2011 Technology Demand case declines by 5.0 percent, to 101.9 thousand Btu per square foot of commercial floorspace in 2035. In comparison, intensity decreases faster in the High Demand Technology case (16.0 percent) and fastest in the Best Available Demand Technology case (20.0 percent).

Efficiency standards reduce electric energy intensity in commercial buildings

Figure 79. Energy intensity of selected commercial electric end uses, 2010 and 2035 (thousand Btu per square foot)

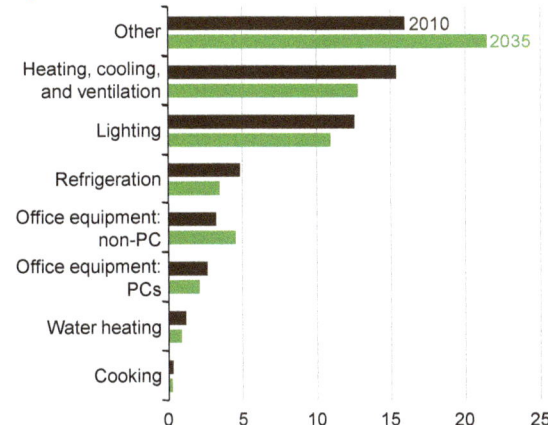

Electricity, which accounted for 52 percent of total commercial delivered energy use in 2010, increases to 56 percent in 2035 in the *AEO2012* Reference case, as commercial floorspace grows at an average annual rate of 1 percent and new electric end uses become more prevalent. Despite such growth, improved efficiency of commercial equipment slows the growth of purchased electricity over the projection period.

Commercial energy intensity in this figure, defined as the ratio of energy consumption in these appliances to floorspace, decreases for most electric end uses from 2010 to 2035 in the Reference case (Figure 79). Electricity intensity decreases by 1.3 percent annually for both cooking and refrigeration, by 0.5 percent annually for lighting, and by 0.7 percent annually for space conditioning (heating, cooling, and ventilation).

End uses such as space heating and cooling, water heating, refrigeration, and lighting are covered by Federal efficiency standards that act to limit growth in energy consumption to less than the growth in commercial floorspace. "Other" electric end uses, some of which are not subject to standards, account for much of the growth in commercial electricity consumption in the Reference case. Electricity consumption for "other" electrical end uses—including video displays and medical devices—increases by an average of 2.2 percent per year and in 2035 accounts for 38 percent of total commercial electricity consumption. Energy consumption for "other" office equipment—including servers and mainframe computers—increases by 2.3 percent per year from 2010 to 2035, as demand for high-speed networks and internet connectivity continues to grow.

Commercial sector energy demand

Technologies for major energy applications lead efficiency gains in commercial sector

Figure 80. Efficiency gains for selected commercial equipment in three cases, 2035 (percent change from 2010 installed stock efficiency)

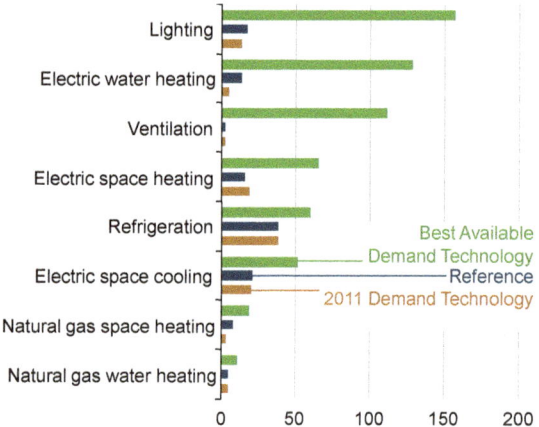

Investment tax credits could increase distributed generation in commercial sector

Figure 81. Additions to electricity generation capacity in the commercial sector in two cases, 2010-2035 (gigawatts)

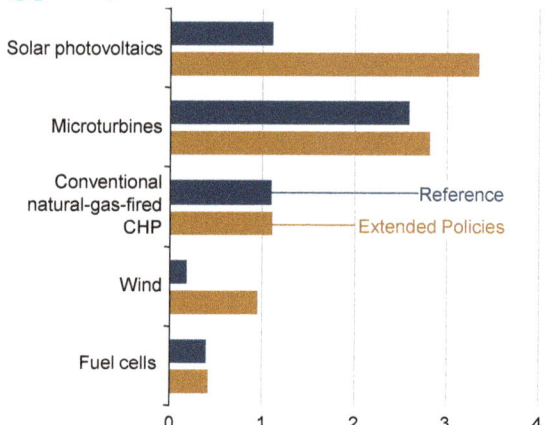

Delivered energy consumption for space heating, ventilation, air conditioning, water heating, lighting, cooking, and refrigeration uses in the commercial sector grows by an average of 0.2 percent per year from 2010 to 2035 in the AEO2012 Reference case, compared with 1.0-percent annual growth in commercial floorspace. The core end uses, which frequently have been the focus of energy efficiency standards, accounted for just over 60 percent of commercial delivered energy demand in 2010. In 2035, their share falls to 53 percent. Energy consumption for all the remaining end uses grows by 1.3 percent per year, led by office equipment other than computers and other electric end uses.

The percentage gains in efficiency in the Reference case are highest for refrigeration, as a result of provisions in the Energy Policy Act of 2005 and EISA2007. Electric space cooling shows the next-largest percentage improvement, followed by lighting and electric space heating (Figure 80).

The Best Available Demand Technology case demonstrates significant potential for further improvement—especially in electric equipment, led by lighting, water heating, and ventilation. In the Best Available Demand Technology case, the share of total commercial delivered energy use in the core end uses falls to 49 percent in 2035, with significant efficiency gains coming from high-efficiency variable air volume ventilation systems, LED lighting, ground-source heat pumps, high-efficiency rooftop heat pumps, centrifugal chillers, and solar water heaters. Those technologies are relatively costly, however, and thus unlikely to gain wide adoption in commercial applications without improved economics. Additional efficiency improvements could also come from an expansion of standards to include some of the rapidly growing miscellaneous electric applications.

ITCs have a major impact on the growth of renewable DG in the commercial sector. Although most ITCs are set to expire at the end of 2016, the tax credit for solar PV installations reverts from 30 percent to 10 percent and continues indefinitely. Commercial PV capacity increases by 2.7 percent annually from 2010 through 2035 in the AEO2012 Reference Case. Extending the ITCs to all DG technologies through 2035 in the AEO2012 Extended Policies case causes PV capacity to increase at an average annual rate of 5.7 percent (Figure 81).

Growth in small-scale wind capacity more than doubles in the Extended Policies case relative to the Reference case, increasing at an average annual rate of 11.4 percent from 2010 to 2035. Wind accounts for 9.2 percent of the 11.1 gigawatts of total commercial DG capacity in 2035 in the Extended Policies case, and PV accounts for 40.6 percent. In the Extended Policies case, renewable energy accounts for 53 percent of all commercial DG capacity, compared with about 37 percent in the Reference case.

Although ITCs affect the rate of adoption of renewable DG by offsetting a portion of capital costs, their potential effects on nonrenewable DG technologies are offset by rising natural gas prices. In the Reference case, microturbine capacity using natural gas grows by an average of 18.1 percent per year from 42 megawatts in 2010 to 2.6 gigawatts in 2035, and the growth rate in the Extended Policies case is only slightly higher, at 18.4 percent. In the Extended Policies case, the microturbine share of total DG capacity in 2035 is 25.6 percent, as compared with 33.4 percent in the Reference case.

Industrial sector energy demand

Manufacturing heat and power energy consumption increases modestly

Figure 82. Industrial delivered energy consumption by application, 2010-2035 (quadrillion Btu)

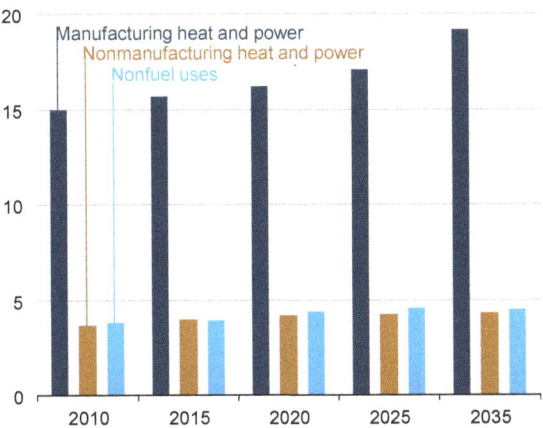

Reliance on natural gas and natural gas liquids rises as industrial energy use grows

Figure 83. Industrial energy consumption by fuel, 2010, 2025 and 2035 (quadrillion Btu)

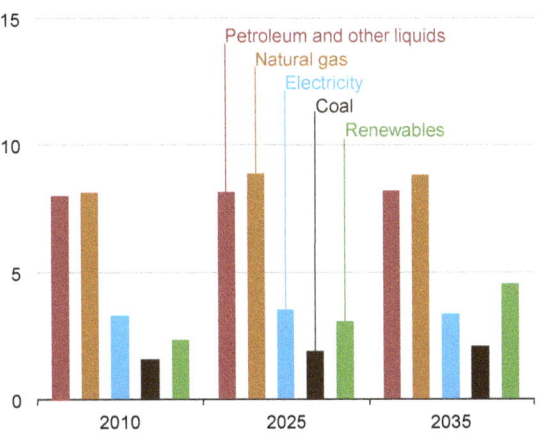

Despite a 49-percent increase in industrial shipments, industrial delivered energy consumption increases by only 15 percent from 2010 to 2035 in the AEO2012 Reference case, reflecting a shift in the share of shipments from energy-intensive manufacturing industries (which include bulk chemicals, petroleum refineries, paper products, iron and steel, food products, aluminum, cement, and glass) to other, less energy-intensive industries, such as plastics, computers, and transportation equipment. Although energy use for most of the energy-intensive industries continues to grow after 2012, with the stronger growth in refining, declines in the energy intensity of heat and power production offset some the growth in their energy use.

The share of industrial delivered energy consumption used for heat and power in manufacturing increases from 64 percent in 2010 to 71 percent in 2035 (Figure 82). The increase in heat and power energy consumption in manufacturing in the Reference case is primarily a result of a large increase (2 quadrillion Btu) in total energy use in the petroleum refining industry, including production increases for CTL, coal- and biomass-to-liquids (CBTL), and biomass pyrolysis oil production.

Heat and power consumption in the nonmanufacturing industries (agriculture, mining, and construction) is flat in the Reference case projection, accounting for about 16 percent of total industrial energy consumption over the 2010-2035 period. The remaining consumption consists of nonfuel uses of energy—primarily, feedstocks for chemical manufacturing and asphalt for construction. The share of total industrial energy consumption represented by nonfuel use increases by 1.6 percent from 2010 to 2020 as a result of increased shipments of organic chemicals, then declines as competition from foreign producers slows the growth of domestic production.

Led by increasing use of natural gas, total delivered industrial energy consumption grows at an annual rate of 0.6 percent from 2010 through 2035 in the Reference case. The mix of fuels changes slowly, reflecting limited capability for fuel switching with the current capital stock (Figure 83).

Industrial natural gas use grows by 8 percent from 2010 to 2035, reflecting relatively low natural gas prices. As a result, 33 percent of delivered industrial energy consumption is met with natural gas in 2035. The second-largest share is met by petroleum and other liquids (30 percent) and the remainder by renewables, electricity, and coal (37 percent). NGL, an increasingly valuable liquid component of natural gas processing, are consumed as a feedstock in the bulk chemicals industry and also are used for heat in other sectors. Industrial use of all petroleum and other liquids increases slightly from 2010 to 2035, and in 2035 the chemical industries use nearly one-half of the total as feedstock.

Coal use in the industrial sector for boilers and for smelting in steelmaking declines as more boilers are fired with natural gas and less metallurgical coal is used for steelmaking. After 2016, increased use of coal for CTL and CBTL production fully offsets the decline in the steel industry and boiler fuel use.

A decline in the electricity share of industrial energy consumption reflects modest growth in combined heat and power (CHP), which offsets purchased electricity requirements, as well as efficiency improvements across industries, primarily as a result of rising standards for motor efficiency. With growth in lumber, paper, and other industries that consume biomass-based byproducts, the renewable share of industrial energy use expands.

Industrial sector energy demand

Iron and steel and cement industries are most sensitive to economic growth rate

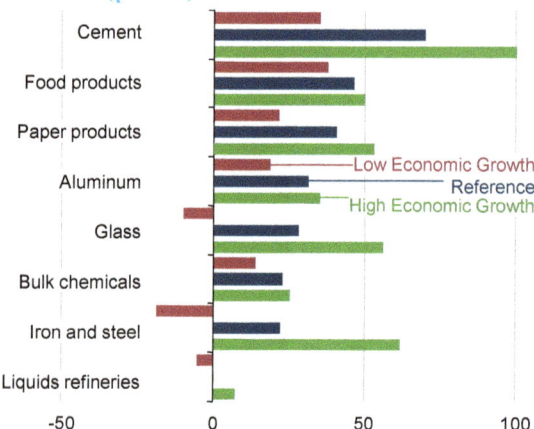

Figure 84. Cumulative growth in value of shipments from energy-intensive industries in three cases, 2010-2035 (percent)

Total shipments from the energy-intensive industries grow by an average of 1 percent per year from 2010 to 2035 in the Reference case, as compared with 0.6 percent in the Low Economic Growth case and 1.2 percent in the High Economic Growth case. The post-recession recovery in shipments is uneven among the industrial subsectors. Paper, bulk chemicals, aluminum, and cement all show strong short-term recoveries from 2010 levels, while shipments from the liquids refinery industry lag. The iron and steel and glass industries show flat to moderate growth in the near term.

Among the energy-intensive industries, the value of shipments in the bulk chemicals, paper, and aluminum take less than 10 years to return to their 2006-2007 pre-recession levels. Others, including cement, iron and steel, and glass, take longer. Shipments from the liquids refinery industry do not reach pre-recession levels by 2035, because demand for transportation fuels is moderated by increasing vehicle efficiencies. Food shipments, which grow in proportion to population and are resistant to recessions, have not shown the same recession-related decline as the other industries. Shipments of bulk chemicals, especially organic chemicals, grow sharply from 2012 to 2025 with the increased use of NGL as feedstock. After 2025, shipments from the bulk chemical industry level off as a result of foreign competition.

The energy-intensive iron and steel and cement industries show the greatest variability in shipments across the three cases (Figure 84), because they supply downstream industries that are sensitive to GDP growth. Construction is a downstream industry for both iron and steel and cement, and the metal-based durables industry is a downstream industry for iron and steel. Shipments in the metal durables industry levels off after 2020, following a decline in iron and steel shipments.

Energy use reflects output and efficiency trends in energy-intensive industries

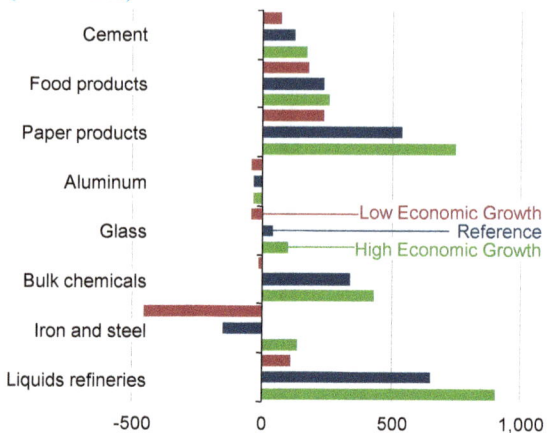

Figure 85. Change in delivered energy for energy-intensive industries in three cases, 2010-2035 (trillion Btu)

Changes in energy consumption from 2010 to 2035 in the energy-intensive industries ranges from almost nothing in the Low Economic Growth case to 0.8 percent per year or 5 quadrillion Btu in the High Economic Growth case (Figure 85). Changes in energy consumption by the industrial subsector largely reflect the corresponding changes in gross shipments. Energy efficiency improvements and changes in manufacturing methods and requirements, however, also affect energy consumption.

Starting from low levels of economic activity in 2010, shipments from all industries grow over the projection period. For example, steel industry shipments grow by 23 percent in the *AEO2012* Reference case from 2010 to 2035, but energy use declines by 12 percent due to a shift from the use of blast furnace steel production to the use of recycled products and electric arc furnaces. The continued decline of primary aluminum production and concurrent rise in less energy-intensive secondary production lead to a similar decline in aluminum industry energy use despite an increase in shipments. The paper industry shows a far less noticeable improvement in energy efficiency because of greater demand for more energy-intensive products such as paperboard by consumers.

The only industrial subsector that shows an increase in energy intensity is refining. In each of the three Economic Growth cases (Reference, Low Growth, and High Growth), the increase in liquids refinery industry energy consumption exceeds the growth in shipments over the projection period as a result of increased use of coal after 2015 for CTL and CBTL production. Production of alternative fuels is inherently more energy-intensive than production of traditional fuels, because they are refined from solids with relatively low energy densities.

Industrial sector energy demand

Transportation equipment shows strongest growth in non-energy-intensive shipments

Figure 86. Cumulative growth in value of shipments from non-energy-intensive industries in three cases, 2010-2035 (percent)

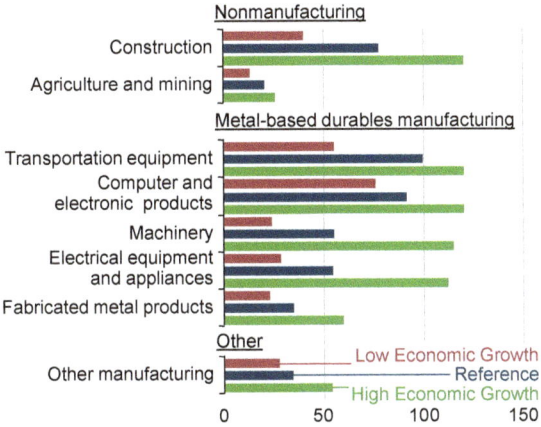

Nonmanufacturing and transportation equipment lead energy efficiency gains

Figure 87. Change in delivered energy for non-energy-intensive industries in three cases, 2010-2035 (trillion Btu)

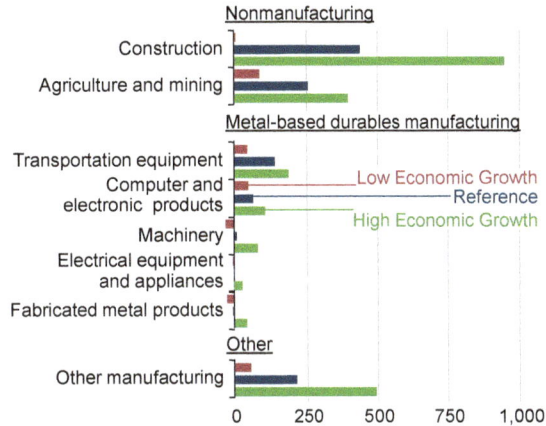

In 2035, non-energy-intensive manufacturing and nonmanufacturing industrial subsectors account for $6.7 trillion (2005 dollars) in shipments in the Reference case—a 57-percent increase from 2010. From 2010 to 2035, growth in those shipments averages 1.2 percent per year in the Low Economic Growth case and 2.5 percent in the High Economic Growth case, compared with 1.8 percent in the Reference case (Figure 86). Non-energy-intensive manufacturing and nonmanufacturing are segments of the industrial sector that primarily consume fuels for thermal or electrical needs, not as raw materials or feedstocks.

In the three cases, shipments from the two subsectors grow at roughly twice the annual rate projected for energy-intensive manufacturing, based on production of high-tech, high-value goods and strong supply chain linkages between energy-intensive manufacturing and many non-energy-intensive manufacturing industries (such as machinery and transportation equipment produced for the metals industries). Recovery in the two subsectors from 2010 to 2015 is rapid because of increased U.S. competiveness in the transportation equipment and machinery industries, as well as a recovering construction industry, which saw residential starts bottom out in 2010. After 2015, the growth is more moderate.

In the Reference case, shipments from the non-energy-intensive manufacturing and nonmanufacturing industries generally exceed pre-recession levels by 2017, reflecting a slow and extended economic recovery. Pre-recession shipment levels are exceeded in 2015 and 2024 in the High Economic Growth and Low Economic Growth cases, respectively.

From 2010 to 2035, total energy consumption in the non-energy-intensive manufacturing and nonmanufacturing industrial subsectors changes by 2 percent or 178 trillion Btu in the Low Economic Growth case, 15 percent or 1,134 trillion Btu in the Reference case, and 30 percent or 2,282 trillion Btu in the High Economic Growth case (Figure 87). In each of the three cases, those industries together account for more than 40 percent of the projected increase in total industrial natural gas consumption.

The transportation equipment and construction industries account for roughly 20 percent of the projected increase in energy use but approximately 40 percent of the projected growth in total industrial shipments in all cases. The transportation equipment industry, in particular, shows a rapid decline in energy intensity from 2010 to 2035. Energy consumption increases by 37 percent from 2010 to 2035 and production doubles, yielding an annualized decline in energy intensity of 1.3 percent per year in the transportation equipment industry over the projection period in the *AEO2012* Reference case.

Overall, the combined energy intensity of the non-energy-intensive manufacturing and nonmanufacturing industries declines by 25 percent in the Low Economic Growth case and 29 percent in the High Economic Growth case. The more rapid decline in the High Economic Growth case is consistent with an expectation that energy intensity will fall more rapidly when stronger economic growth facilitates additional investment in more energy-efficient equipment.

Transportation sector energy demand

Transportation energy use grows slowly in comparison with historical trend

Figure 88. Delivered energy consumption for transportation by mode in two cases, 2010 and 2035 (quadrillion Btu)

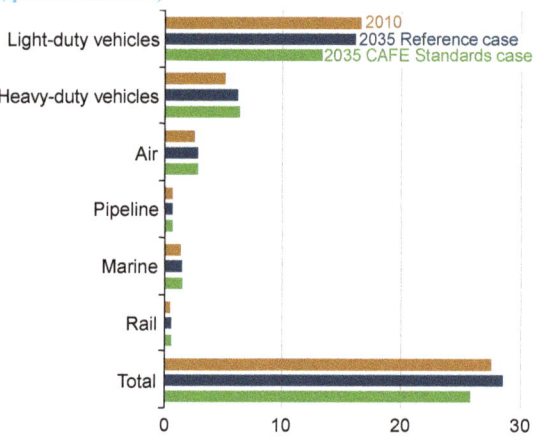

CAFE and greenhouse gas emissions standards boost vehicle fuel economy

Figure 89. Average fuel economy of new light-duty vehicles in two cases, 1980-2035 (miles per gallon)

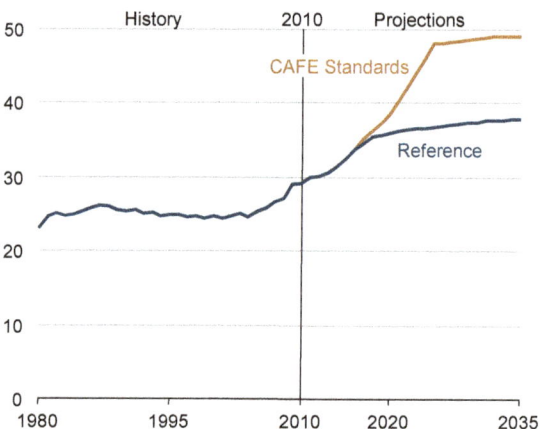

Transportation sector energy consumption grows at an average annual rate of 0.1 percent from 2010 to 2035 (from 27.6 quadrillion Btu to 28.6 quadrillion Btu), much slower than the 1.2-percent average from 1975 to 2010. The slower growth results primarily from improvement in fuel economy for both LDVs and heavy-duty vehicles (HDVs), as well as relatively modest growth in demand for personal travel.

LDV energy demand falls by 3.2 percent (0.5 quadrillion Btu) from 2010 to 2035 (Figure 88). Personal travel demand rises more slowly than in recent history, with the increase more than offset by existing GHG standards for model year (MY) 2012 to 2016 and by EISA2007 fuel economy standards for MY 2017 to 2020. Inclusion of the proposed standards for MY 2017-2025, which are not included in the Reference case, reduce LDV energy demand by 20.0 percent (3.2 quadrillion Btu) from 2010 to 2035.

Energy demand for HDVs (including tractor trailers, buses, vocational vehicles, and heavy-duty pickups and vans) increases by 21 percent, or 1.1 quadrillion Btu, from 2010 to 2035, as a result of increases in vehicle miles traveled (VMT) as economic output recovers. Fuel efficiency and GHG emissions standards temper growth in energy demand even as more miles are traveled overall.

Energy demand for aircraft increases by 11 percent, or 0.3 quadrillion Btu from 2010 to 2035. Higher incomes and moderate growth in fuel costs encourage more personal air travel, the resulting increase in energy use offset by gains in aircraft fuel efficiency. Air freight use of energy grows as a result of export growth. Energy consumption for marine and rail travel also increases, as industrial output grows and more coal is transported. Energy use for pipelines also increases, even though more natural gas production occurs closer to end-use markets.

The introduction of Corporate Average Fuel Economy (CAFE) standards for LDVs in 1978 resulted in an increase in fuel economy from 19.9 miles per gallon (mpg) in 1978 to 26.2 mpg in 1987. Over the two decades that followed, despite improvements in LDV technology, fuel economy fell to between 24 and 26 mpg as sales of light-duty trucks increased from 20 percent of new LDV sales in 1980 to almost 55 percent in 2004 [124]. The subsequent rise in fuel prices and reduction in sales of light-duty trucks, coupled with tighter CAFE standards for light-duty trucks starting with MY 2008, led to a rise in LDV fuel economy to 29.2 mpg in 2010.

The National Highway Traffic Safety Administration (NHTSA) introduced attribute-based CAFE standards for MY 2011 LDVs in 2009 and, together with the U.S. Environmental Protection Agency (EPA), in 2010 announced CAFE and GHG emissions standards for MY 2012 to MY 2016. EISA2007 further requires that LDVs achieve an average fuel economy of 35 mpg by MY 2020 [125]. In the AEO2012 Reference case, the fuel economy of new LDVs [126] rises to 30.0 mpg in 2011, 33.8 mpg in 2016, and 35.9 mpg in 2020 (Figure 89). After 2020, CAFE standards remain constant, with LDV fuel economy increasing moderately to 37.9 mpg in 2035 as a result of more widespread adoption of fuel-saving technologies.

In December 2011, NHTSA and EPA proposed more stringent attribute-based CAFE and GHG emissions standards for MYs 2017 to 2025 [127]. The proposal calls for a projected average LDV CAFE of 49.6 mpg by 2025 together with a GHG standard equivalent to 54.5 mpg. With the inclusion of the proposed LDV CAFE standards, LDV fuel economy in the CAFE Standards case increases by nearly 30 percent in 2035 compared to the Reference case.

Transportation sector energy demand

Travel demand for personal vehicles increases more slowly than in the past

Figure 90. Vehicle miles traveled per licensed driver, 1970-2035 (thousand miles)

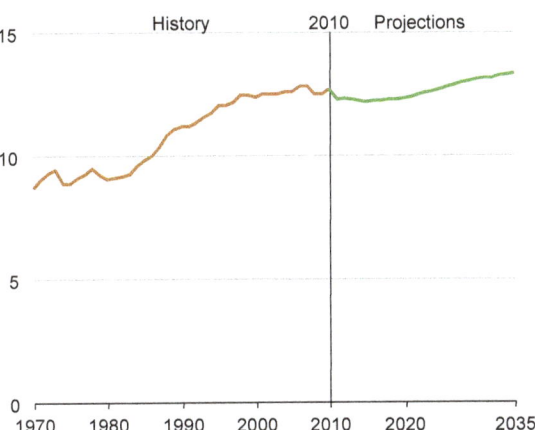

Personal vehicle travel demand, measured as VMT per licensed driver, grew at an average annual rate of 1.1 percent from 1970 to 2007, from about 8,700 miles per driver in 1970 to 12,800 miles per driver in 2007. Increased travel was supported by rising incomes, declining costs of driving per mile (determined by fuel economy and fuel price), and demographic changes (such as women entering the workforce). Between 2007 and 2010, VMT per licensed driver declined to around 12,700 miles per driver because of a spike in the cost of driving per mile and the economic downturn. In the AEO2012 Reference case, VMT per licensed driver grows by an average of 0.2 percent per year, to 13,350 miles per driver in 2035 (Figure 90).

Although the real price of motor gasoline in the transportation sector increases by 48 percent from 2010 to 2035 in the Reference case, VMT per licensed driver still grows as real disposable personal income climbs by 81 percent. Faster growth in income than in fuel prices ensures that travel demand continues to rise by reducing the percentage of income spent on fuel. In addition, the effect of rising fuel costs is moderated by a 30-percent improvement in new vehicle fuel economy following the implementation of more stringent GHG and CAFE standards for LDVs.

Several demographic forces play a role in moderating the growth in VMT per licensed driver despite the rise in real disposable income. Although LDV sales increase through 2035, the number of vehicles per licensed driver remains relatively constant (at just over 1 per licensed driver). Also, unemployment remains above pre-recession levels in the Reference case until later in the projection, further tempering the increase in personal travel demand.

Sales of alternative fuel, fuel flexible, and hybrid vehicles rise

Figure 91. Sales of light-duty vehicles using non-gasoline technologies by fuel type, 2010, 2020, and 2035 (million vehicles sold)

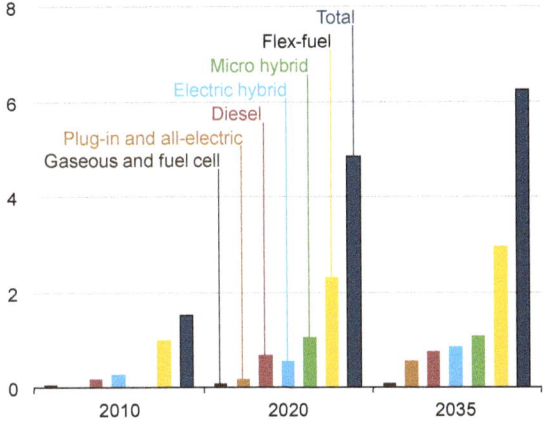

LDVs that use diesel, other alternative fuels, hybrid-electric, or all-electric systems play a significant role in meeting more stringent GHG emissions and fuel economy standards, as well as offering fuel savings in the face of higher fuel prices. Sales of such vehicles increase from 14 percent of all new LDV sales in 2010 to 35 percent in 2035 in the AEO2012 Reference case. Sales would be even higher with consideration of the proposed fuel economy standards covering MYs 2017 through 2025 that are not included in the Reference case (see discussion in "Issues in focus").

Flex-fuel vehicles (FFVs), which can use blends of ethanol up to 85 percent, represent the largest share of vehicles, at 17 percent of all new vehicle sales. Manufacturers selling FFVs currently receive incentives in the form of fuel economy credits earned for CAFE compliance through MY 2016. FFVs also play a critical role in meeting the RFS for biofuels.

Sales of hybrid electric and all-electric vehicles that use stored electric energy grow considerably in the Reference case (Figure 91). Micro hybrids, which use start/stop technology to manage engine operation while at idle, account for 6 percent of total LDV sales in 2035, which is the largest share for vehicles that use electric storage. Gasoline-electric and diesel-electric hybrid vehicles account for 5 percent of total LDV sales in 2035; and plug-in and all-electric hybrid vehicles account for 3 percent of LDV sales and 9 percent of sales of vehicles using diesel, alternative fuels, hybrid, or all-electric systems.

Sales of diesel vehicles also increase, to 4 percent of total LDV sales in 2035. Light-duty gaseous and fuel cell vehicles account for less than 0.5 percent of new vehicle sales throughout the projection because of the limited availability of a fueling infrastructure and their high incremental cost.

Electricity demand

Heavy-duty vehicle energy demand continues to grow but slows from historical rates

Figure 92. Heavy-duty vehicle energy consumption, 1995-2035 (quadrillion Btu)

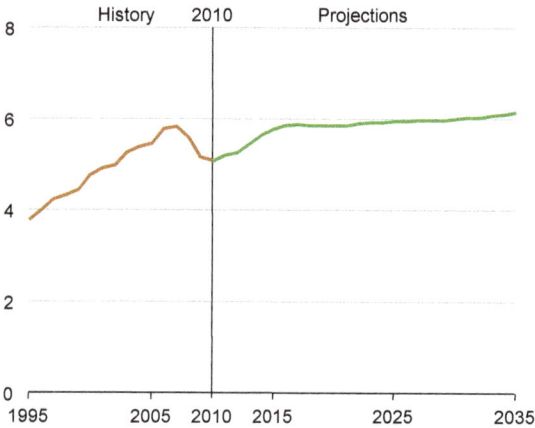

Energy demand for HDVs—including tractor trailers, vocational vehicles, heavy-duty pickups and vans, and buses—increases from 5.1 quadrillion Btu in 2010 to 6.2 quadrillion Btu in 2035, at an average annual growth rate of 0.8 percent, which is the highest among transportation modes. Still, the increase in energy demand for HDVs is lower than the 2-percent annual average from 1995 to 2010, as increases in VMT are offset by improvements in fuel economy following the recent introduction of new standards for HDV fuel efficiency and GHG emissions.

The total number of miles traveled annually by all HDVs grows by 48 percent from 2010 to 2035, from 234 billion miles to 345 billion miles, for an average annual increase of 1.6 percent. The rise in VMT is supported by rising economic output over the projection period and an increase in the number of trucks on the road, from 8.9 million in 2010 to 12.5 million in 2035.

Higher fuel economy for HDVs partially offsets the increase in their VMT, as average new vehicle fuel economy increases from 6.6 mpg in 2010 to 8.2 mpg in 2035. The gain in fuel economy is primarily a consequence of the new GHG emissions and fuel efficiency standards enacted by EPA and NHTSA that begin in MY 2014 and reach the most stringent levels in MY 2018 [128]. Fuel economy continues to improve moderately after 2018, as fuel-saving technologies continue to be adopted for economic reasons (Figure 92).

Residential and commercial sectors dominate electricity demand growth

Figure 93. U.S. electricity demand growth, 1950-2035 (percent, 3-year moving average)

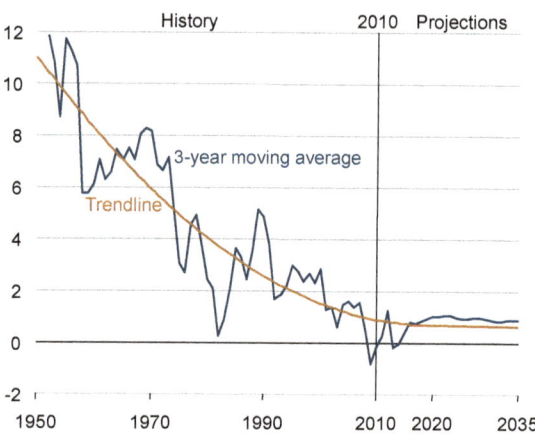

Electricity demand (including retail sales and direct use) growth has slowed in each decade since the 1950s, from a 9.8-percent annual rate of growth from 1949 to 1959 to only 0.7 percent per year in the first decade of the 21st century. In the AEO2012 Reference case, electricity demand growth rebounds somewhat from those low levels but remains relatively slow, as growing demand for electricity services is offset by efficiency gains from new appliance standards and investments in energy-efficient equipment (Figure 93).

Electricity demand grows by 22 percent in the AEO2012 Reference case, from 3,877 billion kilowatthours in 2010 to 4,716 billion kilowatthours in 2035. Residential demand grows by 18 percent over the same period, to 1,718 billion kilowatthours in 2035, spurred by population growth, rising disposable income, and continued population shifts to warmer regions with greater cooling requirements. Commercial sector electricity demand increases by 28 percent, to 1,699 billion kilowatthours in 2035, led by demand in the service industries. In the industrial sector, electricity demand has been generally declining since 2000, and it grows by only 2 percent from 2010 to 2035, slowed by increased competition from overseas manufacturers and a shift of U.S. manufacturing toward consumer goods that require less energy to produce. Electricity demand in the transportation sector is small, but it is expected to more than triple from 7 billion kilowatthours in 2010 to 22 billion kilowatthours in 2035 as sales of electric plug-in LDVs increase.

Average annual electricity prices (in 2010 dollars) increase by 3 percent from 2010 to 2035 in the Reference case, generally falling through 2020 in response to lower fuel prices used to generate electricity. After 2020, rising fuel costs more than offset lower costs for transmission and distribution.

Electricity generation

Coal-fired plants continue to be the largest source of U.S. electricity generation

Figure 94. Electricity generation by fuel, 2010, 2020, and 2035 (billion kilowatthours)

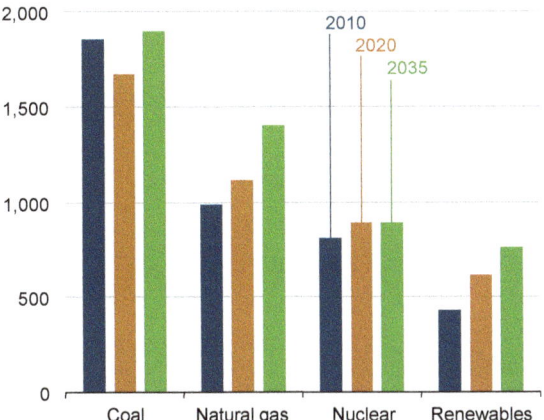

Coal remains the dominant fuel for electricity generation in the AEO2012 Reference case (Figure 94), but its share declines significantly. In 2010, coal accounted for 45 percent of total U.S. generation; in 2020 and 2035 its projected share of total generation is 39 percent and 38 percent, respectively. Competition from natural gas and renewables is a key factor in the decline. Overall, coal-fired generation in 2035 is 2 percent higher than in 2010 but still 6 percent below the 2007 pre-recession level.

Generation from natural gas grows by 42 percent from 2010 to 2035, and its share of total generation increases from 24 percent in 2010 to 28 percent in 2035. The relatively low cost of natural gas makes the dispatching of existing natural gas plants more competitive with coal plants and, in combination with relatively low capital costs, makes natural gas the primary choice to fuel new generation capacity.

Generation from renewable sources grows by 77 percent in the Reference case, raising its share of total generation from 10 percent in 2010 to 15 percent in 2035. Most of the growth in renewable electricity generation comes from wind and biomass facilities, which benefit from State RPS requirements, Federal tax credits, and, in the case of biomass, the availability of low-cost feedstocks and the RFS.

Generation from U.S. nuclear power plants increases by 10 percent from 2010 to 2035, but the share of total generation declines from 20 percent in 2010 to 18 percent in 2035. Although new nuclear capacity is added by new reactors and uprates of older ones, total generation grows faster and the nuclear share falls. Nuclear capacity grows from 101 gigawatts in 2010 to 111 gigawatts in 2035, with 7.3 gigawatts of additional uprates and 8.5 gigawatts of new capacity between 2010 and 2035. Some older nuclear capacity is retired, which reduces overall nuclear generation.

Most new capacity additions use natural gas and renewables

Figure 95. Electricity generation capacity additions by fuel type, including combined heat and power, 2011-2035 (gigawatts)

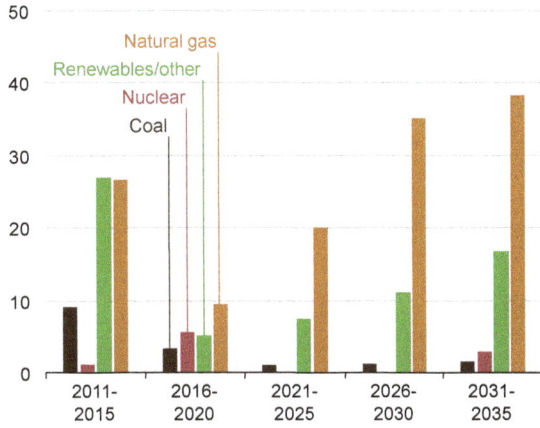

Decisions to add capacity, and the choice of fuel for new capacity, depend on a number of factors [129]. With growing electricity demand and the retirement of 88 gigawatts of existing capacity, 235 gigawatts of new generating capacity (including end-use combined heat and power) are projected to be added between 2011 and 2035 (Figure 95).

Natural-gas-fired plants account for 60 percent of capacity additions between 2011 and 2035 in the Reference case, compared with 29 percent for renewables, 7 percent for coal, and 4 percent for nuclear. Escalating construction costs have the largest impact on capital-intensive technologies, which include nuclear, coal, and renewables. However, Federal tax incentives, State energy programs, and rising prices for fossil fuels increase the competitiveness of renewable and nuclear capacity. Current Federal and State environmental regulations also affect fossil fuel use, particularly coal. Uncertainty about future limits on GHG emissions and other possible environmental programs also reduces the competitiveness of coal-fired plants (reflected in AEO2012 by adding 3 percentage points to the cost of capital for new coal-fired capacity).

Uncertainty about demand growth and fuel prices also affects capacity planning. Total capacity additions from 2011 to 2035 range from 166 gigawatts in the Low Economic Growth case to 305 gigawatts in the High Economic Growth case. In the AEO2012 Low Tight Oil and Shale Gas Resource case, natural gas prices are higher than in the Reference case and new natural gas fired capacity from 2011 to 2035 accounts for 102 gigawatts, which represents 47 percent of total additions. In the High Tight Oil and Shale Gas Resource case, delivered natural gas prices are lower than in the Reference case and natural gas-fired capacity additions by 2035 are 155 gigawatts, or 66 percent of total new capacity.

Electricity sales

Additions to power plant capacity slow after 2012 but accelerate beyond 2020

Figure 96. Additions to electricity generating capacity, 1985-2035 (gigawatts)

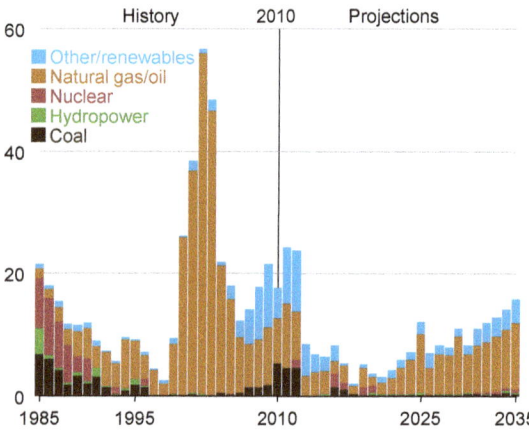

Typically, investments in electricity generation capacity have gone through "boom and bust" cycles. Periods of slower growth have been followed by strong growth in response to changing expectations for future electricity demand and fuel prices, as well as changes in the industry, such as restructuring (Figure 96). A construction boom in the early 2000s saw capacity additions averaging 35 gigawatts a year from 2000 to 2005, much higher than had been seen before. Since then, average annual builds have dropped to 17 gigawatts per year from 2006 to 2010.

In the AEO2012 Reference case, capacity additions between 2011 and 2035 total 235 gigawatts, including new plants built not only in the power sector but also by end-use generators. Annual additions in 2011 and 2012 remain relatively high, averaging 24 gigawatts per year [130]. Of those early builds, about 40 percent are renewable plants built to take advantage of Federal tax incentives and to meet State renewable standards.

Annual builds drop significantly after 2012 and remain below 9 gigawatts per year until 2025. During that period, existing capacity is adequate to meet growth in demand in most regions, given the earlier construction boom and relatively slow growth in electricity demand after the economic recession. Between 2025 and 2035, average annual builds increase to 11 gigawatts per year, as excess capacity is depleted and the rate of total capacity growth is more consistent with electricity demand growth. More than 70 percent of the capacity additions from 2025 to 2035 are natural gas fired, given the higher construction costs for other capacity types and uncertainty about the prospects for future limits on GHG emissions.

Growth in generating capacity parallels rising demand for electricity

Figure 97. Electricity sales and power sector generating capacity, 1949-2035 (index, 1949 = 1.0)

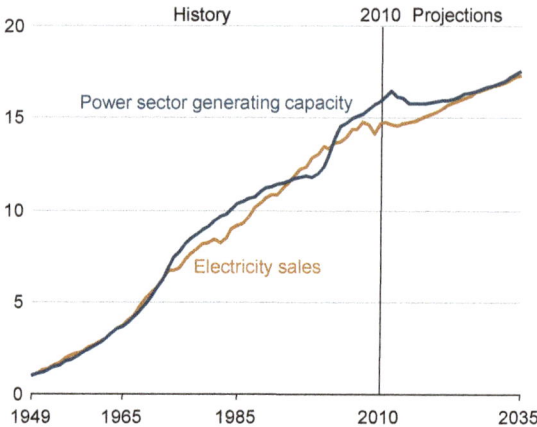

Over the long term, growth in electricity generating capacity parallels the growth in end-use demand for electricity. However, unexpected shifts in demand or dramatic changes affecting capacity investment decisions can cause imbalances that can take years to work out.

Figure 97 shows indexes summarizing relative changes in total generating capacity and electricity demand. During the 1950s and 1960s, the capacity and demand indexes tracked closely. The energy crises of the 1970s and 1980s, together with other factors, slowed electricity demand growth, and capacity growth outpaced demand for more than 10 years thereafter, as planned units continued to come on line. Demand and capacity did not align again until the mid-1990s. Then, in the late 1990s, uncertainty about deregulation of the electricity industry caused a downturn in capacity expansion, and another period of imbalance followed, with growth in electricity demand exceeding capacity growth.

In 2000, a boom in construction of new natural gas fired plants began, quickly bringing capacity back into balance with demand and, in fact, creating excess capacity. Construction of new intermittent wind capacity that sometimes needs backup capacity also began to grow after 2000. More recently, the 2008-2009 economic recession caused a significant drop in electricity demand, which has recovered only partially in the post-recession period. In combination with slow near-term growth in electricity demand, the slow economic recovery creates excess generating capacity in the AEO2012 Reference case. Capacity currently under construction is completed in the Reference case, but only a limited amount of additional capacity is built before 2025, while older capacity is retired. In 2025, capacity growth and demand growth are in balance again, and they grow at similar rates through 2035.

Electricity capacity

Costs and regulatory uncertainties vary across options for new capacity

Figure 98. Levelized electricity costs for new power plants, excluding subsidies, 2020 and 2035 (2010 cents per kilowatthour)

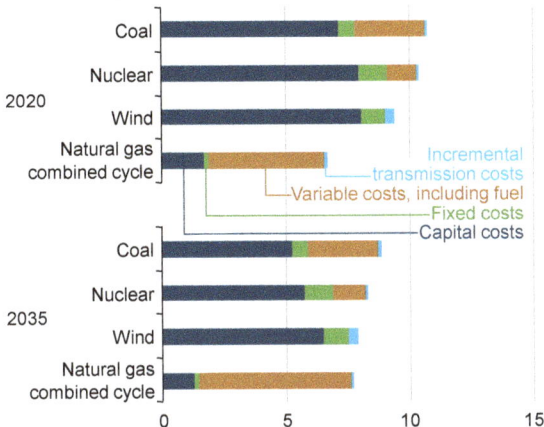

Technology choices for new generating capacity are based largely on capital, operating, and transmission costs. Coal, nuclear, and renewable plants are capital-intensive (Figure 98), whereas operating (fuel) expenditures make up most of the costs for natural gas capacity [131]. Capital costs depend on such factors as equipment costs, interest rates, and cost recovery periods. Fuel costs vary with operating efficiency, fuel price, and transportation costs.

In addition to considerations of levelized costs [132], some technologies and fuels receive subsidies, such as production tax credits and ITCs. Also, new plants must satisfy local and Federal emissions standards and must be compatible with the utility's load profile.

Regulatory uncertainty also affects capacity planning. New coal plants may require carbon control and sequestration equipment, resulting in higher material, labor, and operating costs. Alternatively, coal plants without carbon controls could incur higher costs for siting and permitting. Because nuclear and renewable power plants (including wind plants) do not emit GHGs, their costs are not directly affected by regulatory uncertainty in this area.

Capital costs can decline over time as developers gain technology experience, with the largest rate of decline in new technologies. In the AEO2012 Reference case, the capital costs of new technologies are adjusted upward initially to compensate for the optimism inherent in early estimates of project costs, then decline as project developers gain experience. The decline continues at a progressively slower rate as more units are built. Operating efficiencies also are assumed to improve over time, resulting in reduced variable costs unless increases in fuel costs exceed the savings from efficiency gains.

Nuclear power plant capacity grows slowly through uprates and new builds

Figure 99. Electricity generating capacity at U.S. nuclear power plants in three cases, 2010, 2025, and 2035 (gigawatts)

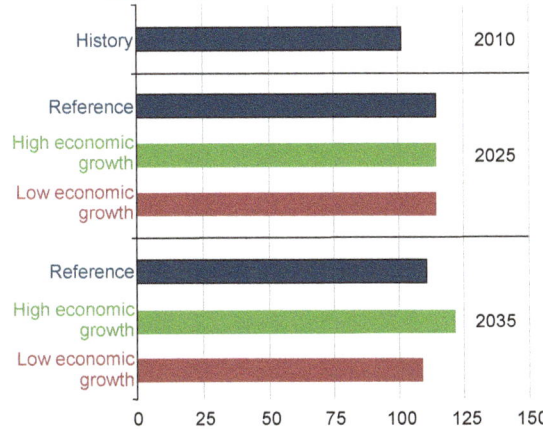

In the AEO2012 Reference case, nuclear power capacity increases from 101.2 gigawatts in 2010 to a high of 114.7 gigawatts in 2025, before declining to 110.9 gigawatts in 2035 (Figure 99), largely as a result of plant retirements. The capacity increase through 2025 includes 7.3 gigawatts of expansion at existing plants and 6.8 gigawatts of new capacity, which includes completion of two conventional reactors at the Watts Bar and Bellefonte sites. Four advanced reactors, reported as under construction, are also assumed to be brought online by 2020 and to be eligible for Federal financial incentives. High construction costs for nuclear plants, especially relative to natural gas fired plants, make additional options for new nuclear capacity uneconomical until the later years of the projection, when an additional 1.8 gigawatts is added. Nuclear capacity additions vary with assumptions about overall demand for electricity. Across the Economic Growth cases, nuclear capacity additions from 2011 to 2035 range from 6.8 gigawatts in the Low Economic Growth case to 19.2 gigawatts in the High Economic Growth case.

One nuclear unit, Oyster Creek, is expected to be retired at the end of 2019, as announced by Exelon in December 2010. An additional 5.5 gigawatts of nuclear capacity is assumed to be retired by 2035. All other existing nuclear units continue to operate through 2035 in the Reference case, which assumes that they will apply for and receive operating license renewals, including in some cases a second 20-year extension after 60 years of operation (for more discussion, see "Issues in focus"). With costs for natural gas fired generation rising in the Reference case and uncertainty about future regulation of GHG emissions, the economics of keeping existing nuclear power plants in operation are favorable.

Renewable capacity

Wind dominates renewable capacity growth, but solar and biomass gain market share

Figure 100. Nonhydropower renewable electricity generation capacity by energy source, including end-use capacity, 2010-2035 (gigawatts)

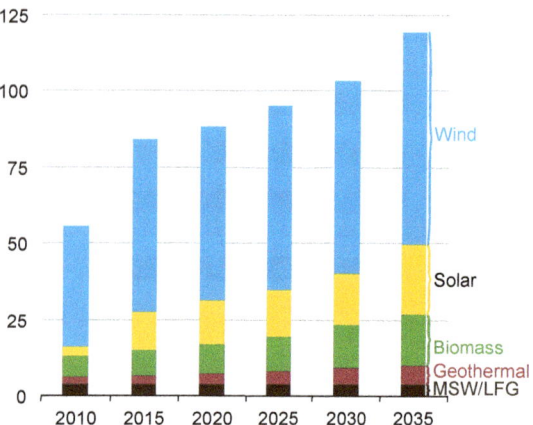

Nonhydropower renewable generation surpasses hydropower by 2020

Figure 101. Hydropower and other renewable electricity generation, including end-use generation, 2010-2035 (billion kilowatthours)

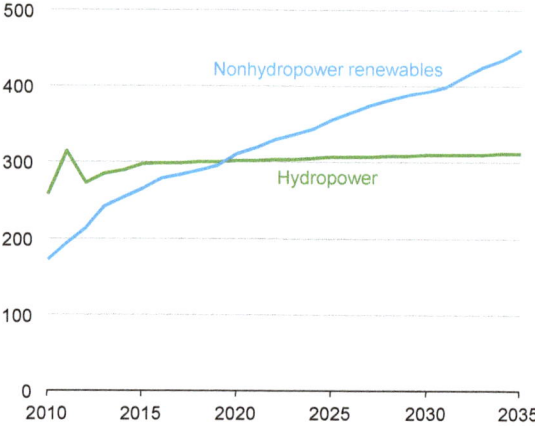

From 2010 to 2035, total nonhydropower renewable generating capacity more than doubles in the *AEO2012* Reference case (Figure 100). Wind accounts for the largest share of that new capacity, increasing from 39 gigawatts in 2010 to 70 gigawatts in 2035. Both solar capacity and biomass capacity grow at faster rates than wind capacity, but they start from smaller levels.

Excluding new projects already under construction, PV accounts for nearly all solar capacity additions both in the end-use sectors (where 11 gigawatts of PV capacity is added from 2010 to 2035) and in the electric power sector (8 gigawatts added from 2010 to 2035). While end-use solar capacity grows throughout the projection, the growth of solar capacity in the electric power sector is concentrated primarily in the last decade of the projection period (2025-2035) when the technology becomes more cost-competitive. Geothermal capacity nearly triples over the projection period, but in 2035 it still accounts for only about 5 percent of total nonhydropower renewable generating capacity.

Renewable capacity additions are supported by State RPS programs, the Federal RFS, and Federal tax credits. Total renewable capacity—particularly, wind and solar—grows rapidly in the near term in the *AEO2012* Reference case. There is, however, relatively little projected need for new generation capacity of any type, including renewables, for the remainder of the current decade, primarily because there is an abundance of existing natural gas fired capacity that can be operated at higher capacity factors. After 2020 there is a need for new generation capacity in the Reference case, resulting in a resurgence in renewable capacity growth.

In the *AEO2012* Reference case, nonhydropower renewable generation grows at an average annual rate of 3.9 percent, nearly tripling from 2010 to 2035. Generation from nonhydropower renewable sources has been small historically in comparison with hydroelectric generation; however, nonhydropower renewable generation surpasses hydroelectric generation in 2020 in the Reference case (Figure 101).

The share of the total electricity generation accounted for by nonhydropower renewable generation increases from about 4 percent in 2010 to 9 percent in 2035. Although wind remains the largest source of nonhydropower renewable generation through 2035, both solar and biomass generation grow at faster annual rates. Solar generation increases by an average of nearly 10 percent per year, and biomass generation increases by 6 percent per year.

Both solar and wind energy are intermittent resources, and as a result their contributions to the generation mix are less than their contribution to the capacity mix. Biomass-fired generation, on the other hand, is dispatchable and grows to levels approaching wind generation by the end of the projection, at 145 billion kilowatthours in 2035, as compared with 194 billion kilowatthours for wind-powered generation. Most of the growth in biomass generation comes from CHP units used in the production of biomass-based liquid fuels, primarily in response to the Federal RFS. Biomass co-firing and end-use generation play an important role in satisfying State RPS mandates, particularly from 2010 to 2020, when overall capacity growth is modest.

Natural gas prices

State renewable portfolio standards increase renewable electricity generation

Figure 102. Regional growth in nonhydropower renewable electricity generation, including end-use generation, 2010-2035 (billion kilowatthours)

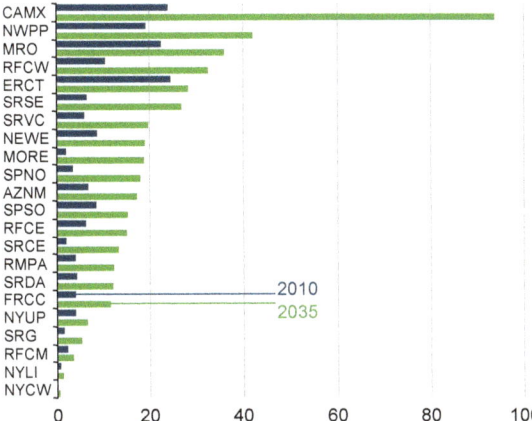

Natural gas prices are expected to rise with the marginal cost of production

Figure 103. Annual average Henry Hub spot natural gas prices, 1990-2035 (2010 dollars per million Btu)

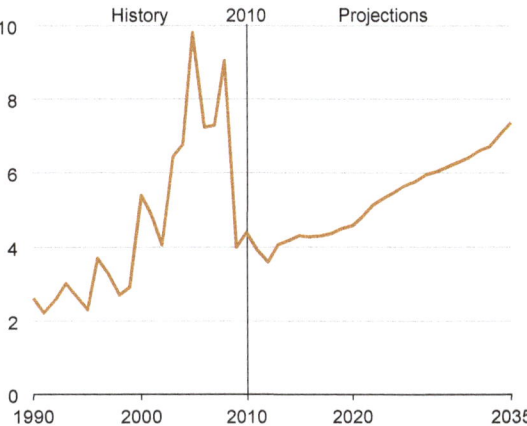

Regional growth in renewable electricity generation is based largely on two factors: availability of renewable energy resources and the existence of State RPS programs that require the use of renewable generation. After a period of robust RPS enactments in several States, the past few years have been relatively quiet in terms of State program expansions, primarily due to the subdued economic climate.

The highest level of nonhydroelectric renewable generation in 2035, 93.9 billion kilowatthours, occurs in the WECC California (CAMX) region (Figure 102), whose area approximates the California State boundaries. (For a map of the electricity regions presented, see Appendix F.) The three largest contributors to the total are wind, solar, and geothermal generation. The region encompassing the Pacific Northwest has more overall renewable generation, the vast majority of which comes from hydroelectric sources.

Although the Western and Southwestern States have the most projected solar installations, State RPS programs heavily influence the growth of solar capacity in the eastern States, where both the Reliability First Corporation/East (RFCE) and the Reliability First Corporation/West (RFCW) regions have large amounts of end-use solar generation, with 1.7 billion kilowatthours and 1.9 billion kilowatthours, respectively. The two regions are not known for a strong solar resource base, and the installations are in response to the ITC as well as solar requirements embedded in State RPS programs. Most biomass capacity—confined largely to the end-use sectors—is built at the sites of cellulosic ethanol plants, many of which are in the Southeast.

U.S. natural gas prices are determined largely by supply and demand conditions in North American markets. At current (2012) price levels, natural gas prices are below average replacement cost. However, over time natural gas prices rise with the cost of developing incremental production capacity (Figure 103). After 2017, natural gas prices rise in the AEO2012 Reference case more rapidly than crude oil prices, but oil prices remain at least three times higher than natural gas prices through the end of the projection (Figure 104).

As of January 1, 2010, total proved and unproved natural gas resources are estimated at 2,203 trillion cubic feet. Development costs for natural gas wells are expected to grow slowly. Henry Hub spot prices for natural gas rise by 2.1 percent per year from 2010 through 2035 in the Reference case, to an annual average of $7.37 per million Btu (2010 dollars) in 2035.

Figure 104. Ratio of low-sulfur light crude oil price to Henry Hub natural gas price on energy equivalent basis, 1990-2035

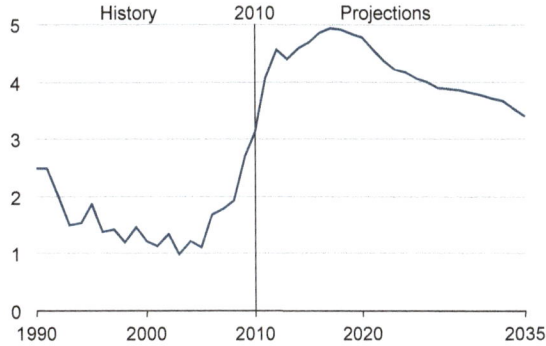

Natural gas production

Natural gas prices vary with economic growth and shale gas well recovery rates

Figure 105. Annual average Henry Hub spot natural gas prices in five cases, 1990-2035 (2010 dollars per million Btu)

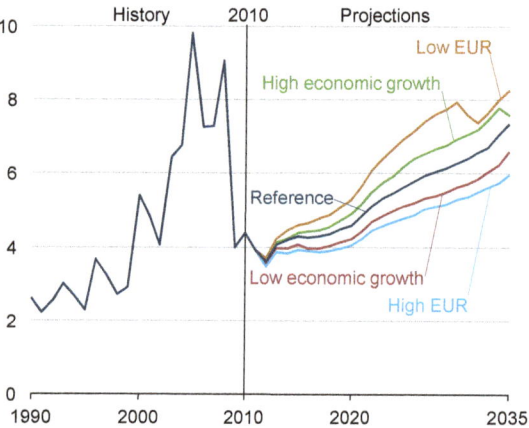

The rate at which natural gas prices change in the future can vary, depending on a number of factors. Two important factors are the future rate of macroeconomic growth and the expected cumulative production of shale gas wells over their lifetimes—the estimated ultimate recovery (EUR) per well. Alternative cases with different assumptions for these factors are shown in Figure 105.

Higher rates of economic growth lead to increased consumption of natural gas, causing more rapid depletion of natural gas resources and a more rapid increase in the cost of developing new incremental natural gas production. Conversely, lower rates of economic growth lead to lower levels of natural gas consumption and, ultimately, a slower increase in the cost of developing new production.

In the High and Low EUR cases, the EUR per shale gas well is increased and decreased by 50 percent, respectively. Future shale gas well recovery rates are an important determinant of future prices. Changes in well recovery rates affect the long-run marginal cost of shale gas production, which in turn affects both natural gas prices and the volumes of new shale gas production developed (further analysis and discussion are included in the "Issues in focus" section of this report). In the Low EUR case, an Alaska gas pipeline starts operating in 2031, accompanied by a dip in natural gas prices. A recent proposal to build a natural gas pipeline along the route of the Alyeska oil pipeline with an LNG export facility could speed up construction. In the High Economic Growth case, the pipeline begins operation in 2035, with a similar effect on prices.

With rising domestic production, the United States become a net exporter of natural gas

Figure 106. Total U.S. natural gas production, consumption, and net imports, 1990-2035 (trillion cubic feet)

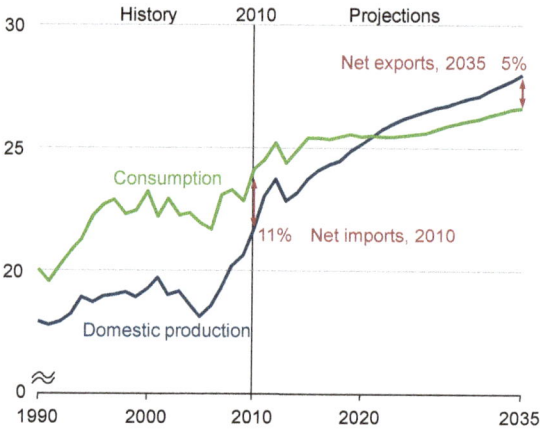

The United States consumed more natural gas than it produced in 2010, importing 2.6 trillion cubic feet from other countries. In the *AEO2012* Reference case, domestic natural gas production grows more quickly than consumption. As a result, the United States becomes a net exporter of natural gas by around 2022, and in 2035 net exports of natural gas from the United States total about 1.4 trillion cubic feet (Figure 106).

U.S. natural gas consumption grows at a rate of 0.4 percent per year from 2010 to 2035 in the Reference case, or by a total of 2.5 trillion cubic feet, to 26.6 trillion cubic feet in 2035. Growth in domestic natural gas consumption depends on many factors, including the rate of economic growth and the delivered prices of natural gas and other fuels. Natural gas consumption in the commercial and industrial sectors grows by less than 0.5 percent per year through 2035, and consumption for electric power generation grows by 0.8 percent per year. Residential natural gas consumption declines over the same period, by a total of 0.3 trillion cubic feet from 2010 to 2035.

U.S. natural gas production grows by 1.0 percent per year, to 27.9 trillion cubic feet in 2035, more than enough to meet domestic needs for consumption, which allows for exports. The prospects for future U.S. natural gas exports are highly uncertain and depend on many factors that are difficult to anticipate, such as the development of new natural gas production capacity in foreign countries, particularly from deepwater reservoirs, shale gas deposits, and the Arctic.

Shale gas provides largest source of growth in U.S. natural gas supply

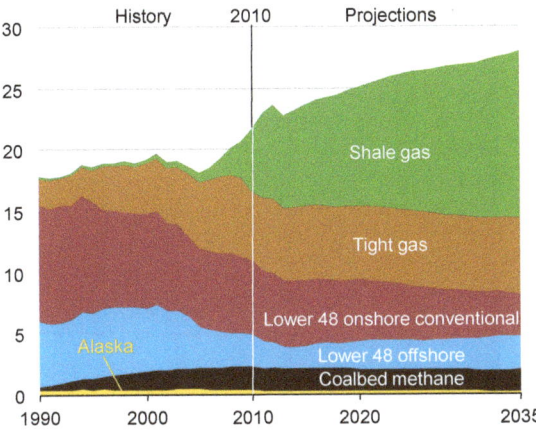

Figure 107. Natural gas production by source, 1990-2035 (trillion cubic feet)

In most U.S. regions, natural gas production growth is led by shale gas development

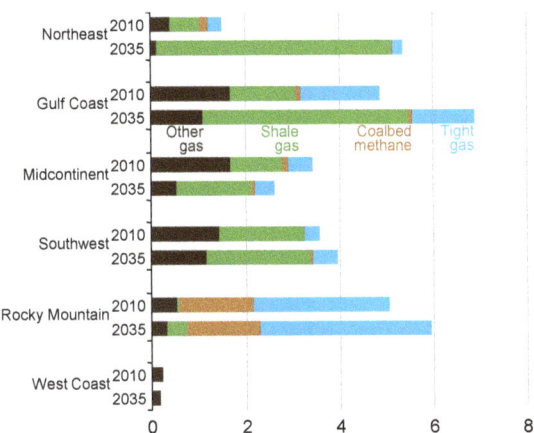

Figure 108. Lower 48 onshore natural gas production by region, 2010 and 2035 (trillion cubic feet)

The increase in natural gas production from 2010 to 2035 in the *AEO2012* Reference case results primarily from the continued development of shale gas resources (Figure 107). Shale gas is the largest contributor to production growth; there is relatively little change in production levels from tight formations, coalbed methane deposits, and offshore fields.

Shale gas accounts for 49 percent of total U.S. natural gas production in 2035, more than double its 23-percent share in 2010. In the Reference case, estimated proved and unproved shale gas resources amount to a combined 542 trillion cubic feet, out of a total U.S. resource of 2,203 trillion cubic feet. Estimates of shale gas resources and well productivity remain uncertain (see "Issues in focus" for discussion).

Tight gas produced from low permeability sandstone and carbonate reservoirs is the second-largest source of domestic supply in the Reference case, averaging 6.1 trillion cubic feet of production per year from 2010 to 2035. Coalbed methane production remains relatively constant throughout the projection, averaging 1.8 trillion cubic feet per year.

Offshore natural gas production declines by 0.8 trillion cubic feet from 2010 through 2014, following the 2010 moratorium on offshore drilling, as exploration and development activities in the Gulf of Mexico focus on oil-directed activity. After 2014 offshore production continues to rise throughout the remainder of the projection period.

Shale gas production, which more than doubles from 2010 to 2035, is the largest contributor to the projected growth in total U.S. natural gas production in the Reference case. Regional production growth largely reflects expected increases in production from shale beds. See Figure F4 in Appendix F for a map of U.S. natural gas supply regions.

In the Northeast, natural gas production grows by an average of 5.2 percent per year, or a total of 3.9 trillion cubic feet from 2010 to 2035 (Figure 108). The Marcellus shale, which accounts for 3.0 trillion cubic feet of the expected increase, is particularly attractive for development because of its large resource base, its proximity to major natural gas consumption markets, and the extensive pipeline infrastructure that already exists in the Northeast.

In the Gulf Coast region, natural gas production grows by 2.0 trillion cubic feet from 2010 to 2035, at an average rate of 1.4 percent per year. Natural gas production from the Haynesville/Bossier and Eagle Ford formations increases by 2.8 trillion cubic feet over the period, but declines in production from other natural gas fields in the region offset some of the gains, so that the net increase in production for the region as a whole is only about 2 trillion cubic feet.

In the Rocky Mountain region, natural gas production grows by 0.9 trillion cubic feet from 2010 through 2035, with tight sandstone and carbonate production increasing by 0.8 trillion cubic feet and shale gas production by 0.4 trillion cubic feet. As in the Gulf Coast region, production growth in the Rocky Mountain region is offset in part by production declines in the region's other natural gas fields.

Petroleum and other liquids consumption

The U.S. becomes a net natural gas exporter

Figure 109. U.S. net imports of natural gas by source, 1990-2035 (trillion cubic feet)

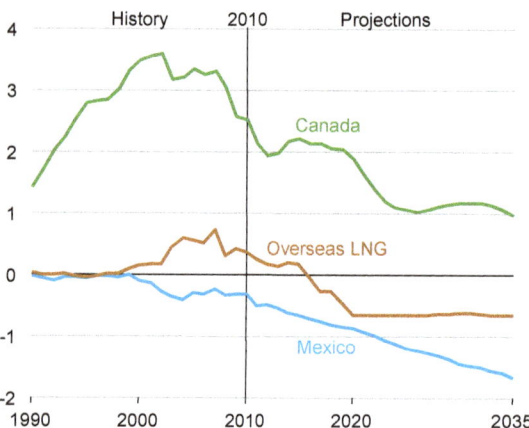

Transportation uses lead growth in consumption of petroleum and other liquids

Figure 110. Consumption of petroleum and other liquids by sector, 1990-2035 (million barrels per day)

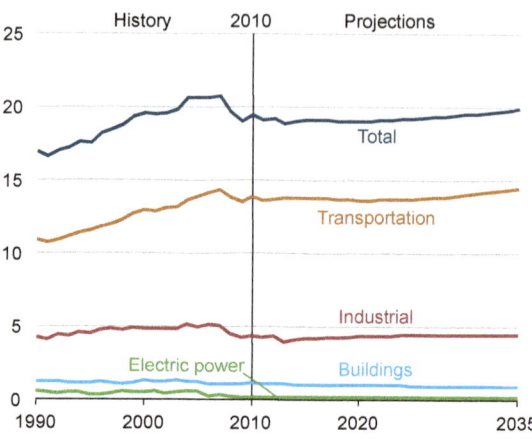

In 2010, the United States imported 11 percent of its total natural gas supply. In the *AEO2012* Reference case, U.S. natural gas production grows faster than consumption, so that early in the next decade exports exceed imports. In 2035, U.S. net natural gas exports are about 1.4 trillion cubic feet (about 4 billion cubic feet per day), half of which is exported overseas as liquefied natural gas (LNG). The other half is transported by pipelines, primarily to Mexico.

U.S. LNG exports supplied from lower 48 natural gas production are assumed to start when LNG export capacity of 1.1 billion cubic feet per day goes into operation in 2016. An additional 1.1 billion cubic feet per day of capacity is expected to come on line in 2019. At full capacity, the facilities could ship 0.8 trillion cubic feet of LNG to overseas consumers per year. Net U.S. LNG exports are somewhat lower than those figures imply, however, because LNG imports to the New England region are projected to continue. In general, future U.S. exports of LNG depend on a number of factors that are difficult to anticipate and thus are highly uncertain.

Net natural gas imports from Canada decline over the next decade in the Reference case and then stabilize at about 1.1 trillion cubic feet per year (Figure 109), when natural gas prices in the U.S. lower 48 States become high enough to motivate Canadian producers to expand their production of shale gas and tight gas. In Mexico, natural gas consumption shows robust growth through 2035, while Mexico's production grows at a slower rate. As a result, increasing volumes of imported natural gas from the United States fill the growing gap between Mexico's production and consumption.

U.S. consumption of petroleum and other liquids totals 19.9 million barrels per day in 2035 in the *AEO2012* Reference case, an increase of 0.7 million barrels per day over the 2010 total (Figure 110). With the exception of the transportation sector, where consumption grows by about 0.6 million barrels per day from 2010 through 2035, petroleum and other liquids consumption remains relatively flat. The transportation sector accounts for 72 percent of total petroleum and other liquids consumption in 2035. Proposed fuel economy standards covering MYs 2017 through 2025 that are not included in the Reference case would further reduce projected petroleum use (see "Issues in focus").

Motor gasoline, ultra-low-sulfur diesel fuel, and jet fuel are the primary transportation fuels, supplemented by biofuels such as ethanol and biodiesel. Petroleum-based motor gasoline consumption drops by approximately 0.9 million barrels per day from 2010 to 2035 in the Reference case, displaced by increased ethanol use in the form of higher blends in gasoline and by E85 consumption, which increases from virtually zero in 2010 to 0.8 million barrels per day in 2035. Diesel fuel consumption increases from 3.3 million barrels per day in 2010 to 4.1 million barrels per day in 2035.

Biodiesel and a number of next-generation biofuels account for a large share of the increase in petroleum and other liquids consumption (excluding ethanol) for transportation from 2010 to 2035 (about 0.7 million barrels per day). The growth in biofuels consumption (including ethanol) is attributable to the EISA2007 RFS mandates, as well as high crude oil prices. The growth in diesel fuel use results primarily from increased sales of light-duty diesel vehicles needed to meet more stringent CAFE standards, with a corresponding increase in domestic production of diesel fuel.

Petroleum and other liquids supply

Biofuels and natural gas liquids lead growth in total petroleum and other liquids supply

Figure 111. U.S. production of petroleum and other liquids by source, 2010-2035 (million barrels per day)

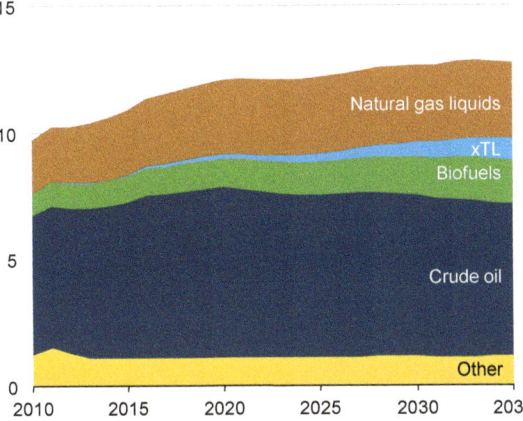

U.S. crude oil production increases, led by lower 48 onshore production

Figure 112. Domestic crude oil production by source, 1990-2035 (million barrels per day)

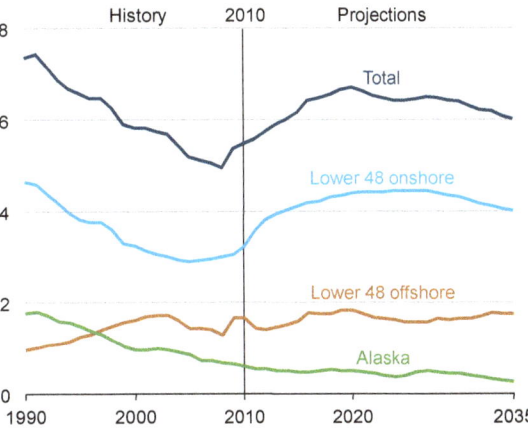

In the *AEO2012* Reference case, domestic production of petroleum and other liquids grows by 3.1 million barrels per day from 2010 to 2035 (Figure 111). Total production grows rapidly, from 9.7 million barrels per day in 2010 to 12.1 million barrels per day in 2020, as production of crude oil and NGL from tight oil formations (including shale plays) increases sharply. After 2020, total U.S. production of petroleum and other liquids grows more slowly, to 12.7 million barrels per day in 2035, as tight oil production levels off despite continued increases in crude oil prices. As production of other liquid fuels increases, the crude oil share of total domestic petroleum and other liquids production declines from 56 percent in 2010 to 47 percent in 2035. NGL production increases by more than 0.9 million barrels per day, to 3.0 million barrels per day in 2035, mainly as a result of strong growth in production of both tight oil and shale gas, which contain significant volumes of NGLs.

Biofuels production grows by 0.8 million barrels per day from 2010 to 2035 as a result of the EISA2007 RFS, with ethanol and biodiesel accounting for 0.7 and 0.1 million barrels per day, respectively, of the increase in the Reference case. The increase in domestic ethanol production reduces consumption of petroleum-based motor gasoline components by about 6 percent in 2035 on an energy-equivalent basis. In the early years of the projection, ethanol is used primarily for blending in E10 (motor gasoline blends containing up to 10 percent ethanol) and E15 (15 percent ethanol). In 2035, 37 percent of domestic ethanol production is used in E85 (85 percent ethanol) and 63 percent in E10 and E15 blends. In addition, growth in next-generation "xTL" production, which includes both biomass-to-liquids and CTL, contributes significantly to the growth in total U.S. petroleum and other liquids production, particularly after 2020, adding about 0.6 and 0.3 million barrels per day of production, respectively, from 2010 to 2035.

As world oil prices increase in the *AEO2012* Reference case, U.S. production of tight oil (liquid oil embedded in low-permeable sandstone, carbonate, and shale rock) and production using carbon dioxide-enhanced oil recovery (CO_2-EOR) techniques add to the projected increase in domestic crude oil production from 2010 to 2035 (Figure 112). Growth in lower 48 onshore crude oil production comes primarily from the continued development of tight oil resources, mostly from the Bakken and Eagle Ford formations. Tight oil production surpasses 1.3 million barrels per day in 2027 and then declines to about 1.2 million barrels per day in 2035 as "sweet spots" are depleted. *AEO2012* also includes six other tight formations in the projections for tight oil production: the Austin Chalk, Avalon/Bone Springs, Monterey, Niobrara, Spraberry, and Woodford formations. Additional tight oil resources are likely to be identified in the future as more work is completed to identify currently producing reservoirs that may be better categorized as tight formations, and as new tight oil plays are identified and incorporated (see next column).

Crude oil production using CO_2-EOR increases significantly after 2020, when oil prices are higher, the more profitable tight oil deposits are depleted, and affordable anthropogenic sources of carbon dioxide (CO_2) are available. It plateaus at about 650,000 barrels per day from 2032 to 2035, when its profitability is limited by reservoir quality and CO_2 availability. From 2011 through 2035, CO_2-EOR production exceeds 4 billion barrels of oil.

Lower 48 offshore oil production remains relatively constant in the Reference case. The decline in currently producing fields is offset primarily by exploration and development of new fields in the deep waters of the Gulf of Mexico and, after 2029, in the Pacific Outer Continental Shelf.

Petroleum and other liquids supply

U.S. crude oil production varies with price and resource assumptions

Figure 113. Total U.S. crude oil production in six cases, 1990-2035 (million barrels per day)

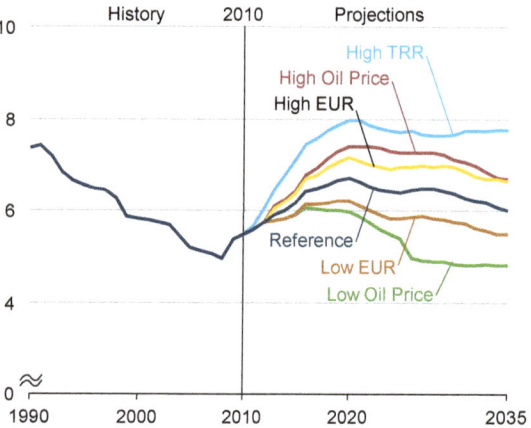

U.S. crude oil production varies with changes in assumptions about the extent of productivity improvement and well spacing in emerging tight oil resources examined in the High Technically Recoverable Resources (TRR) case and in the High and Low EUR cases (see discussion in "Issues in focus") and with changes in assumptions about crude oil prices in the Low and High Crude Oil Price cases (Figure 113). In the High TRR case, assumptions for tight oil allow for more rapid growth in crude oil production in the short and long term than in the Reference case, with production reaching nearly 8 million barrels per day in 2020. In the Low EUR case there is very little growth in domestic crude oil production over the projection period.

Higher oil prices lead to an increase in the level of investment in new oil projects. However, the returns from increased investment diminish as the average size and quality of available reservoirs decline. For example, in the High Oil Price case tight oil production is, on average, 225,000 barrels per day higher from 2020 to 2030 than in the Reference case but returns to Reference case levels in 2035. In contrast, low oil prices result in less investment in new oil projects and encourage producers to plug and abandon existing fields at earlier dates. For example, in the Low Oil Price case, oil production from the Alaska North Slope is shut down by around 2025, when the projected operating costs exceed wellhead production revenues (see "Issues in focus"). From 2020 to 2035, tight oil production is, on average, roughly 300,000 barrels per day lower in the Low Oil Price case than in the Reference case.

U.S. net imports of petroleum and other liquids fall in the Reference case

Figure 114. Net import share of U.S. petroleum and other liquids consumption in three cases, 1990-2035 (percent)

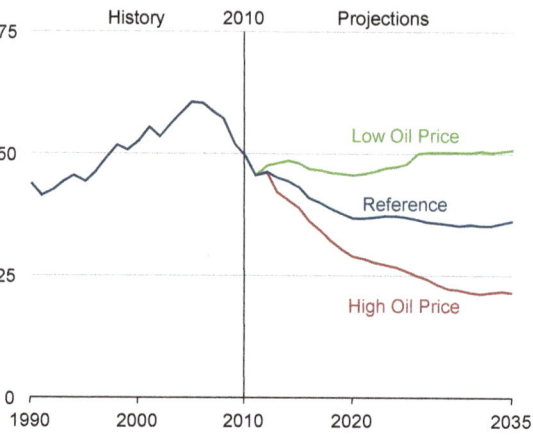

U.S. imports of petroleum and other liquids (including crude oil, petroleum liquids, and liquids derived from nonpetroleum sources) grew steadily from the mid-1980s to 2005 but have declined since then. In the *AEO2012* Reference and High Oil Price cases, U.S. imports of petroleum and other liquids continue to decline from 2010 to 2035, even as they provide a major part of total U.S. supply. Tighter fuel efficiency standards, increased use of biofuels, and greater production of domestic petroleum and other liquids contribute to the decrease in the share of imports. The combination of higher prices and renewable fuel mandates leads to more domestic production of petroleum and biofuels, which, combined with declines in the petroleum share of finished products after 2015, results in sustained net product exports.

The net import share of U.S. petroleum and other liquids consumption, which fell from 60 percent in 2005 to 50 percent in 2010, continues to decline in the Reference case, with the net import share falling to 36 percent in 2035 (Figure 114). In the High Oil Price case, the net import share falls even lower to a 22-percent share in 2035. In the Low Oil Price case, the net import share remains flat in the near term but rises to 51 percent in 2035, as domestic demand increases and imports become cheaper than crude oil produced domestically.

As a result of increased domestic production and slow growth in consumption, the United States becomes a net exporter of petroleum products, with net exports in the Reference case increasing from 0.18 million barrels per day in 2011 to 0.34 million barrels per day in 2035. In the High Oil Price case, net exports of petroleum products increase to 0.9 million barrels per day in 2035.

Petroleum and other liquids supply

U.S. consumption of cellulosic biofuels exceeds renewable fuels standard in 2035

Figure 115. EISA2007 RFS credits earned in selected years, 2010-2035 (billion credits)

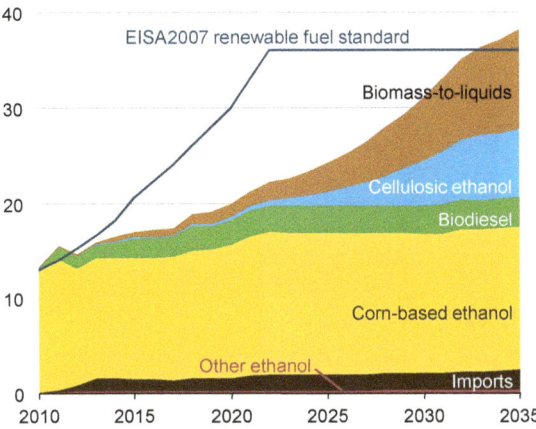

Infrastructure hurdles limit near-term growth in consumption of E15 and E85 fuels

Figure 116. U.S. ethanol use in blended gasoline and E85, 2000-2035 (billion gallons per year)

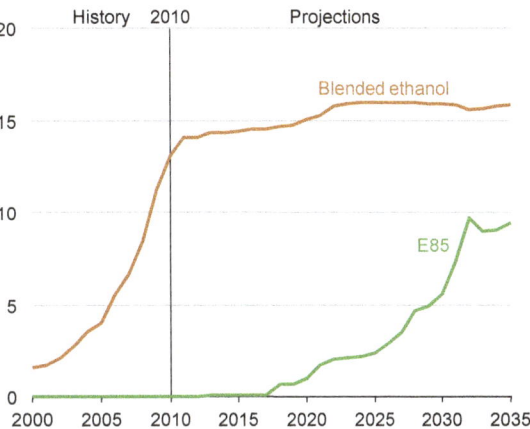

Although biofuel production increases substantially in the *AEO2012* Reference case, it does not meet the mandated RFS of 36 billion gallons in 2022 (Figure 115). Financial and technological hurdles delay the start of many advanced biofuel projects, particularly cellulosic biofuel projects. Three consecutive years of substantial reductions in the cellulosic biofuels mandate [133, 134, 135] have significantly reduced the possibility that the original RFS levels mandated in EISA2007 will be reached by 2022.

Between 2012 and 2022, it is expected that the EPA will evaluate the status of biofuel capacity annually and revise the production mandates for the following year, according to provisions in the RFS [136]. In 2011, after the EPA reduced the cellulosic biofuel mandate for both 2010 and 2011 from 100 million and 250 million gallons, respectively, to approximately 6 million gallons in both years, it also reduced the 2012 mandate from 500 million gallons to about 8 million gallons. Taking into account those modifications and anticipated future changes, only 22.1 billion of RFS credits are generated in 2022 in the Reference case, with 15 billion gallons of credits coming from domestic production of corn-based ethanol.

In the Reference case, the remainder of the biofuel supply consists of imported ethanol, biodiesel, cellulosic ethanol, and smaller volumes of next-generation biofuels. U.S. consumption of cellulosic ethanol grows from 0.6 billion gallons in 2022 to 7.2 billion gallons in 2035, when imports of ethanol and biodiesel total 2.2 billion gallons and 0.2 billion gallons, respectively.

A number of factors have recently limited the amount of ethanol that can be consumed domestically. Currently, given the limited availability of E85, the primary use of ethanol is as a blendstock for gasoline. With rapid growth in ethanol capacity and production in recent years, ethanol consumption in 2010 approached the legal gasoline blending limit of 10 percent (E10). As of January 2011, the EPA increased the blending limit to 15 percent for vehicles built in 2001 and later [137]. Once the final requirements are put in place, blenders will no longer be prohibited from blending beyond 10 percent for the general stock; however, a number of issues are expected to limit the rate at which terminals and retail outlets choose to take advantage of the option.

Liability from potential misfueling and infrastructure problems is one of the top concerns expected to slow the widespread adoption of E15. Retailers are hesitant to sell E15, even with the EPA's warning label, if they are not relieved of responsibility for damage to consumers' vehicles that may result from misfueling with the higher ethanol blend or from malfunctions of storage equipment or infrastructure. Consumer acceptance of the new fuel blend will also play a part, and warning labels may deter customers from risking potential damage from the use of E15, which potentially could void vehicle warranties.

In light of those potential issues, ethanol blending in gasoline increases slowly in the Reference case, from 13.2 billion gallons in 2010 (about 9 percent of the gasoline pool) to 15.0 billion gallons in 2020 (about 11 percent) and 15.8 billion gallons in 2035 (12.5 percent). Given the blending limitations, the remaining growth in ethanol use is in E85, which grows from about 0.6 billion gallons in 2018 to 9.5 billion gallons in 2035 (Figure 116).

Coal production

Shifts in fuel consumption guide future investment decisions for refiners

Figure 117. U.S. motor gasoline and diesel fuel consumption, 2000-2035 (million barrels per day)

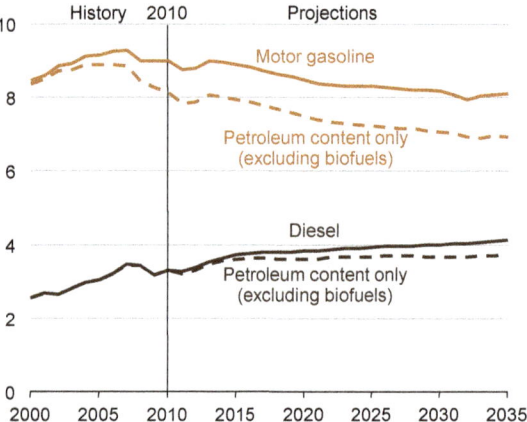

Early declines in coal production are more than offset by growth after 2015

Figure 118. Coal production by region, 1970-2035 (quadrillion Btu)

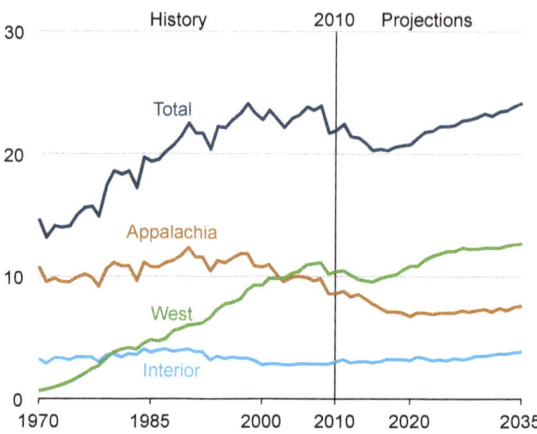

Tighter vehicle efficiency standards for LDVs require new LDVs to average 35 mpg by 2020, and newly issued regulations require increased use of ethanol. The Reference case does not include the proposed fuel economy standards covering MYs 2017 through 2025 that would raise vehicle efficiency standards even higher. Demand for motor gasoline declines in the Reference case. In combination with a tighter market for diesel fuel, the decrease in gasoline consumption leads to a shift in refinery outputs and investments. As some smaller and less integrated refineries begin to idle capacity as a result of higher costs, new refinery projects are focused on shifting production from gasoline to distillate fuels. The restructuring results in a net reduction in refinery capacity of 2.4 million barrels per day over the projection period.

In the Reference case, new capacity that was planned before the economic downturn of 2008-2009 comes on line early in the projection period, adding approximately 400,000 barrels per day of new refining distillation capacity from 2010 to 2015. As a result of refinery economics and concerns about the potential for enactment of legislation that could constrain carbon emissions, raise refiners' costs, and limit the growth in demand for petroleum and other liquids, no additional refinery capacity is built after 2015 until around 2030. Total refining capacity in the United States declines gradually after 2015 as additional capacity is idled.

Motor gasoline consumption and diesel fuel consumption (either including or excluding biofuels) trend in opposite directions in the Reference case (Figure 117). Consumption of diesel fuel increases by approximately 0.8 million barrels per day from 2010 to 2035, while motor gasoline consumption falls by 0.9 million barrels per day.

Although higher coal exports provide some support in 2011, U.S. coal production declines for four years thereafter as a result of low natural gas prices, rising coal prices, lack of growth in electricity demand, and increasing generation from renewables. In addition, new requirements to control emissions of nitrogen oxides (NO_X), sulfur dioxide (SO_2), and air toxics (such as mercury and acid gases), result in the retirement of some coal-fired generating capacity, contributing to the reduction in demand for coal. After 2015, coal production grows at an average annual rate of 1.0 percent through 2035, with coal use for electricity generation increasing as electricity demand grows and natural gas prices rise. More coal is also used for production of synthetic liquids, and coal exports increase.

Western coal production grows through 2035 (Figure 118) but at a much slower rate than in the past, as demand growth continues to slow. Low-cost supplies of coal from the West satisfy much of the additional need for fuel at coal-fired power plants east of the Mississippi River and supply most of the coal used at new CTL and CBTL plants.

Coal production in the Interior region, which has trended downward slightly since the early 1990s, recovers to near historic highs in the AEO2012 Reference case. Additional production from the Interior region originates from mines tapping into the substantial reserves of mid- and high-sulfur bituminous coal in Illinois, Indiana, and western Kentucky and from lignite mines in Texas and Louisiana. Appalachian coal production declines substantially from current levels, as coal produced from the extensively mined, higher cost reserves of Central Appalachia is supplanted by lower cost coal from other supply regions. An expected increase in production from the northern part of the Appalachia basin, however, moderates the overall production decline in Appalachia.

Coal production and prices

U.S. coal production is affected by actions to cut GHG emissions from existing power plants

Figure 119. U.S. total coal production in six cases, 2010, 2020, and 2035 (quadrillion Btu)

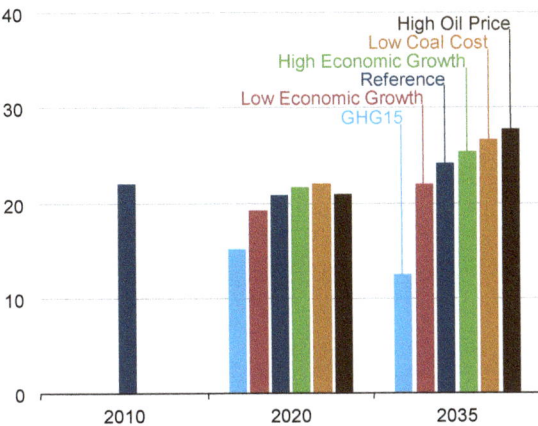

U.S. coal production varies across the *AEO2012* cases, reflecting different assumptions about the costs of producing and transporting coal, the outlook for economic growth, the outlook for world oil prices, and possible restrictions on GHG emissions (Figure 119). As shown in the GHG15 case, where a CO_2 emissions price that grows to $44 per metric ton in 2035 is assumed, actions to restrict or reduce GHG emissions can significantly affect the outlook for U.S. coal production.

Assumptions about economic growth primarily affect the projections for overall electricity demand, which in turn determine the need for coal-fired electricity generation. In contrast, assumptions about the costs of producing and transporting coal primarily affect the choice of technologies for electricity generation, with coal capturing a larger share of the U.S. electricity market in the Low Coal Cost case. In the High Oil Price case, higher oil prices stimulate the demand for coal-based synthetic liquids, leading to more coal use at CTL and CBTL plants. Production of coal-based synthetic liquids totals 1.3 million barrels per day in 2035 in the High Oil Price case, more than four times the amount in the Reference case.

From 2010 to 2035, changes in total annual coal production across the cases (excluding the GHG case) range from a decrease of 1 percent to an increase of 26 percent. In the earlier years of the projections, coal production is lower than in 2010 in most cases, as other sources of electricity generation displace coal-fired generation. From 2010 to 2020, changes in coal production across the cases (excluding the GHG case) range from a decline of 13 percent to virtually no change, with a 6-percent decline projected in the *AEO2012* Reference case.

Average minemouth price continues to rise, but at a slower pace than in recent years

Figure 120. Average annual minemouth coal prices by region, 1990-2035 (2010 dollars per million Btu)

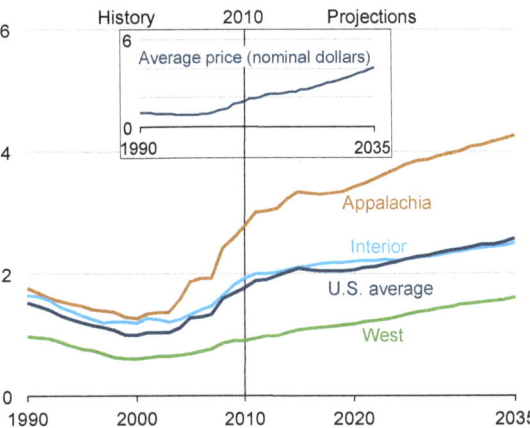

In the *AEO2012* Reference case, the average real minemouth price for U.S. coal increases by 1.5 percent per year, from $1.76 per million Btu in 2010 to $2.56 in 2035, continuing the upward trend in coal prices that began in 2000 (Figure 120). A key factor underlying the higher coal prices in the projection is an expectation that coal mining productivity will continue to decline, but at slower rates than during the 2000s.

In the Appalachian region, the average minemouth coal price increases by 1.7 percent per year from 2010 to 2035. In addition to continued declines in coal mining productivity, the higher price outlook for the Appalachian region reflects a shift to higher-value coking coal, resulting from the combination of growing exports of coking coal and declining shipments of steam/thermal coal to domestic markets. Recent increases in the average price of Appalachian coal, from $1.28 per million Btu in 2000 to $2.77 per million Btu in 2010, in part a result of significant declines in mining productivity over the past decade, have substantially reduced the competitiveness of Appalachian coal with coal from other regions.

In the Western and Interior coal supply regions, declines in mining productivity, combined with increasing production, lead to increases in the real minemouth price of coal, averaging 2.3 percent per year for the Western region and 1.0 percent per year for the Interior region from 2010 to 2035.

Emissions from energy use

Concerns about future GHG policies affect investments in emissions-intensive capacity

Figure 121. Cumulative coal-fired generating capacity additions by sector in two cases, 2011-2035 (gigawatts)

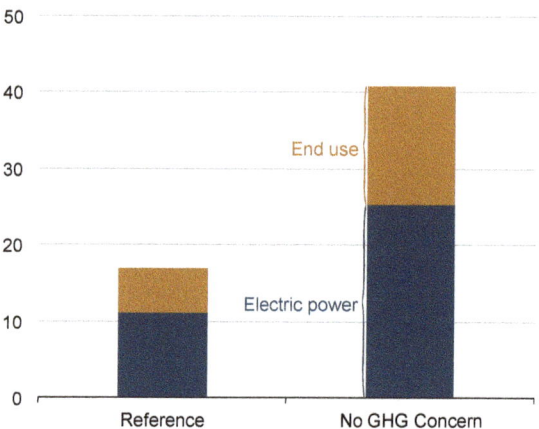

Projected energy-related carbon dioxide emissions remain below their 2005 level

Figure 122. U.S. energy-related carbon dioxide emissions by sector and fuel, 2005 and 2035 (million metric tons)

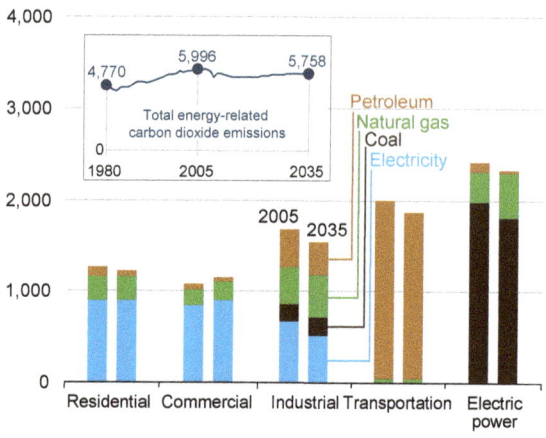

In the *AEO2012* Reference case, the cost of capital for investments in GHG-intensive technologies—including new coal-fired power plants without carbon capture and storage (CCS), new CTL and CBTL plants, and capital investment projects at existing coal-fired power plants (excluding CCS)—is increased by 3 percentage points to reflect the behavior of utilities, other energy companies, and regulators concerning the possible enactment of GHG legislation that could require owners to purchase emissions allowances, invest in CCS, or invest in other projects to offset their emissions in the future. The No GHG Concern case illustrates the potential impact on energy investments when the additional 3 percentage points added to the cost of capital for GHG-intensive technologies is removed.

In the No GHG Concern case, the lower cost of capital leads to 40 gigawatts of new coal-fired capacity additions from 2011 to 2035, up from 17 gigawatts in the Reference case (Figure 121). As a result, additions of both natural gas and renewable generating capacity are lower in the No GHG Concern case than in the Reference case. In the end-use sectors, all new coal-fired capacity additions in the No GHG Concern case are at CTL and CBTL plants, where part of the electricity is used to produce synthetic liquids and the remaining portion is sold to the grid. As a result, production of coal-based synthetic liquids totals 0.7 million barrels per day in 2035, compared with 0.3 million barrels per day in the Reference case. Total coal consumption (including coal converted to synthetic fuels) increases to 24.3 quadrillion Btu in 2035 in the No GHG Concern case, 2.6 quadrillion Btu (12 percent) higher than in the Reference case. Energy-related CO_2 emissions in 2035 are 5,900 million metric tons in the No GHG Concern case, about 2 percent higher than in the Reference case and 2 percent lower than their 2005 level.

On average, energy-related CO_2 emissions in the *AEO2012* Reference case decline by 0.1 percent per year from 2005 to 2035, as compared with an average increase of 0.9 percent per year from 1980 to 2005. Reasons for the decline include an expected slow and extended recovery from the recession of 2008-2009, growing use of renewable technologies and fuels, efficiency improvements, slower growth in electricity demand, and more use of natural gas, which is less carbon-intensive than other fossil fuels. In the Reference case, energy-related CO_2 emissions remain below 2005 levels through 2035, when they total 5,758 million metric tons—238 million metric tons (4.0 percent) below their 2005 level (Figure 122).

Petroleum remains the largest source of U.S. CO_2 emissions over the projection period, but its share falls to 40 percent in 2035 from 44 percent in 2005. CO_2 emissions from petroleum use, mainly in the transportation sector, were at relatively low levels in 2009. Although they increase somewhat from 2025 to 2035, emissions from petroleum use remain fairly stable, as improvements in transportation fuel economy and the expanded use of ethanol and other biofuels outweigh expected increases in travel demand. CO_2 emissions from petroleum would be even lower if proposed fuel economy standards covering MYs 2017 through 2025 were included in the Reference case.

Emissions from coal, the second largest source of CO_2 emissions, remain below 2005 levels through 2035 in the Reference case. Coal's share of total U.S. CO_2 emissions remains relatively unchanged through 2035, because the percentage decline in emissions from coal combustion is roughly the same as the percentage decline in total CO_2 emissions over the period. The natural gas share of CO_2 emissions increases from just under 20 percent in 2005 to 25 percent in 2035 as the use of natural gas to fuel electricity generation and industrial applications increases.

Emissions from energy use

Power plant emissions of sulfur dioxide are reduced by further environmental controls

Figure 123. Sulfur dioxide emissions from electricity generation, 1990-2035 (million short tons)

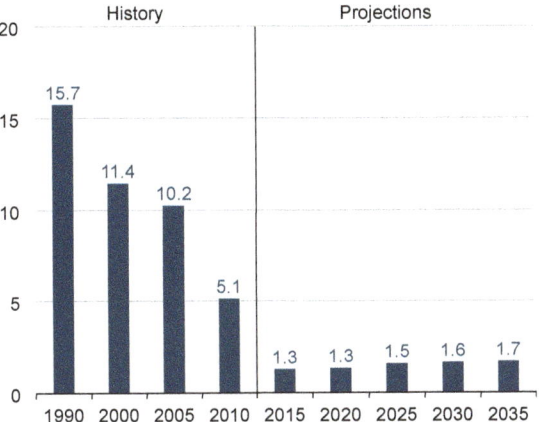

Nitrogen oxide emissions show little change from 2010 to 2035 in the Reference case

Figure 124. Nitrogen oxide emissions from electricity generation, 1990-2035 (million short tons)

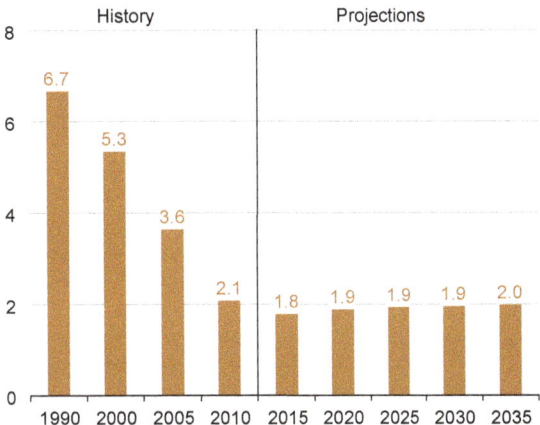

In the *AEO2012* Reference case, SO_2 emissions from the U.S. electric power sector fall from 5.1 million short tons in 2010 to a range of 1.3 to 1.7 million short tons in the 2015-2035 projection period. The reduction occurs in response to the EPA's Cross-State Air Pollution Rule (CSAPR) and Mercury and Air Toxics Standards (MATS) [138]. Although SO_2 is not directly regulated by the MATS, the reductions are achieved as a result of the technology requirements for acid gas and non-mercury metal controls on coal-fired power plants. *AEO2012* assumes that, in order to continue operating, coal plants must have either flue gas desulfurization (FGD) or dry sorbent injection (DSI) systems installed by 2015. Both technologies, which are used to reduce acid gas emissions, also reduce SO_2 emissions.

EIA assumes a 95-percent SO_2 removal efficiency for FGD units and a 70-percent SO_2 removal efficiency for DSI systems. DSI systems can achieve 70-percent efficiency when they include a baghouse filter, which also is assumed to be needed for compliance with the non-mercury metal component of the MATS.

From 2010 to 2035, approximately 48 gigawatts of coal-fired capacity is retrofitted with FGD units in the Reference case, and another 58 gigawatts is retrofitted with DSI systems. By 2015, all operating coal-fired power plants are assumed to have either DSI or FGD systems installed on units larger than 25 megawatts. As a result, after a 75-percent decrease from 2010 to 2015, SO_2 emissions increase slowly from 2016 to 2035 (Figure 123), as total electricity generation from coal-fired power plants increases.

Annual emissions of NO_X from the electric power sector, which totaled 2.1 million short tons in 2010, range between 1.8 and 2.0 million short tons from 2015 to 2035 (Figure 124). Annual NO_X emissions from electricity generation dropped by 43 percent from 2005 to 2010 due to implementation of the Clean Air Interstate Rule (CAIR), which led to the installation of additional NO_X pollution control equipment.

In the *AEO2012* Reference case, NO_X emissions are 5 percent below 2010 levels in 2035, despite a 2-percent increase in coal-fired electricity generation over the same period. The drop in emissions is a result primarily of CSAPR [139], which includes both annual and seasonal cap-and-trade systems for NO_X in 28 States. A slight rise in NO_X emissions after 2015 corresponds to a recovery in coal-fired generation as natural gas prices rise in the later years of the projection period.

The MATS does not have a direct effect on NO_X emissions, because none of the potential technologies required to comply with MATS has a significant impact on NO_X emissions. However, because MATS contributes to a reduction in coal-fired generation overall, it indirectly reduces NO_X emissions in the power sector in States without CSAPR where coal- and oil-fired units are used.

Coal-fired power plants can be retrofitted with one of three types of NO_X control technologies: selective catalytic reduction (SCR), selective noncatalytic reduction (SNCR), or low-NO_X burners. The type of retrofit used depends on the specific characteristics of the plant, including the boiler configuration and the type of coal used. From 2010 to 2035, 28 gigawatts of coal-fired capacity is retrofitted with NO_X controls in the Reference case: 69 percent with SCR, 3 percent with SNCR, and 29 percent with low-NO_X burners.

Endnotes for Market trends

Links current as of June 2012

121. In the recessions highlighted in Figure 46, percentage changes in annual GDP relative to the previous year were negative.
122. The industrial sector includes manufacturing, agriculture, construction, and mining. The energy-intensive manufacturing sectors include food, paper, bulk chemicals, petroleum refining, glass, cement, steel, and aluminum.
123. Energy expenditures relative to GDP are not the energy share of GDP, because they include energy as an intermediate product. The energy share of GDP corresponds to the share of value added by domestic energy-producing sectors, excluding the value of energy as an intermediate product.
124. S.C. Davis, S.W. Diegel, and R.G. Boundy, *Transportation Energy Databook: Edition 30*, ORNL-6986 (Oak Ridge, TN: June 2011), Chapter 4, "Light Vehicles and Characteristics," website cta.ornl.gov/data/index.shtml.
125. The *AEO2012* Reference case does not include the proposed LDV GHG and fuel economy standards published by the EPA and NHTSA in December 2011. (See "2017 and Later Model Year Light-Duty Vehicle Greenhouse Gas Emissions and Corporate Average Fuel Economy Standards," website www.nhtsa.gov/fuel-economy.)
126. LDV fuel economy includes AFVs and banked credits toward compliance.
127. U.S. Environmental Protection Agency and National Highway Transportation Safety Administration, "2017 and Later Model Year Light-Duty Vehicle Greenhouse Gas Emissions and Corporate Average Fuel Economy Standards; Proposed Rule," Federal Register, Vol. 76, No. 231 (Washington, DC, December 1, 2011), website www.nhtsa.gov/staticfiles/rulemaking/pdf/cafe/2017-25_CAFE_NPRM.pdf. 49 CFR Parts 523, 531, 533, 536, and 537.
128. U.S. Environmental Protection Agency and National Highway Traffic Safety Administration, "Greenhouse Gas Emissions Standards and Fuel Efficiency Standards for Medium- and Heavy-Duty Engines and Vehicles; Final Rule," Federal Register, Vol. 76, No. 179 (Washington, DC: September 15, 2011), pp. 57106-57513, website www.gpo.gov/fdsys/pkg/FR-2011-09-15/html/2011-20740.htm.
129. The factors that influence decisionmaking on capacity additions include electricity demand growth, the need to replace inefficient plants, the costs and operating efficiencies of different generation options, fuel prices, State RPS programs, and the availability of Federal tax credits for some technologies.
130. The 24 gigawatts include the 1.12 gigawatt Watts Bar 2 unit in 2012 that was subsequently delayed by TVA until 2015 due to cost overruns; www.tva.gov/news/releases/aprjun12/0426_board.htm.
131. Unless otherwise noted, the term "capacity" in the discussion of electricity generation indicates utility, nonutility, and CHP capacity. Costs reflect the average of regional costs.
132. For detailed discussion of levelized costs, see U.S. Energy Information Administration, "Levelized Cost of New Generation Resources in the Annual Energy Outlook 2012," website www.eia.gov/forecasts/aeo/electricity_generation.cfm.
133. U.S. Environmental Protection Agency, "EPA Finalizes Regulations for the National Renewable Fuel Standard Program for 2010 and Beyond," EPA-420-F-10-007 (Washington, DC: February 2010), website www.epa.gov/otaq/renewablefuels/420f10007.pdf.
134. U.S. Environmental Protection Agency, "EPA Finalizes 2011 Renewable Fuel Standards," EPA-420-F-10-056 (Washington, DC: November 2010), website www.epa.gov/oms/fuels/renewablefuels/420f10056.pdf.
135. U.S. Environmental Protection Agency, "EPA Finalizes 2012 Renewable Fuel Standards," EPA-420-F-11-044 (Washington, DC: December 2011), website www.epa.gov/otaq/fuels/renewablefuels/documents/420f11044.pdf.
136. EISA2007, Section 211(o)(7) of the Clean Air Act.
137. U.S. Environmental Protection Agency, "E15 (a blend of gasoline and ethanol)," website www.epa.gov/otaq/regs/fuels/additive/e15.
138. U.S. Environmental Protection Agency, "Mercury and Air Toxics Standards," website www.epa.gov/mats.
139. U.S. Environmental Protection Agency, "Cross-State Air Pollution Rule (CSAPR)," website epa.gov/airtransport.

Comparison with other projections

Energy Information Administration (EIA) and other contributors have endeavored to make these projections as objective, reliable, and useful as possible; however, they should serve as an adjunct to, not a substitute for, a complete and focused analysis of public policy initiatives. None of the EIA or any of the other contributors shall be responsible for any loss sustained due to reliance on the information included in this report.

Comparison with other projections

Only IHS Global Insight (IHSGI) produces a comprehensive energy projection with a time horizon similar to that of the *Annual Energy Outlook 2012 (AEO2012)*. Other organizations, however, address one or more aspects of the U.S. energy market. The most recent projection from IHSGI, as well as others that concentrate on economic growth, international oil prices, energy consumption, electricity, natural gas, petroleum, and coal, are compared here with the *AEO2012* Reference case.

1. Economic growth

The range of projected economic growth in the outlooks included in the comparison tends to be wider over the first 5 years of the projection period than over a longer period, because the group of variables—such as population, productivity, and labor force growth—that are used to influence long-run economic growth is smaller than the group of variables that affect projections of short-run growth. The average annual rate of growth of real gross domestic product (GDP) from 2010 to 2015 (in 2005 dollars) ranges from 2.4 percent to 3.4 percent (Table 22). From 2010 to 2020, the 10-year average annual growth rate ranges from 2.5 percent to 3.1 percent.

From 2010 to 2015, real GDP is projected to grow at a 2.5-percent average annual rate in the *AEO2012* Reference case, lower than projected by the Office of Management and Budget (OMB), Congressional Budget Office (CBO), Blue Chip Consensus (Blue Chip), Social Security Administration (in *The 2011 Annual Report of the Board of Trustees of the Federal Old-Age and Survivors Insurance and Federal Disability Insurance Trust Funds*), ExxonMobil, and the Interindustry Forecasting Project at the University of Maryland (INFORUM) and higher than projected by Strategic Energy and Economic Research, Inc. (SEER). The *AEO2012* projection of GDP growth is similar to the IHSGI average annual rate of 2.5 percent over the same period.

The average annual GDP growth of 2.5 percent in the *AEO2012* Reference case from 2010 to 2020 is at the low end of the range of outlooks, with OMB, INFORUM, and the Social Security Administration projecting the strongest recovery from the 2008-2009 recession. INFORUM projects average annual GDP growth of 3.1 percent from 2010 to 2020, while OMB and the Social Security Administration project annual average growth of 3.0 percent over the same period. The CBO, ExxonMobil, Blue Chip, the International Energy Agency's (IEA) November 2011 *World Energy Outlook* Current Policies Scenario, and SEER also project higher growth than the *AEO2012* Reference case from 2010 to 2020, ranging between 2.6 and 2.8 percent per year over the next 10 years.

There are few public or private projections of GDP growth for the United States that extend to 2035. The *AEO2012* Reference case projects 2.5-percent average annual GDP growth from 2010 to 2035, consistent with trends in labor force and productivity growth. IHSGI, ExxonMobil, and the Social Security Administration project GDP growth averaging 2.5 percent per year from 2010 to 2035, and INFORUM (at 2.7 percent) and SEER (at 2.8 percent) project higher GDP growth than in the *AEO2012* Reference Case over the same period. IEA projects a slightly lower rate of 2.4 percent per year from 2010 to 2035.

2. Oil prices

In the *AEO2012* Reference case, oil prices [West Texas Intermediate (WTI)] rise from $79 per barrel in 2010 to about $117 per barrel in 2015 and $127 per barrel in 2020 (Table 23). From the 2020 level, prices increase slowly to $145 per barrel in 2035. This price trend is slightly higher than the trend shown in last year's *AEO2011* Reference case.

Table 22. Projections of average annual economic growth, 2010-2035

	Average annual percentage growth rates			
Projection	2010-2015	2010-2020	2020-2035	2010-2035
AEO2012 (Reference case)	2.5	2.5	2.6	2.5
AEO2011 (Reference case)	3.0	2.8	2.6	2.7
IHSGI (November 2011)	2.5	2.5	2.5	2.5
OMB (January 2012)[a]	3.1	3.0	--	--
CBO (January 2012)[a]	2.7	2.8	--	--
INFORUM (January 2012)	3.4	3.1	2.4	2.7
Social Security Administration (August 2011)	3.3	3.0	2.1	2.5
IEA (2011)[b]	--	2.6	2.4	2.4
Blue Chip Consensus (October 2011)[a]	2.6	2.6	--	--
ExxonMobil	2.7	2.7	2.3	2.5
SEER	2.4	2.7	2.8	2.8

-- = not reported.
[a]OMB, CBO, and Blue Chip forecasts end in 2022, and growth rates cited are for 2010-2022.
[b]IEA publishes U.S. growth rates for certain intervals: 2009-2020 growth is 2.6 percent, and 2009-2035 growth rate is 2.4 percent.

Market volatility and different assumptions about the future of the world economy are reflected in the range of price projections for both the short term and the long term; however, most projections show prices rising over the entire course of the projection period. The projections range from $82 per barrel to $117 per barrel in 2015 (a span of $35 per barrel) and from $98 per barrel to $145 per barrel in 2035 (a span of $47 per barrel). The wide range underscores the uncertainty inherent in the projections. The range of the projections is encompassed in the range of the *AEO2012* Low and High Oil Price cases, from $58 per barrel to $182 per barrel in 2015 and from $62 per barrel to $200 per barrel in 2035.

The measure of oil prices is, by and large, comparable across projections. EIA reports the price of low-sulfur, light crude oil, approximately the same as the WTI price widely cited in the trade press. The only series that do not report projections in WTI terms are IEA, with prices in the Current Policies Scenario expressed as the price of imported crude oil, and INFORUM, with prices expressed as the average U.S. refiner acquisition cost (RAC) of imported crude oil.

3. Total energy consumption

Five projections by other organizations—INFORUM, IHSGI, ExxonMobil, IEA, and BP—include energy consumption by sector. To allow comparison with the IHSGI projection, the *AEO2012* Reference case was adjusted to remove coal-to-liquids (CTL) heat and power, biofuels heat and co-products, and natural gas feedstock use. To allow comparison with the ExxonMobil projection, electricity consumption in each sector was removed from the *AEO2012* Reference case projections. To allow comparison with the IEA and BP projections, the *AEO2012* Reference case projections for the residential and commercial sectors were combined to produce a buildings sector projection. BP does not include the electric power sector in its projection for total energy consumption; however, it does include conversion losses that allow comparison on the basis of total energy consumption. The IEA projections have a base year of 2009, as opposed to 2010 in the other projections, and BP's projections extend only through 2030, not 2035.

Total energy consumption is higher in all projection years in both the IHSGI and INFORUM projections than in the *AEO2012* Reference case. ExxonMobil, IEA, and BP show lower total energy consumption in all years (Table 24). ExxonMobil and BP include a cost for carbon dioxide (CO_2) emissions in their outlooks, which helps to explain the lower level of consumption in those outlooks. While the IEA reference case also includes a cost for CO_2 emissions, the IEA Current Policies Scenario (which assumes that no new policies are added to those in place in mid-2011) was used for comparison in this analysis, because it corresponds better with the assumptions in *AEO2012*.

The INFORUM projection of total energy consumption in 2035 is almost 8 quadrillion Btu higher than the *AEO2012* Reference case projection, with the industrial and electric power sectors each about 2 quadrillion Btu higher and the transportation sector about 3 quadrillion Btu higher. For the transportation sector, the difference appears to result from a higher number of light-duty vehicle miles traveled in the INFORUM results, which offsets slightly higher motor gasoline prices in the INFORUM projection. Vehicle efficiency is essentially the same in the INFORUM and *AEO2012* projections. INFORUM also projects higher revenue passenger-miles for air travel than *AEO2012*. Diesel prices are lower in the INFORUM projection, which leads to higher demand (about 1 quadrillion Btu) than in *AEO2012*. In the industrial sector, INFORUM projects industrial shipments in 2035 that are approximately 1.5 times the level of those in the *AEO2012* Reference case, which helps to explain the higher level of industrial energy consumption in the INFORUM projection relative to *AEO2012*.

IHSGI projects significantly higher electricity consumption for all sectors than in the *AEO2012* Reference case, which helps to explain much of the difference in total energy consumption between the two projections. In the IHSGI projection, the electric power sector consumes 13 quadrillion Btu more energy in 2035 than in the *AEO2012* Reference case. The greater use of electricity in the IHSGI projection, including 300 trillion Btu used by electric vehicles, also results in higher electricity prices than in the *AEO2012* Reference case.

Table 23. Projections of oil prices, 2015-2035 (2010 dollars per barrel)

Projection	2015	2020	2025	2030	2035
AEO2012 (Reference case)	116.91	126.68	132.56	138.49	144.98
AEO2011 (Reference case)	95.41	109.05	118.57	124.17	126.03
EVA	82.24	84.75	89.07	94.78	102.11
IEA (Current Policies Scenario)	106.30	118.10	127.30	134.50	140.00
INFORUM	91.78	105.84	113.35	117.83	116.76
IHSGI	99.16	72.89	87.19	95.65	98.08
Purvin & Gertz	98.75	103.77	106.47	107.37	107.37
SEER	94.20	101.57	107.13	111.26	121.94

Although there are differences in energy consumption by sector between the ExxonMobil and BP projections, in both cases total energy consumption declines from 2010 levels and is lower than in the *AEO2012* Reference case. The difference appears to result primarily from the inclusion of a tax on CO_2 emissions in both the ExxonMobil and BP projections, which is not considered in the *AEO2012* projection. Energy consumption in the transportation sector declines from 2010 levels in both the ExxonMobil and BP projections, driven by policy changes and technology improvement; however, BP projects a much larger drop in transportation energy consumption, a total of 4 quadrillion Btu (or four times the decline in the ExxonMobil projection) between 2010 and 2030.

Although energy consumption in all sectors in the IEA projection is higher in 2035 than in 2010, energy consumption in the transportation and industrial sectors declines from 2020 to 2030, by less than 1 quadrillion Btu in each sector.

IEA projects little change for energy use in those two sectors from 2030 to 2035, with industrial energy consumption declining very slowly and transportation energy consumption increasing very slightly. IEA projects total energy consumption that is higher than BP in 2030 and higher than ExxonMobil in 2035 but considerably lower than in the *AEO2012* Reference case.

4. Electricity

Table 25 compares summary results for the electric power sector from the *AEO2012* Reference case with projections by Energy Ventures Analysis (EVA), IHSGI, and INFORUM. In 2015, total electricity sales range from a low of 3,753 billion kilowatthours in the *AEO2012* Reference case to a high of 4,173 billion kilowatthours in the IHSGI projection. IHSGI shows higher sales across

Table 24. Projections of energy consumption by sector, 2010-2035 (quadrillion Btu)

Sector	AEO2012 Reference	INFORUM	IHSGI	ExxonMobil	IEA	BP
			2010			
Residential	11.7	11.4	11.2	--	--	--
Residential excluding electricity	6.7	6.5	6.2	6.0	--	--
Commercial	8.7	8.5	8.6	--	--	--
Commercial excluding electricity	4.2	3.9	4.0	4.0	--	--
Buildings sector	20.4	20.0	19.8	10.0	19.1[a]	21.8
Industrial	23.4	23.1	--	--	22.9[a]	23.0
Industrial excluding electricity	20.1	19.9	--	20.0	--	--
Losses[b]	0.8	--	--	--	--	--
Natural gas feedstocks	0.5	--	--	--	--	--
Industrial removing losses and feedstocks	22.0	--	21.4	--	--	--
Transportation	27.6	27.4	26.6	27.0	22.9[a]	22.8
Electric power	39.6	40.1	40.8	37.0	35.6[a]	--
Less: electricity demand[c]	12.8	12.8	12.8	--	14.3[a]	--
Electric power losses	26.8	27.3	--	--	--	23.1
Total primary energy	**98.2**	**97.8**	--	**94.0**	**85.7**[a]	**90.7**
Excluding losses[b] and feedstocks	96.8	--	95.8	--	--	--
			2020			
Residential	11.4	11.2	11.8	--	--	--
Residential excluding electricity	6.4	6.4	5.8	6.0	--	--
Commercial	9.2	9.5	9.5	--	--	--
Commercial excluding electricity	4.3	4.3	4.0	4.0	--	--
Buildings sector	20.5	20.7	21.3	9.0	20.4	21.9
Industrial	24.6	27.4	--	--	24.8	23.4
Industrial excluding electricity	21.2	23.9	--	20.0	--	--
Losses[b]	1.2	--	--	--	--	--
Natural gas feedstocks	0.5	--	--	--	--	--
Industrial removing losses and feedstocks	22.9	--	22.5	--	--	--
Transportation	27.3	29.0	27.4	28.0	23.8	21.0
Electric power	40.2	41.6	48.6	39.0	39.3	--
Less: electricity demand[c]	13.3	13.6	15.7	--	16.4	--
Electric power losses	26.9	28.0	--	--	--	23.7
Total primary energy	**99.3**	**105.1**	--	**96.0**	**91.4**	**90.1**
Excluding losses[b] and feedstocks	97.6	--	104.1	--	--	--

-- = not reported.
See notes at end of table.

(continued on next page)

all sectors in 2015 in comparison with the other projections. Total electricity sales in 2035 in the IHSGI projection (5,652 billion kilowatthours) are higher than in the others: 4,415 billion kilowatthours in the *AEO2012* Reference case, 4,483 billion kilowatthours in the INFORUM projection, and 4,726 billion kilowatthours in the EVA projection. Although IHSGI projects higher electricity sales in all sectors in 2035, the largest percentage differences between the IHSGI and other projections are in the industrial sector. Electricity sales in the industrial sector in 2035 in the IHSGI projection are 1,387 billion kilowatthours, as compared with 977 billion kilowatthours in the *AEO2012* Reference case, 941 billion kilowatthours in the EVA projection, and 968 billion kilowatthours in the INFORUM projection.

Table 24. Projections of energy consumption by sector, 2010-2035 (quadrillion Btu) (continued)

Sector	AEO2012 Reference	INFORUM	IHSGI	ExxonMobil	IEA	BP
			2030			
Residential	11.7	11.6	12.6	--	--	--
Residential excluding electricity	6.2	6.3	5.7	5.0	--	--
Commercial	9.9	10.6	10.4	--	--	--
Commercial excluding electricity	4.4	4.5	4 0	4.0	--	--
Buildings sector	21.6	22.1	23 0	9.0	22.0	23.0
Industrial	26.1	28.8	--	--	24.1	23.2
Industrial excluding electricity	22.7	25.3	--	19.0	--	--
Losses[b]	2.4	--	--	--	--	--
Natural gas feedstocks	0.5	--	--	--	--	--
Industrial removing losses and feedstocks	23.3	--	23 0	--	--	--
Transportation	27.9	30.7	27 5	26.0	22.9	18.5
Electric power	43.2	45.0	54 3	41.0	41.6	--
Less: electricity demand[c]	14.5	14.8	18.1	--	17.9	--
Electric power losses	28.7	30.1	--	--	--	24.1
Total primary energy	**104.3**	**111.8**	--	**94.0**	**92.3**	**88.9**
Excluding losses[b] and feedstocks	101.5	--	109.7	--	--	--
			2035			
Residential	11.9	11.7	13 0	--	--	--
Residential excluding electricity	6.1	6.2	5 5	5.0	--	--
Commercial	10.3	11.1	10 8	--	--	--
Commercial excluding electricity	4.5	4.6	4 0	3.0	--	--
Buildings sector	22.2	22.8	23 8	8.0	22.9	--
Industrial	26.9	29.1	--	--	23.9	--
Industrial excluding electricity	23.6	25.7	--	18.0	--	--
Losses[b]	3.2	--	--	--	--	--
Natural gas feedstocks	0.4	--	--	--	--	--
Industrial removing losses and feedstocks	23.3	--	23 3	--	--	--
Transportation	28.6	31.9	27 8	25.0	23.1	--
Electric power	44.2	46.2	57 2	40.0	42.5	--
Less: electricity demand[c]	15.1	15.3	19 3	--	18.6	--
Electric power losses	29.2	30.8	--	--	--	--
Total primary energy	**106.9**	**114.7**	--	**92.0**	**93.4**	--
Excluding losses[b] and feedstocks	103.3	--	112.7	--	--	--

-- = not reported.
[a]IEA data are for 2009.
[b]Losses in CTL and biofuel production.
[c]Energy consumption in the sectors includes electricity demand purchases from the electric power sector, which are subtracted to avoid double counting in deriving total primary energy consumption.

Table 25. Comparison of electricity projections, 2015, 2025, and 2035 (billion kilowatthours, except where noted)

Projection	2010	AEO2012 Reference case	Other projections		
			EVA	IHSGI	INFORUM
2015					
Average end-use price (2010 cents per kilowatthour)[a]	9.8	9.7	--	10.2	--
Residential	11.5	11.8	12.8	12.0	10.5
Commercial	10.1	9.9	11.5	10.7	9.3
Industrial	6.7	6.5	7.9	7.0	6.2
Total generation plus imports	**4,152**	**4,181**	**4,053**	**4,611**	--
Coal	1,851	1,581	1,591	1,905	--
Petroleum	37	28	--	45	--
Natural gas[b]	982	1,130	1,090	1,223	--
Nuclear	807	830	827	839	--
Hydroelectric/other[c]	449	583	515	576	--
Net imports	26	29	29	24	--
Electricity sales	3,749	3,753	3,921	4,173	3,854
Residential	1,451	1,392	1,481	1,563	1,365
Commercial/other[d]	1,336	1,354	1,414	1,489	1,438
Industrial	962	1,008	1,025	1,121	1,051
Capacity, including CHP (gigawatts)[e]	1,036	1,042	1,094	1,101	--
Coal	318	286	289	309	--
Oil and natural gas	459	464	514	491	--
Nuclear	101	104	106	104	--
Hydroelectric/other[f]	158	188	185	197	--
2025					
Average end-use price (2010 cents per kilowatthour)[a]	9.8	9.7	--	10.9	--
Residential	11.5	11.6	13.2	12.8	10.5
Commercial	10.1	9.9	11.7	11.4	9.3
Industrial	6.7	6.7	8.0	7.4	6.2
Total generation plus imports	**4,152**	**4,578**	**4,514**	**5,417**	--
Coal	1,851	1,786	1,653	1,774	--
Petroleum	37	29	--	45	--
Natural gas[b]	982	1,140	1,335	1,760	--
Nuclear	807	917	870	918	--
Hydroelectric/other[c]	449	683	629	896	--
Net imports	26	22	27	25	--
Electricity sales	3,749	4,090	4,298	4,942	4,167
Residential	1,451	1,533	1,650	1,887	1,468
Commercial/other[d]	1,336	1,525	1,679	1,793	1,660
Industrial	962	1,032	969	1,261	1,039
Capacity, including CHP (gigawatts)[e]	1,036	1,091	1,119	1,274	--
Coal	318	282	267	283	--
Oil and natural gas	459	493	518	566	--
Nuclear	101	115	110	114	--
Hydroelectric/other[f]	158	201	224	312	--

-- = not reported.
See notes at end of table.

(continued on next page)

Comparison with other projections

Only IHSGI and the *AEO2012* Reference case provide average electricity price projections through 2035. Average electricity prices in the *AEO2012* Reference case are 9.8 cents per kilowatthour in 2010 and 9.7 cents per kilowatthour in 2015 and 2025 before reaching 10.1 cents per kilowatthour in 2035. In the IHSGI projection, the average electricity price rises continuously (with the exception of a small decrease from 2017 to 2018), from 9.8 cents per kilowatthour in 2010 to 10.2 cents in 2015, 10.9 cents in 2025, and 12.1 cents per kilowatthour in 2035.

In all the projections, average electricity prices by sector follow patterns similar to changes in the weighted average electricity price across all sectors (including transportation services). The lowest prices by sector in 2015 are in the INFORUM projection (10.5 cents per kilowatthour in the residential sector, 9.3 cents per kilowatthour in the commercial sector, and 6.2 cents per kilowatthour in the industrial sector). The highest average electricity prices by sector in 2015 are in the EVA projection (12.8 cents per kilowatthour in the residential sector, 11.5 cents per kilowatthour in the commercial sector, and 7.9 cents per kilowatthour in the industrial sector).

In the *AEO2012* Reference case, electricity prices for the residential sector are 11.8 cents per kilowatthour in both 2015 and 2035, electricity prices for the commercial sector increase from 9.9 cents per kilowatthour in 2015 to 10.1 cents per kilowatthour in 2035, and electricity prices for the industrial sector increase from 6.5 cents per kilowatthour in 2015 to 7.1 cents per kilowatthour in 2035. When compared with the *AEO2012* Reference case prices in 2035, the largest difference is with the IHSGI projection. The IHSGI price projections are much higher than those in the *AEO2012* Reference case. IHSGI shows real electricity prices rising to 14.3 cents per kilowatthour for the residential sector, 12.5 cents per kilowatthour for the commercial sector, and 8.1 cents per kilowatthour for the industrial sector in 2035.

Table 25. Comparison of electricity projections, 2015, 2025, and 2035 (billion kilowatthours, except where noted) (continued)

Projection	2010	AEO2012 Reference case	Other projections EVA	IHSGI	INFORUM
			2035		
Average end-use price (2010 cents per kilowatthour)[a]	9.8	10.1	--	12.1	--
Residential	11.5	11.8	12.9	14.3	10.5
Commercial	10.1	10.1	11.3	12.5	9.3
Industrial	6.7	7.1	7.6	8.1	6.2
Total generation plus imports	4,152	5,004	--	6,199	--
Coal	1,851	1,897	--	1,618	--
Petroleum	37	30	--	45	--
Natural gas[b]	982	1,398	--	2,354	--
Nuclear	807	887	--	1,030	--
Hydroelectric/other[c]	449	780	--	1,124	--
Net imports	26	12	--	28	--
Electricity sales	3,749	4,415	4,726	5,652	4,483
Residential	1,451	1,718	1,778	2,178	1,611
Commercial/other[d]	1,336	1,721	2,008	2,088	1,904
Industrial	962	977	941	1,387	968
Capacity, including CHP (gigawatts)[e]	1,036	1,190	--	1,450	--
Coal	318	285	--	262	--
Oil and natural gas	459	568	--	665	--
Nuclear	101	111	--	128	--
Hydroelectric/other[f]	158	226	--	396	--

-- = not reported.
[a] Average end-use price includes the transportation sector.
[b] Includes supplemental gaseous fuels. For EVA, represents total oil and natural gas.
[c] "Other" includes conventional hydroelectric, pumped storage, geothermal, wood, wood waste, municipal waste, other biomass, solar and wind power, batteries, chemicals, hydrogen, pitch, purchased steam, sulfur, petroleum coke, and miscellaneous technologies.
[d] "Other" includes sales of electricity to government and other transportation services.
[e] EIA capacity is net summer capacity, including CHP plants.
[f] "Other" includes conventional hydro, geothermal, wood, wood waste, all municipal waste, landfill gas, other biomass, solar, wind power, pumped storage, and fuel cells.

Total electricity generation plus imports in 2015 ranges from a low of 4,053 billion kilowatthours in the EVA projection to a high of 4,611 billion kilowatthours in the IHSGI projection, compared with 4,181 billion kilowatthours in the *AEO2012* Reference case. Although coal represents the largest share of generation in 2015 in all the projections, the natural gas share of total generation grows from 2015 to 2035 in all the projections, particularly IHSGI. In the IHSGI projection, coal has a 33-percent share of total generation in 2025, and the natural gas share is 32 percent. IHSGI shows natural gas overtaking coal as a share of total generation by 2035 as a result of the carbon tax assumed in the IHSGI projection and the need to replace existing units that are uneconomical or are being retired for various regulatory or environmental reasons. In 2035, the coal share in the IHSGI projection is 26 percent of total generation, and the natural gas share is 38 percent. In the *AEO2012* Reference case, which does not include a carbon tax, the coal share also decreases but only to 38 percent of total generation, while the natural gas share increases to 28 percent.

Nuclear generation in 2015 ranges from a low of 827 billion kilowatthours in the EVA projection to a high of 839 billion kilowatthours in the IHSGI projection. From 2015 to 2025, EVA projects a 5-percent increase in nuclear generation, to 870 billion kilowatthours. IHSGI and *AEO2012* project increases of 9 percent and 10 percent, respectively. In the IHSGI projection, nuclear generation totals 1,030 billion kilowatthours in 2035, a 12-percent increase from 2025. The *AEO2012* Reference case shows nuclear generation declining to 887 billion kilowatthours in 2035, a 3-percent decrease from 2025, as units are retired when they reach the end of their useful generation lifetimes.

Total generating capacity by fuel in 2015 is relatively similar across the projections, ranging from 1,042 gigawatts in the *AEO2012* Reference case to 1,101 gigawatts in the IHSGI projection, but IHSGI shows a much larger decrease in capacity in 2025. IHSGI projects more aggressive growth in total generating capacity, due to what appears to be a much higher demand projection. Natural gas and oil-fired capacity grows to 566 gigawatts in 2025 in the IHSGI projection, compared with 493 gigawatts in *AEO2012* and 518 gigawatts in the EVA projections. Hydroelectric/other capacity grows to 312 gigawatts in 2025 in the IHSGI projection, higher than the 201 gigawatts in *AEO2012*. The faster growth in natural gas and hydroelectric/other capacity in the IHSGI projection continues through 2035. Natural gas and oil-fired capacity grows to 665 gigawatts in 2035, and hydroelectric/other capacity grows to 396 gigawatts in 2035 in the IHSGI projection. By comparison, natural gas and oil-fired capacity grows to 568 gigawatts and hydroelectric/other capacity grows to 226 gigawatts in the *AEO2012* Reference case in 2035.

5. Natural gas

The projections of natural gas consumption, production, imports, and prices (Table 26) vary significantly as a result of differences in assumptions. For example, the *AEO2012* Reference case assumes that current laws and regulations remain unchanged throughout the projection period (including the implication that laws which include sunset dates do, in fact, become ineffective at the time of those sunset dates), whereas the other projections may include anticipated policy developments over the next 25 years. In particular, the *AEO2012* Reference case does not assume changes in CO_2 emissions policies.

Each of the projections shows an increase in overall natural gas consumption from 2010 to 2035, with the IHSGI projection showing the largest increase, 39 percent. The ExxonMobil projection includes an increase of around 20 percent. The EVA projection shows an increase of 26 percent from 2010 to 2030 (EVA does not extend to 2035). Total natural gas consumption in the *AEO2012*, Deloitte, and SEER projections increases from 2010 to 2035, with total natural gas consumption growing from 4 to 31 percent. IHSGI shows the largest increase and INFORUM the smallest. The IHSGI projection for total natural gas consumption in 2035 is 36 percent higher than the INFORUM projection. In the *AEO2012* Reference case, total natural gas consumption grows by 5 percent from 2015 to 2035.

The IHSGI and ExxonMobil projections for natural gas consumption by electricity generators are much higher than the other projections shown in Table 26. In 2035, natural gas consumption by electricity generators in the IHGSI projection is more than double the consumption projected by INFORUM, and the ExxonMobil projection is 77 percent higher than the INFORUM projection. The *AEO2012* Reference case, SEER, and INFORUM projections show similar levels of natural gas consumption in the electricity generation sector in 2035, with average annual growth of 1 percent or less across the projection period, while consumption grows by an average of 3 percent in the ExxonMobil and IHSGI projections. The slower rate of growth in the *AEO2012* Reference case reflects relatively slower growth in electricity consumption and faster growth in renewable energy consumption than in the other projections.

Industrial natural gas consumption is similar across the projections, but with more rapid growth projected by EVA, Deloitte, and INFORUM. Natural gas consumption increases by 23 percent from 2010 to 2030 in the EVA projection and by 23 percent and 11 percent, respectively, from 2010 to 2035 in the INFORUM and Deloitte projections. All of the growth in industrial natural gas consumption in the Deloitte and INFORUM projections is between 2010 and 2015. In the *AEO2012* Reference case, in contrast, industrial natural gas consumption grows by 6 percent from 2010 to 2035. In the ExxonMobil projection, industrial natural gas consumption remains constant over the projection period; in the IHSGI projection industrial natural gas consumption falls from 2010 to 2035; and in the INFORUM, SEER, and Deloitte projections, after an initial increase, industrial natural gas consumption declines from 2015 to 2035.

The levels of commercial sector natural gas consumption are similar across the projections, but projections for the residential sector vary significantly [140]. Three of the seven projections (INFORUM, Deloitte, and EVA) show similar growth in residential consumption through 2030, and INFORUM and Deloitte are similar through 2035; however, the IHSGI and *AEO2012* projections show larger declines in residential consumption of natural gas from 2010 to 2035 (11 percent and 6 percent, respectively). The SEER projection for residential natural gas consumption shows a decrease of 4 percent from 2015 to 2025, then a partial recovery by 2035.

Table 26. Comparison of natural gas projections, 2015, 2025, and 2035 (trillion cubic feet, except where noted)

Projection	2010	AEO2012 Reference case	IHSGI	EVA	Deloitte	SEER	ExxonMobil	INFORUM
					2015			
Dry gas production[a]	21.58	23.65	23.81	23.80	24.52	23.66	24 00	24.29
Net imports	**2.58**	**1.73**	**1.62**	**2.20**	**1.30**	**1.73**	**1.20**	**--**
Pipeline	2.21	1.56	--	1.80	1.22	1.56	--	--
LNG	0.37	0.16	--	0.40	0.08	0.16	--	--
Consumption	**24.13**	**25.39**	**25.52**	**26.60**	**24.07[b]**	**26.05**	**25.00[c]**	**23.61[b]**
Residential	4.94	4.85	4.64	4.90	4.86	4.91	8.00[d]	4.87
Commercial	3.20	3.33	3.10	3.20	3.23	3.41	--	3.43
Industrial[e]	6.60	7.01	6.64	7.00	7.51	7.64	8 00	8.19
Electricity generators[f]	7.38	8.08	9.02	9.30	8.46	8.06	9 00	7.12
Others[g]	2.01	2.12	2.11	2.20	--	2.04	--	--
Henry Hub spot market price (2010 dollars per million Btu)	4.39	4.29	4.75	4.07	4.25	4.28		
End-use prices (2010 dollars per thousand cubic feet)								
Residential	11.36	10.56	11.82	--	--	11.68	--	--
Commercial	9.32	8.82	9.88	--	--	8.31	--	--
Industrial[h]	5.65	5.00	6.95	--	--	4.63	--	--
Electricity generators	5.25	4.65	5.20	--	--	5.17	--	--
					2025			
Dry gas production[a]	21.58	26.28	27.23	26.70	27.32	25.88	27 00	27.57
Net imports	**2.58**	**-0.79**	**2.13**	**1.30**	**0.38**	**0.29**	**1.50**	**--**
Pipeline	2.21	-0.13	--	0.90	0.29	1.03	--	--
LNG	0.37	-0.66	--	0.40	0.09	-0.74	--	--
Consumption	**24.13**	**25.53**	**29.39**	**29.00**	**26.36[b]**	**27.10**	**29.00[c]**	**23.43[b]**
Residential	4.94	4.76	4.53	5.00	5.05	4.71	8.00[d]	4.90
Commercial	3.20	3.44	3.15	3.30	3.46	3.53	--	3.60
Industrial[e]	6.60	7.14	6.52	7.70	7.58	7.47	8 00	8.20
Electricity generators[f]	7.38	7.87	12.78	10.50	10.27	9.27	13 00	6.74
Others[g]	2.01	2.31	2.42	2.50	--	2.12	--	--
Henry Hub spot market price (2010 dollars per million Btu)	4.39	5.63	4.82	6.47	5.80	6.29		
End-use prices (2010 dollars per thousand cubic feet)								
Residential	11.36	12.33	11.70	--	--	14.40	--	--
Commercial	9.32	10.27	9.81	--	--	10.68	--	--
Industrial[h]	5.65	6.19	6.99	--	--	6.96	--	--
Electricity generators	5.25	5.73	5.28	--	--	7.47	--	--

-- = not reported.
See notes at end of table.

(continued on next page)

With the exception of ExxonMobil, which shows a decline in U.S. production of domestic natural gas between 2030 and 2035, all the projections show increasing U.S. production of domestic natural gas over the projection period, although at different rates. The highest level of natural gas production is projected by IHSGI, exceeding the ExxonMobil projection by 21 percent in 2035. Coupled with a significant decline in net pipeline imports, SEER, INFORUM, and the AEO2012 Reference case project a strong increase in the share of total U.S. natural gas supply accounted for by domestic production. The other projections show relatively stable and similar percentages for the contribution of domestic natural gas production to total supply, with the exception of IHSGI, which shows a notable increase in net imports after 2015. In all the projections, with the exception of EVA, net LNG imports remain below the 2010 level of 0.4 trillion cubic feet throughout the projection period. In all the projections, however, net pipeline imports decline from 2010 levels, with AEO2012, SEER, and Deloitte projecting more severe declines than EVA (only through 2030 since EVA does not show 2035).

The AEO2012 Reference case and SEER show similar levels of natural gas production and Henry Hub spot prices, both with increasing production and prices over time. EVA shows similar levels of natural gas production as the AEO2012 Reference case through 2025, but higher Henry Hub spot prices. IHSGI projects a larger increase in natural gas production but at relatively stable prices. In 2015, the Henry Hub spot price in the IHSGI projection is 11 percent higher than the price in the SEER projection; however, the SEER Henry Hub spot price quickly surpasses the IHSGI price, and it is 50 percent higher in 2035. Deloitte, ExxonMobil, and INFORUM did not include price projections.

Only IHSGI and SEER included delivered natural gas prices that can be compared with those in the AEO2012 Reference case [141]. However, there appear to be definitional differences in the projections, based on an examination of 2010 price levels. In particular,

Table 26. Comparison of natural gas projections, 2015, 2025, and 2035 (trillion cubic feet, except where noted) (continued)

Projection	2010	AEO2012 Reference case	Other projections					
			IHSGI	EVA	Deloitte	SEER	ExxonMobil	INFORUM
					2035			
Dry gas production[a]	21.58	27.93	31.35	--	27.87	27.00	26 00	30.71
Net imports	2.58	-1.36	2.36	--	0.14	-0.46	2.50	--
Pipeline	2.21	-0.70	--	--	0.07	0.28	--	--
LNG	0.37	-0.66	--	--	0.08	-0.74	--	--
Consumption	24.13	26.63	33.54	--	27.30[b]	27.24	29.00[c]	24.66[b]
Residential	4.94	4.64	4.38	--	5.03	4.80	7.00[d]	4.83
Commercial	3.20	3.60	3.18	--	3.60	3.64	--	3.83
Industrial[e]	6.60	7.00	6.35	--	7.31	7.30	8 00	8.09
Electricity generators[f]	7.38	8.96	16.90	--	11.37	9.37	14 00	7.90
Others[g]	2.01	2.43	2.72	--	--	2.13	--	--
Henry Hub spot market price (2010 dollars per million Btu)	4.39	7.37	5.13	7.26	6.63	7.70	--	--
End-use prices (2010 dollars per thousand cubic feet)								
Residential	11.36	14.33	11.81	--	--	17.15	--	--
Commercial	9.32	11.93	9.99	--	--	13.09	--	--
Industrial[h]	5.65	7.73	7.22	--	--	9.20	--	--
Electricity generators	5.25	7.37	5.62	--	--	9.75	--	--

-- = not reported.
[a]Does not include supplemental fuels.
[b]Does not includes lease, plant, and pipeline fuel and fuel consumed in natural gas vehicles.
[c]Does not includes lease, plant, and pipeline fuel.
[d]Natural gas consumed in the residential and commercial sectors.
[e]Includes consumption for industrial combined heat and power (CHP) plants and a small number of industrial electricity-only plants, and natural gas-to-liquids heat/power production; excludes consumption by nonutility generators.
[f]Includes consumption of energy by electricity-only and CHP plants whose primary business is to sell electricity, or electricity and heat, to the public. Includes electric utilities, small power producers, and exempt wholesale generators.
[g]Includes lease, plant, and pipeline fuel and fuel consumed in natural gas vehicles.
[h]The 2010 industrial natural gas price for IHSGI is $6.53.

the IHSGI industrial delivered natural gas price is difficult to compare. The industrial delivered natural gas price for 2010 in the IHSGI projection is $0.88 higher than the industrial price for 2010 in the *AEO2012* Reference case and $1.13 higher than the 2010 industrial price in the SEER projection (all prices in 2010 dollars per thousand cubic feet). From 2010 to 2035, the delivered price for electricity generators increases by 7 percent in the IHSGI projection, by 40 percent in the *AEO2012* Reference case, and by 86 percent in the SEER projection. The SEER projection also shows the largest increases in residential and commercial delivered prices, at 51 percent and 40 percent, respectively, over the same period. IHSGI shows the smallest increases in residential and commercial delivered prices over the projection period, at 4 percent and 7 percent, respectively. The *AEO2012* Reference case projects a 26-percent increase in residential delivered natural gas prices and a 28-percent increase in commercial prices.

6. Liquid fuels

In the *AEO2012* Reference case, the U.S. RAC for imported crude oil (in 2010 dollars) increases to $113.97 per barrel in 2015, $121.21 per barrel in 2025, and $132.95 per barrel in 2035 (Table 27). Prices are lower in the INFORUM projection, ranging from $91.78 per barrel in 2015 to $116.76 per barrel in 2035. BP, EVA, and Purvin & Gertz (P&G) did not report projections of RAC prices.

Domestic crude oil production increases from about 5.5 million barrels per day in 2010 to a peak of 6.7 million barrels per day in 2020, then declines to about 6.0 million barrels per day in 2035 in the *AEO2012* Reference case. Overall, the production level in 2035 is more than 9 percent higher than the 2010 level. The INFORUM projection shows a steady increase in production, to 5.8 million barrels per day in 2035. Domestic crude oil production decreases to 3.2 million barrels per day in 2035 in the P&G projection.

Supply from renewable sources increases to about 1.1 million barrels per day in 2015, almost 1.5 million barrels per day in 2025 (38.5 percent higher than the 2015 level), and more than 2.3 million barrels per day in 2035 (120.2 percent higher than the 2015 level) in the *AEO2012* Reference case. In the BP projection, supplies from renewable sources, on an energy-equivalent basis, increase by 49.5 percent from 2015 to 2025. BP does not report supplies from renewable sources in 2035, and it is not included in the projections by EVA, INFORUM, and P&G.

Prices for both transportation diesel fuel and gasoline increase through 2035 in the *AEO2012* projection, with diesel prices higher than gasoline prices. INFORUM projects rising gasoline prices from 2015 levels but decreasing diesel prices, with the gasoline price consistently higher than the diesel price. The BP, EVA, and P&G projections do not include delivered fuel prices.

7. Coal

Projections from EVA, IHSGI, INFORUM, IEA, ExxonMobil, and BP offer some opportunity to compare other coal outlooks with the *AEO2012* Reference case. Although many of the assumptions used in the other projections are unknown, ExxonMobil does assume a carbon tax, and EVA assumes some additional regulations affecting coal use that are not included in current laws. Such assumptions

Table 27. Comparison of liquids projections, 2015, 2025, and 2035 (million barrels per day, except where noted)

Projection	2010	AEO2012 Reference case	Other projections			
			BP[a]	EVA	INFORUM	P&G
			2015			
Average U.S. imported RAC (2010 dollars per barrel)	75.87	113.97	--	--	91.78	--
Average WTI price (2010 dollars per barrel)	79.39	116.91	--	82.24	--	98.75
Domestic production	**7.55**	**8.71**	**8.56**	**9.60**	**--**	**7.92**
Crude oil	5.47	6.15	--	6.90	5.43	5.43
Alaska	0.60	0.46	--	0.40	--	0.54
NGL	2.07	2.56	--	2.70	--	2.49
Total net imports	**9.56**	**8.27**	**8.20**	**--**	**9.81**	**--**
Crude oil	9.17	8.52	--	--	8.59	9.69
Products	0.39	-0.25	--	--	1.22	--
Liquids consumption	19.17	19.10	18.26	--	20.04[b]	17.69
Net petroleum import share of liquids supplied (percent)	50	43	45	--	--	--
Supply from renewable sources	0.90	1.05	1.24	--	--	--
Transportation product prices (2010 dollars per gallon)						
Gasoline	2.76	3.54	--	--	3.85	--
Diesel	3.00	3.78	--	--	3.60	--

-- = not reported.
See notes at end of table.

(continued on next page)

probably contribute to lower coal consumption levels compared with historical levels and the *AEO2012* Reference case. BP, EVA, ExxonMobil, and IHSGI have the most pessimistic views of coal use, with consumption declining over their respective projection horizons. In contrast, both the *AEO2012* and INFORUM projections show rising coal consumption after an initial decline. INFORUM's projection for coal consumption in 2035 is the highest—12 percent higher than in the *AEO2012* Reference case (Table 28).

Because most coal consumed in the United States is used for electricity generation, the outlooks with the largest declines in total coal consumption also show similar declines in coal use for electric power generation. The *AEO2012* Reference case has the most pessimistic outlook for coal consumption in the power sector in 2015; however, while coal use in the electric power sector recovers after 2015 in the *AEO2012* Reference case, it continues to decline in the EVA, IHSGI, ExxonMobil, and BP projections. ExxonMobil—which includes a carbon tax—shows the largest decline in coal use for electricity generation compared with the other projections,

Table 27. Comparison of liquids projections, 2015, 2025, and 2035 (million barrels per day, except where noted) (continued)

Projection	2010	AEO2012 Reference case	Other projections			
			BP[a]	EVA	INFORUM	P&G
			2025			
Average U.S. imported RAC (2010 dollars per barrel)	75.87	121.21	--	--	113.35	--
Average WTI price (2010 dollars per barrel)	79.39	132.56	--	89.07	--	106.47
Domestic production	**7.55**	**9.41**	**9.20**	**11.10**	--	**7.37**
Crude oil	5.47	6.40	--	7.10	5.74	4.26
Alaska	0.60	0.40	--	0.00	--	0.45
NGL	2.07	3.01	--	4.00	--	3.11
Total net imports	**9.56**	**7.12**	**5.87**	--	**9.89**	--
Crude oil	9.17	7.24	--	--	8.31	10.71
Products	0.39	-0.12	--	--	1.58	--
Liquids consumption	19.17	19.20	17.30	--	20.38[b]	17.39
Net petroleum import share of liquids supplied (percent)	50	37	34	--	--	--
Supply from renewable sources	0.90	1.45	1.85	--	--	--
Transportation product prices (2010 dollars per gallon)						
Gasoline	2.76	3.85	--	--	4.36	--
Diesel	3.00	4.17	--	--	3.46	--
			2035			
Average U.S. imported RAC (2010 dollars per barrel)	75.87	132.95	--	--	116.76	--
Average WTI price (2010 dollars per barrel)	79.39	144.98	--	102.11	--	107.37
Domestic production	**7.55**	**9.00**	--	--	--	--
Crude oil	5.47	5.99	--	--	5.80	3.23
Alaska	0.60	0.27	--	--	--	0.41
NGL	2.07	3.01	--	--	--	--
Total net imports	**9.56**	**7.18**	--	--	**10.36**	--
Crude oil	9.17	7.52	--	--	8.49	11.68
Products	0.39	-0.34	--	--	1.88	--
Liquids consumption	19.17	19.90	--	--	21.31[b]	17.38
Net petroleum import share of liquids supplied (percent)	50	36	--	--	--	--
Supply from renewable sources	0.90	2.31	--	--	--	--
Transportation product prices (2010 dollars per gallon)						
Gasoline	2.76	4.03	--	--	4.49	--
Diesel	3.00	4.44	--	--	3.30	--

-- = not reported.
[a]For BP, liquids production data were converted from million metric tons to barrels at 8.067817 barrels per metric ton, and liquids demand data were converted at 8.162674 barrels per metric ton. One metric ton equals 1,000 kilograms.
[b]For INFORUM, liquids demand data were converted from quadrillion Btus to barrels at 187.84572 million barrels per quadrillion Btu.

Comparison with other projections

and coal consumption in the BP outlook also declines from 2010 levels. The EVA projection for coal consumption in the electric power sector in 2030 is 13 percent lower than the 2010 level, whereas coal consumption returns to 2010 levels in 2030 in the *AEO2012* Reference case. The IEA projection for coal consumption in the electric power sector in 2035, at 19.2 quadrillion Btu, is similar to the *AEO2012* Reference case projection.

EVA, IHSGI, and the *AEO2012* Reference case all project declining use of coal at coking plants through 2030, with EVA including the most pessimistic outlook. INFORUM's industrial coal consumption figure, which appears to include both coking coal consumption

Table 28. Comparison of coal projections, 2015, 2025, 2030, and 2035 (million short tons, except where noted)

Projection	2010	AEO2012 Reference case		Other projections					
		(million short tons)	(quadrillion Btu)	EVA[a]	IHSGI	INFORUM	IEA[b]	Exxon-Mobil[c]	BP[b]
				(million short tons)			(quadrillion Btu)		
2015									
Production	1,084	993	20.24	1,017	1,144	970	--	--	22.00
East of the Mississippi	446	407	--	411	--	--	--	--	--
West of the Mississippi	638	586	--	606	--	--	--	--	--
Consumption									
Electric power	975	839	16.15	871	1,002	--	--	17.00	18.68
Coke plants	21	22	--	20	21	--	--	--	--
Coal-to-liquids	0	0	--	--	--	--	--	--	--
Other industrial/buildings	55	53	1.66[d]	42	50	1.81[d]	--	--	--
Total consumption (quadrillion Btu)[e]	20.76	--	17.80	--	--	--	--	19.00	20.53
Total consumption (million short tons)	1,051	914	--	933	1,073	916[f]	--	--	--
Net coal exports	64	95	2.38	100	70	54	--	--	1.48
Exports	82	110	2.73	104	89	70	--	--	1.48
Imports	18	15	0.35	4	19	16	--	--	0.00[g]
Minemouth price									
2010 dollars per ton	35.61	42.08	--	--	--	32.80	--	--	--
2010 dollars per Btu	1.76	2.08	--	--	--	--	--	--	--
Average delivered price to electricity generators									
2010 dollars per ton	44.27	45.17	--	--	--	42.72	--	--	--
2010 dollars per Btu	2.26	2.35	--	--	2.39	--	--	--	--
2025									
Production	1,084	1,118	22.25	995	1,038	1,114	--	--	19.40
East of the Mississippi	446	383	--	403	--	--	--	--	--
West of the Mississippi	638	735	--	592	--	--	--	--	--
Consumption									
Electric power	975	952	18.06	847	927	--	--	15.00	16.16
Coke plants	21	19	--	17	19	--	--	--	--
Coal-to-liquids	0	38	--	--	--	--	--	--	--
Other industrial/buildings	55	55	1.63[d]	33	39	2.07[d]	--	--	--
Total consumption (quadrillion Btu)[e]	20.76	--	20.02	--	--	--	--	15.00	17.70
Total consumption (million short tons)	1,051	1,063	--	897	986	1,072[f]	--	--	--
Net coal exports	64	71	1.79	113	53	42	--	--	1.70
Exports	82	115	2.82	118	73	75	--	--	1.70
Imports	18	44	1.03	4	20	33	--	--	0.00[g]
Minemouth price									
2010 dollars per ton	35.61	44.05	--	--	--	33.43	--	--	--
2010 dollars per Btu	1.76	2.23	--	--	--	--	--	--	--
Average delivered price to electricity generators									
2010 dollars per ton	44.27	48.13	--	--	--	43.58	--	--	--
2010 dollars per Btu	2.26	2.54	--	--	2.48	--	--	--	--

-- = not reported.
See notes at end of table.

(continued on next page)

Table 28. Comparison of coal projections, 2015, 2025, 2030, and 2035 (million short tons, except where noted) (continued)

Projection	2010	AEO2012 Reference case		Other projections					
		(million short tons)	(quadrillion Btu)	EVA[a]	IHSGI	INFORUM	IEA[b]	Exxon-Mobil[c]	BP[b]
				(million short tons)			(quadrillion Btu)		
2030									
Production	1,084	1,166	23.22	992	984	1,177	--	--	17.99
East of the Mississippi	446	409	--	396	--	--	--	--	--
West of the Mississippi	638	757	--	596	--	--	--	--	--
Consumption									
Electric power	975	975	18.55	847	885	--	19.2	13.00	14.76
Coke plants	21	18	--	16	19	--	--	--	--
Coal-to-liquids	0	51	--	--	--	--	--	--	--
Other industrial/buildings	55	55	1.60[d]	31	35	2.37[d]	1.1[b]	--	--
Total consumption (quadrillion Btu)[e]	20.76	--	20.59	--	--	--	--	13.00	16.18
Total consumption (million short tons)	1,051	1,099	--	894	938	1,156[f]	--	--	--
Net coal exports	64	83	2.08	113	47	41	--	--	1.81
Exports	82	117	2.85	118	68	74	--	--	1.81
Imports	18	33	0.77	5	20	53	--	--	0.00[g]
Minemouth price									
2010 dollars per ton	35.61	47.28	--	--	--	33.21	--	--	--
2010 dollars per Btu	1.76	2.39	--	--	--	--	--	--	--
Average delivered price to electricity generators									
2010 dollars per ton	44.27	50.56	--	--	--	43.31	--	--	--
2010 dollars per Btu	2.26	2.66	--	--	2.52	--	--	--	--
2035									
Production	1,084	1,212	24.14	--	926	1,284	--	--	--
East of the Mississippi	446	431	--	--	--	--	--	--	--
West of the Mississippi	638	781	--	--	--	--	--	--	--
Consumption									
Electric power	975	998	19.03	--	837	--	19.2	11.00	--
Coke plants	21	17	--	--	18	--	--	--	--
Coal-to-liquids	0	67	--	--	--	--	--	--	--
Other industrial/buildings	55	56	1.58[d]	--	31	2.70[d]	1.1	--	--
Total consumption (quadrillion Btu)[e]	20.76	--	21.15	--	--	--	--	11.00	--
Total consumption (million short tons)	1,051	1,137	--	--	886	1,277[f]	--	--	--
Net coal exports	64	94	2.31	--	42	8	--	--	--
Exports	82	129	3.13	--	63	71	--	--	--
Imports	18	36	0.82	--	20	64	--	--	--
Minemouth price									
2010 dollars per ton	35.61	50.52	--	--	--	33.06	--	--	--
2010 dollars per Btu	1.76	2.56	--	--	--	--	--	--	--
Average delivered price to electricity generators									
2010 dollars per ton	44.27	53.31	--	--	--	43.13	--	--	--
2010 dollars per Btu	2.26	2.80	--	--	2.54	--	--	--	--

-- = not reported.
[a]Regulations known to be accounted for in the EVA projections include MATS, CSAPR, regulations for cooling-water intake structures under Section 316(b) of the Clean Water Act, and regulations for coal combustion residuals under authority of the Resource Conservation and Recovery Act.
[b]For IEA and BP, data were converted from millions of tons oil equivalent (toe) at 39.683 million Btu per toe.
[c]ExxonMobil projections include a carbon tax.
[d]Coal consumption in quadrillion Btu. INFORUM's value appears to include coal consumption at coke plants. To facilitate comparison the AEO2012 value also includes coal consumption at coke plants.
[e]For AEO2012, excludes coal converted to coal-based synthetic liquids.
[f]Calculated as consumption = (production - exports + imports).
[g]Calculated as imports = (consumption - production + exports).

and coal use at industrial steam plants, is higher than projected in the *AEO2012* Reference case. EVA and IHSGI show declines in coal use in the industrial/buildings sector (excluding the coking sector), whereas the *AEO2012* outlook is more stable. According to ExxonMobil's projection, coal is consumed only for electricity generation after 2015, as implied consumption in all other sectors drops to zero. The *AEO2012* Reference case appears to be the only projection that includes coal use in CTL production.

Only EVA provides regional production information for comparison with the *AEO2012* Reference case. Despite much lower total coal consumption than in *AEO2012*, EVA's estimate of coal production east of the Mississippi is similar to that in the *AEO2012* Reference case. The differences in coal production are primarily in basins west of the Mississippi, where *AEO2012* projects 161 million more tons of coal production in 2030 than projected by EVA.

With respect to exports, two broad consensus groups are identifiable among the projections. The most optimistic projections are EVA and *AEO2012*, which show exports remaining above 100 million tons through 2030. However, EVA and *AEO2012* do differ, in that the *AEO2012* Reference case projects stronger growth for coking coal exports, and EVA projects stronger growth for thermal coal exports. The second group of projections, including BP, INFORUM, and IHSGI, shows a less optimistic outlook for U.S. coal exports. Coal exports in 2030 in the *AEO2012* Reference case are 1.0 quadrillion Btu higher than projected by BP. If BP's average heat rate for exports is assumed to be similar to that in *AEO2012*, BP's projected coal exports in 2030 are about 70 million tons, similar to the INFORUM and IHSGI projections for the same year. IHSGI's projection of exports is the lowest of this group, peaking in 2025 and then falling to 63 million tons in 2035.

The outlook for coal imports varies considerably across the projections, with little consensus. In the EVA projection, imports drop to a negligible 4 million tons early on and remain at that level for the balance of the projection; and in the BP projection, there are no coal imports to the United States after 2015. In the IHSGI projection, coal imports vary little through 2035. In 2035, coal imports in the *AEO2012* Reference case are just over one-half those in the INFORUM outlook.

Coal price comparisons can be made only for the *AEO2012*, IHSGI, and INFORUM projections. *AEO2012* includes the highest minemouth coal prices, which rise by 42 percent from 2010 to 2035. IHSGI and the *AEO2012* Reference case do project similar delivered coal prices to the electricity sector through 2020, but after 2020 IHSGI's prices change little, whereas prices in the *AEO2012* Reference case continue to rise. The difference may indicate that IHSGI's more pessimistic coal consumption outlook has less to do with high coal prices than with other factors. Similarly, INFORUM's delivered coal price to the electricity sector falls and then remains constant at around 2015 levels through 2035, lower than the price in 2010.

Endnotes for Comparison with other projections

Links current as of June 2012

140. ExxonMobil's projection for residential consumption includes commercial consumption.
141. SEER's prices include a carbon tax.

List of acronyms

AB	Assembly Bill
AB32	California Assembly Bill 32
ACI	Activated carbon injection
AEO	*Annual Energy Outlook*
AEO2012	*Annual Energy Outlook 2012*
ANWR	Arctic National Wildlife Refuge
ARRA2009	American Recovery and Reinvestment Act of 2009
ASHRAE	American Society of Heating, Refrigerating, and Air-Conditioning Engineers
Blue Chip	Blue Chip Consensus
BTL	Biomass-to-liquids
Btu	British thermal unit
CAFE	Corporate average fuel economy
CAIR	Clean Air Interstate Rule
CARB	California Air Resources Board
CBO	Congressional Budget Office
CBTL	Coal- and biomass-to-liquids
CCS	Carbon capture and storage
CHP	Combined heat and power
CI	Carbon intensity
CMM	Coal Market Module
CNG	Compressed natural gas
CO_2	Carbon dioxide
CO_2-EOR	Carbon dioxide-enhanced oil recovery
CSAPR	Cross-State Air Pollution Rule
CTL	Coal-to-liquids
DG	Distributed generation
dge	Diesel gallon equivalent
DOE	U.S. Department of Energy
DSI	Direct sorbent injection
E10	Motor gasoline blend containing up to 10 percent ethanol
E15	Motor gasoline blend containing up to 15 percent ethanol
E85	Motor fuel containing up to 85 percent ethanol
EERE	Energy Efficiency and Renewable Energy
EIA	U.S. Energy Information Administration
EIEA2008	Energy Improvement and Extension Act of 2008
EISA2007	Energy Independence and Security Act of 2007
EOR	Enhanced oil recovery
EPA	U.S. Environmental Protection Agency
EPACT05	Energy Policy Act of 2005
EUR	Estimated ultimate recovery
EV	Electric vehicle
EVA	Energy Ventures Analysis
FEMP	Federal Energy Management Program
FFV	Flex-fuel vehicle
FGD	Flue gas desulfurization
GDP	Gross domestic product
GHG	Greenhouse gas
GTL	Gas-to-liquids
GVWR	Gross vehicle weight rating
HAP	Hazardous air pollutant
HB	House Bill
HCl	Hydrogen chloride
HD	Heavy-duty
HDV	Heavy-duty vehicle
HEV	Hybrid electric vehicle
Hg	Mercury
ICE	Internal combustion engine
IDM	Industrial Demand Module
IEA	International Energy Agency
IECC2006	2006 International Energy Conversion Code
IEM	International Energy Module
IHSGI	IHS Global Insight
INFORUM	Interindustry Forecasting Project at the University of Maryland
IOU	Invester-owned utility
IREC	Interstate Renewable Energy Council
ITC	Investment tax credit
LCFS	Low Carbon Fuel Standard
LDV	Light-duty vehicle
LED	Light-emitting diode
LFMM	Liquid Fuels Market Module
LNG	Liquefied natural gas
MATS	Mercury and Air Toxics Standards
MAM	Macroeconomic Activity Module
mmt	Million metric tons
$MMTCO_2e$	Million metric tons carbon dioxide equivalent
mpg	Miles per gallon
MSRP	Manufacturer's suggested retail price
MY	Model year
NAICS	North American Industry Classification System
NEMS	National Energy Modeling System
NERC	North American Electric Reliability Corporation
NGL	Natural gas liquids
NGPL	Natural gas plant liquids
NGTDM	Natural Gas Transmission and Distribution Module
NGV	Natural gas vehicle
NHTSA	National Highway Traffic Safety Administration
NO_x	Nitrogen oxides
NRC	U.S. Nuclear Regulatory Commission
OECD	Organization for Economic Cooperation and Development
OMB	Office of Management and Budget
OPEC	Organization of the Petroleum Exporting Countries
P&G	Purvin & Gertz
PADD	Petroleum Administration for Defense District
PCs	Personal computers
PHEV	Plug-in hybrid electric vehicle
PM	Particulate matter
$PM_{2.5}$	Particulate matter less than 2.5 microns diameter
PMM	Petroleum Market Module
PTC	Production tax credit
PV	Solar photovoltaic
RAC	U.S. Refiner Acquisition Cost
RECS	Residential Energy Consumption Survey
RFM	Renewable Fuels Module
RFS	Renewable fuel standard
RGGI	Regional Greenhouse Gas Initiative
RPS	Renewable portfolio standard
SB	Senate Bill
SCR	Selective catalytic reduction
SEER	Strategic Energy and Economic Research, Inc.
SEIA	Solar Energy Industries Association
SNCR	Selective noncatalytic reduction
SO_2	Sulfur dioxide
STEO	Short-Term Energy Outlook
TAPS	Trans-Alaska Pipeline System
TRR	Technically recoverable resource
UEC	Unit energy consumption
UPS	Uninterruptible power supply
USGS	United States Geological Survey
VIUS	Vehicle Inventory and Use Survey
VMT	Vehicle miles traveled
WTI	West Texas Intermediate

Notes and sources

Table notes and sources

Table 1. HD National Program vehicle regulatory categories: U.S. Environmental Protection Agency and National Highway Traffic Safety Administration, "Greenhouse Gas Emissions Standards and Fuel Efficiency Standards for Medium- and Heavy-Duty Engines and Vehicles: Final Rule," *Federal Register*, Vol. 76, No. 179 (Washington, DC: September 15, 2011), pp. 57106-57513, website www.gpo.gov/fdsys/pkg/FR-2011-09-15/html/2011-20740.htm.

Table 2. HD National Program standards for combination tractor greenhouse gas emissions and fuel consumption: U.S. Environmental Protection Agency and National Highway Traffic Safety Administration, *Greenhouse Gas Emissions Standards and Fuel Efficiency Standards for Medium- and Heavy-Duty Engines and Vehicles*, 49 CFR Parts 523, 534, and 535, RIN 2060-AP61; 2127-AK74, Federal Register Notice Vol. 76, No. 179, Thursday, September 15, 2011.

Table 3. HD National Program standards for vocational vehicle greenhouse gas emissions and fuel consumption: U.S. Environmental Protection Agency and National Highway Traffic Safety Administration, *Greenhouse Gas Emissions Standards and Fuel Efficiency Standards for Medium- and Heavy-Duty Engines and Vehicles*, 49 CFR Parts 523, 534, and 535, RIN 2060-AP61; 2127-AK74, Federal Register Notice Vol. 76, No. 179, Thursday, September 15, 2011.

Table 4. Renewable portfolio standards in the 30 States with current mandates: U.S. Energy Information Administration, Office of Energy Analysis. Based on a review of enabling legislation and regulatory actions from the various States of policies identified by the Database of States Incentives for Renewable Energy as of January 1, 2012, website www.dsireuse.org.

Table 5. Key analyses of interest from "Issues in focus" in recent *AEO*s: U.S. Energy Information Administration, *Annual Energy Outlook 2011*, DOE/EIA-0383(2011) (Washington, DC, April 2011); U.S. Energy Information Administration, *Annual Energy Outlook 2010*, DOE/EIA-0383(2010) (Washington, DC, April 2010); and U.S. Energy Information Administration, *Annual Energy Outlook 2009*, DOE/EIA-0383(2009) (Washington, DC, March 2009).

Table 6. Key assumptions for the residential sector in the *AEO2012* Integrated Demand Technology case: Projections: AEO2012 National Energy Modeling System, runs FROZTECH.D030812A, HIGHTECH.D032812A, and BESTTECH.D032812A.

Table 7. Key assumptions for the commercial sector in the *AEO2012* Integrated Demand Technology case: Projections: AEO2012 National Energy Modeling System, runs FROZTECH.D030812A, HIGHTECH.D032812A, and BESTTECH.D032812A.

Table 8. Estimated average fuel economy and greenhouse gas emissions standards proposed for light-duty vehicles, model years 2017-2025: U.S. Environmental Protection Agency and National Highway Transportation Safety Administration, "2017 and Later Model Year Light-Duty Vehicle Greenhouse Gas Emissions and Corporate Average Fuel Economy Standards: Proposed Rule," *Federal Register*, Vol. 76, No. 231 (Washington, DC: December 1, 2011), website www.nhtsa.gov/staticfiles/rulemaking/pdf/cafe/2017-25_CAFE_NPRM.pdf.

Table 9. Vehicle types that do not rely solely on a gasoline internal combustion engine for motive and accessory power: U.S. Energy Information Administration, Office of Energy Analysis.

Table 10. Description of battery-powered electric vehicles: U.S. Energy Information Administration, Office of Energy Analysis.

Table 11. Comparison of operating and incremental costs of battery electric vehicles and conventional gasoline vehicles: U.S. Energy Information Administration, Office of Energy Analysis.

Table 12. Summary of key results from the Reference, High Nuclear, and Low Nuclear cases, 2010-2035: History: U.S. Energy Information Administration, *Annual Energy Review 2010*, DOE/EIA-0384 (Washington, DC, October 2011). **Projections:** AEO2012 National Energy Modeling System, runs REF2012.D020112C, HINUC12.D022312A and LOWNUC12.D022312b.

Table 13. Alaska North Slope wells completed during 2010 in selected oil fields: Alaska Oil and Gas Conservation Commission, Public Databases Website at doa.alaska.gov/ogc/publicdb.html. The North Slope well total includes exploration wells, water disposal wells, service wells, etc. The Alpine field is the primary field within the Colville River Unit.

Table 14. Unproved technically recoverable resource assumption by basin: U.S. Energy Information Administration, Office of Energy Analysis.

Table 15. *AEO2012* unproved technically recoverable resources for selected shale gas plays as of January 1, 2010: U.S. Energy Information Administration, Office of Energy Analysis. **Note:** Average well spacing, percent of area untested, and percent of area with potential have been rounded to the nearest unit.

Table 16. *AEO2012* unproved technically recoverable tight oil resources as of January 1, 2010: U.S. Energy Information Administration, Office of Energy Analysis. **Note:** Average well spacing, percent of area untested, and percent of area with potential have been rounded to the nearest unit.

Table 17. Estimated ultimate recovery for selected shale gas plays in three *AEO*s: Projections: AEO2012 National Energy Modeling System, runs REF2012.D020112C, AEO2011 National Energy Modeling System, runs REF2011.D0209A, and AEO2010 National Energy Modeling System, runs REF2010.D111809A.

Table 18. Petroleum supply, consumption, and prices in four cases, 2020 and 2035: **History:** Crude oil lower 48 average wellhead prices: U.S. Energy Information Administration, *Petroleum Marketing Annual 2009*, DOE/EIA-0487(2009) (Washington, DC, August 2010). Lower 48 onshore, lower 48 offshore, and Alaska crude oil production: U.S. Energy Information Administration, *Petroleum Supply Annual 2010*, DOE/EIA-0340(2010)/1 (Washington, DC, July 2011). **Projections:** AEO2012 National Energy Modeling System, runs REF2012.D020112C, REF2012.LEUR12.D022112A, REF2012.HEUR12.D022112A, and HTRR12.D050412A.

Table 19. Natural gas prices, supply, and consumption in four cases, 2020 and 2035: **History:** Alaska and Lower 48 natural gas production, net imports, and other consumption: U.S. Energy Information Administration, *Natural Gas Monthly*, DOE/EIA-0130(2011/07) (Washington, DC, July 2011). Other production: U.S. Energy Information Administration, Office of Energy Analysis. Consumption by sector based on: U.S. Energy Information Administration, *Annual Energy Review 2010*, DOE/EIA-0384(2010) (Washington, DC, October 2011). Henry Hub natural gas prices: U.S. Energy Information Administration, Short-Term Energy Outlook Query System, Monthly Natural Gas Data, Variable NGHHUUS. **Projections:** AEO2012 National Energy Modeling System, runs REF2012.D020112C, REF2012.LEUR12.D022112A, REF2012.HEUR12.D022112A, and HTRR12.D050412A.

Table 20. Marcellus unproved technically recoverable resources in *AEO2012* (as of January 1, 2010): U.S. Energy Information Administration, Office of Energy Analysis. **Note:** Average well spacing, percent of area untested, and percent of area with potential have been rounded to the nearest unit.

Table 21. Marcellus unproved technically recoverable resources: *AEO2011*, USGS 2011, and *AEO2012*: **Projections:** *AEO2011*: AEO2011 National Energy Modeling System, run REF2011.D0209A; USGS 2011: USGS 2011 Open-File Report 2011-1298, website pubs.usgs.gov/of/2011/1298; and Fact Sheet 2011-3092, website pubs.usgs.gov/fs/2011/3092; *AEO2012*: AEO2012 National Energy Modeling System, run REF2012.D020112C. **Note:** Average well spacing, percent of area untested, and percent of area with potential have been rounded to the nearest unit.

Table 22. Projections of average annual economic growth, 2010-2035: *AEO2012* **(Reference case):** AEO2012 National Energy Modeling System, run AEO2012.REF2012.D020112C. *AEO2011* **(Reference case):** AEO2011 National Energy Modeling System, run AEO2011.REF2011.D020911A. **IHSGI:** IHS Global Insight, *30-year U.S. and Regional Economic Forecast* (Lexington, MA, November 2011), website www.ihs.com/products/global-insight/index.aspx (subscription site). **OMB:** Office of Management and Budget, *Fiscal Year 2013 Budget of the U.S. Government* (Washington, DC, February 13, 2012), website www.whitehouse.gov/sites/default/files/omb/budget/fy2013/assets/budget.pdf. **CBO:** Congressional Budget Office, *The Budget and Economic Outlook: Fiscal Years 2012 to 2022* (Washington, DC, January 31, 2012), website www.cbo.gov/publication/42905. **INFORUM:** "Inforum Lift (Long-term Interindustry Forecasting Tool) Model" (College Park, MD, February 2012), website inforumweb.umd.edu/services/models/lift.html. **SSA:** Social Security Administration, *The 2011 Annual Report of the Board of Trustees of the Federal Old-Age And Survivors Insurance And Federal Disability Insurance Trust Funds* (U.S. Government Printing Office: Washington, DC, May 13, 2011), website www.ssa.gov/OACT/TR/2011/tr2011.pdf. **IEA (2011):** International Energy Agency, *World Energy Outlook 2011* (Paris, France, November 2011), website www.worldenergyoutlook.org. **Blue Chip Consensus:** *Blue Chip Economic Indicators* (Aspen Publishers, October 2011), website www.aspenpublishers.com/Topics/Banking-Law-Finance-Economic-Forecast/. **ExxonMobil:** ExxonMobil Corporation, *The Outlook for Energy: A View to 2040* (Irving, TX, 2012), website www.exxonmobil.com/Corporate/energy_outlook.aspx. **SEER:** Strategic Energy and Economic Research, Inc., e-mail from Ron Denhardt (February 21, 2012).

Table 23. Projections of oil prices, 2015-2035: *AEO2012* **(Reference case):** AEO2012 National Energy Modeling System, run AEO2012.REF2012.D020112C. *AEO2011* **(Reference case):** AEO2011 National Energy Modeling System, run AEO2011.REF2011.D020911A. **EVA:** Energy Ventures Analysis, Inc., e-mail from Anthony Petruzzo (January 26, 2012). **IEA (Current Policies Scenario):** International Energy Agency, *World Energy Outlook 2011* (Paris, France, November 2011), website www.worldenergyoutlook.org. **INFORUM:** "Inforum Lift (Long-term Interindustry Forecasting Tool) Model" (College Park, MD, February 2012), website inforumweb.umd.edu/services/models/lift.html. **IHSGI:** IHS Global Insight, *30-year U.S. and Regional Economic Forecast* (Lexington, MA, November 2011), website www.ihs.com/products/global-insight/index.aspx (subscription site). **P&G:** Purvin and Gertz, Inc., *Global Petroleum Market Outlook 2011* (Houston, TX, March 2011), website www.purvingertz.com/pubs.cfm?Area=1 (subscription site). **SEER:** Strategic Energy & Economic Research, Inc., e-mail from Ron Denhardt (February 21, 2012).

Table 24. Projections of energy consumption by sector, 2010-2035: *AEO2012* **(Reference case):** AEO2012 National Energy Modeling System, run AEO2012.REF2012.D020112C. **INFORUM:** "Inforum Lift (Long-term Interindustry Forecasting Tool) Model" (College Park, MD, February 2012), website inforumweb.umd.edu/services/models/lift.html. **IHSGI:** IHS Global Insight, *30-year U.S. and Regional Economic Forecast* (Lexington, MA, November 2011), website www.ihs.com/products/global-insight/index.aspx (subscription site). **ExxonMobil:** ExxonMobil Corporation, *The Outlook for Energy: A View to 2040* (Irving, TX, 2012), website www.exxonmobil.com/Corporate/energy_outlook.aspx. **IEA:** International Energy Agency, *World Energy Outlook 2011* (Paris, France, November 2011), website www.worldenergyoutlook.org. **BP:** BP, Inc., e-mail from Mark Finley (January 15, 2012).

Table 25. Comparison of electricity projections, 2010, 2015, 2025, and 2035: *AEO2012* **(Reference case):** AEO2012 National Energy Modeling System, run AEO2012.REF2012.D020112C. **EVA:** Energy Ventures Analysis, Inc., e-mail from Anthony Petruzzo (January 26, 2012). **IHSGI:** IHS Global Insight, *30-year U.S. and Regional Economic Forecast* (Lexington, MA, November 2011), website www.ihs.com/products/global-insight/index.aspx (subscription site). **INFORUM:** "Inforum Lift (Long-term Interindustry Forecasting Tool) Model" (College Park, MD, February 2012), website inforumweb.umd.edu/services/models/lift.html.

Table 26. Comparison of natural gas projections, 2010, 2015, 2025, and 2035: *AEO2012* **(Reference case):** AEO2012 National Energy Modeling System, run AEO2012.REF2012.D020112C. **IHSGI:** IHS Global Insight, *30-year U.S. and Regional Economic Forecast* (Lexington, MA, November 2011), website www.ihs.com/products/global-insight/index.aspx (subscription site). **EVA:** Energy Ventures Analysis, Inc., e-mail from Anthony Petruzzo (January 26, 2012). **Deloitte:** Deloitte LLP, e-mail from Tom Choi (January 26, 2012). **SEER:** Strategic Energy and Economic Research, Inc., e-mail from Ron Denhardt (February 21, 2012). **ExxonMobil:** ExxonMobil Corporation, *The Outlook for Energy: A View to 2040* (Irving, TX, 2012), website www.exxonmobil.com/Corporate/energy_outlook.aspx. **INFORUM:** "Inforum Lift (Long-term Interindustry Forecasting Tool) Model" (College Park, MD, February 2012), website inforumweb.umd.edu/services/models/lift.html.

Table 27. Comparison of liquids projections, 2010, 2015, 2025, and 2035: *AEO2012* **(Reference case):** AEO2012 National Energy Modeling System, run AEO2012.REF2012.D020112C. **BP:** BP, Inc., e-mail from Mark Finley (January 15, 2012). **EVA:** Energy Ventures Analysis, Inc., e-mail from Anthony Petruzzo (January 26, 2012). **IHSGI:** IHS Global Insight, *30-year U.S. and Regional Economic Forecast* (Lexington, MA, November 2011), website www.ihs.com/products/global-insight/index.aspx (subscription site). **INFORUM:** "Inforum Lift (Long-term Interindustry Forecasting Tool) Model" (College Park, MD, February 2012), website inforumweb.umd.edu/services/models/lift.html. **P&G:** Purvin and Gertz, Inc., *Global Petroleum Market Outlook 2011* (Houston, TX, March 2011), website www.purvingertz.com/pubs.cfm?Area=1 (subscription site).

Table 28. Comparison of coal projections, 2010, 2015, 2025, and 2035: *AEO2012* **(Reference case):** AEO2012 National Energy Modeling System, run AEO2012.REF2012.D020112C. **EVA:** Energy Ventures Analysis, Inc., e-mail from Anthony Petruzzo (January 26, 2012). **IHSGI:** IHS Global Insight, *30-year U.S. and Regional Economic Forecast* (Lexington, MA, November 2011), website www.ihs.com/products/global-insight/index.aspx (subscription site). **INFORUM:** "Inforum Lift (Long-term Interindustry Forecasting Tool) Model" (College Park, MD, February 2012), website inforumweb.umd.edu/services/models/lift.html. **IEA:** International Energy Agency, *World Energy Outlook 2011* (Paris, France, November 2011), website www.worldenergyoutlook.org. **BP:** BP, Inc., e-mail from Mark Finley (January 15, 2012). **ExxonMobil:** ExxonMobil Corporation, *The Outlook for Energy: A View to 2040* (Irving, TX, 2012), website www.exxonmobil.com/Corporate/energy_outlook.aspx. **BP:** BP, Inc., e-mail from Mark Finley (January 15, 2012).

Figure notes and sources

Figure 1. Energy use per capita and per dollar of gross domestic product, 1980-2035: History: U.S. Energy Information Administration, *Annual Energy Review 2010*, DOE/EIA-0384(2010) (Washington, DC, October 2011). **Projections:** AEO2012 National Energy Modeling System, run REF2012.D020112C.

Figure 2. U.S. production of tight oil in four cases, 2000-2035: History: U.S. Energy Information Administration, *Annual Energy Review 2010*, DOE/EIA-0384(2010) (Washington, DC, October 2011). **Projections:** AEO2012 National Energy Modeling System, runs REF2012.D020112C, REF2012.LEUR12.D02212A, REF2012.HEUR12.D02212A, and REF2012.HTRR12.D050412A.

Figure 3. U.S. dependence on imported petroleum and other liquids, 1970-2035: U.S. Energy Information Administration, *Annual Energy Review 2010*, DOE/EIA-0384(2010) (Washington, DC, October 2011). **Projections:** AEO2012 National Energy Modeling System, runs REF2012.D020112C.

Figure 4. Total U.S. natural gas production, consumption, and net imports, 1990-2035: History: U.S. Energy Information Administration, *Annual Energy Review 2010*, DOE/EIA-0384(2010) (Washington, DC, October 2011). **Projection:** AEO2012 National Energy Modeling System, runs REF2012.D020112C.

Figure 5. Cumulative retirements of coal-fired generating capacity by NERC region in nine cases, 2010-2035: Projection: AEO2012 National Energy Modeling System, runs REF2012.D020112C, REF_R05.D030712A, REF2012.HEUR12.D022112A, REF2012.LEUR12.D022112A, HEUR12_R05.D022312A, HCCST12.D031312A, LCCST12.D031312A, HM2012.D022412A, and LM2012.D022412A.

Figure 6. U.S. energy-related carbon dioxide emissions by sector and fuel, 2005 and 2035: History: U.S. Energy Information Administration, *Annual Energy Review 2010*, DOE/EIA-0384(2010) (Washington, DC, October 2011). **Projection:** AEO2012 National Energy Modeling System, runs REF2012.D020112C.

Figure 7. HD National Program model year standards for diesel pickup and van greenhouse gas emissions and fuel consumption, 2014-2018: U.S. Environmental Protection Agency and National Highway Traffic Safety Administration, *Greenhouse Gas Emissions Standards and Fuel Efficiency Standards for Medium- and Heavy-Duty Engines and Vehicles*, 49 CFR Parts 523, 534, and 535, RIN 2060-AP61; 2127-AK74, Federal Register Notice Vol. 76, No. 179, Thursday, September 15, 2011.

Figure 8. HD National Program model year standards for gasoline pickup and van greenhouse gas emissions and fuel consumption, 2014-2018: U.S. Environmental Protection Agency and National Highway Traffic Safety Administration, *Greenhouse Gas Emissions Standards and Fuel Efficiency Standards for Medium- and Heavy-Duty Engines and Vehicles*, 49 CFR Parts 523, 534, and 535, RIN 2060-AP61; 2127-AK74, Federal Register Notice Vol. 76, No. 179, Thursday, September 15, 2011.

Figure 9. States covered by CSAPR limits on emissions of sulfur dioxide and nitrogen oxides: U.S. Environmental Protection Agency, *Cross-State Air Pollution Fact Sheet* (Washington, DC, July 2011), website www.epa.gov/airtransport/pdfs/CSAPRFactsheet.pdf.

Figure 10. Total combined requirements for State renewable portfolio standards, 2015-2035: Projections: AEO2012 National Energy Modeling System, runs REF2012.D020112C.

Figure 11. Total energy consumption in three cases, 2005-2035: **History:** U.S. Energy Information Administration, *Annual Energy Review 2010*, DOE/EIA-0384(2010) (Washington, DC, October 2011). **Projections:** AEO2012 National Energy Modeling System, runs REF2012.D020112C, NOSUNSET.D032112A, and EXTENDED.D050612B.

Figure 12. Consumption of petroleum and other liquids for transportation in three cases, 2005-2035: **History:** U.S. Energy Information Administration, *Annual Energy Review 2010*, DOE/EIA-0384(2010) (Washington, DC, October 2011). **Projections:** AEO2012 National Energy Modeling System, runs REF2012.D020112C, NOSUNSET.D032112A, and EXTENDED.D050612B.

Figure 13. Renewable electricity generation in three cases, 2005-2035: **History:** U.S. Energy Information Administration, *Annual Energy Review 2010*, DOE/EIA-0384(2010) (Washington, DC, October 2011). **Projections:** AEO2012 National Energy Modeling System, runs REF2012.D020112C, NOSUNSET.D032112A, and EXTENDED.D050612B.

Figure 14. Electricity generation from natural gas in three cases, 2005-2035: **History:** U.S. Energy Information Administration, *Annual Energy Review 2010*, DOE/EIA-0384(2010) (Washington, DC, October 2011). **Projections:** AEO2012 National Energy Modeling System, runs REF2012.D020112C, NOSUNSET.D032112A, and EXTENDED.D050612B.

Figure 15. Energy-related carbon dioxide emissions in three cases, 2005-2035: **History:** U.S. Energy Information Administration, *Annual Energy Review 2010*, DOE/EIA-0384(2010) (Washington, DC, October 2011). **Projections:** AEO2012 National Energy Modeling System, runs REF2012.D020112C, NOSUNSET.D032112A, and EXTENDED.D050612B.

Figure 16. Natural gas wellhead prices in three cases, 2005-2035: **History:** U.S. Energy Information Administration, *Annual Energy Review 2010*, DOE/EIA-0384(2010) (Washington, DC, October 2011). **Projections:** AEO2012 National Energy Modeling System, runs REF2012.D020112C, NOSUNSET.D032112A, and EXTENDED.D050612B.

Figure 17. Average electricity prices in three cases, 2005-2035: **History:** U.S. Energy Information Administration, *Annual Energy Review 2010*, DOE/EIA-0384(2010) (Washington, DC, October 2011). **Projections:** AEO2012 National Energy Modeling System, runs REF2012.D020112C, NOSUNSET.D032112A, and EXTENDED.D050612B.

Figure 18. Average annual oil prices in three cases, 1980-2035: **History:** U.S. Energy Information Administration, *Annual Energy Review 2010*, DOE/EIA-0384(2010) (Washington, DC, October 2011). **Projections:** AEO2012 National Energy Modeling System, runs REF2012.D020112C, LP2012.D022112A, and HP2012.D022112A.

Figure 19. World petroleum and other liquids production, 2000-2035: **History:** U.S. Energy Information Administration, *Annual Energy Review 2010*, DOE/EIA-0384(2010) (Washington, DC, October 2011). **Projections:** AEO2012 National Energy Modeling System, run REF2012.D020112C.

Figure 20. Residential and commercial delivered energy consumption in four cases, 2010-2035: **Projections:** AEO2012 National Energy Modeling System, runs REF2012.D020112C, FROZTECH.D030812A, HIGHTECH.D032812A, and BESTTECH.D032812A.

Figure 21. Cumulative reductions in residential energy consumption relative to the Integrated 2011 Demand Technology case, 2011-2035: **Projection:** AEO2012 National Energy Modeling System, run FROZTECH.D030812A, HIGHTECH.D032812A, and BESTTECH.D032812A.

Figure 22. Cumulative reductions in commercial energy consumption relative to the Integrated 2011 Demand Technology case, 2011-2035: **Projection:** AEO2012 National Energy Modeling System, run FROZTECH.D030812A, HIGHTECH.D032812A, and BESTTECH.D032812A.

Figure 23. Light-duty vehicle market shares by technology type in two cases, model year 2025: **Projections:** AEO2012 National Energy Modeling System, runs REF2012.D020112C and CAFEY.D032112A.

Figure 24. On-road fuel economy of the light-duty vehicle stock in two cases, 2005-2035: **History:** U.S. Energy Information Administration, *Annual Energy Review 2010*, DOE/EIA-0384(2010) (Washington, DC, October 2011). **Projections:** AEO2012 National Energy Modeling System, run REF2012.D020112C and CAFEY.D032112A.

Figure 25. Total transportation consumption of petroleum and other liquids in two cases, 2005-2035: **History:** U.S. Energy Information Administration, *Annual Energy Review 2010*, DOE/EIA-0384(2010) (Washington, DC, October 2011). **Projections:** AEO2012 National Energy Modeling System, run REF2012.D020112C and CAFEY.D032112A.

Figure 26. Total carbon dioxide emissions from transportation energy use in two cases, 2005-2035: **History:** U.S. Energy Information Administration, *Annual Energy Review 2010*, DOE/EIA-0384(2010) (Washington, DC, October 2011). **Projections:** AEO2012 National Energy Modeling System, run REF2012.D020112C and CAFEY.D032112A.

Figure 27. Cost of electric vehicle battery storage to consumers in two cases, 2012-2035: **Projections:** AEO2012 National Energy Modeling System, run REF2012.D020112C and BATTECH.D032112A. Note: U.S. Department of Energy Office of Energy Efficiency and Renewable Energy high-energy battery cost goal includes mark-up of 1.5 for retail price equivalency

Figure 28. Costs of electric drivetrain nonbattery systems to consumers in two cases, 2012-2035: **Projections:** AEO2012 National Energy Modeling System, run REF2012.D020112C and BATTECH.D032112A.

Figure 29. Total prices to consumers for compact passenger cars in two cases, 2015 and 2035: **Projections:** AEO2012 National Energy Modeling System, run REF2012.D020112C and BATTECH.D032112A.

Notes and sources

Figure 30. Total prices to consumers for small sport utility vehicles in two cases, 2015 and 2035: Projections: AEO2012 National Energy Modeling System, run REF2012.D020112C and BATTECH.D032112A.

Figure 31. Sales of new light-duty vehicles in two cases, 2015 and 2035: Projections: AEO2012 National Energy Modeling System, run REF2012.D020112C and BATTECH.D032112A.

Figure 32. Consumption of petroleum and other liquids, electricity, and total energy by light-duty vehicles in two cases, 2000-2035: History: Derived from U.S. Energy Information Administration, *Annual Energy Review 2010*, DOE/EIA-0384(2010) (Washington, DC, October 2011), Oak Ridge National Laboratory, *Transportation Energy Data Book*, Edition 30 and Annual (Oak Ridge, TN: 2011). **Projections:** AEO2012 National Energy Modeling System, runs REF2012.D020112C and BATTECH.D032112A.

Figure 33. Energy-related carbon dioxide emissions from light-duty vehicles in two cases, 2005-2035: History: Derived from U.S. Energy Information Administration, *Annual Energy Review 2010*, DOE/EIA-0384(2010) (Washington, DC: October 2011). **Projections:** AEO2012 National Energy Modeling System, runs REF2012.D020112C and BATTECH.D032112A.

Figure 34. U.S. spot market prices for crude oil and natural gas, 1997-2012: History: U.S. Energy Information Administration, Office of Energy Analysis based on Reuters data.

Figure 35. Distribution of annual vehicle-miles traveled by light-medium (Class 3) and heavy (Class 7 and 8) heavy-duty vehicles, 2002: Derived from U.S. Census Bureau, Vehicle Inventory and Use Survey, 2002, website www.census.gov/svsd/www/vius/2002.html.

Figure 36. Diesel and natural gas transportation fuel prices in the HDV Reference case, 2005-2035: History: Prices for diesel based on U.S. Energy Information Administration, *Petroleum Marketing Annual 2009*, DOE/EIA-0487(2009) (Washington, DC: August 2010). **Historical prices for natural gas transportation fuel and projections:** AEO2012 National Energy Modeling System, run NOSUBNGV12.D050412A.

Figure 37. Sales of new heavy-duty natural gas vehicles in two cases, 2008-2035: Projections: AEO2012 National Energy Modeling System, runs RFNGV12.D050412A and NOSUBNGV12.D050412A.

Figure 38. Natural gas fuel use by heavy-duty vehicles in tow cases, 2008-2035: Projections: AEO2012 National Energy Modeling System, runs RFNGV12.D050412A and NOSUBNGV12.D050412A.

Figure 39. Reduction in petroleum and other liquid fuels use by heavy-duty vehicles in the HD NGV Potential case compared with the HDV Reference case, 2010-2035: Projections: AEO2012 National Energy Modeling System, runs RFNGV12.D050412A and NOSUBNGV12.D050412A.

Figure 40. Diesel and natural gas transportation fuel prices in two cases, 2035: Projections: AEO2012 National Energy Modeling System, runs RFNGV12.D050412A and NOSUBNGV12.D050412A.

Figure 41. U.S. liquids fuels production industry: U.S. Energy Information Administration, Office of Energy Analysis.

Figure 42. Mass-based overview of the U.S. liquids fuels production industry in the LFMM case, 2000, 2011, and 2035: History: EIA, *Petroleum Supply Annual 2010*, DOE/EIA-0340(2010)/1 (Washington, DC, July 2011). **Projections:** AEO2012 National Energy Modeling System runs REF2012.D121011B and REF_LFMM.D050312A.

Figure 43. New regional format for EIA's Liquid Fuels Market Module: U.S. Energy Information Administration, Office of Energy Analysis.

Figure 44. RFS mandated consumption of renewable fuels, 2009-2022: *Federal Register*, "Regulation of Fuels and Fuel Additives: Changes to Renewable Fuel Standard Program", EPA Final Rule, March 26, 2010, website www.gpo.gov/fdsys/pkg/FR-2010-03-26/pdf/2010-3851.pdf.

Figure 45. Natural gas delivered prices to the electric power sector in three cases, 2010-2035: Projections: AEO2012 National Energy Modeling System, runs REF2012.D020112C, REF2012.LEUR12.D022112A, and REF2012.HEUR12.D022112A.

Figure 46. U.S. electricity demand in three cases, 2010-2035: Projections: AEO2012 National Energy Modeling System, runs REF2012.D020112C, LM2012.D022412A and HM2012.D022412A.

Figure 47. Cumulative retirements of coal-fired generating capacity by NERC region in nine cases, 2010-2035: Projection: AEO2012 National Energy Modeling System, runs REF2012.D020112C, REF_R05.D030712A, REF2012.HEUR12.D022112A, REF2012.LEUR12.D022112A, HEUR12_R05.D022312A, HCCST12.D031312A, LCCST12.D031312A, HM2012.D022412A, and LM2012.D022412A.

Figure 48. Electricity generation by fuel in eleven cases, 2010 and 2020: History: U.S. Energy Information Administration, *Annual Energy Review 2010*, DOE/EIA-0384(2010) (Washington, DC, October 2011). **Projections:** AEO2012 National Energy Modeling System, runs REF2012.D020112C, REF_R05.D030712A, REF2012.HEUR12.D022112A, REF2012.LEUR12.D022112A, HEUR12_R05.D022312A, HCCST12.D031312A, LCCST12.D031312A, HM2012.D022412A, and LM2012.D022412A.

Figure 49. Electricity generation by fuel in eleven cases, 2010 and 2035: History: U.S. Energy Information Administration, *Annual Energy Review 2010*, DOE/EIA-0384(2010) (Washington, DC, October 2011). **Projections:** AEO2012 National Energy Modeling

System, runs REF2012.D020112C, REF_R05.D030712A, REF2012.HEUR12.D022112A, REF2012.LEUR12.D022112A, HEUR12_R05.D022312A, HCCST12.D031312A, LCCST12.D031312A, HM2012.D022412A, and LM2012.D022412A.

Figure 50. Cumulative retrofits of generating capacity with scrubbers and dry sorbent injection for emissions control, 2011-2020: **Projections:** AEO2012 National Energy Modeling System, runs REF2012.D020112C, REF_R05.D030712A, REF2012.HEUR12.D022112A, REF2012.LEUR12.D022112A, HEUR12_R05.D022312A, HCCST12.D031312A, LCCST12.D031312A, HM2012.D022412A, and LM2012.D022412A.

Figure 51. Nuclear power plant retirements by NERC region in the Low Nuclear case, 2010-2035: **Projections:** AEO2011 National Energy Modeling System, run LOWNUC12.D022312B.

Figure 52. Alaska North Slope oil production in three cases, 2010-2035: **Projections:** AEO2012 National Energy Modeling System, runs REF2012.D020112C, HP2012.D022112A, and LP2012.D022112A.

Figure 53. Alaska North Slope wellhead oil revenue in three cases, assuming no minimum revenue requirement, 2010-2035: **Projections:** AEO2012 National Energy Modeling System, runs REF2012.D020112C, HP2012.D022112A, and LP2012.D022112A.

Figure 54. Average production profiles for shale gas wells in major U.S. shale plays by years of operation: U.S. Energy Information Administration, analysis of well-level production from HPDI database; and Pennsylvania Department of Environmental Protection Oil & Gas Reporting, website www.paoilandgasreporting.state.pa.us/publicreports/Modules/DataExports/DataExports.aspx (accessed October 2011).

Figure 55. U.S. production of tight oil in four cases, 2000-2035: **History:** U.S. Energy Information Administration, *Annual Energy Review 2010*, DOE/EIA-0384(2010) (Washington, DC, October 2011). **Projections:** AEO2012 National Energy Modeling System, runs REF2012.D020112C, REF2012.LEUR12.D02212A, REF2012.HEUR12.D02212A, and REF2012.HTRR12.D050412A.

Figure 56. U.S. production of shale gas in four cases, 2000-2035: **History:** U.S. Energy Information Administration, *Annual Energy Review 2010*, DOE/EIA-0384(2010) (Washington, DC, October 2011). **Projections:** AEO2012 National Energy Modeling System, runs REF2012.D020112C, REF2012.LEUR12.D02212A, REF2012.HEUR12.D02212A, and REF2012.HTRR12.D050412A.

Figure 57. United States Geological Survey Marcellus Assessment Units: U.S Energy Information Administration, Office of Energy Analysis based on image published by the USGS in their Marcellus assessment fact sheet (USGS Fact Sheet 2011-3092, pubs.usgs.gov/fs/2011/3092/pdf/fs2011-3092.pdf).

Figure 58. Average annual growth rates of real GDP, labor force, and nonfarm labor productivity in three cases, 2010-2035: AEO2012 National Energy Modeling System, runs REF2012.D020112C, HM2012.D022412A, and LM2012.D022412A.

Figure 59. Average annual growth rates over 5 years following troughs of U.S. recessions in 1975, 1982, 1991, and 2008: **History:** Bureau of Economic Analysis, Bureau of Labor Statistics (unemployment rate). **Projections:** AEO2012 National Energy Modeling System, run REF2011.D020112C.

Figure 60. Average annual growth rates for real output and its major components in three cases, 2010-2035: AEO2012 National Energy Modeling System, runs REF2012.D020112C, HM2012.D022412A, and LM2012.D022412A.

Figure 61. Sectoral composition of industrial output growth rates in three cases, 2010-2035: AEO2012 National Energy Modeling System, runs REF2012.D020112C, HM2012.D022412A, and LM2012.D022412A.

Figure 62. Energy end-use expenditures as a share of gross domestic product, 1970-2035: **History:** U.S. Energy Information Administration, *Annual Energy Review 2010*, DOE/EIA-0384(2010) (Washington, DC, October 2011). **Projections:** AEO2012 National Energy Modeling System, run REF2012.D020112C.

Figure 63. Energy end-use expenditures as a share of gross output, 1987-2035: **History:** U.S. Energy Information Administration, *Annual Energy Review 2010*, DOE/EIA-0384(2010) (Washington, DC, October 2011). **Projections:** AEO2012 National Energy Modeling System, run REF2012.D020112C.

Figure 64. Average annual oil prices in three cases, 1980-2035: **History:** U.S. Energy Information Administration, *Annual Energy Review 2010*, DOE/EIA-0384(2010) (Washington, DC, October 2011). **Projections:** AEO2012 National Energy Modeling System, runs REF2012.D020112C. HP2012.D022112A, and LP2012.D022112A.

Figure 65. World petroleum and other liquids supply and demand by region in three cases, 2010 and 2035: **History:** U.S. Energy Information Administration, *Annual Energy Review 2010*, DOE/EIA-0384(2010) (Washington, DC, October 2011). **Projections:** AEO2012 National Energy Modeling System, runs REF2012.D020112C. HP2012.D022112A, and LP2012.D022112A.

Figure 66. Total world production of nonpetroleum liquids, bitumen, and extra-heavy oil in three cases, 2010 and 2035: **History:** Derived from U.S. Energy Information Administration, International Energy Statistics database (as of January 2012), website www.eia.gov/ies. **Projections:** Generate World Oil Balance (GWOB) Model and AEO2012 National Energy Modeling System, runs REF2012.D020112C, LP2012.D022112A, and HP2012.D022112A.

Figure 67. North American natural gas trade, 2010-2035: AEO2012 National Energy Modeling System, run REF2012.D020112C.

Notes and sources

Figure 68. World energy consumption by region, 1990-2035: History: U.S. Energy Information Administration, International Energy Statistics database (as of January, 2012), website www.eia.gov/ies. **Projections:** U.S. Energy Information Administration, World Energy Projections System Plus (2012) model.

Figure 69. Installed nuclear capacity in OECD and non-OECD countries, 2010 and 2035: U.S. Energy Information Administration, World Energy Projections System Plus (2012) model.

Figure 70. World renewable electricity generation by source, excluding hydropower, 2005-2035: History: U.S. Energy Information Administration, International Energy Statistics database (as of January, 2012), website www.eia.gov/ies. **Projections:** U.S. Energy Information Administration, World Energy Projections System Plus (2012) model.

Figure 71. Energy use per capita and per dollar of gross domestic product, 1980-2035: History: U.S. Energy Information Administration, Annual Energy Review 2010, DOE/EIA-0384(2010) (Washington, DC, October 2011). **Projections:** AEO2012 National Energy Modeling System, run REF2012.D020112C.

Figure 72. Primary energy use by end-use sector, 2010-2035: History: U.S. Energy Information Administration, *Annual Energy Review 2010*, DOE/EIA-0384(2010) (Washington, DC, October 2011). **Projections:** AEO2012 National Energy Modeling System, run REF2012.D020112C.

Figure 73. Primary energy use by fuel, 1980-2035: History: U.S. Energy Information Administration, *Annual Energy Review 2010*, DOE/EIA-0384(2010) (Washington, DC, October 2011). **Projections:** AEO2012 National Energy Modeling System, run REF2012.D020112C.

Figure 74. Residential delivered energy intensity in four cases, 2005-2035: History: U.S. Energy Information Administration, *Annual Energy Review 2010*, DOE/EIA-0384(2010) (Washington, DC, October 2011). **Projections:** AEO2012 National Energy Modeling System, runs REF2012.D020112C, FROZTECH.D030812A, BESTTECH.D032812A, and HIGHTECH.D032812A.

Figure 75. Change in residential electricity consumption for selected end uses in the Reference case, 2010-2035: AEO2012 National Energy Modeling System, run REF2012.D020112C.

Figure 76. Ratio of residential delivered energy consumption for selected end uses: AEO2012 National Energy Modeling System, runs REF2012.D020112C, BESTTECH.D032812A, HIGHTECH.D032812A, and EXTENDED.D050612B.

Figure 77. Residential market penetration by renewable technologies in two cases, 2010, 2020, and 2035: AEO2012 National Energy Modeling System, runs REF2012.D020112C and EXTENDED.D050612B.

Figure 78. Commercial delivered energy intensity in four cases, 2005-2035: History: U.S. Energy Information Administration, *Annual Energy Review 2010*, DOE/EIA-0384(2010) (Washington, DC, October 2011). **Projections:** AEO2012 National Energy Modeling System, runs REF2012.D020112C, FROZTECH.D030812A, BESTTECH.D032812A, and HIGHTECH.D032812A.

Figure 79. Energy intensity of selected commercial electric end uses, 2010 and 2035: AEO2012 National Energy Modeling System, runs REF2012.D020112C.

Figure 80. Efficiency gains for selected commercial equipment in three cases, 2035: AEO2012 National Energy Modeling System, runs REF2012.D020112C, FROZTECH.D030812A, and BESTTECH.D032812A.

Figure 81. Additions to electricity generation capacity in the commercial sector in two cases, 2010-2035: AEO2012 National Energy Modeling System, runs REF2012.D020112C and EXTENDED.D050612B.

Figure 82. Industrial delivered energy consumption by application, 2010-2035: AEO2012 National Energy Modeling System, run REF2012.D020112C.

Figure 83. Industrial energy consumption by fuel, 2010, 2025 and 2035: AEO2012 National Energy Modeling System, runs REF2012.D020112C.

Figure 84. Cumulative growth in value of shipments from energy-intensive industries in three cases, 2010-2035: AEO2012 National Energy Modeling System, runs REF2012.D020112C, HM2012.D022412A, and LM2012.D022412A.

Figure 85. Change in delivered energy for energy-intensive industries in three cases, 2010-2035: AEO2012 National Energy Modeling System, runs REF2012.D020112C, HM2012.D022412A, and LM2012.D022412A.

Figure 86. Cumulative growth in value of shipments from non-energy-intensive industries in three cases, 2010-2035: AEO2012 National Energy Modeling System, runs REF2012.D020112C, HM2012.D022412A, and LM2012.D022412A.

Figure 87. Change in delivered energy for non-energy-intensive industries in three cases, 2010-2035: AEO2012 National Energy Modeling System, runs REF2012.D020112C, HM2012.D022412A, and LM2012.D022412A.

Figure 88. Delivered energy consumption for transportation by mode in two cases, 2010 and 2035: AEO2012 National Energy Modeling System, runs REF2012.D020112C and CAFEY.D032112C.

Figure 89. Average fuel economy of new light-duty vehicles in two cases, 1980-2035: History: Oak Ridge National Laboratory, *Transportation Energy Data Book*, Edition 30 and Annual (Oak Ridge, TN: 2011). **Projections:** AEO2012 National Energy Modeling System, runs REF2012.D020112C and CAFEY.D032112C.

Figure 90. **Vehicle miles traveled per licensed driver, 1970-2035: History:** Derived from U.S. Department of Transportation, Federal Highway Administration, *Highway Statistics 2010* (Washington, DC: 2012), website www.fhwa.dot.gov/policyinformation/statistics/2010. **Projections:** AEO2012 National Energy Modeling System, run REF2012.D020112C.

Figure 91. **Sales of light-duty vehicles using non-gasoline technologies by fuel type, 2010, 2020, and 2035:** AEO2012 National Energy Modeling System, runs REF2012.D020112C.

Figure 92. **Heavy-duty vehicle energy consumption, 1995-2035: History:** Derived from U.S. Energy Information Administration, *Annual Energy Review 2010*, DOE/EIA-0384(2010) (Washington, DC: October 2011); and Oak Ridge National Laboratory, *Transportation Energy Data Book*, Edition 30 and Annual (Oak Ridge, TN: 2011); and U.S. Department of Transportation, Federal Highway Administration, *Highway Statistics 2010* (Washington, DC: 2012), website www.fhwa.dot.gov/policyinformation/statistics/2010. **Projections:** AEO2012 National Energy Modeling System, run REF2012.D020112C.

Figure 93. **U.S. electricity demand growth, 1950-2035: History:** U.S. Energy Information Administration, *Annual Energy Review 2010*, DOE/EIA-0384(2010) (Washington, DC, October 2011). **Projections:** AEO2012 National Energy Modeling System, runs REF2012.D020112C.

Figure 94. **Electricity generation by fuel, 2010, 2020, and 2035:** AEO2012 National Energy Modeling System, runs REF2012.D020112C.

Figure 95. **Electricity generation capacity additions by fuel type, including combined heat and power, 2011-2035:** AEO2012 National Energy Modeling System, runs REF2012.D020112C.

Figure 96. **Additions to electricity generation capacity, 1985-2035: History:** Energy Information Administration, Form EIA-860, "Annual Electric Generator Report." **Projections:** AEO2012 National Energy Modeling System, runs REF2012.D020112C.

Figure 97. **Electricity sales and power sector generating capacity, 1949-2035: History:** U.S. Energy Information Administration, *Annual Energy Review 2010*, DOE/EIA-0384(2010) (Washington, DC, October 2011). **Projections:** AEO2012 National Energy Modeling System, run REF2012.D020112C.

Figure 98. **Levelized electricity costs for new power plants, excluding subsidies, 2020 and 2035:** AEO2012 National Energy Modeling System, run REF2012.D020112C.

Figure 99. **Electricity generating capacity at U.S. nuclear power plants in three cases, 2010, 2025, and 2035:** AEO2012 National Energy Modeling System, runs REF2012.D020112C, LM2012.D022412A, and HM2012.D022412A.

Figure 100. **Nonhydropower renewable electricity generation capacity by energy source, including end-use capacity, 2010-2035:** AEO2012 National Energy Modeling System, runs REF2012.D020112.

Figure 101. **Hydropower and other renewable electricity generation, including end-use generation, 2010-2035:** AEO2012 National Energy Modeling System, runs REF2012.D020112C.

Figure 102. **Regional growth in nonhydroelectric renewable electricity generation, including end-use generation, 2010-2035:** AEO2012 National Energy Modeling System, runs REF2012.D020112C.

Figure 103. **Annual average Henry Hub spot natural gas prices, 1990-2035: History:** U.S. Energy Information Administration, *Short-Term Energy Outlook* Query System, Monthly Natural Gas Data, Variable NGHHUUS. **Projections:** AEO2012 National Energy Modeling System, run REF2012.D020112C.

Figure 104. **Ratio of low-sulfur light crude oil price to Henry Hub natural gas price on an energy equivalent basis, 1990-2035: History:** U.S. Energy Information Administration, *Short-Term Energy Outlook* Query System, Monthly Natural Gas Data, Variable NGHHUUS, and U.S. Energy Information Administration, Form EIA-856, "Monthly Foreign Crude Oil Acquisition Report." **Projections:** AEO2012 National Energy Modeling System, run REF2012.D020112C.

Figure 105. **Annual average Henry Hub spot natural gas prices in seven cases, 1990-2035: History:** U.S. Energy Information Administration, *Natural Gas Annual 2010*, DOE/EIA-0131(2010) (Washington, DC, December 2011). **Projections:** AEO2012 National Energy Modeling System, runs REF2012.D020112C, REF2012.HEUR12.D022112A, REF2012.LEUR12.D022112A, LM2012.D022412A, and HM2012.D022412A.

Figure 106. **Natural gas production, consumption, and net imports, 1990-2035: History:** U.S. Energy Information Administration, *Natural Gas Annual 2010*, DOE/EIA-0131(2010) (Washington, DC, December 2011). **Projections:** AEO2012 National Energy Modeling System, runs REF2012.D020112C.

Figure 107. **Natural gas production by source, 1990-2035: History:** U.S. Energy Information Administration, *Natural Gas Annual 2010*, DOE/EIA-0131(2010) (Washington, DC, December 2011). **Projections:** AEO2012 National Energy Modeling System, runs REF2012.D020112C.

Figure 108. **Lower 48 onshore natural gas production by region, 2010 and 2035:** AEO2012 National Energy Modeling System, runs REF2012.D020112C.

Notes and sources

Figure 109. U.S. net imports of natural gas by source, 1990-2035: History: U.S. Energy Information Administration, *Natural Gas Annual 2010*, DOE/EIA-0131(2010) (Washington, DC, December 2011). **Projections:** AEO2012 National Energy Modeling System, runs REF2012.D020112C.

Figure 110. Consumption of petroleum and other liquids by sector, 1990-2035: History: U.S. Energy Information Administration, *Annual Energy Review 2010*, DOE/EIA-0384(2010) (Washington, DC, October 2011). **Projections:** AEO2012 National Energy Modeling System, run REF2012.D020112C.

Figure 111. U.S. production of petroleum and other liquids by source, 2010-2035: AEO2012 National Energy Modeling System, run REF2012.D020112C.

Figure 112. Domestic crude oil production by source, 1990-2035: History: U.S. Energy Information Administration, *Petroleum Supply Annual 2010*, DOE/EIA-0340(2010)/1 (Washington, DC, July 2011). **Projections:** AEO2012 National Energy Modeling System, run REF2012.D020112C.

Figure 113. Total U.S. crude oil production in six cases, 1990-2035: History: U.S. Energy Information Administration, *Annual Energy Review 2010*, DOE/EIA-0384(2010) (Washington, DC, October 2011). **Projections:** AEO2012 National Energy Modeling System, run REF2012.D020112C, LP2012.D022112A, HP2012.D022112A, REF2012.HEUR12.D022112A, REF2012.LEUR.D022112A, and HTRR12.D050412A.

Figure 114. Net import share of U.S. petroleum and other liquids consumption in three cases, 1990-2035: History: U.S. Energy Information Administration, *Annual Energy Review 2010*, DOE/EIA-0384(2010) (Washington, DC, October 2011). **Projections:** AEO2012 National Energy Modeling System, run REF2012.D020112C, LP2012.D022112A, and HP2012.D022112A.

Figure 115. EISA2007 RFS credits earned in selected years, 2010-2035: AEO2012 National Energy Modeling System, run REF2012.D020112C.

Figure 116. U.S. ethanol use in blended gasoline and E85, 2000-2035: History: U.S. Energy Information Administration, *Annual Energy Review 2010*, DOE/EIA-0384(2010) (Washington, DC, October 2011). **Projections:** AEO2012 National Energy Modeling System, run REF2012.D020112C.

Figure 117. U.S. motor gasoline and diesel fuel consumption, 2000-2035: History: U.S. Energy Information Administration, *Annual Energy Review 2010*, DOE/EIA-0384(2010) (Washington, DC, October 2011). **Projections:** AEO2012 National Energy Modeling System, run REF2012.D020112C.

Figure 118. Coal production by region, 1970-2035: History (short tons): 1970-1990: U.S. Energy Information Administration, *The U.S. Coal Industry, 1970-1990: Two Decades of Change*, DOE/EIA-0559 (Washington, DC, November 2002). **1991-2000:** U.S. Energy Information Administration, *Coal Industry Annual*, DOE/EIA-0584 (various years). **2001-2010:** U.S. Energy Information Administration, *Annual Coal Report 2010*, DOE/EIA-0584(2010) (Washington, DC, November 2011), and previous issues. **History (conversion to quadrillion Btu): 1970-2010: Estimation Procedure:** Estimates of average heat content by region and year are based on coal quality data collected through various energy surveys (see sources) and national-level estimates of U.S. coal production by year in units of quadrillion Btu, published in EIA's *Annual Energy Review*. **Sources:** U.S. Energy Information Administration, *Annual Energy Review 2010*, DOE/EIA-0384(2010) (Washington, DC, October 2011), Table 1.2; Form EIA-3, "Quarterly Coal Consumption and Quality Report, Manufacturing and Transformation/Processing Coal Plants and Commercial and Institutional Coal Users"; Form EIA-5, "Quarterly Coal Consumption and Quality Report, Coke Plants"; Form EIA-6A, "Coal Distribution Report"; Form EIA-7A, "Annual Coal Production and Preparation Report"; Form EIA-423, "Monthly Cost and Quality of Fuels for Electric Plants Report"; Form EIA-906, "Power Plant Report"; Form EIA-920, "Combined Heat and Power Plant Report"; Form EIA-923, "Power Plant Operations Report"; U.S. Department of Commerce, Bureau of the Census, "Monthly Report EM 545"; and Federal Energy Regulatory Commission, Form 423, "Monthly Report of Cost and Quality of Fuels for Electric Plants." **Projections:** AEO2012 National Energy Modeling System, run REF2012.D020112C. Note: For 1989-2035, coal production includes waste coal.

Figure 119. U.S. total coal production in six cases, 2010, 2020, and 2035: AEO2012 National Energy Modeling System, run REF2012.D020112C, LCCST12.D031312A, HP2012.D022112A, HM2012.D022412A, LM2012.D022412A, and CO2FEE15.D031312A. **Note:** Coal production includes waste coal.

Figure 120. Average annual minemouth coal prices by region, 1990-2035: History (dollars per short ton): 1990-2000: U.S. Energy Information Administration, *Coal Industry Annual*, DOE/EIA-0584 (various years). **2001-2010:** U.S. Energy Information Administration, *Annual Coal Report 2010*, DOE/EIA-0584(2010) (Washington, DC, November 2011), and previous issues. **History (conversion to dollars per million Btu): 1970-2009: Estimation Procedure:** Estimates of average heat content by region and year based on coal quality data collected through various energy surveys (see sources) and national-level estimates of U.S. coal production by year in units of quadrillion Btu published in EIA's *Annual Energy Review*. **Sources:** U.S. Energy Information Administration, *Annual Energy Review 2010*, DOE/EIA-0384(2010) (Washington, DC, October 2011), Table 1.2; Form EIA-3, "Quarterly Coal Consumption and Quality Report, Manufacturing and Transformation/Processing Coal Plants and Commercial and Institutional Coal Users"; Form EIA-5, "Quarterly Coal Consumption and Quality Report, Coke Plants"; Form EIA-6A, "Coal Distribution Report"; Form EIA-7A, "Annual Coal Production and Preparation Report"; Form EIA-423, "Monthly Cost and Quality of Fuels for Electric Plants Report"; Form EIA-906, "Power Plant Report"; and Form EIA-920, "Combined Heat and Power Plant

Report"; Form EIA-923, "Power Plant Operations Report"; U.S. Department of Commerce, Bureau of the Census, "Monthly Report EM 545"; and Federal Energy Regulatory Commission, Form 423, "Monthly Report of Cost and Quality of Fuels for Electric Plants." **Projections:** AEO2012 National Energy Modeling System, run REF2012.D020112C. **Note:** Includes reported prices for both open-market and captive mines.

Figure 121. Cumulative coal-fired generating capacity additions by sector in two cases, 2011-2035: AEO2012 National Energy Modeling System, run REF2012.D020112C and NOGHGCONCERN.D031212A.

Figure 122. U.S. energy-related carbon dioxide emissions by sector and fuel, 2005 and 2035: AEO2012 National Energy Modeling System, run REF2012.D020112C.

Figure 123. Sulfur dioxide emissions from electricity generation, 1990-2035: 1990, 2000, 2005: U.S. Environmental Protection Agency, *National Air Pollutant Emissions Trends, 1990-1998*, EPA-454/R-00-002 (Washington, DC, March 2000); U.S. Environmental Protection Agency, *Acid Rain Program Preliminary Summary Emissions Report, Fourth Quarter 2004*, website ampd.epa.gov/ampd/. **2010 and Projections:** AEO2012 National Energy Modeling System, run REF2012.D020112C.

Figure 124. Nitrogen oxide emissions from electricity generation, 1990-2035: History: 1990, 2000, 2005: U.S. Environmental Protection Agency, *National Air Pollutant Emissions Trends, 1990-1998*, EPA-454/R-00-002 (Washington, DC, March 2000); U.S. Environmental Protection Agency, *Acid Rain Program Preliminary Summary Emissions Report, Fourth Quarter 2004*, website ampd.epa.gov/ampd/. **2010 and Projections:** AEO2012 National Energy Modeling System, run REF2012.D020112C.

This page intentionally left blank

Appendix A
Reference case

Table A1. Total energy supply, disposition, and price summary
(quadrillion Btu per year, unless otherwise noted)

Supply, disposition, and prices	Reference case							Annual growth 2010-2035 (percent)
	2009	2010	2015	2020	2025	2030	2035	
Production								
Crude oil and lease condensate	11.35	11.59	13.23	14.40	13.77	13.71	12.89	0.4%
Natural gas plant liquids	2.57	2.78	3.33	3.79	3.93	3.98	3.94	1.4%
Dry natural gas	21.09	22.10	24.22	25.69	26.91	27.58	28.60	1.0%
Coal[1]	21.63	22.06	20.24	20.74	22.25	23.22	24.14	0.4%
Nuclear / uranium[2]	8.36	8.44	8.68	9.28	9.60	9.56	9.28	0.4%
Hydropower	2.67	2.51	2.90	2.95	2.99	3.02	3.04	0.8%
Biomass[3]	3.72	4.05	4.45	5.26	6.26	7.60	9.07	3.3%
Other renewable energy[4]	1.11	1.34	1.99	2.04	2.22	2.41	2.81	3.0%
Other[5]	0.47	0.64	0.60	0.64	0.69	0.79	0.91	1.4%
Total	**72.97**	**75.50**	**79.64**	**84.80**	**88.61**	**91.87**	**94.67**	**0.9%**
Imports								
Crude oil	19.70	20.14	18.87	16.00	16.23	16.04	16.90	-0.7%
Liquid fuels and other petroleum[6]	5.40	5.02	4.32	4.03	4.08	4.04	4.14	-0.8%
Natural gas[7]	3.85	3.81	3.73	3.49	2.75	3.00	2.84	-1.2%
Other imports[8]	0.61	0.52	0.44	0.72	1.07	0.78	0.81	1.8%
Total	**29.56**	**29.49**	**27.37**	**24.25**	**24.14**	**23.86**	**24.69**	**-0.7%**
Exports								
Liquid fuels and other petroleum[9]	4.20	4.81	5.00	4.39	4.46	4.67	4.95	0.1%
Natural gas[10]	1.08	1.15	1.93	3.09	3.51	3.86	4.17	5.3%
Coal	1.51	2.10	2.73	2.36	2.82	2.85	3.13	1.6%
Total	**6.79**	**8.06**	**9.66**	**9.84**	**10.79**	**11.38**	**12.25**	**1.7%**
Discrepancy[11]	1.04	-1.23	-0.08	-0.10	-0.03	0.04	0.18	--
Consumption								
Liquid fuels and other petroleum[12]	36.50	37.25	36.72	36.38	36.58	36.99	37.70	0.0%
Natural gas	23.43	24.71	26.00	26.07	26.14	26.72	27.26	0.4%
Coal[13]	19.62	20.76	17.80	18.73	20.02	20.59	21.15	0.1%
Nuclear / uranium[2]	8.36	8.44	8.68	9.28	9.60	9.56	9.28	0.4%
Hydropower	2.67	2.51	2.90	2.95	2.99	3.02	3.04	0.8%
Biomass[14]	2.72	2.88	3.04	3.58	4.17	4.78	5.44	2.6%
Other renewable energy[4]	1.11	1.34	1.99	2.04	2.22	2.41	2.81	3.0%
Other[15]	0.32	0.29	0.30	0.29	0.28	0.25	0.24	-0.6%
Total	**94.71**	**98.16**	**97.43**	**99.32**	**101.99**	**104.32**	**106.93**	**0.3%**
Prices (2010 dollars per unit)								
Petroleum (dollars per barrel)								
Low sulfur light crude oil	62.37	79.39	116.91	126.68	132.56	138.49	144.98	2.4%
Imported crude oil[16]	59.72	75.87	113.97	115.74	121.21	126.51	132.95	2.3%
Natural gas (dollars per million Btu)								
at Henry hub	4.00	4.39	4.29	4.58	5.63	6.29	7.37	2.1%
at the wellhead[17]	3.75	4.06	3.84	4.10	5.00	5.56	6.48	1.9%
Natural gas (dollars per thousand cubic feet)								
at the wellhead[17]	3.85	4.16	3.94	4.19	5.12	5.69	6.64	1.9%
Coal (dollars per ton)								
at the minemouth[18]	33.62	35.61	42.08	40.96	44.05	47.28	50.52	1.4%
Coal (dollars per million Btu)								
at the minemouth[18]	1.68	1.76	2.08	2.06	2.23	2.39	2.56	1.5%
Average end-use[19]	2.32	2.38	2.56	2.58	2.70	2.81	2.94	0.9%
Average electricity (cents per kilowatthour)	9.9	9.8	9.7	9.6	9.7	9.8	10.1	0.1%

Table A1. Total energy supply, disposition, and price summary (continued)
(quadrillion Btu per year, unless otherwise noted)

Supply, disposition, and prices	Reference case							Annual growth 2010-2035 (percent)
	2009	2010	2015	2020	2025	2030	2035	
Prices (nominal dollars per unit)								
Petroleum (dollars per barrel)								
Low sulfur light crude oil	61.65	79.39	125.97	148.87	170.09	197.10	229.55	4.3%
Imported crude oil[16]	59.04	75.87	122.81	136.02	155.52	180.06	210.51	4.2%
Natural gas (dollars per million Btu)								
at Henry hub	3.95	4.39	4.62	5.39	7.23	8.95	11.67	4.0%
at the wellhead[17]	3.71	4.06	4.14	4.81	6.42	7.92	10.26	3.8%
Natural gas (dollars per thousand cubic feet)								
at the wellhead[17]	3.80	4.16	4.24	4.93	6.57	8.11	10.51	3.8%
Coal (dollars per ton)								
at the minemouth[18]	33.24	35.61	45.34	48.13	56.52	67.28	80.00	3.3%
Coal (dollars per million Btu)								
at the minemouth[18]	1.66	1.76	2.24	2.42	2.86	3.41	4.05	3.4%
Average end-use[19]	2.30	2.38	2.76	3.03	3.47	4.01	4.66	2.7%
Average electricity (cents per kilowatthour)	9.8	9.8	10.4	11.3	12.5	13.9	16.0	2.0%

[1] Includes waste coal.
[2] These values represent the energy obtained from uranium when it is used in light water reactors. The total energy content of uranium is much larger, but alternative processes are required to take advantage of it.
[3] Includes grid-connected electricity from wood and wood waste; biomass, such as corn, used for liquid fuels production; and non-electric energy demand from wood. Refer to Table A17 for details.
[4] Includes grid-connected electricity from landfill gas; biogenic municipal waste; wind; photovoltaic and solar thermal sources; and non-electric energy from renewable sources, such as active and passive solar systems. Excludes electricity imports using renewable sources and nonmarketed renewable energy. See Table A17 for selected nonmarketed residential and commercial renewable energy data.
[5] Includes non-biogenic municipal waste, liquid hydrogen, methanol, and some domestic inputs to refineries.
[6] Includes imports of finished petroleum products, unfinished oils, alcohols, ethers, blending components, and renewable fuels such as ethanol.
[7] Includes imports of liquefied natural gas that is later re-exported.
[8] Includes coal, coal coke (net), and electricity (net). Excludes imports of fuel used in nuclear power plants.
[9] Includes crude oil, petroleum products, ethanol, and biodiesel.
[10] Includes re-exported liquefied natural gas.
[11] Balancing item. Includes unaccounted for supply, losses, gains, and net storage withdrawals.
[12] Includes petroleum-derived fuels and non-petroleum derived fuels, such as ethanol and biodiesel, and coal-based synthetic liquids. Petroleum coke, which is a solid, is included. Also included are natural gas plant liquids and crude oil consumed as a fuel. Refer to Table A17 for detailed renewable liquid fuels consumption.
[13] Excludes coal converted to coal-based synthetic liquids and natural gas.
[14] Includes grid-connected electricity from wood and wood waste, non-electric energy from wood, and biofuels heat and coproducts used in the production of liquid fuels, but excludes the energy content of the liquid fuels.
[15] Includes non-biogenic municipal waste, liquid hydrogen, and net electricity imports.
[16] Weighted average price delivered to U.S. refiners.
[17] Represents lower 48 onshore and offshore supplies.
[18] Includes reported prices for both open market and captive mines.
[19] Prices weighted by consumption; weighted average excludes residential and commercial prices, and export free-alongside-ship (f.a.s.) prices.
Btu = British thermal unit.
- - = Not applicable.
Note: Totals may not equal sum of components due to independent rounding. Data for 2009 and 2010 are model results and may differ slightly from official EIA data reports.
Sources: 2009 natural gas supply values: U.S. Energy Information Administration (EIA), *Natural Gas Annual 2009*, DOE/EIA-0131(2009) (Washington, DC, December 2010). 2010 natural gas supply values and natural gas wellhead price: EIA, *Natural Gas Monthly*, DOE/EIA-0130(2011/07) (Washington, DC, July 2011). 2009 natural gas wellhead price: U.S. Department of the Interior, Office of Natural Resources Revenue; and EIA, *Natural Gas Annual 2009*, DOE/EIA-0131(2009) (Washington, DC, December 2010). 2009 and 2010 coal minemouth and delivered coal prices: EIA, *Annual Coal Report 2010*, DOE/EIA-0584(2010) (Washington, DC, November 2011). 2010 petroleum supply values and 2009 crude oil and lease condensate production: EIA, *Petroleum Supply Annual 2010*, DOE/EIA-0340(2010)/1 (Washington, DC, July 2011). Other 2009 petroleum supply values: EIA, *Petroleum Supply Annual 2009*, DOE/EIA-0340(2009)/1 (Washington, DC, July 2010). 2009 and 2010 low sulfur light crude oil price: EIA, Form EIA-856, "Monthly Foreign Crude Oil Acquisition Report." Other 2009 and 2010 coal values: *Quarterly Coal Report, October-December 2010*, DOE/EIA-0121(2010/4Q) (Washington, DC, May 2011). Other 2009 and 2010 values: EIA, *Annual Energy Review 2010*, DOE/EIA-0384(2010) (Washington, DC, October 2011). **Projections:** EIA, AEO2012 National Energy Modeling System run REF2012.D020112C.

Table A2. Energy consumption by sector and source
(quadrillion Btu per year, unless otherwise noted)

Sector and source	Reference case							Annual growth 2010-2035 (percent)
	2009	2010	2015	2020	2025	2030	2035	
Energy consumption								
Residential								
Liquefied petroleum gases	0.51	0.56	0.51	0.50	0.50	0.51	0.51	-0.4%
Kerosene	0.03	0.03	0.02	0.02	0.02	0.02	0.02	-1.7%
Distillate fuel oil	0.60	0.63	0.55	0.48	0.43	0.38	0.35	-2.3%
Liquid fuels and other petroleum subtotal	1.14	1.22	1.08	1.01	0.95	0.91	0.87	-1.3%
Natural gas	4.90	5.06	4.97	4.95	4.88	4.84	4.76	-0.2%
Coal	0.01	0.01	0.01	0.01	0.01	0.01	0.01	-1.1%
Renewable energy[1]	0.43	0.42	0.43	0.43	0.43	0.43	0.43	0.1%
Electricity	4.66	4.95	4.75	4.96	5.23	5.55	5.86	0.7%
Delivered energy	**11.13**	**11.66**	**11.24**	**11.36**	**11.51**	**11.73**	**11.93**	**0.1%**
Electricity related losses	9.80	10.39	9.58	10.01	10.52	10.95	11.35	0.4%
Total	**20.93**	**22.05**	**20.81**	**21.36**	**22.02**	**22.68**	**23.28**	**0.2%**
Commercial								
Liquefied petroleum gases	0.13	0.14	0.14	0.14	0.15	0.15	0.16	0.3%
Motor gasoline[2]	0.05	0.05	0.05	0.05	0.05	0.06	0.06	0.4%
Kerosene	0.00	0.00	0.00	0.00	0.00	0.01	0.01	0.7%
Distillate fuel oil	0.41	0.43	0.35	0.34	0.33	0.33	0.32	-1.2%
Residual fuel oil	0.08	0.08	0.08	0.08	0.08	0.08	0.08	-0.0%
Liquid fuels and other petroleum subtotal	0.68	0.72	0.62	0.62	0.62	0.62	0.62	-0.5%
Natural gas	3.20	3.28	3.41	3.51	3.53	3.60	3.69	0.5%
Coal	0.07	0.06	0.06	0.06	0.06	0.06	0.06	-0.0%
Renewable energy[3]	0.11	0.11	0.11	0.11	0.11	0.11	0.11	0.0%
Electricity	4.46	4.54	4.59	4.88	5.16	5.48	5.80	1.0%
Delivered energy	**8.51**	**8.70**	**8.80**	**9.18**	**9.48**	**9.87**	**10.28**	**0.7%**
Electricity related losses	9.39	9.52	9.27	9.85	10.38	10.82	11.23	0.7%
Total	**17.90**	**18.22**	**18.06**	**19.03**	**19.86**	**20.69**	**21.50**	**0.7%**
Industrial[4]								
Liquefied petroleum gases	2.00	2.00	1.83	2.06	2.17	2.18	2.15	0.3%
Motor gasoline[2]	0.24	0.25	0.28	0.30	0.30	0.30	0.30	0.8%
Distillate fuel oil	1.11	1.16	1.25	1.18	1.19	1.17	1.18	0.1%
Residual fuel oil	0.11	0.12	0.09	0.08	0.08	0.08	0.08	-1.3%
Petrochemical feedstocks	0.90	0.94	1.01	1.20	1.29	1.31	1.30	1.3%
Other petroleum[5]	3.57	3.59	3.44	3.18	3.11	3.09	3.19	-0.5%
Liquid fuels and other petroleum subtotal	7.93	8.05	7.89	7.99	8.13	8.13	8.21	0.1%
Natural gas	6.32	6.76	7.19	7.26	7.32	7.21	7.18	0.2%
Natural-gas-to-liquids heat and power	0.00	0.00	0.00	0.00	0.00	0.00	0.00	- -
Lease and plant fuel[6]	1.31	1.37	1.43	1.55	1.57	1.59	1.63	0.7%
Natural gas subtotal	7.63	8.14	8.62	8.80	8.89	8.80	8.81	0.3%
Metallurgical coal	0.40	0.55	0.57	0.48	0.49	0.46	0.43	-1.0%
Other industrial coal	0.94	1.01	1.03	1.04	1.08	1.08	1.08	0.3%
Coal-to-liquids heat and power	0.00	0.00	0.00	0.26	0.36	0.48	0.60	- -
Net coal coke imports	-0.02	-0.01	-0.01	-0.02	-0.03	-0.04	-0.06	9.3%
Coal subtotal	1.32	1.56	1.59	1.76	1.90	1.98	2.06	1.1%
Biofuels heat and coproducts	0.82	0.84	0.81	0.96	1.27	1.92	2.57	4.6%
Renewable energy[7]	1.37	1.50	1.61	1.67	1.82	1.87	1.95	1.1%
Electricity	3.13	3.28	3.44	3.46	3.52	3.44	3.33	0.1%
Delivered energy	**22.20**	**23.37**	**23.96**	**24.64**	**25.53**	**26.14**	**26.94**	**0.6%**
Electricity related losses	6.59	6.89	6.94	6.97	7.09	6.80	6.46	-0.3%
Total	**28.79**	**30.26**	**30.90**	**31.61**	**32.61**	**32.93**	**33.39**	**0.4%**

Table A2. Energy consumption by sector and source (continued)
(quadrillion Btu per year, unless otherwise noted)

Sector and source	Reference case							Annual growth 2010-2035 (percent)
	2009	2010	2015	2020	2025	2030	2035	
Transportation								
Liquefied petroleum gases	0.05	0.04	0.04	0.04	0.04	0.05	0.05	0.5%
E85[8]	0.00	0.00	0.01	0.13	0.30	0.72	1.22	27.0%
Motor gasoline[2]	16.84	16.91	16.13	15.31	14.90	14.69	14.53	-0.6%
Jet fuel[9]	2.98	3.07	3.03	3.09	3.19	3.27	3.33	0.3%
Distillate fuel oil[10]	5.53	5.77	6.55	6.80	7.03	7.20	7.44	1.0%
Residual fuel oil	0.81	0.90	0.91	0.92	0.93	0.93	0.94	0.2%
Other petroleum[11]	0.16	0.17	0.17	0.17	0.17	0.17	0.17	0.0%
Liquid fuels and other petroleum subtotal	26.36	26.88	26.83	26.46	26.57	27.02	27.67	0.1%
Pipeline fuel natural gas	0.61	0.65	0.68	0.67	0.67	0.68	0.69	0.2%
Compressed / liquefied natural gas	0.04	0.04	0.06	0.09	0.11	0.14	0.16	5.7%
Liquid hydrogen	0.00	0.00	0.00	0.00	0.00	0.00	0.00	--
Electricity	0.02	0.02	0.03	0.03	0.04	0.06	0.07	4.8%
Delivered energy	**27.04**	**27.59**	**27.60**	**27.25**	**27.40**	**27.90**	**28.60**	**0.1%**
Electricity related losses	0.05	0.05	0.05	0.06	0.08	0.11	0.14	4.5%
Total	**27.09**	**27.63**	**27.65**	**27.32**	**27.49**	**28.01**	**28.75**	**0.2%**
Delivered energy consumption for all sectors								
Liquefied petroleum gases	2.69	2.75	2.51	2.74	2.86	2.88	2.86	0.2%
E85[8]	0.00	0.00	0.01	0.13	0.30	0.72	1.22	27.0%
Motor gasoline[2]	17.13	17.21	16.46	15.66	15.25	15.04	14.88	-0.6%
Jet fuel[9]	2.98	3.07	3.03	3.09	3.19	3.27	3.33	0.3%
Kerosene	0.04	0.04	0.03	0.03	0.03	0.03	0.03	-1.2%
Distillate fuel oil	7.65	7.99	8.69	8.81	8.99	9.08	9.29	0.6%
Residual fuel oil	0.99	1.11	1.08	1.08	1.09	1.09	1.11	0.0%
Petrochemical feedstocks	0.90	0.94	1.01	1.20	1.29	1.31	1.30	1.3%
Other petroleum[12]	3.72	3.76	3.61	3.34	3.27	3.26	3.36	-0.4%
Liquid fuels and other petroleum subtotal	36.10	36.87	36.43	36.08	36.28	36.68	37.38	0.1%
Natural gas	14.46	15.15	15.64	15.81	15.85	15.79	15.79	0.2%
Natural-gas-to-liquids heat and power	0.00	0.00	0.00	0.00	0.00	0.00	0.00	--
Lease and plant fuel[6]	1.31	1.37	1.43	1.55	1.57	1.59	1.63	0.7%
Pipeline natural gas	0.61	0.65	0.68	0.67	0.67	0.68	0.69	0.2%
Natural gas subtotal	16.38	17.17	17.75	18.03	18.09	18.06	18.11	0.2%
Metallurgical coal	0.40	0.55	0.57	0.48	0.49	0.46	0.43	-1.0%
Other coal	1.01	1.08	1.09	1.10	1.14	1.14	1.15	0.3%
Coal-to-liquids heat and power	0.00	0.00	0.00	0.26	0.36	0.48	0.60	--
Net coal coke imports	-0.02	-0.01	-0.01	-0.02	-0.03	-0.04	-0.06	9.3%
Coal subtotal	1.39	1.62	1.65	1.82	1.96	2.04	2.12	1.1%
Biofuels heat and coproducts	0.82	0.84	0.81	0.96	1.27	1.92	2.57	4.6%
Renewable energy[13]	1.91	2.03	2.15	2.21	2.36	2.41	2.50	0.8%
Liquid hydrogen	0.00	0.00	0.00	0.00	0.00	0.00	0.00	--
Electricity	12.27	12.79	12.81	13.33	13.96	14.53	15.06	0.7%
Delivered energy	**68.87**	**71.32**	**71.59**	**72.43**	**73.92**	**75.64**	**77.75**	**0.3%**
Electricity related losses	25.83	26.84	25.84	26.89	28.07	28.67	29.18	0.3%
Total	**94.71**	**98.16**	**97.43**	**99.32**	**101.99**	**104.32**	**106.93**	**0.3%**
Electric power[14]								
Distillate fuel oil	0.07	0.08	0.08	0.09	0.09	0.09	0.09	0.5%
Residual fuel oil	0.32	0.30	0.21	0.21	0.22	0.22	0.23	-1.1%
Liquid fuels and other petroleum subtotal	0.39	0.38	0.29	0.30	0.31	0.31	0.32	-0.7%
Natural gas	7.04	7.54	8.25	8.05	8.04	8.66	9.16	0.8%
Steam coal	18.23	19.13	16.15	16.91	18.06	18.55	19.03	-0.0%
Nuclear / uranium[15]	8.36	8.44	8.68	9.28	9.60	9.56	9.28	0.4%
Renewable energy[16]	3.77	3.85	4.96	5.40	5.75	5.87	6.22	1.9%
Electricity imports	0.12	0.09	0.10	0.09	0.08	0.05	0.04	-2.9%
Total[17]	**38.10**	**39.63**	**38.64**	**40.22**	**42.03**	**43.20**	**44.24**	**0.4%**

Table A2. Energy consumption by sector and source (continued)
(quadrillion Btu per year, unless otherwise noted)

Sector and source	Reference case							Annual growth 2010-2035 (percent)
	2009	2010	2015	2020	2025	2030	2035	
Total energy consumption								
Liquefied petroleum gases	2.69	2.75	2.51	2.74	2.86	2.88	2.86	0.2%
E85[8]	0.00	0.00	0.01	0.13	0.30	0.72	1.22	27.0%
Motor gasoline[2]	17.13	17.21	16.46	15.66	15.25	15.04	14.88	-0.6%
Jet fuel[9]	2.98	3.07	3.03	3.09	3.19	3.27	3.33	0.3%
Kerosene	0.04	0.04	0.03	0.03	0.03	0.03	0.03	-1.2%
Distillate fuel oil	7.72	8.07	8.78	8.89	9.07	9.17	9.38	0.6%
Residual fuel oil	1.32	1.41	1.29	1.29	1.31	1.32	1.34	-0.2%
Petrochemical feedstocks	0.90	0.94	1.01	1.20	1.29	1.31	1.30	1.3%
Other petroleum[12]	3.72	3.76	3.61	3.34	3.27	3.26	3.36	-0.4%
Liquid fuels and other petroleum subtotal	36.50	37.25	36.72	36.38	36.58	36.99	37.70	0.0%
Natural gas	21.51	22.69	23.89	23.85	23.89	24.45	24.94	0.4%
Natural-gas-to-liquids heat and power	0.00	0.00	0.00	0.00	0.00	0.00	0.00	--
Lease and plant fuel[6]	1.31	1.37	1.43	1.55	1.57	1.59	1.63	0.7%
Pipeline natural gas	0.61	0.65	0.68	0.67	0.67	0.68	0.69	0.2%
Natural gas subtotal	23.43	24.71	26.00	26.07	26.14	26.72	27.26	0.4%
Metallurgical coal	0.40	0.55	0.57	0.48	0.49	0.46	0.43	-1.0%
Other coal	19.23	20.21	17.24	18.01	19.20	19.69	20.18	-0.0%
Coal-to-liquids heat and power	0.00	0.00	0.00	0.26	0.36	0.48	0.60	--
Net coal coke imports	-0.02	-0.01	-0.01	-0.02	-0.03	-0.04	-0.06	9.3%
Coal subtotal	19.62	20.76	17.80	18.73	20.02	20.59	21.15	0.1%
Nuclear / uranium[15]	8.36	8.44	8.68	9.28	9.60	9.56	9.28	0.4%
Biofuels heat and coproducts	0.82	0.84	0.81	0.96	1.27	1.92	2.57	4.6%
Renewable energy[18]	5.68	5.88	7.11	7.61	8.11	8.29	8.71	1.6%
Liquid hydrogen	0.00	0.00	0.00	0.00	0.00	0.00	0.00	--
Electricity imports	0.12	0.09	0.10	0.09	0.08	0.05	0.04	-2.9%
Total	**94.71**	**98.16**	**97.43**	**99.32**	**101.99**	**104.32**	**106.93**	**0.3%**
Energy use and related statistics								
Delivered energy use	68.87	71.32	71.59	72.43	73.92	75.64	77.75	0.3%
Total energy use	94.71	98.16	97.43	99.32	101.99	104.32	106.93	0.3%
Ethanol consumed in motor gasoline and E85	0.96	1.11	1.22	1.35	1.55	1.82	2.15	2.7%
Population (millions)	307.84	310.83	326.16	342.01	358.06	374.09	390.09	0.9%
Gross domestic product (billion 2005 dollars)	12703	13088	14803	16740	19185	21725	24539	2.5%
Carbon dioxide emissions (million metric tons)	5424.8	5633.6	5407.2	5434.4	5552.5	5647.3	5757.9	0.1%

[1]Includes wood used for residential heating. See Table A4 and/or Table A17 for estimates of nonmarketed renewable energy consumption for geothermal heat pumps, solar thermal water heating, and electricity generation from wind and solar photovoltaic sources.
[2]Includes ethanol (blends of 15 percent or less) and ethers blended into gasoline.
[3]Excludes ethanol. Includes commercial sector consumption of wood and wood waste, landfill gas, municipal waste, and other biomass for combined heat and power. See Table A5 and/or Table A17 for estimates of nonmarketed renewable energy consumption for solar thermal water heating and electricity generation from wind and solar photovoltaic sources.
[4]Includes energy for combined heat and power plants, except those whose primary business is to sell electricity, or electricity and heat, to the public.
[5]Includes petroleum coke, asphalt, road oil, lubricants, still gas, and miscellaneous petroleum products.
[6]Represents natural gas used in well, field, and lease operations, and in natural gas processing plant machinery.
[7]Includes consumption of energy produced from hydroelectric, wood and wood waste, municipal waste, and other biomass sources. Excludes ethanol blends (15 percent or less) in motor gasoline.
[8]E85 refers to a blend of 85 percent ethanol (renewable) and 15 percent motor gasoline (nonrenewable). To address cold starting issues, the percentage of ethanol varies seasonally. The annual average ethanol content of 74 percent is used for this forecast.
[9]Includes only kerosene type.
[10]Diesel fuel for on- and off- road use.
[11]Includes aviation gasoline and lubricants.
[12]Includes unfinished oils, natural gasoline, motor gasoline blending components, aviation gasoline, lubricants, still gas, asphalt, road oil, petroleum coke, and miscellaneous petroleum products.
[13]Includes electricity generated for sale to the grid and for own use from renewable sources, and non-electric energy from renewable sources. Excludes ethanol and nonmarketed renewable energy consumption for geothermal heat pumps, buildings photovoltaic systems, and solar thermal water heaters.
[14]Includes consumption of energy by electricity-only and combined heat and power plants whose primary business is to sell electricity, or electricity and heat, to the public.
[15]These values represent the energy obtained from uranium when it is used in light water reactors. The total energy content of uranium is much larger, but alternative processes are required to take advantage of it.
[16]Includes conventional hydroelectric, geothermal, wood and wood waste, biogenic municipal waste, other biomass, wind, photovoltaic, and solar thermal sources. Excludes net electricity imports.
[17]Includes non-biogenic municipal waste not included above.
[18]Includes conventional hydroelectric, geothermal, wood and wood waste, biogenic municipal waste, other biomass, wind, photovoltaic, and solar thermal sources. Excludes ethanol, net electricity imports, and nonmarketed renewable energy consumption for geothermal heat pumps, buildings photovoltaic systems, and solar thermal water heaters.

Btu = British thermal unit.
- - = Not applicable.
Note: Totals may not equal sum of components due to independent rounding. Data for 2009 and 2010 are model results and may differ slightly from official EIA data reports.
Sources: 2009 and 2010 consumption based on: U.S. Energy Information Administration (EIA), *Annual Energy Review 2010*, DOE/EIA-0384(2010) (Washington, DC, October 2011). 2009 and 2010 population and gross domestic product: IHS Global Insight Industry and Employment models, August 2011. 2009 and 2010 carbon dioxide emissions: EIA, *Monthly Energy Review, October 2011* DOE/EIA-0035(2011/10) (Washington, DC, October 2011). Projections: EIA, AEO2012 National Energy Modeling System run REF2012.D020112C.

Table A3. Energy prices by sector and source
(2010 dollars per million Btu, unless otherwise noted)

Sector and source	Reference case							Annual growth 2010-2035 (percent)
	2009	2010	2015	2020	2025	2030	2035	
Residential								
Liquefied petroleum gases	24.84	27.02	30.70	31.07	32.27	33.29	34.64	1.0%
Distillate fuel oil	18.35	21.21	27.26	28.81	30.15	31.42	32.73	1.8%
Natural gas	11.95	11.08	10.31	10.84	12.03	12.76	13.98	0.9%
Electricity	34.01	33.69	34.59	33.87	34.08	34.06	34.58	0.1%
Commercial								
Liquefied petroleum gases	21.76	23.52	27.42	27.78	28.97	29.96	31.30	1.1%
Distillate fuel oil	16.16	20.77	23.98	25.49	26.86	27.98	29.18	1.4%
Residual fuel oil	13.66	11.07	16.18	17.60	18.24	19.04	18.90	2.2%
Natural gas	9.82	9.10	8.60	8.98	10.02	10.60	11.64	1.0%
Electricity	30.06	29.73	29.03	28.69	29.00	28.68	29.48	-0.0%
Industrial[1]								
Liquefied petroleum gases	20.05	21.80	27.43	27.76	29.24	30.48	32.18	1.6%
Distillate fuel oil	16.74	21.32	24.20	25.73	27.22	28.39	29.53	1.3%
Residual fuel oil	12.16	10.92	19.21	20.53	21.23	21.71	21.65	2.8%
Natural gas[2]	5.33	5.51	4.88	5.12	6.04	6.57	7.54	1.3%
Metallurgical coal	5.49	5.84	7.22	7.58	8.11	8.61	9.11	1.8%
Other industrial coal	2.99	2.71	3.27	3.30	3.38	3.50	3.64	1.2%
Coal to liquids	--	--	1.26	2.05	2.08	2.22	2.38	--
Electricity	20.05	19.63	18.91	18.95	19.60	19.81	20.78	0.2%
Transportation								
Liquefied petroleum gases[3]	25.84	26.88	31.93	32.21	33.38	34.37	35.74	1.1%
E85[4]	20.76	25.21	29.03	29.91	28.81	30.75	31.96	1.0%
Motor gasoline[5]	19.52	22.70	29.26	30.77	32.10	33.03	33.61	1.6%
Jet fuel[6]	12.75	16.22	23.74	25.26	26.45	27.58	29.13	2.4%
Diesel fuel (distillate fuel oil)[7]	18.02	21.87	27.56	28.98	30.42	31.38	32.40	1.6%
Residual fuel oil	10.61	10.42	18.32	19.58	20.62	20.76	20.95	2.8%
Natural gas[8]	14.17	13.20	12.40	12.50	13.29	13.68	14.51	0.4%
Electricity	35.71	32.99	30.50	29.74	31.53	32.54	33.82	0.1%
Electric power[9]								
Distillate fuel oil	14.54	18.73	22.77	24.18	25.35	26.43	27.80	1.6%
Residual fuel oil	8.98	11.89	23.00	24.38	25.40	25.55	25.72	3.1%
Natural gas	4.85	5.14	4.55	4.72	5.60	6.21	7.21	1.4%
Steam coal	2.22	2.26	2.35	2.41	2.54	2.66	2.80	0.9%
Average price to all users[10]								
Liquefied petroleum gases	16.13	17.28	22.99	23.06	24.19	25.23	26.63	1.7%
E85[4]	20.76	25.21	29.03	29.91	28.81	30.75	31.96	1.0%
Motor gasoline[5]	19.47	22.59	29.26	30.77	32.10	33.03	33.61	1.6%
Jet fuel	12.75	16.22	23.74	25.26	26.45	27.58	29.13	2.4%
Distillate fuel oil	17.73	21.65	26.87	28.36	29.81	30.87	31.91	1.6%
Residual fuel oil	10.51	10.82	19.01	20.31	21.31	21.53	21.68	2.8%
Natural gas	7.37	7.16	6.45	6.77	7.74	8.30	9.30	1.1%
Metallurgical coal	5.49	5.84	7.22	7.58	8.11	8.61	9.11	1.8%
Other coal	2.26	2.29	2.41	2.47	2.59	2.71	2.85	0.9%
Coal to liquids	--	--	1.26	2.05	2.08	2.22	2.38	--
Electricity	29.02	28.68	28.38	28.09	28.54	28.65	29.56	0.1%
Non-renewable energy expenditures by sector (billion 2010 dollars)								
Residential	240.88	251.69	246.72	251.77	266.75	280.17	298.72	0.7%
Commercial	177.13	179.08	177.92	187.57	201.89	212.88	231.98	1.0%
Industrial	184.40	198.98	223.88	239.75	261.92	268.58	282.31	1.4%
Transportation	479.66	573.78	746.84	770.94	803.52	829.88	856.65	1.6%
Total non-renewable expenditures	1082.08	1203.54	1395.36	1450.04	1534.08	1591.52	1669.66	1.3%
Transportation renewable expenditures	0.07	0.08	0.25	3.77	8.74	22.00	38.86	28.2%
Total expenditures	**1082.15**	**1203.62**	**1395.61**	**1453.81**	**1542.81**	**1613.52**	**1708.52**	**1.4%**

Table A3. Energy prices by sector and source (continued)
(nominal dollars per million Btu, unless otherwise noted)

Sector and source	Reference case							Annual growth 2010-2035 (percent)
	2009	2010	2015	2020	2025	2030	2035	
Residential								
Liquefied petroleum gases	24.55	27.02	33.08	36.51	41.41	47.38	54.86	2.9%
Distillate fuel oil	18.14	21.21	29.38	33.86	38.68	44.72	51.82	3.6%
Natural gas	11.82	11.08	11.11	12.74	15.43	18.16	22.14	2.8%
Electricity	33.62	33.69	37.27	39.80	43.72	48.47	54.76	2.0%
Commercial								
Liquefied petroleum gases	21.51	23.52	29.54	32.65	37.17	42.65	49.56	3.0%
Distillate fuel oil	15.97	20.77	25.83	29.95	34.47	39.82	46.20	3.2%
Residual fuel oil	13.51	11.07	17.43	20.68	23.41	27.10	29.93	4.1%
Natural gas	9.70	9.10	9.27	10.56	12.86	15.08	18.43	2.9%
Electricity	29.71	29.73	31.28	33.71	37.21	40.82	46.67	1.8%
Industrial[1]								
Liquefied petroleum gases	19.82	21.80	29.56	32.63	37.51	43.38	50.95	3.5%
Distillate fuel oil	16.55	21.32	26.08	30.24	34.93	40.40	46.76	3.2%
Residual fuel oil	12.02	10.92	20.70	24.13	27.24	30.89	34.28	4.7%
Natural gas[2]	5.27	5.51	5.26	6.02	7.75	9.35	11.93	3.1%
Metallurgical coal	5.43	5.84	7.78	8.91	10.40	12.26	14.42	3.7%
Other industrial coal	2.96	2.71	3.52	3.87	4.34	4.98	5.77	3.1%
Coal to liquids	- -	- -	1.36	2.41	2.67	3.16	3.78	- -
Electricity	19.83	19.63	20.38	22.27	25.15	28.20	32.90	2.1%
Transportation								
Liquefied petroleum gases[3]	25.55	26.88	34.41	37.85	42.83	48.91	56.59	3.0%
E85[4]	20.52	25.21	31.28	35.15	36.97	43.77	50.61	2.8%
Motor gasoline[5]	19.29	22.70	31.53	36.17	41.19	47.01	53.22	3.5%
Jet fuel[6]	12.61	16.22	25.58	29.68	33.94	39.25	46.12	4.3%
Diesel fuel (distillate fuel oil)[7]	17.82	21.87	29.69	34.06	39.03	44.66	51.29	3.5%
Residual fuel oil	10.49	10.42	19.74	23.01	26.45	29.55	33.18	4.7%
Natural gas[8]	14.01	13.20	13.36	14.69	17.05	19.47	22.97	2.2%
Electricity	35.31	32.99	32.86	34.95	40.46	46.31	53.55	2.0%
Electric power[9]								
Distillate fuel oil	14.37	18.73	24.53	28.42	32.52	37.61	44.02	3.5%
Residual fuel oil	8.88	11.89	24.78	28.66	32.59	36.37	40.73	5.0%
Natural gas	4.80	5.14	4.90	5.55	7.19	8.84	11.42	3.2%
Steam coal	2.19	2.26	2.53	2.83	3.25	3.78	4.43	2.7%

Table A3. Energy prices by sector and source (continued)
(nominal dollars per million Btu, unless otherwise noted)

Sector and source	Reference case							Annual growth 2010-2035 (percent)
	2009	2010	2015	2020	2025	2030	2035	
Average price to all users[10]								
Liquefied petroleum gases	15.94	17.28	24.78	27.10	31.04	35.90	42.17	3.6%
E85[4]	20.52	25.21	31.28	35.15	36.97	43.77	50.61	2.8%
Motor gasoline[5]	19.25	22.59	31.53	36.16	41.19	47.01	53.22	3.5%
Jet fuel	12.61	16.22	25.58	29.68	33.94	39.25	46.12	4.3%
Distillate fuel oil	17.53	21.65	28.96	33.33	38.24	43.94	50.52	3.4%
Residual fuel oil	10.39	10.82	20.48	23.87	27.34	30.64	34.33	4.7%
Natural gas	7.28	7.16	6.95	7.96	9.93	11.81	14.73	2.9%
Metallurgical coal	5.43	5.84	7.78	8.91	10.40	12.26	14.42	3.7%
Other coal	2.23	2.29	2.60	2.90	3.32	3.86	4.51	2.8%
Coal to liquids	--	--	1.36	2.41	2.67	3.16	3.78	--
Electricity	28.68	28.68	30.58	33.01	36.62	40.77	46.80	2.0%
Non-renewable energy expenditures by sector (billion nominal dollars)								
Residential	238.13	251.69	265.85	295.89	342.26	398.75	472.99	2.6%
Commercial	175.11	179.08	191.71	220.43	259.04	302.97	367.31	2.9%
Industrial	182.29	198.98	241.24	281.75	336.06	382.26	447.01	3.3%
Transportation	474.19	573.78	804.75	906.02	1030.98	1181.11	1356.41	3.5%
Total non-renewable expenditures	1069.72	1203.54	1503.55	1704.09	1968.35	2265.08	2643.72	3.2%
Transportation renewable expenditures	0.07	0.08	0.27	4.43	11.21	31.31	61.53	30.6%
Total expenditures	**1069.78**	**1203.62**	**1503.82**	**1708.52**	**1979.56**	**2296.40**	**2705.26**	**3.3%**

[1]Includes energy for combined heat and power plants, except those whose primary business is to sell electricity, or electricity and heat, to the public.
[2]Excludes use for lease and plant fuel.
[3]Includes Federal and State taxes while excluding county and local taxes.
[4]E85 refers to a blend of 85 percent ethanol (renewable) and 15 percent motor gasoline (nonrenewable). To address cold starting issues, the percentage of ethanol varies seasonally. The annual average ethanol content of 74 percent is used for this forecast.
[5]Sales weighted-average price for all grades. Includes Federal, State and local taxes.
[6]Kerosene-type jet fuel. Includes Federal and State taxes while excluding county and local taxes.
[7]Diesel fuel for on-road use. Includes Federal and State taxes while excluding county and local taxes.
[8]Natural gas used as a vehicle fuel. Includes estimated motor vehicle fuel taxes and estimated dispensing costs or charges.
[9]Includes electricity-only and combined heat and power plants whose primary business is to sell electricity, or electricity and heat, to the public.
[10]Weighted averages of end-use fuel prices are derived from the prices shown in each sector and the corresponding sectoral consumption.
Btu = British thermal unit.
- - = Not applicable.
Note: Data for 2009 and 2010 are model results and may differ slightly from official EIA data reports.
Sources: 2009 and 2010 prices for motor gasoline, distillate fuel oil, and jet fuel are based on prices in the U.S. Energy Information Administration (EIA), *Petroleum Marketing Annual 2009*, DOE/EIA-0487(2009) (Washington, DC, August 2010). 2009 residential and commercial natural gas delivered prices: EIA,*Natural Gas Annual 2009*, DOE/EIA-0131(2009) (Washington, DC, December 2010). 2010 residential and commercial natural gas delivered prices: EIA, *Natural Gas Monthly*, DOE/EIA-0130(2011/07) (Washington, DC, July 2011). 2009 and 2010 industrial natural gas delivered prices are estimated based on: EIA, *Manufacturing Energy Consumption Survey* and industrial and wellhead prices from the *Natural Gas Annual 2009*, DOE/EIA-0131(2009) (Washington, DC, December 2010) and the *Natural Gas Monthly*, DOE/EIA-0130(2011/07) (Washington, DC, July 2011). 2009 transportation sector natural gas delivered prices are based on: EIA, *Natural Gas Annual 2009*, DOE/EIA-0131(2009) (Washington, DC, December 2010) and estimated State taxes, Federal taxes, and dispensing costs or charges. 2010 transportation sector natural gas delivered prices are model results. 2009 and 2010 electric power sector distillate and residual fuel oil prices: EIA, *Monthly Energy Review*, DOE/EIA-0035(2010/09) (Washington, DC, September 2010). 2009 and 2010 electric power sector natural gas prices: EIA, *Electric Power Monthly*, DOE/EIA-0226, April 2010 and April 2011, Table 4.2, and EIA, *State Energy Data Report 2009*, DOE/EIA-0214(2009) (Washington, DC, June 2011). 2009 and 2010 coal prices based on: EIA, *Quarterly Coal Report, October-December 2010*, DOE/EIA-0121(2010/4Q) (Washington, DC, May 2011) and EIA, AEO2012 National Energy Modeling System run REF2012.D020112C. 2009 and 2010 electricity prices: EIA, *Annual Energy Review 2010*, DOE/EIA-0384(2010) (Washington, DC, October 2011). 2009 and 2010 E85 prices derived from monthly prices in the Clean Cities Alternative Fuel Price Report. **Projections:** EIA, AEO2012 National Energy Modeling System run REF2012.D020112C.

Table A4. Residential sector key indicators and consumption
(quadrillion Btu per year, unless otherwise noted)

Key indicators and consumption	Reference case							Annual growth 2010-2035 (percent)
	2009	2010	2015	2020	2025	2030	2035	
Key indicators								
Households (millions)								
Single-family	81.73	82.11	85.49	89.94	94.26	98.56	102.54	0.9%
Multifamily	25.41	25.52	26.98	29.31	31.47	33.70	35.96	1.4%
Mobile homes	6.65	6.56	6.25	6.56	6.86	7.04	7.14	0.3%
Total	**113.78**	**114.19**	**118.73**	**125.82**	**132.60**	**139.30**	**145.64**	**1.0%**
Average house square footage	1646	1653	1684	1705	1725	1743	1759	0.2%
Energy intensity								
(million Btu per household)								
Delivered energy consumption	97.8	102.1	94.6	90.3	86.8	84.2	81.9	-0.9%
Total energy consumption	184.0	193.1	175.3	169.8	166.1	162.8	159.9	-0.8%
(thousand Btu per square foot)								
Delivered energy consumption	59.4	61.8	56.2	52.9	50.3	48.3	46.6	-1.1%
Total energy consumption	111.8	116.8	104.1	99.6	96.3	93.4	90.9	-1.0%
Delivered energy consumption by fuel								
Electricity								
Space heating	0.28	0.30	0.28	0.30	0.31	0.33	0.34	0.5%
Space cooling	0.81	1.08	1.01	1.06	1.12	1.18	1.24	0.6%
Water heating	0.44	0.45	0.47	0.50	0.52	0.53	0.53	0.7%
Refrigeration	0.38	0.37	0.37	0.38	0.39	0.41	0.43	0.6%
Cooking	0.11	0.11	0.11	0.12	0.13	0.14	0.15	1.4%
Clothes dryers	0.19	0.19	0.19	0.18	0.18	0.17	0.18	-0.3%
Freezers	0.08	0.08	0.08	0.08	0.09	0.09	0.09	0.3%
Lighting	0.70	0.69	0.52	0.48	0.46	0.46	0.47	-1.5%
Clothes washers[1]	0.03	0.03	0.03	0.03	0.02	0.02	0.02	-1.2%
Dishwashers[1]	0.10	0.10	0.10	0.10	0.10	0.10	0.11	0.4%
Color televisions and set-top boxes	0.32	0.33	0.32	0.34	0.37	0.40	0.43	1.1%
Personal computers and related equipment	0.17	0.17	0.19	0.22	0.24	0.26	0.27	1.8%
Furnace fans and boiler circulation pumps	0.14	0.13	0.14	0.14	0.14	0.15	0.15	0.4%
Other uses[2]	0.90	0.92	0.92	1.03	1.16	1.31	1.44	1.8%
Delivered energy	**4.66**	**4.95**	**4.75**	**4.96**	**5.23**	**5.55**	**5.86**	**0.7%**
Natural gas								
Space heating	3.31	3.50	3.39	3.34	3.27	3.24	3.19	-0.4%
Space cooling	0.00	0.00	0.00	0.00	0.00	0.00	0.00	-0.3%
Water heating	1.32	1.29	1.31	1.33	1.33	1.31	1.27	-0.1%
Cooking	0.22	0.22	0.22	0.22	0.22	0.23	0.23	0.3%
Clothes dryers	0.05	0.06	0.06	0.06	0.06	0.06	0.07	0.7%
Delivered energy	**4.90**	**5.06**	**4.97**	**4.95**	**4.88**	**4.84**	**4.76**	**-0.2%**
Distillate fuel oil								
Space heating	0.50	0.53	0.48	0.42	0.38	0.34	0.31	-2.1%
Water heating	0.10	0.10	0.07	0.06	0.05	0.04	0.04	-3.9%
Delivered energy	**0.60**	**0.63**	**0.55**	**0.48**	**0.43**	**0.38**	**0.35**	**-2.3%**
Liquefied petroleum gases								
Space heating	0.26	0.30	0.26	0.25	0.24	0.23	0.22	-1.1%
Water heating	0.08	0.07	0.05	0.04	0.04	0.04	0.03	-3.0%
Cooking	0.03	0.03	0.03	0.03	0.03	0.03	0.02	-0.9%
Other uses[3]	0.14	0.16	0.17	0.18	0.20	0.21	0.22	1.3%
Delivered energy	**0.51**	**0.56**	**0.51**	**0.50**	**0.50**	**0.51**	**0.51**	**-0.4%**
Marketed renewables (wood)[4]	0.43	0.42	0.43	0.43	0.43	0.43	0.43	0.1%
Other fuels[5]	0.04	0.04	0.03	0.03	0.03	0.03	0.03	-1.6%

Reference case

Table A4. Residential sector key indicators and consumption (continued)
(quadrillion Btu per year, unless otherwise noted)

Key indicators and consumption	Reference case							Annual growth 2010-2035 (percent)
	2009	2010	2015	2020	2025	2030	2035	
Delivered energy consumption by end use								
Space heating	4.81	5.08	4.86	4.78	4.67	4.60	4.52	-0.5%
Space cooling	0.81	1.08	1.01	1.06	1.12	1.18	1.24	0.6%
Water heating	1.94	1.91	1.90	1.92	1.94	1.91	1.88	-0.1%
Refrigeration	0.38	0.37	0.37	0.38	0.39	0.41	0.43	0.6%
Cooking	0.35	0.35	0.36	0.37	0.38	0.39	0.40	0.5%
Clothes dryers	0.25	0.25	0.25	0.25	0.24	0.24	0.25	-0.0%
Freezers	0.08	0.08	0.08	0.08	0.09	0.09	0.09	0.3%
Lighting	0.70	0.69	0.52	0.48	0.46	0.46	0.47	-1.5%
Clothes washers[1]	0.03	0.03	0.03	0.03	0.02	0.02	0.02	-1.2%
Dishwashers[1]	0.10	0.10	0.10	0.10	0.10	0.10	0.11	0.4%
Color televisions and set-top boxes	0.32	0.33	0.32	0.34	0.37	0.40	0.43	1.1%
Personal computers and related equipment	0.17	0.17	0.19	0.22	0.24	0.26	0.27	1.8%
Furnace fans and boiler circulation pumps	0.14	0.13	0.14	0.14	0.14	0.15	0.15	0.4%
Other uses[6]	1.04	1.08	1.09	1.21	1.36	1.52	1.67	1.8%
Delivered energy	11.13	11.66	11.24	11.36	11.51	11.73	11.93	0.1%
Electricity related losses	9.80	10.39	9.58	10.01	10.52	10.95	11.35	0.4%
Total energy consumption by end use								
Space heating	5.41	5.70	5.42	5.37	5.29	5.24	5.17	-0.4%
Space cooling	2.52	3.34	3.06	3.19	3.36	3.51	3.65	0.4%
Water heating	2.87	2.85	2.85	2.93	2.98	2.96	2.90	0.1%
Refrigeration	1.17	1.15	1.11	1.14	1.18	1.23	1.28	0.4%
Cooking	0.58	0.58	0.59	0.61	0.64	0.67	0.69	0.7%
Clothes dryers	0.65	0.65	0.64	0.62	0.59	0.58	0.60	-0.4%
Freezers	0.26	0.26	0.25	0.26	0.26	0.26	0.26	0.1%
Lighting	2.18	2.13	1.58	1.45	1.39	1.37	1.37	-1.7%
Clothes washers[1]	0.10	0.10	0.10	0.08	0.07	0.07	0.07	-1.4%
Dishwashers[1]	0.31	0.31	0.30	0.30	0.30	0.31	0.33	0.2%
Color televisions and set-top boxes	1.00	1.02	0.98	1.03	1.10	1.18	1.26	0.9%
Personal computers and related equipment	0.53	0.53	0.57	0.65	0.72	0.76	0.79	1.6%
Furnace fans and boiler circulation pumps	0.42	0.42	0.42	0.43	0.44	0.44	0.44	0.2%
Other uses[6]	2.94	3.01	2.96	3.29	3.70	4.10	4.47	1.6%
Total	20.93	22.05	20.81	21.36	22.02	22.68	23.28	0.2%
Nonmarketed renewables[7]								
Geothermal heat pumps	0.00	0.01	0.01	0.02	0.02	0.02	0.03	6.4%
Solar hot water heating	0.01	0.01	0.02	0.02	0.02	0.02	0.02	2.4%
Solar photovoltaic	0.00	0.00	0.04	0.05	0.05	0.06	0.06	10.7%
Wind	0.00	0.00	0.01	0.01	0.01	0.01	0.01	9.1%
Total	0.02	0.02	0.08	0.10	0.10	0.11	0.11	6.9%
Heating degree days[8]	4408	4382	4208	4172	4136	4101	4067	-0.3%
Cooling degree days[8]	1279	1498	1392	1409	1426	1443	1459	-0.1%

[1]Does not include water heating portion of load.
[2]Includes small electric devices, heating elements, and motors not listed above. Electric vehicles are included in the transportation sector.
[3]Includes such appliances as outdoor grills and mosquito traps.
[4]Includes wood used for primary and secondary heating in wood stoves or fireplaces as reported in the *Residential Energy Consumption Survey 2005*.
[5]Includes kerosene and coal.
[6]Includes all other uses listed above.
[7]Represents delivered energy displaced.
[8]See Table A5 for regional detail.
Btu = British thermal unit.
- - = Not applicable.
Note: Totals may not equal sum of components due to independent rounding. Data for 2009 and 2010 are model results and may differ slightly from official EIA data reports.
Sources: 2009 and 2010 consumption based on: U.S. Energy Information Administration (EIA), *Annual Energy Review 2010*, DOE/EIA-0384(2010) (Washington, DC, October 2011). 2009 and 2010 degree days based on state-level data from the National Oceanic and Atmospheric Administration's Climatic Data Center and Climate Prediction Center. Projections: EIA, AEO2012 National Energy Modeling System run REF2012.D020112C.

Table A5. Commercial sector key indicators and consumption
(quadrillion Btu per year, unless otherwise noted)

Key indicators and consumption	Reference case							Annual growth 2010-2035 (percent)
	2009	2010	2015	2020	2025	2030	2035	
Key indicators								
Total floorspace (billion square feet)								
Surviving	78.0	79.3	82.4	87.0	91.9	96.2	100.7	1.0%
New additions	2.3	1.8	1.7	2.0	2.0	2.0	2.3	1.0%
Total	80.3	81.1	84.1	89.1	93.9	98.2	103.0	1.0%
Energy consumption intensity (thousand Btu per square foot)								
Delivered energy consumption	106.0	107.3	104.6	103.1	101.0	100.6	99.8	-0.3%
Electricity related losses	117.0	117.3	110.2	110.6	110.6	110.2	109.0	-0.3%
Total energy consumption	223.0	224.5	214.8	213.7	211.5	210.7	208.8	-0.3%
Delivered energy consumption by fuel								
Purchased electricity								
Space heating[1]	0.18	0.18	0.16	0.16	0.16	0.16	0.16	-0.6%
Space cooling[1]	0.47	0.56	0.50	0.50	0.51	0.52	0.53	-0.2%
Water heating[1]	0.09	0.09	0.09	0.09	0.09	0.09	0.08	-0.4%
Ventilation	0.50	0.51	0.53	0.56	0.58	0.61	0.63	0.9%
Cooking	0.02	0.02	0.02	0.02	0.02	0.02	0.02	-0.3%
Lighting	1.03	1.01	1.00	1.03	1.06	1.10	1.13	0.4%
Refrigeration	0.40	0.39	0.35	0.34	0.34	0.34	0.35	-0.4%
Office equipment (PC)	0.22	0.21	0.19	0.19	0.20	0.21	0.21	0.0%
Office equipment (non-PC)	0.25	0.26	0.31	0.37	0.40	0.44	0.46	2.3%
Other uses[2]	1.29	1.30	1.43	1.62	1.80	2.00	2.22	2.2%
Delivered energy	4.46	4.54	4.59	4.88	5.16	5.48	5.80	1.0%
Natural gas								
Space heating[1]	1.61	1.65	1.69	1.73	1.70	1.68	1.64	-0.0%
Space cooling[1]	0.03	0.04	0.04	0.04	0.03	0.03	0.03	-1.1%
Water heating[1]	0.43	0.44	0.48	0.51	0.52	0.53	0.54	0.8%
Cooking	0.17	0.18	0.19	0.20	0.21	0.22	0.22	0.9%
Other uses[3]	0.95	0.98	1.01	1.04	1.07	1.14	1.25	1.0%
Delivered energy	3.20	3.28	3.41	3.51	3.53	3.60	3.69	0.5%
Distillate fuel oil								
Space heating[1]	0.16	0.14	0.12	0.11	0.10	0.10	0.09	-1.7%
Water heating[1]	0.03	0.03	0.03	0.03	0.03	0.03	0.03	0.9%
Other uses[4]	0.22	0.26	0.20	0.20	0.20	0.20	0.19	-1.2%
Delivered energy	0.41	0.43	0.35	0.34	0.33	0.33	0.32	-1.2%
Marketed renewables (biomass)	0.11	0.11	0.11	0.11	0.11	0.11	0.11	0.0%
Other fuels[5]	0.33	0.34	0.33	0.34	0.34	0.35	0.36	0.2%
Delivered energy consumption by end use								
Space heating[1]	1.95	1.97	1.98	2.00	1.96	1.93	1.89	-0.2%
Space cooling[1]	0.50	0.60	0.54	0.54	0.54	0.55	0.57	-0.2%
Water heating[1]	0.55	0.56	0.60	0.63	0.64	0.65	0.66	0.7%
Ventilation	0.50	0.51	0.53	0.56	0.58	0.61	0.63	0.9%
Cooking	0.20	0.20	0.21	0.23	0.23	0.24	0.24	0.8%
Lighting	1.03	1.01	1.00	1.03	1.06	1.10	1.13	0.4%
Refrigeration	0.40	0.39	0.35	0.34	0.34	0.34	0.35	-0.4%
Office equipment (PC)	0.22	0.21	0.19	0.19	0.20	0.21	0.21	0.0%
Office equipment (non-PC)	0.25	0.26	0.31	0.37	0.40	0.44	0.46	2.3%
Other uses[6]	2.90	2.99	3.09	3.30	3.53	3.80	4.13	1.3%
Delivered energy	8.51	8.70	8.80	9.18	9.48	9.87	10.28	0.7%

Table A5. Commercial sector key indicators and consumption (continued)
(quadrillion Btu per year, unless otherwise noted)

Key indicators and consumption	Reference case						Annual growth 2010-2035 (percent)
	2009	2010	2015	2020	2025	2030	2035	
Electricity related losses	9.39	9.52	9.27	9.85	10.38	10.82	11.23	0.7%
Total energy consumption by end use								
Space heating[1]	2.34	2.35	2.31	2.33	2.28	2.24	2.19	-0.3%
Space cooling[1]	1.50	1.77	1.54	1.55	1.57	1.58	1.60	-0.4%
Water heating[1]	0.75	0.75	0.78	0.80	0.81	0.82	0.82	0.4%
Ventilation	1.56	1.57	1.60	1.69	1.75	1.81	1.84	0.6%
Cooking	0.25	0.25	0.26	0.27	0.27	0.28	0.29	0.5%
Lighting	3.21	3.14	3.01	3.12	3.21	3.27	3.32	0.2%
Refrigeration	1.24	1.21	1.06	1.02	1.02	1.02	1.04	-0.6%
Office equipment (PC)	0.67	0.66	0.57	0.58	0.59	0.61	0.63	-0.2%
Office equipment (non-PC)	0.77	0.81	0.95	1.10	1.21	1.30	1.36	2.1%
Other uses[6]	5.62	5.71	5.98	6.56	7.15	7.75	8.42	1.6%
Total	17.90	18.22	18.06	19.03	19.86	20.69	21.50	0.7%
Nonmarketed renewable fuels[7]								
Solar thermal	0.03	0.03	0.03	0.03	0.03	0.04	0.04	1.4%
Solar photovoltaic	0.00	0.01	0.01	0.01	0.01	0.01	0.01	2.8%
Wind	0.00	0.00	0.00	0.00	0.00	0.00	0.00	5.3%
Total	0.03	0.03	0.04	0.04	0.04	0.05	0.05	1.7%
Heating Degree Days								
New England	6649	5944	6349	6351	6355	6358	6360	0.3%
Middle Atlantic	5798	5453	5588	5587	5586	5585	5583	0.1%
East North Central	6542	6209	6215	6215	6215	6215	6215	0.0%
West North Central	6837	6585	6456	6461	6463	6466	6468	-0.1%
South Atlantic	2839	3183	2728	2703	2677	2651	2625	-0.8%
East South Central	3599	4003	3474	3480	3485	3491	3496	-0.5%
West South Central	2198	2503	2156	2149	2143	2137	2131	-0.6%
Mountain	4852	4808	4780	4749	4713	4677	4641	-0.1%
Pacific	3188	3202	3130	3135	3138	3140	3143	-0.1%
United States	4408	4382	4208	4172	4136	4101	4067	-0.3%
Cooling Degree Days								
New England	363	655	518	518	517	517	516	-0.9%
Middle Atlantic	587	997	783	783	783	784	784	-1.0%
East North Central	547	978	779	780	780	781	781	-0.9%
West North Central	720	1123	976	975	974	973	973	-0.6%
South Atlantic	2047	2289	2103	2118	2134	2149	2165	-0.2%
East South Central	1491	1999	1668	1665	1662	1658	1655	-0.8%
West South Central	2582	2755	2602	2607	2611	2615	2619	-0.2%
Mountain	1551	1489	1578	1595	1617	1637	1658	0.4%
Pacific	967	746	891	888	887	885	883	0.7%
United States	1279	1498	1392	1409	1426	1443	1459	-0.1%

[1]Includes fuel consumption for district services.
[2]Includes miscellaneous uses, such as service station equipment, automated teller machines, telecommunications equipment, and medical equipment.
[3]Includes miscellaneous uses, such as pumps, emergency generators, combined heat and power in commercial buildings, and manufacturing performed in commercial buildings.
[4]Includes miscellaneous uses, such as cooking, emergency generators, and combined heat and power in commercial buildings.
[5]Includes residual fuel oil, liquefied petroleum gases, coal, motor gasoline, and kerosene.
[6]Includes miscellaneous uses, such as service station equipment, automated teller machines, telecommunications equipment, medical equipment, pumps, emergency generators, combined heat and power in commercial buildings, manufacturing performed in commercial buildings, and cooking (distillate), plus residual fuel oil, liquefied petroleum gases, coal, motor gasoline, and kerosene.
[7]Represents delivered energy displaced.
Btu = British thermal unit.
PC = Personal computer.
Note: Totals may not equal sum of components due to independent rounding. Data for 2009 and 2010 are model results and may differ slightly from official EIA data reports.
Sources: 2009 and 2010 consumption based on: U.S. Energy Information Administration (EIA), *Annual Energy Review 2010*, DOE/EIA-0384(2010) (Washington, DC, October 2011). 2009 and 2010 degree days based on state-level data from the National Oceanic and Atmospheric Administration's Climatic Data Center and Climate Prediction Center. Projections: EIA, AEO2012 National Energy Modeling System run REF2012.D020112C.

Table A6. Industrial sector key indicators and consumption

Key indicators and consumption	Reference case							Annual growth 2010-2035 (percent)
	2009	2010	2015	2020	2025	2030	2035	
Key indicators								
Value of shipments (billion 2005 dollars)								
Manufacturing	4052	4260	4857	5260	5745	6023	6285	1.6%
Nonmanufacturing	1615	1578	1873	2103	2228	2305	2407	1.7%
Total	**5667**	**5838**	**6730**	**7363**	**7973**	**8328**	**8692**	**1.6%**
Energy prices								
(2010 dollars per million Btu)								
Liquefied petroleum gases	20.05	21.80	27.43	27.76	29.24	30.48	32.18	1.6%
Motor gasoline	16.79	16.77	29.20	30.72	32.06	33.01	33.55	2.8%
Distillate fuel oil	16.74	21.32	24.20	25.73	27.22	28.39	29.53	1.3%
Residual fuel oil	12.16	10.92	19.21	20.53	21.23	21.71	21.65	2.8%
Asphalt and road oil	6.59	5.59	9.30	9.94	10.37	10.45	10.69	2.6%
Natural gas heat and power	4.59	4.78	4.16	4.41	5.33	5.88	6.89	1.5%
Natural gas feedstocks	6.16	6.32	5.68	5.93	6.83	7.36	8.33	1.1%
Metallurgical coal	5.49	5.84	7.22	7.58	8.11	8.61	9.11	1.8%
Other industrial coal	2.99	2.71	3.27	3.30	3.38	3.50	3.64	1.2%
Coal for liquids	--	--	1.26	2.05	2.08	2.22	2.38	--
Electricity	20.05	19.63	18.91	18.95	19.60	19.81	20.78	0.2%
(nominal dollars per million Btu)								
Liquefied petroleum gases	19.82	21.80	29.56	32.63	37.51	43.38	50.95	3.5%
Motor gasoline	16.60	16.77	31.46	36.10	41.14	46.98	53.12	4.7%
Distillate fuel oil	16.55	21.32	26.08	30.24	34.93	40.40	46.76	3.2%
Residual fuel oil	12.02	10.92	20.70	24.13	27.24	30.89	34.28	4.7%
Asphalt and road oil	6.52	5.59	10.02	11.68	13.30	14.87	16.93	4.5%
Natural gas heat and power	4.54	4.78	4.49	5.19	6.84	8.37	10.91	3.4%
Natural gas feedstocks	6.09	6.32	6.12	6.96	8.77	10.48	13.18	3.0%
Metallurgical coal	5.43	5.84	7.78	8.91	10.40	12.26	14.42	3.7%
Other industrial coal	2.96	2.71	3.52	3.87	4.34	4.98	5.77	3.1%
Coal for liquids	--	--	1.36	2.41	2.67	3.16	3.78	--
Electricity	19.83	19.63	20.38	22.27	25.15	28.20	32.90	2.1%
Energy consumption (quadrillion Btu)[1]								
Industrial consumption excluding refining								
Liquefied petroleum gases heat and power	0.45	0.41	0.36	0.39	0.41	0.41	0.40	-0.0%
Liquefied petroleum gases feedstocks	1.54	1.58	1.45	1.65	1.75	1.76	1.74	0.4%
Motor gasoline	0.24	0.25	0.28	0.30	0.30	0.30	0.30	0.8%
Distillate fuel oil	1.11	1.15	1.25	1.18	1.19	1.17	1.18	0.1%
Residual fuel oil	0.10	0.11	0.09	0.08	0.08	0.08	0.08	-1.1%
Petrochemical feedstocks	0.90	0.94	1.01	1.20	1.29	1.31	1.30	1.3%
Petroleum coke	0.28	0.16	0.20	0.19	0.15	0.12	0.13	-1.1%
Asphalt and road oil	0.87	0.88	1.00	1.00	0.98	0.94	0.94	0.3%
Miscellaneous petroleum[2]	0.38	0.52	0.14	0.12	0.12	0.11	0.12	-5.8%
Petroleum subtotal	5.87	6.00	5.78	6.11	6.27	6.20	6.19	0.1%
Natural gas heat and power	4.48	4.84	5.23	5.22	5.27	5.23	5.23	0.3%
Natural gas feedstocks	0.47	0.48	0.48	0.51	0.50	0.47	0.44	-0.3%
Lease and plant fuel[3]	1.31	1.37	1.43	1.55	1.57	1.59	1.63	0.7%
Natural gas subtotal	6.25	6.69	7.14	7.27	7.34	7.29	7.31	0.4%
Metallurgical coal and coke[4]	0.38	0.55	0.56	0.46	0.46	0.42	0.38	-1.5%
Other industrial coal	0.88	0.95	0.97	0.98	1.02	1.02	1.02	0.3%
Coal subtotal	1.26	1.50	1.53	1.44	1.47	1.44	1.40	-0.3%
Renewables[5]	1.37	1.50	1.61	1.67	1.82	1.87	1.95	1.1%
Purchased electricity	2.94	3.09	3.24	3.26	3.33	3.24	3.12	0.0%
Delivered energy	**17.69**	**18.78**	**19.30**	**19.75**	**20.23**	**20.04**	**19.97**	**0.2%**
Electricity related losses	6.19	6.47	6.55	6.58	6.69	6.39	6.04	-0.3%
Total	**23.88**	**25.25**	**25.84**	**26.33**	**26.92**	**26.44**	**26.01**	**0.1%**

Table A6. Industrial sector key indicators and consumption (continued)

Key indicators and consumption	Reference case							Annual growth 2010-2035 (percent)
	2009	2010	2015	2020	2025	2030	2035	
Refining consumption								
Liquefied petroleum gases heat and power	0 01	0.01	0.01	0.01	0 01	0.01	0.01	0.4%
Distillate fuel oil	0.00	0 00	0.00	0.00	0 00	0 00	0.00	--
Residual fuel oil	0.01	0 01	0.00	0.00	0 00	0 00	0.00	--
Petroleum coke	0.52	0 52	0.53	0.49	0.49	0 51	0.53	0.1%
Still gas	1.50	1 50	1.55	1.36	1 34	1 39	1.45	-0.1%
Miscellaneous petroleum[2]	0.02	0 02	0.02	0.02	0 02	0 02	0.02	1.2%
Petroleum subtotal	2.05	2 05	2.11	1.89	1 86	1 93	2.02	-0.1%
Natural gas heat and power	1.38	1.44	1.48	1.53	1 55	1 51	1.51	0.2%
Natural-gas-to-liquids heat and power	0.00	0 00	0.00	0.00	0 00	0 00	0.00	--
Natural gas subtotal	1.38	1.44	1.48	1.53	1 55	1 51	1.51	0.2%
Other industrial coal	0.06	0 06	0.06	0.06	0 06	0 06	0.06	0.0%
Coal-to-liquids heat and power	0.00	0 00	0.00	0.26	0 36	0 48	0.60	--
Coal subtotal	0.06	0 06	0.06	0.32	0.42	0 54	0.66	10.0%
Biofuels heat and coproducts	0.82	0 84	0.81	0.96	1 27	1 92	2.57	4.6%
Purchased electricity	0.19	0 20	0.20	0.20	0.19	0 20	0.21	0.3%
Delivered energy	**4.51**	**4.60**	**4.66**	**4.89**	**5.30**	**6.10**	**6.97**	**1.7%**
Electricity related losses	0.40	0.41	0.39	0.39	0 39	0.40	0.41	0.0%
Total	**4.91**	**5.01**	**5.05**	**5.28**	**5.69**	**6.50**	**7.39**	**1.6%**
Total industrial sector consumption								
Liquefied petroleum gases heat and power	0.46	0.42	0.38	0.41	0.42	0.42	0.41	-0.0%
Liquefied petroleum gases feedstocks	1.54	1 58	1.45	1.65	1.75	1.76	1.74	0.4%
Motor gasoline	0.24	0 25	0.28	0.30	0 30	0 30	0.30	0.8%
Distillate fuel oil	1.11	1.16	1.25	1.18	1.19	1.17	1.18	0.1%
Residual fuel oil	0.11	0.12	0.09	0.08	0 08	0 08	0.08	-1.3%
Petrochemical feedstocks	0.90	0 94	1.01	1.20	1 29	1 31	1.30	1.3%
Petroleum coke	0.80	0 68	0.73	0.68	0 64	0 63	0.66	-0.1%
Asphalt and road oil	0.87	0 88	1.00	1.00	0 98	0 94	0.94	0.3%
Still gas	1.50	1 50	1.55	1.36	1 34	1 39	1.45	-0.1%
Miscellaneous petroleum[2]	0.40	0 54	0.17	0.14	0.14	0.13	0.14	-5.3%
Petroleum subtotal	7.93	8 05	7.89	7.99	8.13	8.13	8.21	0.1%
Natural gas heat and power	5.86	6 28	6.71	6.75	6 82	6.74	6.74	0.3%
Natural-gas-to-liquids heat and power	0.00	0 00	0.00	0.00	0 00	0 00	0.00	--
Natural gas feedstocks	0.47	0.48	0.48	0.51	0 50	0.47	0.44	-0.3%
Lease and plant fuel[3]	1.31	1 37	1.43	1.55	1 57	1 59	1.63	0.7%
Natural gas subtotal	7.63	8.14	8.62	8.80	8 89	8 80	8.81	0.3%
Metallurgical coal and coke[4]	0.38	0 55	0.56	0.46	0.46	0.42	0.38	-1.5%
Other industrial coal	0.94	1 01	1.03	1.04	1 08	1 08	1.08	0.3%
Coal-to-liquids heat and power	0.00	0 00	0.00	0.26	0 36	0 48	0.60	--
Coal subtotal	1.32	1 56	1.59	1.76	1 90	1 98	2.06	1.1%
Biofuels heat and coproducts	0.82	0 84	0.81	0.96	1 27	1 92	2.57	4.6%
Renewables[5]	1.37	1 50	1.61	1.67	1 82	1 87	1.95	1.1%
Purchased electricity	3.13	3 28	3.44	3.46	3 52	3.44	3.33	0.1%
Delivered energy	**22.20**	**23.37**	**23.96**	**24.64**	**25.53**	**26.14**	**26.94**	**0.6%**
Electricity related losses	6.59	6 89	6.94	6.97	7 09	6 80	6.46	-0.3%
Total	**28.79**	**30.26**	**30.90**	**31.61**	**32.61**	**32.93**	**33.39**	**0.4%**

Table A6. Industrial sector key indicators and consumption (continued)

Key indicators and consumption	Reference case							Annual growth 2010-2035 (percent)
	2009	2010	2015	2020	2025	2030	2035	
Energy consumption per dollar of shipments (thousand Btu per 2005 dollar)								
Liquid fuels and other petroleum	1.40	1.38	1.17	1.09	1.02	0.98	0.94	-1.5%
Natural gas	1.35	1.39	1.28	1.20	1.11	1.06	1.01	-1.3%
Coal	0.23	0.27	0.24	0.24	0.24	0.24	0.24	-0.5%
Renewable fuels[5]	0.39	0.40	0.36	0.36	0.39	0.45	0.52	1.0%
Purchased electricity	0.55	0.56	0.51	0.47	0.44	0.41	0.38	-1.5%
Delivered energy	**3.92**	**4.00**	**3.56**	**3.35**	**3.20**	**3.14**	**3.10**	**-1.0%**
Industrial combined heat and power								
Capacity (gigawatts)	25.08	25.64	30.38	35.48	40.71	48.10	55.79	3.2%
Generation (billion kilowatthours)	130.57	141.07	168.00	201.40	235.62	287.62	341.40	3.6%

[1]Includes energy for combined heat and power plants, except those whose primary business is to sell electricity, or electricity and heat, to the public.
[2]Includes lubricants and miscellaneous petroleum products.
[3]Represents natural gas used in well, field, and lease operations, and in natural gas processing plant machinery.
[4]Includes net coal coke imports.
[5]Includes consumption of energy produced from hydroelectric, wood and wood waste, municipal waste, and other biomass sources.
Btu = British thermal unit.
- - = Not applicable.
Note: Totals may not equal sum of components due to independent rounding. Data for 2009 and 2010 are model results and may differ slightly from official EIA data reports.
Sources: 2009 and 2010 prices for motor gasoline and distillate fuel oil are based on: U.S. Energy Information Administration (EIA), *Petroleum Marketing Annual 2009*, DOE/EIA-0487(2009) (Washington, DC, August 2010). 2009 and 2010 petrochemical feedstock and asphalt and road oil prices are based on: EIA, *State Energy Data Report 2009*, DOE/EIA-0214(2009) (Washington, DC, June 2011). 2009 and 2010 coal prices are based on: EIA, *Quarterly Coal Report, October-December 2010*, DOE/EIA-0121(2010/4Q) (Washington, DC, May 2011) and EIA, AEO2012 National Energy Modeling System run REF2012.D020112C. 2009 and 2010 electricity prices: EIA, *Annual Energy Review 2010*, DOE/EIA-0384(2010) (Washington, DC, October 2011). 2009 and 2010 natural gas prices are based on: EIA, *Manufacturing Energy Consumption Survey* and industrial and wellhead prices from the *Natural Gas Annual 2009*, DOE/EIA-0131(2009) (Washington, DC, December 2010) and the *Natural Gas Monthly*, DOE/EIA-0130(2011/07) (Washington, DC, July 2011). 2009 refining consumption values are based on: *Petroleum Supply Annual 2009*, DOE/EIA-0340(2009)/1 (Washington, DC, July 2010). 2010 refining consumption based on: *Petroleum Supply Annual 2010*, DOE/EIA-0340(2010)/1 (Washington, DC, July 2011). Other 2009 and 2010 consumption values are based on: EIA, *Annual Energy Review 2010*, DOE/EIA-0384(2010) (Washington, DC, October 2011). 2009 and 2010 shipments: IHS Global Insight, Global Insight Industry model, August 2011. Projections: EIA, AEO2012 National Energy Modeling System run REF2012.D020112C.

Reference case

Table A7. Transportation sector key indicators and delivered energy consumption

Key indicators and consumption	Reference case							Annual growth 2010-2035 (percent)
	2009	2010	2015	2020	2025	2030	2035	
Key indicators								
Travel indicators								
(billion vehicle miles traveled)								
Light-duty vehicles less than 8,501 pounds	2625	2662	2710	2881	3111	3363	3583	1 2%
Commercial light trucks[1]	58	64	70	76	83	88	92	1.5%
Freight trucks greater than 10,000 pounds	240	234	273	297	317	330	345	1.6%
(billion seat miles available)								
Air	964	999	1028	1075	1120	1164	1208	0.8%
(billion ton miles traveled)								
Rail	1532	1559	1503	1662	1782	1826	1871	0.7%
Domestic shipping	477	522	549	587	604	617	627	0.7%
Energy efficiency indicators								
(miles per gallon)								
New light-duty vehicle CAFE standard[2]	25.4	25.7	32.4	35.0	35.2	35 3	35 3	1.3%
New car[2]	28.2	28.2	37 0	39.9	39.9	39.9	39 9	1.4%
New light truck[2]	23.0	23.4	27 9	29.2	29.2	29.2	29 2	0.9%
Compliance new light-duty vehicle[3]	29.3	29.2	32 5	35.9	36.8	37.4	37 9	1.0%
New car[3]	34.0	33.8	37.4	40.3	41.3	42.2	42 9	1.0%
New light truck[3]	25.4	25.5	27.7	30.6	31.0	31.2	31 5	0.8%
Tested new light-duty vehicle[4]	28.2	28.3	31 5	35.9	36.8	37.4	37 9	1.2%
New car[4]	33.2	33.3	36.4	40.3	41.2	42.2	42 8	1.0%
New light truck[4]	24.2	24.3	26.7	30.6	31.0	31.2	31 5	1.0%
On-road new light-duty vehicle[5]	23.0	22.9	25 6	29.2	30.0	30.5	30 9	1.2%
New car[5]	27.4	27.3	29 9	33.1	33.9	34.7	35 2	1.0%
New light truck[5]	19.5	19.6	21 6	24.7	24.9	25.2	25.4	1.0%
Light-duty stock[6]	20.4	20.4	21 5	23.6	25.6	27.1	28 2	1.3%
New commercial light truck[1]	15.6	15.7	16.7	18.8	18.9	19.0	19.1	0.8%
Stock commercial light truck[1]	14.3	14.4	15 2	16.7	18.0	18.7	19 0	1.1%
Freight truck	6.7	6.7	6 8	7.3	7.7	8.0	8.1	0.8%
(seat miles per gallon)								
Aircraft	62.0	62.3	62 8	63.8	65.2	67.0	69 3	0.4%
(ton miles per thousand Btu)								
Rail	3.4	3.4	3 5	3.5	3.5	3.5	3 5	0.1%
Domestic shipping	2.4	2.4	2.4	2.5	2.5	2.5	2 5	0.2%
Energy use by mode								
(quadrillion Btu)								
Light-duty vehicles	15.89	16 06	15.39	14.84	14.73	15 05	15.46	-0.2%
Commercial light trucks[1]	0.51	0 55	0.58	0.57	0 58	0 59	0.61	0.4%
Bus transportation	0.21	0 25	0.26	0.27	0 29	0 30	0.31	0.9%
Freight trucks	4.95	4 82	5.51	5.57	5 66	5 69	5.84	0.8%
Rail, passenger	0.04	0 05	0.05	0.06	0 06	0 06	0.06	1.2%
Rail, freight	0.36	0.45	0.43	0.48	0 51	0 52	0.53	0.6%
Shipping, domestic	0.17	0 22	0.23	0.24	0 25	0 25	0.25	0.5%
Shipping, international	0.77	0 86	0.87	0.87	0 88	0 88	0.89	0.1%
Recreational boats	0.24	0 25	0.26	0.26	0 27	0 28	0.29	0.5%
Air	2.44	2 52	2.55	2.63	2.71	2.76	2.79	0.4%
Military use	0.71	0.77	0.66	0.64	0 66	0.70	0.74	-0.1%
Lubricants	0.13	0.14	0.13	0.14	0.14	0.14	0.14	0.1%
Pipeline fuel	0.61	0 65	0.68	0.67	0 67	0 68	0.69	0.2%
Total	27.04	27.59	27.60	27.25	27.40	27.90	28.60	0.1%

Table A7. Transportation sector key indicators and delivered energy consumption (continued)

Key indicators and consumption	Reference case							Annual growth 2010-2035 (percent)
	2009	2010	2015	2020	2025	2030	2035	
Energy use by mode (million barrels per day oil equivalent)								
Light-duty vehicles	8.50	8.63	8.30	8.05	8.05	8.31	8.64	0.0%
Commercial light trucks[1]	0.26	0.28	0.30	0.29	0.30	0.30	0.31	0.4%
Bus transportation	0.10	0.12	0.13	0.13	0.14	0.14	0.15	0.9%
Freight trucks	2.39	2.32	2.65	2.68	2.72	2.74	2.81	0.8%
Rail, passenger	0.02	0.02	0.02	0.03	0.03	0.03	0.03	1.2%
Rail, freight	0.17	0.22	0.21	0.23	0.24	0.25	0.25	0.6%
Shipping, domestic	0.08	0.10	0.11	0.11	0.11	0.11	0.12	0.5%
Shipping, international	0.34	0.38	0.38	0.38	0.38	0.39	0.39	0.1%
Recreational boats	0.13	0.14	0.14	0.14	0.15	0.15	0.16	0.5%
Air	1.18	1.22	1.23	1.27	1.31	1.33	1.35	0.4%
Military use	0.34	0.37	0.32	0.31	0.32	0.34	0.36	-0.1%
Lubricants	0.06	0.07	0.06	0.06	0.07	0.07	0.07	0.1%
Pipeline fuel	0.29	0.31	0.32	0.32	0.32	0.32	0.32	0.2%
Total	13.87	14.17	14.17	14.01	14.14	14.48	14.95	0.2%

[1]Commercial trucks 8,501 to 10,000 pounds gross vehicle weight rating.
[2]CAFE standard based on projected new vehicle sales.
[3]Includes CAFE credits for alternative fueled vehicle sales and credit banking.
[4]Environmental Protection Agency rated miles per gallon.
[5]Tested new vehicle efficiency revised for on-road performance.
[6]Combined "on-the-road" estimate for all cars and light trucks.
CAFE = Corporate average fuel economy.
Btu = British thermal unit.
Note: Totals may not equal sum of components due to independent rounding. Data for 2009 and 2010 are model results and may differ slightly from official EIA data reports.
Sources: 2009 and 2010: U.S. Energy Information Administration (EIA), *Natural Gas Annual 2009*, DOE/EIA-0131(2009) (Washington, DC, December 2010); EIA, *Annual Energy Review 2010*, DOE/EIA-0384(2010) (Washington, DC, October 2011); Federal Highway Administration, *Highway Statistics 2009* (Washington, DC, April 2011); Oak Ridge National Laboratory, *Transportation Energy Data Book: Edition 30 and Annual* (Oak Ridge, TN, 2011); National Highway Traffic and Safety Administration, *Summary of Fuel Economy Performance* (Washington, DC, October 28, 2010); U.S. Department of Commerce, Bureau of the Census, "Vehicle Inventory and Use Survey," EC02TV (Washington, DC, December 2004); EIA, Alternatives to Traditional Transportation Fuels 2008 (Part II - User and Fuel Data), April 2010; EIA, *State Energy Data Report 2009*, DOE/EIA-0214(2009) (Washington, DC, June 2011); U.S. Department of Transportation, Research and Special Programs Administration, *Air Carrier Statistics Monthly, December 2010/2009* (Washington, DC, December 2010); EIA, *Fuel Oil and Kerosene Sales 2009*, DOE/EIA-0535(2009) (Washington, DC, February 2011); and United States Department of Defense, Defense Fuel Supply Center, Fact Book (January, 2010). **Projections:** EIA, AEO2012 National Energy Modeling System run REF2012.D020112C.

Reference case

Table A8. Electricity supply, disposition, prices, and emissions
(billion kilowatthours, unless otherwise noted)

Supply, disposition, prices, and emissions	Reference case							Annual growth 2010-2035 (percent)
	2009	2010	2015	2020	2025	2030	2035	
Generation by fuel type								
Electric power sector[1]								
Power only[2]								
Coal	1712	1799	1531	1604	1710	1757	1803	0.0%
Petroleum	32	32	25	26	26	27	27	-0.6%
Natural gas[3]	723	776	903	874	882	983	1074	1.3%
Nuclear power	799	807	830	887	917	914	887	0.4%
Pumped storage/other[4]	2	2	2	2	2	2	2	-1.2%
Renewable sources[5]	384	390	504	544	579	594	630	1.9%
Distributed generation (natural gas)	0	0	0	1	2	3	4	--
Total	3651	3806	3796	3937	4118	4279	4427	0.6%
Combined heat and power[6]								
Coal	29	32	30	30	31	31	31	-0.1%
Petroleum	4	3	1	1	1	1	1	-5.2%
Natural gas	118	122	126	124	124	124	123	0.0%
Renewable sources	5	5	4	5	5	5	4	-0.7%
Total	159	165	160	160	161	160	159	-0.1%
Total electric power sector generation	3810	3971	3956	4097	4279	4439	4586	0.6%
Less direct use	14	16	13	13	13	13	13	-0.7%
Net available to the grid	3796	3955	3942	4084	4265	4426	4572	0.6%
End-use sector[7]								
Coal	15	20	20	38	46	54	63	4.7%
Petroleum	3	3	2	2	2	2	2	-0.7%
Natural gas	80	84	101	113	132	160	198	3.5%
Other gaseous fuels[8]	10	11	16	16	15	15	15	1.2%
Renewable sources[9]	31	34	55	65	78	103	125	5.4%
Other[10]	4	4	3	3	3	3	3	-0.8%
Total end-use sector generation	143	155	197	237	277	338	406	3.9%
Less direct use	107	112	149	180	208	243	288	3.8%
Total sales to the grid	36	43	48	57	69	95	118	4.1%
Total electricity generation by fuel								
Coal	1756	1851	1581	1671	1786	1841	1897	0.1%
Petroleum	39	37	28	28	29	29	30	-0.8%
Natural gas	921	982	1130	1113	1140	1270	1398	1.4%
Nuclear power	799	807	830	887	917	914	887	0.4%
Renewable sources[5,9]	420	429	562	614	662	702	760	2.3%
Other[11]	19	21	21	21	21	21	21	-0.0%
Total electricity generation	3953	4126	4152	4334	4556	4777	4992	0.8%
Net generation to the grid	3832	3998	3990	4141	4335	4521	4691	0.6%
Net imports	34	26	29	26	22	14	12	-2.9%
Electricity sales by sector								
Residential	1364	1451	1392	1454	1533	1626	1718	0.7%
Commercial	1307	1329	1346	1431	1513	1607	1699	1.0%
Industrial	917	962	1008	1013	1032	1009	977	0.1%
Transportation	7	7	8	9	12	16	22	4.8%
Total	3596	3749	3753	3907	4090	4258	4415	0.7%
Direct use	121	128	162	193	221	256	302	3.5%
Total electricity use	3717	3877	3915	4100	4311	4514	4716	0.8%

Table A8. Electricity supply, disposition, prices, and emissions (continued)
(billion kilowatthours, unless otherwise noted)

Supply, disposition, prices, and emissions	Reference case							Annual growth 2010-2035 (percent)
	2009	2010	2015	2020	2025	2030	2035	
End-use prices								
(2010 cents per kilowatthour)								
Residential	11.6	11.5	11.8	11.6	11.6	11.6	11.8	0.1%
Commercial	10.3	10.1	9.9	9.8	9.9	9.8	10.1	-0.0%
Industrial	6.8	6.7	6.5	6.5	6.7	6.8	7.1	0.2%
Transportation	12.2	11.3	10.4	10.1	10.8	11.1	11.5	0.1%
All sectors average	9.9	9.8	9.7	9.6	9.7	9.8	10.1	0.1%
(nominal cents per kilowatthour)								
Residential	11.5	11.5	12.7	13.6	14.9	16.5	18.7	2.0%
Commercial	10.1	10.1	10.7	11.5	12.7	13.9	15.9	1.8%
Industrial	6.8	6.7	7.0	7.6	8.6	9.6	11.2	2.1%
Transportation	12.0	11.3	11.2	11.9	13.8	15.8	18.3	2.0%
All sectors average	9.8	9.8	10.4	11.3	12.5	13.9	16.0	2.0%
Prices by service category								
(2010 cents per kilowatthour)								
Generation	6.1	5.9	5.6	5.7	6.0	6.1	6.4	0.3%
Transmission	1.0	1.0	1.1	1.1	1.1	1.1	1.1	0.3%
Distribution	2.9	2.9	3.0	2.8	2.7	2.6	2.6	-0.5%
(nominal cents per kilowatthour)								
Generation	6.0	5.9	6.0	6.7	7.7	8.7	10.2	2.2%
Transmission	1.0	1.0	1.2	1.3	1.4	1.6	1.8	2.2%
Distribution	2.8	2.9	3.3	3.3	3.4	3.7	4.1	1.4%
Electric power sector emissions[1]								
Sulfur dioxide (million short tons)	5.72	5.11	1.26	1.31	1.55	1.62	1.71	-4.3%
Nitrogen oxide (million short tons)	1.99	2.06	1.79	1.87	1.92	1.94	1.96	-0.2%
Mercury (short tons)	36.25	34.70	6.44	6.74	7.24	7.51	7.86	-5.8%

[1]Includes electricity-only and combined heat and power plants whose primary business is to sell electricity, or electricity and heat, to the public.
[2]Includes plants that only produce electricity.
[3]Includes electricity generation from fuel cells.
[4]Includes non-biogenic municipal waste. The U.S. Energy Information Administration estimates that in 2010 approximately 6 billion kilowatthours of electricity were generated from a municipal waste stream containing petroleum-derived plastics and other non-renewable sources. See U.S. Energy Information Administration, *Methodology for Allocating Municipal Solid Waste to Biogenic and Non-Biogenic Energy*, (Washington, DC, May 2007).
[5]Includes conventional hydroelectric, geothermal, wood, wood waste, biogenic municipal waste, landfill gas, other biomass, solar, and wind power.
[6]Includes combined heat and power plants whose primary business is to sell electricity and heat to the public (i.e., those that report North American Industry Classification System code 22).
[7]Includes combined heat and power plants and electricity-only plants in the commercial and industrial sectors; and small on-site generating systems in the residential, commercial, and industrial sectors used primarily for own-use generation, but which may also sell some power to the grid.
[8]Includes refinery gas and still gas.
[9]Includes conventional hydroelectric, geothermal, wood, wood waste, all municipal waste, landfill gas, other biomass, solar, and wind power.
[10]Includes batteries, chemicals, hydrogen, pitch, purchased steam, sulfur, and miscellaneous technologies.
[11]Includes pumped storage, non-biogenic municipal waste, refinery gas, still gas, batteries, chemicals, hydrogen, pitch, purchased steam, sulfur, and miscellaneous technologies.
- - = Not applicable.
Note: Totals may not equal sum of components due to independent rounding. Data for 2009 and 2010 are model results and may differ slightly from official EIA data reports.
Sources: 2009 and 2010 electric power sector generation; sales to the grid; net imports; electricity sales; and electricity end-use prices: U.S. Energy Information Administration (EIA), *Annual Energy Review 2010*, DOE/EIA-0384(2010) (Washington, DC, October 2011), and supporting databases. 2009 and 2010 emissions: U.S. Environmental Protection Agency, Clean Air Markets Database. 2009 and 2010 electricity prices by service category: EIA, AEO2012 National Energy Modeling System run REF2012.D020112C. **Projections:** EIA, AEO2012 National Energy Modeling System run REF2012.D020112C.

Reference case

Table A9. Electricity generating capacity
(gigawatts)

Net summer capacity[1]	Reference case							Annual growth 2010-2035 (percent)
	2009	2010	2015	2020	2025	2030	2035	
Electric power sector[2]								
Power only[3]								
Coal	305.9	308.1	276.7	269.8	269.8	269.9	270.4	-0.5%
Oil and natural gas steam[4]	109.1	107.4	90 0	89.4	88.9	88.0	87 2	-0.8%
Combined cycle	167.7	171.7	187.4	187.7	197.6	218.3	246 0	1.4%
Combustion turbine/diesel	133.1	134.8	138.7	145.6	152.7	158.6	169 0	0.9%
Nuclear power[5]	101.1	101.2	103 6	111.2	114.7	114.3	110 9	0.4%
Pumped storage	22.2	22.2	22.2	22 2	22.2	22.2	22 2	0.0%
Fuel cells	0.0	0.0	0.0	0 0	0.0	0.0	0 0	2.7%
Renewable sources[6]	120.3	125.2	144.4	145.8	151.2	156.1	169 3	1.2%
Distributed generation[7]	0.0	0.0	0.2	0 5	0.8	1.3	2.1	--
Total	959.5	970.6	963.2	972.1	997.8	1028.7	1077.0	0.4%
Combined heat and power[8]								
Coal	5.3	5.2	4.8	4 8	4.8	4.8	4 8	-0.3%
Oil and natural gas steam[4]	0.7	0.7	0.7	0.7	0.7	0.7	0.7	0.0%
Combined cycle	25.8	26.3	26.3	26 3	26.3	26.3	26 3	-0.0%
Combustion turbine/diesel	2.8	2.8	2 8	2 8	2.8	2.8	2 8	-0.0%
Renewable sources[6]	0.8	0.9	0.9	0 9	0.9	0.9	0 9	0.2%
Total	35.4	35.9	35.5	35.5	35.5	35.5	35.5	-0.0%
Cumulative planned additions[9]								
Coal	0.0	0.0	9.3	9 3	9.3	9.3	9 3	--
Oil and natural gas steam[4]	0.0	0.0	0.0	0 0	0.0	0.0	0 0	--
Combined cycle	0.0	0.0	14.3	14 3	14.3	14.3	14 3	--
Combustion turbine/diesel	0.0	0.0	5 0	5 0	5.0	5.0	5 0	--
Nuclear power	0.0	0.0	1.1	6 8	6.8	6.8	6 8	--
Pumped storage	0.0	0.0	0.0	0 0	0.0	0.0	0 0	--
Fuel cells	0.0	0.0	0.0	0 0	0.0	0.0	0 0	--
Renewable sources[6]	0.0	0.0	14.0	14 0	14.0	14.0	14 0	--
Distributed generation[7]	0.0	0.0	0.0	0 0	0.0	0.0	0 0	--
Total	0.0	0.0	43.7	49.3	49.3	49.3	49.3	--
Cumulative unplanned additions[9]								
Coal	0.0	0.0	0.0	0 9	0.9	1.0	1.7	--
Oil and natural gas steam[4]	0.0	0.0	0.0	0 0	0.0	0.0	0 0	--
Combined cycle	0.0	0.0	1.4	1 9	11.8	32.5	60 2	--
Combustion turbine/diesel	0.0	0.0	5 2	12 9	23.2	30.2	41 5	--
Nuclear power	0.0	0.0	0.0	0 0	0.0	0.1	1 8	--
Pumped storage	0.0	0.0	0.0	0 0	0.0	0.0	0 0	--
Fuel cells	0.0	0.0	0.0	0 0	0.0	0.0	0 0	--
Renewable sources[6]	0.0	0.0	5.7	7 0	12.4	17.4	30 5	--
Distributed generation[7]	0.0	0.0	0.2	0 5	0.8	1.3	2.1	--
Total	0.0	0.0	12.4	23.2	49.1	82.5	137.8	--
Cumulative electric power sector additions	0.0	0.0	56.1	72.5	98.5	131.8	187.1	--
Cumulative retirements[10]								
Coal	0.0	0.0	41.0	48 9	48.9	48.9	49 0	--
Oil and natural gas steam[4]	0.0	0.0	17.4	18 0	18.5	19.4	20 3	--
Combined cycle	0.0	0.0	0.0	0 2	0.2	0.2	0 2	--
Combustion turbine/diesel	0.0	0.0	6.4	7 2	10.4	11.4	12.4	--
Nuclear power	0.0	0.0	0.0	0 6	0.6	1.1	6.1	--
Pumped storage	0.0	0.0	0.0	0 0	0.0	0.0	0 0	--
Fuel cells	0.0	0.0	0.0	0 0	0.0	0.0	0 0	--
Renewable sources[6]	0.0	0.0	0.4	0.4	0.4	0.4	0.4	--
Total	0.0	0.0	65.2	75.2	78.9	81.4	88.4	--
Total electric power sector capacity	994.9	1006.5	998.7	1007.6	1033.3	1064.2	1112.5	0.4%

Table A9. Electricity generating capacity (continued)
(gigawatts)

Net summer capacity[1]	Reference case							Annual growth 2010-2035 (percent)
	2009	2010	2015	2020	2025	2030	2035	
End-use generators[11]								
Coal	3.6	4.3	4.2	6.6	7.7	8.8	9.9	3.4%
Petroleum	0.7	0.7	0.7	0.7	0.7	0.7	0.7	0.3%
Natural gas	14.7	14.7	17.7	19.8	22.9	27.4	33.2	3.3%
Other gaseous fuels[12]	1.8	1.7	2.5	2.5	2.5	2.5	2.5	1.5%
Renewable sources[6]	6.7	7.6	17.6	21.1	23.4	27.1	30.6	5.7%
Other[13]	0.6	0.6	0.6	0.6	0.6	0.6	0.6	0.0%
Total	28.0	29.6	43.3	51.3	57.8	67.1	77.5	3.9%
Cumulative capacity additions[9]	0.0	0.0	13.7	21.7	28.2	37.4	47.9	--

[1]Net summer capacity is the steady hourly output that generating equipment is expected to supply to system load (exclusive of auxiliary power), as demonstrated by tests during summer peak demand.
[2]Includes electricity-only and combined heat and power plants whose primary business is to sell electricity, or electricity and heat, to the public.
[3]Includes plants that only produce electricity. Includes capacity increases (uprates) at existing units.
[4]Includes oil-, gas-, and dual-fired capacity.
[5]Nuclear capacity includes 7.3 gigawatts of uprates through 2035.
[6]Includes conventional hydroelectric, geothermal, wood, wood waste, all municipal waste, landfill gas, other biomass, solar, and wind power. Facilities co-firing biomass and coal are classified as coal.
[7]Primarily peak load capacity fueled by natural gas.
[8]Includes combined heat and power plants whose primary business is to sell electricity and heat to the public (i.e., those that report North American Industry Classification System code 22).
[9]Cumulative additions after December 31, 2010.
[10]Cumulative retirements after December 31, 2010.
[11]Includes combined heat and power plants and electricity-only plants in the commercial and industrial sectors; and small on-site generating systems in the residential, commercial, and industrial sectors used primarily for own-use generation, but which may also sell some power to the grid.
[12]Includes refinery gas and still gas.
[13]Includes batteries, chemicals, hydrogen, pitch, purchased steam, sulfur, and miscellaneous technologies.
- - = Not applicable.
Note: Totals may not equal sum of components due to independent rounding. Data for 2009 and 2010 are model results and may differ slightly from official EIA data reports.
Sources: 2009 and 2010 capacity and projected planned additions: U.S. Energy Information Administration (EIA), Form EIA-860, "Annual Electric Generator Report" (preliminary). **Projections:** EIA, AEO2012 National Energy Modeling System run REF2012.D020112C.

Table A10. Electricity trade
(billion kilowatthours, unless otherwise noted)

Electricity trade	Reference case							Annual growth 2010-2035 (percent)
	2009	2010	2015	2020	2025	2030	2035	
Interregional electricity trade								
Gross domestic sales								
Firm power	232.1	237.5	139.1	104.4	47.1	24.2	24.2	-8.7%
Economy	231.9	137.0	206.3	211.9	235.4	230.1	235.8	2.2%
Total	**464.0**	**374.4**	**345.3**	**316.3**	**282.5**	**254.3**	**260.0**	**-1.4%**
Gross domestic sales (million 2010 dollars)								
Firm power	13923.7	14244.9	8341.5	6259.9	2824.5	1450.4	1450.4	-8.7%
Economy	9065.6	6611.0	8320.2	10576.4	14143.6	13529.2	14541.9	3.2%
Total	**22989.2**	**20855.9**	**16661.8**	**16836.3**	**16968.1**	**14979.5**	**15992.2**	**-1.1%**
International electricity trade								
Imports from Canada and Mexico								
Firm power	19.3	13.7	24.3	17.1	5.2	0.4	0.4	-13.3%
Economy	33.1	31.4	24.7	27.7	34.7	31.0	28.2	-0.4%
Total	**52.4**	**45.1**	**49.0**	**44.8**	**39.9**	**31.4**	**28.6**	**-1.8%**
Exports to Canada and Mexico								
Firm power	3.3	3.7	3.0	2.1	0.6	0.0	0.0	- -
Economy	14.7	15.7	16.9	16.7	17.0	17.0	16.5	0.2%
Total	**18.1**	**19.4**	**19.9**	**18.8**	**17.6**	**17.0**	**16.5**	**-0.7%**

- - = Not applicable.
Note: Totals may not equal sum of components due to independent rounding. Data for 2009 and 2010 are model results and may differ slightly from official EIA data reports. Firm power sales are capacity sales, meaning the delivery of the power is scheduled as part of the normal operating conditions of the affected electric systems. Economy sales are subject to curtailment or cessation of delivery by the supplier in accordance with prior agreements or under specified conditions.
Sources: 2009 and 2010 interregional firm electricity trade data: North American Electric Reliability Council (NERC), Electricity Sales and Demand Database 2007; NERC, 2011 Summer Reliability Assessment (May 2011); and NERC, Winter Reliability Assessment 2011/2012 (November 2011). 2009 and 2010 Mexican electricity trade data: U.S. Energy Information Administration (EIA), Electric Power Annual 2010 DOE/EIA-0348(2010) (Washington, DC, November 2011). 2009 Canadian international electricity trade data: National Energy Board, Electricity Exports and Imports Statistics, 2009. 2010 Canadian international electricity trade data: National Energy Board, Electricity Exports and Imports Statistics, 2010. Projections: EIA, AEO2012 National Energy Modeling System run REF2012.D020112C.

Table A11. Liquid fuels supply and disposition
(million barrels per day, unless otherwise noted)

Supply and disposition	Reference case							Annual growth 2010-2035 (percent)
	2009	2010	2015	2020	2025	2030	2035	
Crude oil								
Domestic crude production[1]	5 36	5.47	6.15	6.70	6.40	6.37	5.99	0.4%
Alaska	0 65	0.60	0.46	0.49	0.40	0.44	0.27	-3.2%
Lower 48 states	4.72	4.87	5.69	6.21	6 00	5.94	5.72	0.6%
Net imports	8 97	9.17	8.52	7.15	7 24	7.14	7.52	-0.8%
Gross imports	9 01	9.21	8.56	7.19	7 27	7.17	7.55	-0.8%
Exports	0 04	0.04	0.03	0.04	0 03	0.03	0.03	-1.1%
Other crude supply[2]	0 01	0.08	0.00	0.00	0 00	0.00	0.00	--
Total crude supply	14.34	14.72	14.67	13.85	13.64	13.52	13.51	-0.3%
Other petroleum supply	3.59	3.50	3.25	3.73	3.80	3.70	3.52	0.0%
Natural gas plant liquids	1 91	2.07	2.56	2.91	3 01	3.05	3.01	1.5%
Net product imports	0.75	0.39	-0.25	-0.12	-0.12	-0.25	-0.34	--
Gross refined product imports[3]	1 27	1.23	0.78	0.73	0.79	0.78	0.82	-1.6%
Unfinished oil imports	0 68	0.61	0.64	0.54	0 51	0.50	0.50	-0.8%
Blending component imports	0.72	0.74	0.66	0.64	0 65	0.65	0.66	-0.5%
Exports	1 92	2.19	2.32	2.03	2 07	2.17	2.31	0.2%
Refinery processing gain[4]	0 98	1.07	0.95	0.94	0 91	0.89	0.85	-0.9%
Product stock withdrawal	-0 04	-0.03	0.00	0.00	0 00	0.00	0.00	--
Other non-petroleum supply	0.81	1.00	1.22	1.52	1.86	2.36	2.96	4.4%
Supply from renewable sources	0.75	0.87	1.05	1.22	1.48	1.89	2.37	4.1%
Ethanol	0.73	0.85	0.94	1.04	1.19	1.40	1.65	2.7%
Domestic production	0.72	0.88	0.94	1.04	1.17	1.37	1.59	2.4%
Net imports	0 01	-0.02	0.00	0.00	0 02	0.03	0.06	--
Biodiesel	0 02	0.01	0.09	0.12	0.12	0.13	0.13	9.2%
Domestic production	0 03	0.02	0.09	0.12	0.12	0.13	0.13	7.9%
Net imports	-0 01	-0.01	0.00	0.00	0 00	0.00	-0.00	--
Other biomass-derived liquids[5]	0 00	0.00	0.03	0.06	0.16	0.36	0.59	23.2%
Liquids from gas	0 00	0.00	0.00	0.00	0 00	0.00	0.00	--
Liquids from coal	0 00	0.00	0.00	0.12	0.17	0.22	0.28	--
Other[6]	0 05	0.13	0.17	0.19	0 21	0.25	0.31	3.6%
Total primary supply[7]	18.74	19.22	19.14	19.10	19.29	19.57	19.99	0.2%
Liquid fuels consumption								
by fuel								
Liquefied petroleum gases	2.13	2.27	1.94	2.11	2 21	2.22	2.21	-0.1%
E85[8]	0 00	0.00	0.01	0.09	0 21	0.49	0.83	27.0%
Motor gasoline[9]	9 00	8.99	8.88	8.48	8 29	8.17	8.09	-0.4%
Jet fuel[10]	1 39	1.43	1.46	1.49	1 54	1.58	1.61	0.5%
Distillate fuel oil[11]	3 63	3.80	4.19	4.24	4 33	4.38	4.48	0.7%
Diesel	3.18	3.32	3.71	3.81	3 92	3.99	4.11	0.9%
Residual fuel oil	0 51	0.54	0.56	0.56	0 57	0.57	0.58	0.3%
Other[12]	2.15	2.14	2.06	2.04	2 06	2.06	2.10	-0.1%
by sector								
Residential and commercial	1 05	1.12	1.00	0.96	0 94	0.92	0.91	-0.9%
Industrial[13]	4 24	4.31	4.17	4.31	4.41	4.41	4.44	0.1%
Transportation	13 54	13.82	13.80	13.62	13.71	14.00	14.41	0.2%
Electric power[14]	0.17	0.17	0.13	0.13	0.14	0.14	0.14	-0.7%
Total	18.81	19.17	19.10	19.02	19.20	19.47	19.90	0.1%
Discrepancy[15]	-0.07	0.05	0.05	0.09	0.10	0.10	0.09	--

Table A11. Liquid fuels supply and disposition (continued)
(million barrels per day, unless otherwise noted)

Supply and disposition	Reference case							Annual growth 2010-2035 (percent)
	2009	2010	2015	2020	2025	2030	2035	
Domestic refinery distillation capacity[16]	17.7	17.6	17.5	15 8	15.5	15.4	15 2	-0.6%
Capacity utilization rate (percent)[17]	83.0	86.0	85.9	89 8	90.1	89.6	90 8	0.2%
Net import share of product supplied (percent)	51.9	49.6	43 2	36.8	37.0	35.4	36 2	-1.2%
Net expenditures for imported crude oil and petroleum products (billion 2010 dollars)	206.18	243 07	373.00	322.55	344.58	353 03	389.97	1.9%

[1] Includes lease condensate.
[2] Strategic petroleum reserve stock additions plus unaccounted for crude oil and crude stock withdrawals minus crude product supplied.
[3] Includes other hydrocarbons and alcohols.
[4] The volumetric amount by which total output is greater than input due to the processing of crude oil into products which, in total, have a lower specific gravity than the crude oil processed.
[5] Includes pyrolysis oils, biomass-derived Fischer-Tropsch liquids, and renewable feedstocks used for the on-site production of diesel and gasoline.
[6] Includes domestic sources of other blending components, other hydrocarbons, and ethers.
[7] Total crude supply plus other petroleum supply plus other non-petroleum supply.
[8] E85 refers to a blend of 85 percent ethanol (renewable) and 15 percent motor gasoline (nonrenewable). To address cold starting issues, the percentage of ethanol varies seasonally. The annual average ethanol content of 74 percent is used for this forecast.
[9] Includes ethanol and ethers blended into gasoline.
[10] Includes only kerosene type.
[11] Includes distillate fuel oil and kerosene from petroleum and biomass feedstocks.
[12] Includes aviation gasoline, petrochemical feedstocks, lubricants, waxes, asphalt, road oil, still gas, special naphthas, petroleum coke, crude oil product supplied, methanol, and miscellaneous petroleum products.
[13] Includes consumption for combined heat and power, which produces electricity and other useful thermal energy.
[14] Includes consumption of energy by electricity-only and combined heat and power plants whose primary business is to sell electricity, or electricity and heat, to the public.
[15] Balancing item. Includes unaccounted for supply, losses, and gains.
[16] End-of-year operable capacity.
[17] Rate is calculated by dividing the gross annual input to atmospheric crude oil distillation units by their operable refining capacity in barrels per calendar day.
- - = Not applicable.
Note: Totals may not equal sum of components due to independent rounding. Data for 2009 and 2010 are model results and may differ slightly from official EIA data reports.
Sources: 2009 and 2010 product supplied based on: U.S. Energy Information Administration (EIA), *Annual Energy Review 2010*, DOE/EIA-0384(2010) (Washington, DC, October 2011). Other 2009 data: EIA, *Petroleum Supply Annual 2009*, DOE/EIA-0340(2009)/1 (Washington, DC, July 2010). Other 2010 data: EIA, *Petroleum Supply Annual 2010*, DOE/EIA-0340(2010)/1 (Washington, DC, July 2011). **Projections:** EIA, AEO2012 National Energy Modeling System run REF2012.D020112C.

Table A12. Petroleum product prices
(2010 dollars per gallon, unless otherwise noted)

Sector and fuel	Reference case							Annual growth 2010-2035 (percent)
	2009	2010	2015	2020	2025	2030	2035	
Crude oil prices (2010 dollars per barrel)								
Low sulfur light	62.37	79.39	116.91	126.68	132.56	138.49	144.98	2.4%
Imported crude oil[1]	59.72	75.87	113.97	115.74	121.21	126.51	132.95	2.3%
Delivered sector product prices								
Residential								
Liquefied petroleum gases	2.10	2.29	2.60	2.63	2.73	2.82	2.93	1.0%
Distillate fuel oil	2.54	2.94	3.78	4.00	4.18	4.36	4.54	1.8%
Commercial								
Distillate fuel oil	2.23	2.87	3.30	3.51	3.70	3.85	4.02	1.4%
Residual fuel oil	2.04	1.66	2.42	2.63	2.73	2.85	2.83	2.2%
Residual fuel oil (2010 dollars per barrel)	85.89	69.58	101.70	110.65	114.70	119.73	118.85	2.2%
Industrial[2]								
Liquefied petroleum gases	1.70	1.85	2.32	2.35	2.48	2.58	2.73	1.6%
Distillate fuel oil	2.31	2.93	3.32	3.53	3.74	3.90	4.05	1.3%
Residual fuel oil	1.82	1.63	2.88	3.07	3.18	3.25	3.24	2.8%
Residual fuel oil (2010 dollars per barrel)	76.47	68.62	120.80	129.07	133.47	136.47	136.12	2.8%
Transportation								
Liquefied petroleum gases	2.19	2.28	2.70	2.73	2.83	2.91	3.03	1.1%
Ethanol (E85)[3]	1.98	2.40	2.77	2.85	2.75	2.93	3.05	1.0%
Ethanol wholesale price	1.59	1.71	2.23	2.54	2.33	2.29	2.16	0.9%
Motor gasoline[4]	2.38	2.76	3.54	3.71	3.86	3.97	4.03	1.5%
Jet fuel[5]	1.72	2.19	3.21	3.41	3.57	3.72	3.93	2.4%
Diesel fuel (distillate fuel oil)[6]	2.47	3.00	3.78	3.97	4.17	4.30	4.44	1.6%
Residual fuel oil	1.59	1.56	2.74	2.93	3.09	3.11	3.14	2.8%
Residual fuel oil (2010 dollars per barrel)	66.71	65.53	115.15	123.09	129.62	130.52	131.73	2.8%
Electric power[7]								
Distillate fuel oil	2.02	2.60	3.16	3.35	3.52	3.67	3.86	1.6%
Residual fuel oil	1.34	1.78	3.44	3.65	3.80	3.83	3.85	3.1%
Residual fuel oil (2010 dollars per barrel)	56.46	74.77	144.60	153.30	159.70	160.65	161.71	3.1%
Refined petroleum product prices[8]								
Liquefied petroleum gases	1.37	1.46	1.95	1.95	2.05	2.14	2.26	1.7%
Motor gasoline[4]	2.37	2.74	3.54	3.71	3.85	3.97	4.03	1.6%
Jet fuel[5]	1.72	2.19	3.21	3.41	3.57	3.72	3.93	2.4%
Distillate fuel oil	2.44	2.97	3.69	3.89	4.09	4.23	4.38	1.6%
Residual fuel oil	1.57	1.62	2.85	3.04	3.19	3.22	3.25	2.8%
Residual fuel oil (2010 dollars per barrel)	66.10	68.00	119.50	127.68	133.95	135.33	136.32	2.8%
Average	**2.17**	**2.53**	**3.32**	**3.46**	**3.60**	**3.72**	**3.83**	**1.7%**

Table A12. Petroleum product prices (continued)
(nominal dollars per gallon, unless otherwise noted)

Sector and fuel	Reference case							Annual growth 2010-2035 (percent)
	2009	2010	2015	2020	2025	2030	2035	
Crude oil prices (nominal dollars per barrel)								
Low sulfur light	61.65	79.39	125.97	148.87	170.09	197.10	229.55	4.3%
Imported crude oil[1]	59.04	75.87	122.81	136.02	155.52	180.06	210.51	4.2%
Delivered sector product prices								
Residential								
Liquefied petroleum gases	2.08	2.29	2.80	3.09	3.51	4.01	4.65	2.9%
Distillate fuel oil	2.52	2.94	4.07	4.70	5.36	6.20	7.19	3.6%
Commercial								
Distillate fuel oil	2.20	2.87	3.56	4.12	4.75	5.48	6.36	3.2%
Residual fuel oil	2.02	1.66	2.61	3.10	3.50	4.06	4.48	4.1%
Residual fuel oil (nominal dollars per barrel)	84.91	69.58	109.59	130.04	147.17	170.40	188.19	4.1%
Industrial[2]								
Liquefied petroleum gases	1.68	1.85	2.50	2.76	3.18	3.67	4.31	3.5%
Distillate fuel oil	2.28	2.93	3.58	4.15	4.80	5.55	6.42	3.2%
Residual fuel oil	1.80	1.63	3.10	3.61	4.08	4.62	5.13	4.7%
Residual fuel oil (nominal dollars per barrel)	75.59	68.62	130.16	151.68	171.25	194.23	215.53	4.7%
Transportation								
Liquefied petroleum gases	2.16	2.28	2.91	3.21	3.63	4.14	4.79	3.0%
Ethanol (E85)[3]	1.96	2.40	2.98	3.35	3.52	4.17	4.82	2.8%
Ethanol wholesale price	1.57	1.71	2.40	2.98	2.99	3.25	3.42	2.8%
Motor gasoline[4]	2.35	2.76	3.81	4.36	4.95	5.64	6.39	3.4%
Jet fuel[5]	1.70	2.19	3.45	4.01	4.58	5.30	6.23	4.3%
Diesel fuel (distillate fuel oil)[6]	2.44	3.00	4.07	4.67	5.35	6.12	7.03	3.5%
Residual fuel oil	1.57	1.56	2.95	3.44	3.96	4.42	4.97	4.7%
Residual fuel oil (nominal dollars per barrel)	65.95	65.53	124.07	144.66	166.32	185.76	208.57	4.7%
Electric power[7]								
Distillate fuel oil	1.99	2.60	3.40	3.94	4.51	5.22	6.11	3.5%
Residual fuel oil	1.33	1.78	3.71	4.29	4.88	5.44	6.10	5.0%
Residual fuel oil (nominal dollars per barrel)	55.81	74.77	155.81	180.16	204.91	228.64	256.05	5.0%
Refined petroleum product prices[8]								
Liquefied petroleum gases	1.35	1.46	2.10	2.30	2.63	3.04	3.57	3.6%
Motor gasoline[4]	2.35	2.74	3.81	4.36	4.95	5.64	6.39	3.4%
Jet fuel[5]	1.70	2.19	3.45	4.01	4.58	5.30	6.23	4.3%
Distillate fuel oil	2.41	2.97	3.97	4.57	5.25	6.03	6.93	3.4%
Residual fuel oil	1.56	1.62	3.07	3.57	4.09	4.59	5.14	4.7%
Residual fuel oil (nominal dollars per barrel)	65.34	68.00	128.77	150.05	171.87	192.61	215.84	4.7%
Average	2.14	2.53	3.57	4.06	4.62	5.29	6.06	3.6%

[1]Weighted average price delivered to U.S. refiners.
[2]Includes energy for combined heat and power plants, except those whose primary business is to sell electricity, or electricity and heat, to the public.
[3]E85 refers to a blend of 85 percent ethanol (renewable) and 15 percent motor gasoline (nonrenewable). To address cold starting issues, the percentage of ethanol varies seasonally. The annual average ethanol content of 74 percent is used for this forecast.
[4]Sales weighted-average price for all grades. Includes Federal, State and local taxes.
[5]Includes only kerosene type.
[6]Diesel fuel for on-road use. Includes Federal and State taxes while excluding county and local taxes.
[7]Includes electricity-only and combined heat and power plants whose primary business is to sell electricity, or electricity and heat, to the public.
[8]Weighted averages of end-use fuel prices are derived from the prices in each sector and the corresponding sectoral consumption.
Note: Data for 2009 and 2010 are model results and may differ slightly from official EIA data reports.
Sources: 2009 and 2010 low sulfur light crude oil price: U.S. Energy Information Administration (EIA), Form EIA-856, "Monthly Foreign Crude Oil Acquisition Report." 2009 and 2010 imported crude oil price: EIA, *Annual Energy Review 2010*, DOE/EIA-0384(2010) (Washington, DC, October 2011). 2009 and 2010 prices for motor gasoline, distillate fuel oil, and jet fuel are based on: EIA, *Petroleum Marketing Annual 2009*, DOE/EIA-0487(2009) (Washington, DC, August 2010). 2009 and 2010 residential, commercial, industrial, and transportation sector petroleum product prices are derived from: EIA, Form EIA-782A, "Refiners'/Gas Plant Operators' Monthly Petroleum Product Sales Report." 2009 and 2010 electric power prices based on: EIA, *Monthly Energy Review*, DOE/EIA-0035(2011/09) (Washington, DC, September 2011). 2009 and 2010 E85 prices derived from monthly prices in the Clean Cities Alternative Fuel Price Report. 2009 and 2010 wholesale ethanol prices derived from Bloomberg U.S. average rack price. **Projections:** EIA, AEO2012 National Energy Modeling System run REF2012.D020112C.

Table A13. Natural gas supply, disposition, and prices
(trillion cubic feet per year, unless otherwise noted)

Supply, disposition, and prices	Reference case							Annual growth 2010-2035 (percent)
	2009	2010	2015	2020	2025	2030	2035	
Production								
Dry gas production[1]	20.58	21.58	23.65	25.09	26.28	26.94	27.93	1.0%
Supplemental natural gas[2]	0.07	0.07	0.06	0.06	0.06	0.06	0.06	-0.2%
Net imports	**2.68**	**2.58**	**1.73**	**0.35**	**-0.79**	**-0.89**	**-1.36**	--
Pipeline[3]	2.26	2.21	1.56	1.01	-0.13	-0.27	-0.70	--
Liquefied natural gas	0.42	0.37	0.16	-0.66	-0.66	-0.62	-0.66	--
Total supply	**23.32**	**24.22**	**25.45**	**25.50**	**25.55**	**26.11**	**26.63**	**0.4%**
Consumption by sector								
Residential	4.78	4.94	4.85	4.83	4.76	4.72	4.64	-0.2%
Commercial	3.12	3.20	3.33	3.43	3.44	3.52	3.60	0.5%
Industrial[4]	6.17	6.60	7.01	7.08	7.14	7.03	7.00	0.2%
Natural-gas-to-liquids heat and power[5]	0.00	0.00	0.00	0.00	0.00	0.00	0.00	--
Natural gas to liquids production[6]	0.00	0.00	0.00	0.00	0.00	0.00	0.00	--
Electric power[7]	6.87	7.38	8.08	7.87	7.87	8.47	8.96	0.8%
Transportation[8]	0.04	0.04	0.06	0.08	0.11	0.14	0.16	5.9%
Pipeline fuel	0.60	0.63	0.67	0.66	0.66	0.66	0.67	0.2%
Lease and plant fuel[9]	1.28	1.34	1.39	1.51	1.53	1.55	1.60	0.7%
Total	**22.85**	**24.13**	**25.39**	**25.47**	**25.53**	**26.10**	**26.63**	**0.4%**
Discrepancy[10]	0.47	0.10	0.05	0.04	0.02	0.01	-0.00	--
Natural gas prices								
(2010 dollars per million Btu)								
Henry hub spot price	4.00	4.39	4.29	4.58	5.63	6.29	7.37	2.1%
Average lower 48 wellhead price[11]	3.75	4.06	3.84	4.10	5.00	5.56	6.48	1.9%
(2010 dollars per thousand cubic feet)								
Average lower 48 wellhead price[11]	3.85	4.16	3.94	4.19	5.12	5.69	6.64	1.9%
Delivered prices								
(2010 dollars per thousand cubic feet)								
Residential	12.25	11.36	10.56	11.11	12.33	13.08	14.33	0.9%
Commercial	10.06	9.32	8.82	9.21	10.27	10.86	11.93	1.0%
Industrial[4]	5.47	5.65	5.00	5.25	6.19	6.73	7.73	1.3%
Electric power[7]	4.97	5.25	4.65	4.83	5.73	6.35	7.37	1.4%
Transportation[12]	14.52	13.53	12.71	12.81	13.62	14.02	14.87	0.4%
Average[13]	**7.55**	**7.33**	**6.60**	**6.93**	**7.93**	**8.50**	**9.52**	**1.1%**

Table A13. Natural gas supply, disposition, and prices (continued)
(trillion cubic feet per year, unless otherwise noted)

Supply, disposition, and prices	Reference case							Annual growth 2010-2035 (percent)
	2009	2010	2015	2020	2025	2030	2035	
Natural gas prices								
(nominal dollars per million Btu)								
Henry hub spot price	3.95	4.39	4.62	5.39	7.23	8.95	11.67	4.0%
Average lower 48 wellhead price[11]	3.71	4.06	4.14	4.81	6.42	7.92	10.26	3.8%
(nominal dollars per thousand cubic feet)								
Average lower 48 wellhead price[11]	3.80	4.16	4.24	4.93	6.57	8.11	10.51	3.8%
Delivered prices								
(nominal dollars per thousand cubic feet)								
Residential	12.11	11.36	11.38	13.06	15.82	18.61	22.69	2.8%
Commercial	9.95	9.32	9.50	10.82	13.18	15.46	18.89	2.9%
Industrial[4]	5.40	5.65	5.39	6.17	7.94	9.58	12.23	3.1%
Electric power[7]	4.92	5.25	5.01	5.67	7.35	9.03	11.67	3.2%
Transportation[12]	14.36	13.53	13.70	15.06	17.48	19.95	23.54	2.2%
Average[13]	7.46	7.33	7.11	8.15	10.17	12.10	15.08	2.9%

[1]Marketed production (wet) minus extraction losses.
[2]Synthetic natural gas, propane air, coke oven gas, refinery gas, biomass gas, air injected for Btu stabilization, and manufactured gas commingled and distributed with natural gas.
[3]Includes any natural gas regasified in the Bahamas and transported via pipeline to Florida, as well as gas from Canada and Mexico.
[4]Includes energy for combined heat and power plants, except those whose primary business is to sell electricity, or electricity and heat, to the public.
[5]Includes any natural gas used in the process of converting natural gas to liquid fuel that is not actually converted.
[6]Includes any natural gas converted into liquid fuel.
[7]Includes consumption of energy by electricity-only and combined heat and power plants whose primary business is to sell electricity, or electricity and heat, to the public.
[8]Natural gas used as vehicle fuel.
[9]Represents natural gas used in well, field, and lease operations, and in natural gas processing plant machinery.
[10]Balancing item. Natural gas lost as a result of converting flow data measured at varying temperatures and pressures to a standard temperature and pressure and the merger of different data reporting systems which vary in scope, format, definition, and respondent type. In addition, 2009 and 2010 values include net storage injections.
[11]Represents lower 48 onshore and offshore supplies.
[12]Natural gas used as a vehicle fuel. Price includes estimated motor vehicle fuel taxes and estimated dispensing costs or charges.
[13]Weighted average prices. Weights used are the sectoral consumption values excluding lease, plant, and pipeline fuel.
- - = Not applicable.
Note: Totals may not equal sum of components due to independent rounding. Data for 2009 and 2010 are model results and may differ slightly from official EIA data reports.
Sources: 2009 supply values; and lease, plant, and pipeline fuel consumption: U.S. Energy Information Administration (EIA), *Natural Gas Annual 2009*, DOE/EIA-0131(2009) (Washington, DC, December 2010). 2010 supply values; lease, plant, and pipeline fuel consumption; and wellhead price: EIA, *Natural Gas Monthly*, DOE/EIA-0130(2011/07) (Washington, DC, July 2011). Other 2009 and 2010 consumption based on: EIA, *Annual Energy Review 2010*, DOE/EIA-0384(2010) (Washington, DC, October 2011). 2009 wellhead price: U.S. Department of the Interior, Office of Natural Resources Revenue; and EIA, *Natural Gas Annual 2009*, DOE/EIA-0131(2009) (Washington, DC, December 2010). 2010 residential and commercial delivered prices: EIA, *Natural Gas Monthly*, DOE/EIA-0130(2011/07) (Washington, DC, July 2011). 2009 and 2010 electric power prices: EIA, *Electric Power Monthly*, DOE/EIA-0226, April 2010 and April 2011, Table 4.2, and EIA, *State Energy Data Report 2009*, DOE/EIA-0214(2009) (Washington, DC, June 2011). 2009 and 2010 industrial delivered prices are estimated based on: EIA, *Manufacturing Energy Consumption Survey* and industrial and wellhead prices from the *Natural Gas Annual 2009*, DOE/EIA-0131(2009) (Washington, DC, December 2010) and the *Natural Gas Monthly*, DOE/EIA-0130(2011/07) (Washington, DC, July 2011). 2009 transportation sector delivered prices are based on: EIA, *Natural Gas Annual 2009*, DOE/EIA-0131(2009) (Washington, DC, December 2010) and estimated state taxes, federal taxes, and dispensing costs or charges. 2010 transportation sector delivered prices are model results. **Projections:** EIA, AEO2012 National Energy Modeling System run REF2012.D020112C.

Table A14. Oil and gas supply

Production and supply	Reference case							Annual growth 2010-2035 (percent)
	2009	2010	2015	2020	2025	2030	2035	
Crude oil								
Lower 48 average wellhead price[1] (2010 dollars per barrel)	57.46	80.46	117.84	124.44	130.30	130.74	137.55	2.2%
Production (million barrels per day)[2]								
United States total	5.36	5.47	6.15	6.70	6.40	6.37	5.99	0.4%
Lower 48 onshore	3.04	3.21	4.09	4.38	4.43	4.29	3.99	0.9%
Tight oil[3]	0.25	0.37	0.97	1.20	1.29	1.32	1.23	4.9%
Carbon dioxide enhanced oil recovery	0.27	0.28	0.26	0.33	0.49	0.61	0.66	3.5%
Other	2.52	2.55	2.86	2.85	2.66	2.36	2.10	-0.8%
Lower 48 offshore	1.68	1.67	1.60	1.83	1.57	1.65	1.74	0.2%
Alaska	0.65	0.60	0.46	0.49	0.40	0.44	0.27	-3.2%
Lower 48 end of year reserves[2] (billion barrels)	18.75	18.33	20.55	23.02	23.64	24.34	24.23	1.1%
Natural gas								
Lower 48 average wellhead price[1] (2010 dollars per million Btu)								
Henry hub spot price	4.00	4.39	4.29	4.58	5.63	6.29	7.37	2.1%
Average lower 48 wellhead price[1]	3.75	4.06	3.84	4.10	5.00	5.56	6.48	1.9%
(2010 dollars per thousand cubic feet)								
Average lower 48 wellhead price[1]	3.85	4.16	3.94	4.19	5.12	5.69	6.64	1.9%
Dry production (trillion cubic feet)[4]								
United States total	20.58	21.58	23.65	25.09	26.28	26.94	27.93	1.0%
Lower 48 onshore	17.50	18.66	21.48	22.48	23.64	24.11	24.97	1.2%
Associated-dissolved[5]	1.40	1.40	1.52	1.54	1.41	1.18	1.00	-1.3%
Non-associated	16.10	17.26	19.96	20.94	22.23	22.93	23.97	1.3%
Tight gas	6.40	5.68	6.08	6.06	6.17	6.07	6.14	0.3%
Shale gas	2.91	4.99	8.24	9.69	11.26	12.42	13.63	4.1%
Coalbed methane	1.99	1.99	1.83	1.79	1.77	1.74	1.76	-0.5%
Other	4.80	4.59	3.82	3.40	3.03	2.70	2.44	-2.5%
Lower 48 offshore	2.70	2.56	1.88	2.34	2.38	2.58	2.72	0.3%
Associated-dissolved[5]	0.70	0.71	0.55	0.75	0.67	0.70	0.73	0.1%
Non-associated	2.00	1.85	1.33	1.59	1.71	1.88	2.00	0.3%
Alaska	0.37	0.36	0.29	0.27	0.25	0.25	0.23	-1.8%
Lower 48 end of year dry reserves[4] (trillion cubic feet)	263.40	260.50	274.79	290.32	299.77	307.17	311.58	0.7%
Supplemental gas supplies (trillion cubic feet)[6]	0.07	0.07	0.06	0.06	0.06	0.06	0.06	-0.2%
Total lower 48 wells drilled (thousands)	34.31	43.19	49.79	53.80	59.42	60.21	65.59	1.7%

[1]Represents lower 48 onshore and offshore supplies.
[2]Includes lease condensate.
[3]Tight oil represents resources in low-permeability reservoirs, including shale and chalk formations. The specific plays included in the tight oil category are Bakken/Three Forks/Sanish, Eagle Ford, Woodford, Austin Chalk, Spraberry, Niobrara, Avalon/Bone Springs, and Monterey.
[4]Marketed production (wet) minus extraction losses.
[5]Gas which occurs in crude oil reservoirs either as free gas (associated) or as gas in solution with crude oil (dissolved).
[6]Synthetic natural gas, propane air, coke oven gas, refinery gas, biomass gas, air injected for Btu stabilization, and manufactured gas commingled and distributed with natural gas.
Note: Totals may not equal sum of components due to independent rounding. Data for 2009 and 2010 are model results and may differ slightly from official EIA data reports.
Sources: 2009 and 2010 crude oil lower 48 average wellhead price: U.S. Energy Information Administration (EIA), *Petroleum Marketing Annual 2009*, DOE/EIA-0487(2009) (Washington, DC, August 2010). 2009 and 2010 lower 48 onshore, lower 48 offshore, and Alaska crude oil production: EIA, *Petroleum Supply Annual 2010*, DOE/EIA-0340(2010)/1 (Washington, DC, July 2011). 2009 U.S. crude oil and natural gas reserves: EIA, *U.S. Crude Oil, Natural Gas, and Natural Gas Liquids Reserves*, DOE/EIA-0216(2009) (Washington, DC, November 2010). 2009 Alaska and total natural gas production, and supplemental gas supplies: EIA, *Natural Gas Annual 2009*, DOE/EIA-0131(2009) (Washington, DC, December 2010). 2009 natural gas lower 48 average wellhead price: U.S. Department of the Interior, Office of Natural Resources Revenue; and EIA, *Natural Gas Annual 2009*, DOE/EIA-0131(2009) (Washington, DC, December 2010). 2010 natural gas lower 48 average wellhead price, Alaska and total natural gas production, and supplemental gas supplies: EIA, *Natural Gas Monthly*, DOE/EIA-0130(2011/07) (Washington, DC, July 2011). Other 2009 and 2010 values: EIA, Office of Energy Analysis. **Projections:** EIA, AEO2012 National Energy Modeling System run REF2012.D020112C.

Reference case

Table A15. Coal supply, disposition, and prices
(million short tons per year, unless otherwise noted)

Supply, disposition, and prices	Reference case							Annual growth 2010-2035 (percent)
	2009	2010	2015	2020	2025	2030	2035	
Production[1]								
Appalachia	343	336	300	262	271	282	291	-0.6%
Interior	147	156	151	159	163	181	198	1.0%
West	585	592	542	613	684	703	722	0.8%
East of the Mississippi	450	446	407	377	383	409	431	-0.1%
West of the Mississippi	625	638	586	657	735	757	781	0.8%
Total	1075	1084	993	1034	1118	1166	1212	0.4%
Waste coal supplied[2]	14	14	15	15	16	17	19	1.4%
Net imports								
Imports[3]	21	18	15	28	44	33	36	2.8%
Exports	59	82	110	95	115	117	129	1.8%
Total	-38	-64	-95	-67	-71	-83	-94	- -
Total supply[4]	1050	1034	914	982	1064	1100	1138	0.4%
Consumption by sector								
Residential and commercial	3	3	3	3	3	3	3	-0.3%
Coke plants	15	21	22	18	19	18	17	-1.0%
Other industrial[5]	45	52	50	51	52	52	53	0.0%
Coal-to-liquids heat and power	0	0	0	13	19	26	34	- -
Coal to liquids production	0	0	0	12	18	25	32	- -
Electric power[6]	934	975	839	885	952	975	998	0.1%
Total	997	1051	914	982	1063	1099	1137	0.3%
Discrepancy and stock change[7]	53	-17	-0	-0	1	0	0	- -
Average minemouth price[8]								
(2010 dollars per short ton)	33.62	35.61	42.08	40.96	44.05	47.28	50.52	1.4%
(2010 dollars per million Btu)	1.68	1.76	2.08	2.06	2.23	2.39	2.56	1.5%
Delivered prices (2010 dollars per short ton)[9]								
Coke plants	144.66	153.59	189.11	198.45	212.18	225.36	238.32	1.8%
Other industrial[5]	65.62	59.28	70.14	70.89	72.77	75.43	78.53	1.1%
Coal to liquids	- -	- -	18.65	40.67	39.03	40.20	41.54	- -
Electric power								
(2010 dollars per short ton)	43.83	44.27	45.17	45.98	48.13	50.56	53.31	0.7%
(2010 dollars per million Btu)	2.22	2.26	2.35	2.41	2.54	2.66	2.80	0.9%
Average	46.41	47.17	49.95	49.99	51.90	54.09	56.48	0.7%
Exports[10]	102.61	120.41	140.89	155.03	163.43	172.39	177.66	1.6%

Table A15. Coal supply, disposition, and prices (continued)
(million short tons per year, unless otherwise noted)

Supply, disposition, and prices	Reference case							Annual growth 2010-2035 (percent)
	2009	2010	2015	2020	2025	2030	2035	
Average minemouth price[8]								
(nominal dollars per short ton)	33.24	35.61	45.34	48.13	56.52	67.28	80.00	3.3%
(nominal dollars per million Btu)	1.66	1.76	2.24	2.42	2.86	3.41	4.05	3.4%
Delivered prices (nominal dollars per short ton)[9]								
Coke plants	143.01	153.59	203.77	233.22	272.25	320.74	377.36	3.7%
Other industrial[5]	64.87	59.28	75.58	83.31	93.37	107.35	124.34	3.0%
Coal to liquids	- -	- -	20.09	47.80	50.08	57.22	65.77	- -
Electric power								
(nominal dollars per short ton)	43.33	44.27	48.68	54.03	61.76	71.96	84.40	2.6%
(nominal dollars per million Btu)	2.19	2.26	2.53	2.83	3.25	3.78	4.43	2.7%
Average	**45.88**	**47.17**	**53.83**	**58.74**	**66.60**	**76.98**	**89.43**	**2.6%**
Exports[10]	101.44	120.41	151.81	182.19	209.70	245.35	281.30	3.5%

[1]Includes anthracite, bituminous coal, subbituminous coal, and lignite.
[2]Includes waste coal consumed by the electric power and industrial sectors. Waste coal supplied is counted as a supply-side item to balance the same amount of waste coal included in the consumption data.
[3]Excludes imports to Puerto Rico and the U.S. Virgin Islands.
[4]Production plus waste coal supplied plus net imports.
[5]Includes consumption for combined heat and power plants, except those plants whose primary business is to sell electricity, or electricity and heat, to the public. Excludes all coal use in the coal-to-liquids process.
[6]Includes all electricity-only and combined heat and power plants whose primary business is to sell electricity, or electricity and heat, to the public.
[7]Balancing item: the sum of production, net imports, and waste coal supplied minus total consumption.
[8]Includes reported prices for both open market and captive mines.
[9]Prices weighted by consumption; weighted average excludes residential and commercial prices, and export free-alongside-ship (f.a.s.) prices.
[10]F.a.s. price at U.S. port of exit.
- - = Not applicable.
Btu = British thermal unit.
Note: Totals may not equal sum of components due to independent rounding. Data for 2009 and 2010 are model results and may differ slightly from official EIA data reports.
Sources: 2009 and 2010 data based on: U.S. Energy Information Administration (EIA), *Annual Coal Report 2010*, DOE/EIA-0584(2010) (Washington, DC, November 2011); EIA, *Quarterly Coal Report, October-December 2010*, DOE/EIA-0121(2010/4Q) (Washington, DC, May 2011); and EIA, AEO2012 National Energy Modeling System run REF2012.D020112C. Projections: EIA, AEO2012 National Energy Modeling System run REF2012.D020112C.

Reference case

Table A16. Renewable energy generating capacity and generation
(gigawatts, unless otherwise noted)

Net summer capacity and generation	Reference case							Annual growth 2010-2035 (percent)
	2009	2010	2015	2020	2025	2030	2035	
Electric power sector[1]								
Net summer capacity								
Conventional hydropower	78.01	78.03	78.55	79.13	80.14	80.66	81.25	0.2%
Geothermal[2]	2.37	2.37	2.86	3.57	4.45	5.48	6.30	4.0%
Municipal waste[3]	3.20	3.30	3.36	3.36	3.36	3.36	3.36	0.1%
Wood and other biomass[4]	2.43	2.45	2.72	2.72	2.72	2.72	2.89	0.7%
Solar thermal	0.47	0.47	1.36	1.36	1.36	1.36	1.36	4.3%
Solar photovoltaic[5]	0.15	0.38	2.02	2.03	2.30	2.97	8.18	13.0%
Wind	34.52	39.05	54.26	54.31	57.57	60.29	66.65	2.2%
Offshore wind	0.00	0.00	0.20	0.20	0.20	0.20	0.20	--
Total electric power sector capacity ...	121.16	126.06	145.34	146.68	152.10	157.05	170.19	1.2%
Generation (billion kilowatthours)								
Conventional hydropower	271.50	255.32	295.43	300.54	305.00	307.40	310.08	0.8%
Geothermal[2]	15.01	15.67	18.68	24.41	31.53	39.89	46.54	4.5%
Biogenic municipal waste[6]	16.10	16.56	14.66	14.67	14.67	14.67	14.67	-0.5%
Wood and other biomass	10.74	11.51	21.28	51.60	63.90	57.08	49.28	6.0%
Dedicated plants	9.68	10.15	10.13	13.16	13.30	11.81	10.37	0.1%
Cofiring	1.06	1.36	11.15	38.44	50.60	45.27	38.92	14.4%
Solar thermal	0.74	0.82	2.86	2.86	2.86	2.86	2.86	5.1%
Solar photovoltaic[5]	0.16	0.46	3.61	3.62	4.37	6.16	20.19	16.4%
Wind	73.88	94.49	150.22	150.34	160.73	169.64	189.92	2.8%
Offshore wind	0.00	0.00	0.75	0.75	0.75	0.75	0.75	--
Total electric power sector generation .	388.11	394.82	507.49	548.78	583.81	598.46	634.30	1.9%
End-use sectors[7]								
Net summer capacity								
Conventional hydropower[8]	0.34	0.33	0.33	0.33	0.33	0.33	0.33	0.0%
Geothermal	0.00	0.00	0.00	0.00	0.00	0.00	0.00	--
Municipal waste[9]	0.36	0.35	0.35	0.35	0.35	0.35	0.35	0.0%
Biomass	4.56	4.56	5.73	6.68	8.44	11.31	13.81	4.5%
Solar photovoltaic[5]	1.22	2.05	8.98	11.19	11.69	12.41	13.33	7.8%
Wind	0.18	0.36	2.25	2.57	2.60	2.65	2.74	8.5%
Total end-use sector capacity	6.66	7.65	17.64	21.12	23.41	27.05	30.57	5.7%
Generation (billion kilowatthours)								
Conventional hydropower[8]	1.94	1.76	1.75	1.75	1.75	1.75	1.75	-0.0%
Geothermal	0.00	0.00	0.00	0.00	0.00	0.00	0.00	--
Municipal waste[9]	2.07	2.02	2.79	2.79	2.79	2.79	2.79	1.3%
Biomass	25.31	26.10	33.30	39.53	52.34	76.03	96.17	5.4%
Solar photovoltaic[5]	1.93	3.21	13.88	17.40	18.22	19.40	20.91	7.8%
Wind	0.24	0.47	2.88	3.31	3.36	3.44	3.56	8.5%
Total end-use sector generation	31.48	33.56	54.59	64.77	78.45	103.40	125.17	5.4%

Table A16. Renewable energy generating capacity and generation (continued)
(gigawatts, unless otherwise noted)

Net summer capacity and generation	Reference case							Annual growth 2010-2035 (percent)
	2009	2010	2015	2020	2025	2030	2035	
Total, all sectors								
Net summer capacity								
Conventional hydropower	78.35	78.36	78.88	79.46	80.47	80.99	81.58	0.2%
Geothermal	2.37	2.37	2.86	3.57	4.45	5.48	6.30	4.0%
Municipal waste	3.57	3.65	3.71	3.71	3.71	3.71	3.71	0.1%
Wood and other biomass[4]	6.99	7.00	8.45	9.40	11.16	14.03	16.71	3.5%
Solar[5]	1.85	2.90	12.37	14.58	15.35	16.74	22.87	8.6%
Wind	34.70	39.41	56.72	57.07	60.37	63.15	69.59	2.3%
Total capacity, all sectors	127.83	133.70	162.98	167.80	175.51	184.10	200.76	1.6%
Generation (billion kilowatthours)								
Conventional hydropower	273.44	257.08	297.18	302.28	306.75	309.15	311.83	0.8%
Geothermal	15.01	15.67	18.68	24.41	31.53	39.89	46.54	4.5%
Municipal waste	18.16	18.59	17.45	17.46	17.46	17.46	17.46	-0.3%
Wood and other biomass	36.05	37.61	54.58	91.13	116.24	133.11	145.45	5.6%
Solar[5]	2.82	4.48	20.35	23.87	25.44	28.42	43.96	9.6%
Wind	74.12	94.95	153.85	154.40	164.84	173.83	194.23	2.9%
Total generation, all sectors	419.59	428.38	562.08	613.55	662.25	701.85	759.46	2.3%

[1] Includes electricity-only and combined heat and power plants whose primary business is to sell electricity, or electricity and heat, to the public.
[2] Includes both hydrothermal resources (hot water and steam) and near-field enhanced geothermal systems (EGS). Near-field EGS potential occurs on known hydrothermal sites, however this potential requires the addition of external fluids for electricity generation and is only available after 2025.
[3] Includes municipal waste, landfill gas, and municipal sewage sludge. Incremental growth is assumed to be for landfill gas facilities. All municipal waste is included, although a portion of the municipal waste stream contains petroleum-derived plastics and other non-renewable sources.
[4] Facilities co-firing biomass and coal are classified as coal.
[5] Does not include off-grid photovoltaics (PV). Based on annual PV shipments from 1989 through 2009, EIA estimates that as much as 245 megawatts of remote electricity generation PV applications (i.e., off-grid power systems) were in service in 2009, plus an additional 558 megawatts in communications, transportation, and assorted other non-grid-connected, specialized applications. See U.S. Energy Information Administration, Annual Energy Review 2010, DOE/EIA-0384(2010) (Washington, DC, October 2011), Table 10.9 (annual PV shipments, 1989-2009). The approach used to develop the estimate, based on shipment data, provides an upper estimate of the size of the PV stock, including both grid-based and off-grid PV. It will overestimate the size of the stock, because shipments include a substantial number of units that are exported, and each year some of the PV units installed earlier will be retired from service or abandoned.
[6] Includes biogenic municipal waste, landfill gas, and municipal sewage sludge. Incremental growth is assumed to be for landfill gas facilities. Only biogenic municipal waste is included. The U.S. Energy Information Administration estimates that in 2010 approximately 6 billion kilowatthours of electricity were generated from a municipal waste stream containing petroleum-derived plastics and other non-renewable sources. See U.S. Energy Information Administration, Methodology for Allocating Municipal Solid Waste to Biogenic and Non-Biogenic Energy (Washington, DC, May 2007).
[7] Includes combined heat and power plants and electricity-only plants in the commercial and industrial sectors; and small on-site generating systems in the residential, commercial, and industrial sectors used primarily for own-use generation, but which may also sell some power to the grid.
[8] Represents own-use industrial hydroelectric power.
[9] Includes municipal waste, landfill gas, and municipal sewage sludge. All municipal waste is included, although a portion of the municipal waste stream contains petroleum-derived plastics and other non-renewable sources.
- - = Not applicable.
Note: Totals may not equal sum of components due to independent rounding. Data for 2009 and 2010 are model results and may differ slightly from official EIA data reports.
Sources: 2009 and 2010 capacity: U.S. Energy Information Administration (EIA), Form EIA-860, "Annual Electric Generator Report" (preliminary). 2009 and 2010 generation: EIA, Annual Energy Review 2010, DOE/EIA-0384(2010) (Washington, DC, October 2011). Projections: EIA, AEO2012 National Energy Modeling System run REF2012.D020112C.

Table A17. Renewable energy consumption by sector and source
(quadrillion Btu per year)

Sector and source	2009	2010	2015	2020	2025	2030	2035	Annual growth 2010-2035 (percent)
Marketed renewable energy[1]								
Residential (wood)	0.43	0.42	0.43	0.43	0.43	0.43	0.43	0.1%
Commercial (biomass)	0.11	0.11	0.11	0.11	0.11	0.11	0.11	0.0%
Industrial[2]	2.19	2.34	2.42	2.63	3.09	3.79	4.52	2.7%
Conventional hydroelectric	0.02	0.02	0.02	0.02	0.02	0.02	0.02	0.0%
Municipal waste[3]	0.16	0.17	0.18	0.18	0.18	0.18	0.18	0.1%
Biomass	1.19	1.31	1.42	1.48	1.62	1.68	1.76	1.2%
Biofuels heat and coproducts	0.82	0.84	0.81	0.96	1.27	1.92	2.57	4.6%
Transportation	0.99	1.14	1.45	1.72	2.16	2.88	3.75	4.9%
Ethanol used in E85[4]	0.00	0.00	0.01	0.08	0.20	0.47	0.80	27.0%
Ethanol used in gasoline blending	0.95	1.10	1.21	1.27	1.35	1.35	1.34	0.8%
Biodiesel used in distillate blending	0.04	0.03	0.18	0.23	0.24	0.25	0.26	9.2%
Liquids from biomass	0.00	0.00	0.03	0.11	0.33	0.78	1.31	--
Renewable diesel and gasoline[5]	0.00	0.01	0.03	0.03	0.03	0.03	0.03	6.2%
Electric power[6]	3.77	3.85	4.96	5.40	5.75	5.87	6.22	1.9%
Conventional hydroelectric	2.65	2.49	2.88	2.93	2.98	3.00	3.03	0.8%
Geothermal	0.15	0.15	0.18	0.24	0.31	0.39	0.45	4.5%
Biogenic municipal waste[7]	0.07	0.08	0.09	0.09	0.09	0.09	0.09	0.6%
Biomass	0.17	0.19	0.27	0.60	0.73	0.64	0.56	4.4%
Dedicated plants	0.16	0.17	0.16	0.21	0.22	0.18	0.16	-0.1%
Cofiring	0.01	0.02	0.11	0.39	0.52	0.46	0.40	11.8%
Solar thermal	0.01	0.01	0.03	0.03	0.03	0.03	0.03	5.1%
Solar photovoltaic	0.00	0.00	0.04	0.04	0.04	0.06	0.20	16.4%
Wind	0.72	0.92	1.47	1.47	1.58	1.66	1.86	2.8%
Total marketed renewable energy	7.49	7.87	9.37	10.29	11.54	13.09	15.03	2.6%
Sources of ethanol								
from corn and other starch	0.94	1.14	1.20	1.32	1.39	1.39	1.46	1.0%
from cellulose	0.00	0.00	0.01	0.03	0.13	0.40	0.61	56.6%
Net imports	0.02	-0.03	0.00	0.00	0.03	0.04	0.08	--
Total	0.95	1.11	1.22	1.35	1.55	1.82	2.15	2.7%

Table A17. Renewable energy consumption by sector and source (continued)
(quadrillion Btu per year)

Sector and source	Reference case							Annual growth 2010-2035 (percent)
	2009	2010	2015	2020	2025	2030	2035	
Nonmarketed renewable energy[8]								
Selected consumption								
Residential	0.02	0.02	0.08	0.10	0.10	0.11	0.11	6.9%
Solar hot water heating	0.01	0.01	0.02	0.02	0.02	0.02	0.02	2.4%
Geothermal heat pumps	0.00	0.01	0.01	0.02	0.02	0.02	0.03	6.4%
Solar photovoltaic	0.00	0.00	0.04	0.05	0.05	0.06	0.06	10.7%
Wind	0.00	0.00	0.01	0.01	0.01	0.01	0.01	9.1%
Commercial	0.03	0.03	0.04	0.04	0.04	0.05	0.05	1.7%
Solar thermal	0.03	0.03	0.03	0.03	0.03	0.04	0.04	1.4%
Solar photovoltaic	0.00	0.01	0.01	0.01	0.01	0.01	0.01	2.8%
Wind	0.00	0.00	0.00	0.00	0.00	0.00	0.00	5.3%

[1]Includes nonelectric renewable energy groups for which the energy source is bought and sold in the marketplace, although all transactions may not necessarily be marketed, and marketed renewable energy inputs for electricity entering the marketplace on the electric power grid. Excludes electricity imports; see Table A2.
[2]Includes all electricity production by industrial and other combined heat and power for the grid and for own use.
[3]Includes municipal waste, landfill gas, and municipal sewage sludge. All municipal waste is included, although a portion of the municipal waste stream contains petroleum-derived plastics and other non-renewable sources.
[4]Excludes motor gasoline component of E85.
[5]Renewable feedstocks for the on-site production of diesel and gasoline.
[6]Includes consumption of energy by electricity-only and combined heat and power plants whose primary business is to sell electricity, or electricity and heat, to the public. Actual heat rates used to determine fuel consumption for all renewable fuels except hydropower, geothermal, solar, and wind. Consumption at hydroelectric, geothermal, solar, and wind facilities determined by using the fossil fuel equivalent of 9,760 Btu per kilowatthour.
[7]Includes biogenic municipal waste, landfill gas, and municipal sewage sludge. Incremental growth is assumed to be for landfill gas facilities. Only biogenic municipal waste is included. The U.S. Energy Information Administration estimates that in 2010 approximately 0.3 quadrillion Btus were consumed from a municipal waste stream containing petroleum-derived plastics and other non-renewable sources. See U.S. Energy Information Administration, *Methodology for Allocating Municipal Solid Waste to Biogenic and Non-Biogenic Energy* (Washington, DC, May 2007).
[8]Includes selected renewable energy consumption data for which the energy is not bought or sold, either directly or indirectly as an input to marketed energy. The U.S. Energy Information Administration does not estimate or project total consumption of nonmarketed renewable energy.
- - = Not applicable.
Btu = British thermal unit.
Note: Totals may not equal sum of components due to independent rounding. Data for 2009 and 2010 are model results and may differ slightly from official EIA data reports.
Sources: 2009 and 2010 ethanol: U.S. Energy Information Administration (EIA), *Annual Energy Review 2010*, DOE/EIA-0384(2010) (Washington, DC, October 2011). 2009 and 2010 electric power sector: EIA, Form EIA-860, "Annual Electric Generator Report" (preliminary). Other 2009 and 2010 values: EIA, Office of Energy Analysis. **Projections:** EIA, AEO2012 National Energy Modeling System run REF2012.D020112C.

Table A18. Energy-related carbon dioxide emissions by sector and source
(million metric tons, unless otherwise noted)

Sector and source	Reference case							Annual growth 2010-2035 (percent)
	2009	2010	2015	2020	2025	2030	2035	
Residential								
Petroleum	81	85	74	69	65	61	59	-1.5%
Natural gas	259	267	264	263	259	257	252	-0.2%
Coal	1	1	1	1	1	1	1	-1.3%
Electricity[1]	819	879	746	769	816	862	907	0.1%
Total residential	1159	1232	1084	1101	1141	1181	1218	-0.0%
Commercial								
Petroleum	49	51	44	44	44	44	44	-0.6%
Natural gas	169	173	181	186	187	191	196	0.5%
Coal	6	6	6	6	6	6	6	0.0%
Electricity[1]	785	805	721	757	806	852	897	0.4%
Total commercial	1009	1035	952	993	1043	1093	1142	0.4%
Industrial[2]								
Petroleum	339	344	364	350	351	351	358	0.2%
Natural gas[3]	383	408	445	454	459	455	456	0.4%
Coal	128	157	154	170	183	190	197	0.9%
Electricity[1]	551	583	540	536	550	535	516	-0.5%
Total industrial	1401	1492	1503	1509	1542	1531	1527	0.1%
Transportation								
Petroleum[4]	1818	1836	1825	1785	1778	1791	1814	-0.0%
Natural gas[5]	34	36	39	40	42	44	45	0.9%
Electricity[1]	4	4	4	5	7	9	12	4.2%
Total transportation	1856	1876	1868	1831	1827	1843	1871	-0.0%
Electric power[6]								
Petroleum	34	33	23	23	24	24	25	-1.1%
Natural gas	373	399	438	427	427	459	485	0.8%
Coal	1741	1828	1539	1606	1717	1763	1809	-0.0%
Other[7]	12	12	12	12	12	12	12	0.0%
Total electric power	2159	2271	2011	2067	2179	2258	2330	0.1%
Total by fuel								
Petroleum[3]	2320	2349	2329	2271	2261	2271	2300	-0.1%
Natural gas	1218	1283	1367	1370	1374	1405	1435	0.4%
Coal	1876	1990	1699	1781	1906	1959	2012	0.0%
Other[7]	12	12	12	12	12	12	12	0.0%
Total	5425	5634	5407	5434	5552	5647	5758	0.1%
Carbon dioxide emissions (tons per person)	17.6	18.1	16.6	15.9	15.5	15.1	14.8	-0.8%

[1]Emissions from the electric power sector are distributed to the end-use sectors.
[2]Fuel consumption includes energy for combined heat and power plants, except those plants whose primary business is to sell electricity, or electricity and heat, to the public.
[3]Includes lease and plant fuel.
[4]This includes carbon dioxide from international bunker fuels, both civilian and military, which are excluded from the accounting of carbon dioxide emissions under the United Nations convention. From 1990 through 2009, international bunker fuels accounted for 90 to 126 million metric tons annually.
[5]Includes pipeline fuel natural gas and natural gas used as vehicle fuel.
[6]Includes electricity-only and combined heat and power plants whose primary business is to sell electricity, or electricity and heat, to the public.
[7]Includes emissions from geothermal power and nonbiogenic emissions from municipal waste.
Note: By convention, the direct emissions from biogenic energy sources are excluded from energy-related carbon dioxide emissions. The release of carbon from these sources is assumed to be balanced by the uptake of carbon when the feedstock is grown, resulting in zero net emissions over some period of time. If, however, increased use of biomass energy results in a decline in terrestrial carbon stocks, a net positive release of carbon may occur. See "Energy-Related Carbon Dioxide Emissions by End Use" for the emissions from biogenic energy sources as an indication of the potential net release of carbon dioxide in the absence of offsetting sequestration. Totals may not equal sum of components due to independent rounding. Data for 2009 and 2010 are model results and may differ slightly from official EIA data reports.
Sources: 2009 and 2010 emissions and emission factors: U.S. Energy Information Administration (EIA), *Monthly Energy Review, October 2011* DOE/EIA-0035(2011/10) (Washington, DC, October 2011). Projections: EIA, AEO2012 National Energy Modeling System run REF2012.D020112C.

Table A19. Energy-related carbon dioxide emissions by end use
(million metric tons)

Sector and end use	Reference case							Annual growth 2010-2035 (percent)
	2009	2010	2015	2020	2025	2030	2035	
Residential								
Space heating	280.90	298.51	277.05	272.48	267.41	264.17	259.97	-0.6%
Space cooling	142.72	191.18	159.32	164.10	174.13	183.61	192.21	0.0%
Water heating	160.15	159.68	151.53	154.46	157.58	156.73	154.55	-0.1%
Refrigeration	66.17	66.06	57.91	58.63	61.36	64.38	67.24	0.1%
Cooking	32.01	32.25	30.98	32.26	33.88	35.40	36.82	0.5%
Clothes dryers	36.78	37.23	33.43	31.76	30.86	30.58	31.50	-0.7%
Freezers	14.50	14.62	13.14	13.17	13.46	13.61	13.81	-0.2%
Lighting	123.36	122.27	81.97	74.77	72.02	71.52	72.33	-2.1%
Clothes washers[1]	5.87	5.79	4.96	4.18	3.86	3.64	3.74	-1.7%
Dishwashers[1]	17.70	17.75	15.48	15.32	15.33	16.16	17.28	-0.1%
Color televisions and set-top boxes	56.62	58.20	50.98	53.06	57.14	61.62	66.45	0.5%
Personal computers and related equipment	29.75	30.47	29.70	33.59	37.07	39.80	41.67	1.3%
Furnace fans and boiler circulation pumps	23.80	23.93	21.88	22.19	22.63	22.80	23.00	-0.2%
Other uses	167.37	173.46	155.66	171.03	194.05	216.69	237.60	1.3%
Discrepancy[2]	1.73	0.16	0.00	-0.00	0.00	0.00	0.00	--
Total residential	**1159.44**	**1231.57**	**1083.99**	**1101.00**	**1140.80**	**1180.73**	**1218.17**	**-0.0%**
Commercial								
Space heating[3]	129.16	129.68	124.70	124.97	122.24	120.61	118.00	-0.4%
Space cooling[3]	84.66	101.34	80.33	79.94	81.20	82.60	84.17	-0.7%
Water heating[3]	41.32	41.44	41.47	42.83	43.45	44.00	44.04	0.2%
Ventilation	88.64	90.04	83.19	86.87	90.94	94.43	97.04	0.3%
Cooking	13.27	13.58	13.68	14.20	14.47	14.84	15.13	0.4%
Lighting	181.96	180.09	156.69	160.17	166.24	171.06	174.62	-0.1%
Refrigeration	70.13	69.16	55.15	52.64	52.71	53.53	54.79	-0.9%
Office equipment (PC)	38.00	37.69	29.68	29.85	30.75	32.11	33.19	-0.5%
Office equipment (non-PC)	43.86	46.44	49.41	56.62	62.87	67.77	71.49	1.7%
Other uses[4]	317.61	325.18	317.95	345.09	378.20	411.92	449.71	1.3%
Total commercial	**1008.62**	**1034.63**	**952.26**	**993.16**	**1043.07**	**1092.87**	**1142.18**	**0.4%**
Industrial								
Manufacturing								
Refining	261.44	265.88	268.04	278.94	288.94	303.58	322.94	0.8%
Food products	100.97	105.04	98.92	104.00	108.26	111.71	113.98	0.3%
Paper products	77.15	76.70	71.83	71.82	73.13	71.21	69.81	-0.4%
Bulk chemicals	221.74	234.55	213.65	229.11	233.13	225.47	215.77	-0.3%
Glass	18.92	18.59	19.05	20.00	21.33	21.21	20.50	0.4%
Cement manufacturing	25.91	25.67	33.19	35.70	37.08	36.48	37.41	1.5%
Iron and steel	91.87	116.74	117.01	110.23	114.88	107.91	99.25	-0.6%
Aluminum	27.63	30.89	28.68	27.66	26.37	24.89	23.14	-1.1%
Fabricated metal products	36.69	36.14	36.43	36.81	37.90	35.62	33.25	-0.3%
Machinery	22.80	23.76	24.32	24.32	26.46	25.49	23.73	-0.0%
Computers and electronics	30.67	33.07	32.16	33.69	36.48	36.57	36.74	0.4%
Transportation equipment	43.77	45.62	56.18	54.82	54.85	57.23	58.87	1.0%
Electrical equipment	7.86	8.17	8.23	8.25	9.10	8.85	8.55	0.2%
Wood products	16.74	16.90	19.68	19.99	20.46	19.14	18.50	0.4%
Plastics	37.47	38.26	34.96	35.35	34.86	34.29	33.32	-0.6%
Balance of manufacturing	142.01	142.62	133.94	136.85	138.25	133.50	129.25	-0.4%
Total manufacturing	1163.64	1218.60	1196.68	1227.54	1261.49	1253.14	1245.00	0.1%
Nonmanufacturing								
Agriculture	73.84	73.82	69.73	68.13	68.31	67.95	68.29	-0.3%
Construction	76.16	69.67	83.15	91.08	92.27	91.23	91.95	1.1%
Mining	43.45	46.03	44.37	44.16	43.79	43.23	42.83	-0.3%
Total nonmanufacturing	193.45	189.52	197.25	203.37	204.37	202.41	203.08	0.3%
Discrepancy[2]	43.83	83.41	108.76	78.58	76.09	74.99	78.94	-0.2%
Total industrial	**1400.92**	**1491.53**	**1502.69**	**1509.48**	**1541.94**	**1530.55**	**1527.02**	**0.1%**

Table A19. Energy-related carbon dioxide emissions by end use (continued)
(million metric tons)

Sector and end use	\multicolumn{7}{c	}{Reference case}	Annual growth 2010-2035 (percent)
	2009	2010	2015	2020	2025	2030	2035	
Transportation								
Light-duty vehicles	1068.20	1060.96	1014.74	966.95	945.91	950.30	957.76	-0.4%
Commercial light trucks[5]	35.27	38.02	39.58	38.75	38.76	39.51	40.97	0.3%
Bus transportation	14.85	17.67	17.32	17.17	17.13	17.18	17.32	-0.1%
Freight trucks	356.16	348.09	389.50	391.24	396.52	398.85	409.21	0.6%
Rail, passenger	5.41	5.84	5.76	6.02	6.39	6.70	6.98	0.7%
Rail, freight	26.27	32.99	30.95	33.83	36.05	36.73	37.43	0.5%
Shipping, domestic	13.03	16.31	16.75	17.65	17.97	18.15	18.27	0.5%
Shipping, international	60.55	67.51	67.87	68.23	68.70	69.13	69.55	0.1%
Recreational boats	16.45	17.12	17.27	17.53	17.90	18.42	18.94	0.4%
Air	172.79	178.28	180.48	186.23	192.08	195.53	197.54	0.4%
Military use	50.94	54.70	47.05	45.77	47.13	49.65	52.56	-0.2%
Lubricants	4.71	5.19	5.00	5.10	5.19	5.24	5.28	0.1%
Pipeline fuel	32.53	34.34	36.23	35.81	35.79	35.99	36.36	0.2%
Discrepancy[2]	-1.34	-1.15	-0.21	0.45	1.14	1.81	2.39	- -
Total transportation	**1855.81**	**1875.88**	**1868.28**	**1830.73**	**1826.65**	**1843.20**	**1870.57**	**-0.0%**
Biogenic energy combustion[6]								
Biomass	178.16	190.68	208.91	245.80	271.80	268.87	268.81	1.4%
Electric power sector	15.83	18.00	25.42	56.39	68.61	60.49	52.72	4.4%
Other sectors	162.33	172.68	183.49	189.41	203.18	208.37	216.10	0.9%
Biogenic waste	6.56	7.10	8.20	8.21	8.21	8.21	8.21	0.6%
Biofuels heat and coproducts	77.06	79.11	75.91	89.81	119.14	179.75	241.23	4.6%
Ethanol	65.18	75.71	83.37	92.41	106.14	124.29	146.78	2.7%
Biodiesel	3.07	2.11	12.76	16.51	17.69	18.42	19.18	9.2%
Liquids from biomass	0.00	0.00	2.01	7.99	24.22	57.28	95.80	- -
Renewable diesel and gasoline	0.00	0.50	2.23	2.23	2.23	2.23	2.21	6.2%
Total	**330.03**	**355.21**	**393.39**	**462.96**	**549.43**	**659.05**	**782.23**	**3.2%**

[1]Does not include water heating portion of load.
[2]Represents differences between total emissions by end-use and total emissions by fuel as reported in Table A18. Emissions by fuel may reflect benchmarking and other modeling adjustments to energy use and the associated emissions that are not assigned to specific end uses.
[3]Includes emissions related to fuel consumption for district services.
[4]Includes miscellaneous uses, such as service station equipment, automated teller machines, telecommunications equipment, medical equipment, pumps, emergency generators, combined heat and power in commercial buildings, manufacturing performed in commercial buildings, and cooking (distillate), plus emissions from residual fuel oil, liquefied petroleum gases, coal, motor gasoline, and kerosene.
[5]Commercial trucks 8,501 to 10,000 pounds gross vehicle weight rating.
[6]By convention, the direct emissions from biogenic energy sources are excluded from energy-related carbon dioxide emissions. The release of carbon from these sources is assumed to be balanced by the uptake of carbon when the feedstock is grown, resulting in zero net emissions over some period of time. If, however, increased use of biomass energy results in a decline in terrestrial carbon stocks, a net positive release of carbon may occur. Accordingly, the emissions from biogenic energy sources are reported here as an indication of the potential net release of carbon dioxide in the absence of offsetting sequestration.
- - = Not applicable.
Note: Totals may not equal sum of components due to independent rounding. Data for 2009 and 2010 are model results and may differ slightly from official EIA data reports.
Sources: 2009 and 2010 emissions and emission factors: U.S. Energy Information Administration (EIA), *Monthly Energy Review, October 2011* DOE/EIA-0035(2011/10) (Washington, DC, October 2011). Projections: EIA, AEO2012 National Energy Modeling System run REF2012.D020112C.

Table A20. Macroeconomic indicators
(billion 2005 chain-weighted dollars, unless otherwise noted)

Indicators	Reference case							Annual growth 2010-2035 (percent)
	2009	2010	2015	2020	2025	2030	2035	
Real gross domestic product	12703	13088	14803	16740	19185	21725	24539	2.5%
Components of real gross domestic product								
Real consumption	9037	9221	10218	11250	12697	14359	16220	2.3%
Real investment	1454	1715	2457	2888	3472	4063	4836	4.2%
Real government spending	2540	2557	2355	2407	2525	2667	2818	0.4%
Real exports	1494	1663	2289	3096	4235	5484	6953	5.9%
Real imports	1853	2085	2463	2800	3516	4461	5690	4.1%
Energy intensity								
(thousand Btu per 2005 dollar of GDP)								
Delivered energy	5.42	5.45	4.84	4.33	3.85	3.48	3.17	-2.1%
Total energy	7.46	7.50	6.58	5.93	5.32	4.80	4.36	-2.1%
Price indices								
GDP chain-type price index (2005=1.000)	1.097	1.110	1.196	1.304	1.424	1.580	1.758	1.9%
Consumer price index (1982-4=1.00)								
All-urban	2.15	2.18	2.42	2.67	2.95	3.30	3.72	2.2%
Energy commodities and services	1.93	2.12	2.62	2.94	3.36	3.86	4.37	2.9%
Wholesale price index (1982=1.00)								
All commodities	1.73	1.85	2.10	2.23	2.39	2.58	2.81	1.7%
Fuel and power	1.59	1.86	2.29	2.57	3.01	3.50	4.12	3.2%
Metals and metal products	1.87	2.08	2.43	2.50	2.57	2.61	2.64	1.0%
Industrial commodities excluding energy	1.76	1.83	2.04	2.13	2.22	2.32	2.43	1.1%
Interest rates (percent, nominal)								
Federal funds rate	0.16	0.18	3.26	4.07	4.29	4.52	4.30	--
10-year treasury note	3.26	3.21	4.67	5.10	5.06	5.26	5.18	--
AA utility bond rate	5.75	5.24	6.74	7.41	7.17	7.48	7.56	--
Value of shipments (billion 2005 dollars)								
Service sectors	19996	20602	22469	24967	28029	30911	33430	2.0%
Total industrial	5667	5838	6730	7363	7973	8328	8692	1.6%
Nonmanufacturing	1615	1578	1873	2103	2228	2305	2407	1.7%
Manufacturing	4052	4260	4857	5260	5745	6023	6285	1.6%
Energy-intensive	1509	1595	1664	1786	1901	1973	2034	1.0%
Non-energy-intensive	2543	2664	3194	3474	3844	4050	4251	1.9%
Total shipments	**25664**	**26440**	**29199**	**32329**	**36002**	**39239**	**42122**	**1.9%**
Population and employment (millions)								
Population, with armed forces overseas	307.8	310.8	326.2	342.0	358.1	374.1	390.1	0.9%
Population, aged 16 and over	241.8	244.3	256.5	269.4	282.6	296.2	309.6	1.0%
Population, over age 65	39.7	40.4	47.1	55.1	64.2	72.3	77.7	2.6%
Employment, nonfarm	130.7	129.8	139.4	147.3	154.2	162.0	166.8	1.0%
Employment, manufacturing	11.8	11.5	12.1	11.9	11.4	10.3	9.2	-0.9%
Key labor indicators								
Labor force (millions)	154.2	153.9	158.0	163.6	168.6	174.5	181.7	0.7%
Nonfarm labor productivity (1992=1.00)	1.06	1.10	1.16	1.26	1.42	1.57	1.75	1.9%
Unemployment rate (percent)	9.28	9.63	7.51	6.47	5.54	5.40	5.54	--
Key indicators for energy demand								
Real disposable personal income	9883	10062	11035	12472	14286	16268	18217	2.4%
Housing starts (millions)	0.60	0.63	1.75	1.92	1.96	1.90	1.89	4.5%
Commercial floorspace (billion square feet)	80.3	81.1	84.1	89.1	93.9	98.2	103.0	1.0%
Unit sales of light-duty vehicles (millions)	10.40	11.55	16.16	16.40	17.79	18.11	18.64	1.9%

GDP = Gross domestic product.
Btu = British thermal unit.
-- = Not applicable.
Sources: 2009 and 2010: IHS Global Insight, Global Insight Industry and Employment models, August 2011. **Projections:** U.S. Energy Information Administration, AEO2012 National Energy Modeling System run REF2012.D020112C.

Reference case

Table A21. International liquids supply and disposition summary
(million barrels per day, unless otherwise noted)

Supply and disposition	Reference case							Annual growth 2010-2035 (percent)
	2009	2010	2015	2020	2025	2030	2035	
Crude oil prices (2010 dollars per barrel)								
Low sulfur light	62.37	79.39	116.91	126.68	132.56	138.49	144.98	2.4%
Imported crude oil[1]	59.72	75.87	113.97	115.74	121.21	126.51	132.95	2.3%
Crude oil prices (nominal dollars per barrel)								
Low sulfur light	61.65	79.39	125.97	148.87	170.09	197.10	229.55	4.3%
Imported crude oil[1]	59.04	75.87	122.81	136.02	155.52	180.06	210.51	4.2%
Petroleum liquids production[2]								
OPEC[3]								
Middle East	22.30	23.43	25.46	27.16	29.77	32.07	33.94	1.5%
North Africa	3.92	3.89	3.62	3.42	3.37	3.31	3.27	-0.7%
West Africa	4.16	4.45	5.09	5.35	5.40	5.31	5.26	0.7%
South America	2.43	2.29	2.13	1.97	1.92	1.79	1.72	-1.1%
Total OPEC petroleum production	32.80	34.05	36.30	37.91	40.46	42.48	44.19	1.0%
Non-OPEC								
OECD								
United States (50 states)	8.27	8.79	9.82	10.73	10.53	10.57	10.15	0.6%
Canada	1.96	1.91	1.79	1.82	1.82	1.81	1.78	-0.3%
Mexico and Chile	3.00	2.98	2.65	1.97	1.58	1.65	1.68	-2.3%
OECD Europe[4]	4.70	4.36	3.70	3.33	3.15	3.00	2.83	-1.7%
Japan	0.13	0.13	0.14	0.15	0.15	0.15	0.16	0.7%
Australia and New Zealand	0.65	0.62	0.55	0.54	0.54	0.53	0.53	-0.6%
Total OECD petroleum production	18.71	18.80	18.65	18.54	17.78	17.72	17.14	-0.4%
Non-OECD								
Russia	9.93	10.14	10.04	10.54	11.06	11.62	12.16	0.7%
Other Europe and Eurasia[5]	3.12	3.22	3.67	4.01	4.37	4.52	4.54	1.4%
China	3.99	4.27	4.29	4.46	4.79	4.93	4.70	0.4%
Other Asia[6]	3.67	3.77	3.79	3.55	3.38	3.17	3.00	-0.9%
Middle East	1.56	1.58	1.43	1.31	1.18	1.06	0.97	-1.9%
Africa	2.44	2.41	2.40	2.54	2.68	2.70	2.68	0.4%
Brazil	2.08	2.19	2.72	3.34	3.87	4.21	4.45	2.9%
Other Central and South America	1.90	2.01	2.29	2.32	2.47	2.67	2.65	1.1%
Total non-OECD petroleum production	28.69	29.59	30.63	32.07	33.80	34.88	35.15	0.7%
Total petroleum liquids production	80.21	82.44	85.58	88.52	92.04	95.08	96.47	0.6%
Other liquids production[7]								
United States (50 states)	0.75	0.90	1.05	1.34	1.62	2.08	2.59	4.3%
Other North America	1.69	1.93	2.51	3.08	3.75	4.46	5.16	4.0%
OECD Europe[4]	0.22	0.22	0.23	0.24	0.26	0.27	0.28	1.0%
Middle East	0.01	0.01	0.17	0.21	0.24	0.24	0.24	14.5%
Africa	0.21	0.21	0.28	0.37	0.38	0.39	0.40	2.6%
Central and South America	1.14	1.20	1.78	2.31	2.61	2.90	3.17	3.9%
Other	0.12	0.13	0.16	0.28	0.61	0.92	1.18	9.1%
Total other liquids production	4.14	4.61	6.18	7.82	9.47	11.27	13.02	4.2%
Total production	84.35	87.05	91.76	96.33	101.51	106.34	109.50	0.9%

Table A21. International liquids supply and disposition summary (continued)
(million barrels per day, unless otherwise noted)

Supply and disposition	Reference case							Annual growth 2010-2035 (percent)
	2009	2010	2015	2020	2025	2030	2035	
Liquids consumption[8]								
OECD								
United States (50 states)	18.81	19.17	19.10	19.02	19.20	19.47	19.90	0.1%
United States territories	0.27	0.28	0.31	0.32	0.34	0.36	0.36	1.0%
Canada	2.16	2.21	2.15	2.21	2.25	2.29	2.35	0.2%
Mexico and Chile	2.35	2.34	2.39	2.43	2.50	2.60	2.68	0.5%
OECD Europe[4]	14.66	14.58	14.14	14.43	14.65	14.76	14.74	0.0%
Japan	4.39	4.45	4.51	4.60	4.62	4.51	4.42	-0.0%
South Korea	2.15	2.24	2.25	2.35	2.46	2.53	2.56	0.5%
Australia and New Zealand	1.16	1.13	1.11	1.14	1.17	1.21	1.23	0.3%
Total OECD consumption	45.94	46.40	45.95	46.50	47.19	47.72	48.24	0.2%
Non-OECD								
Russia	2.73	2.93	3.02	2.94	2.91	2.94	2.97	0.1%
Other Europe and Eurasia[5]	2.15	2.08	2.30	2.35	2.45	2.55	2.63	0.9%
China	8.33	9.19	12.10	14.36	16.03	17.65	18.50	2.8%
India	3.11	3.18	3.70	4.58	5.40	5.79	5.80	2.4%
Other non-OECD Asia[6]	6.43	6.73	7.28	7.95	8.85	9.40	9.89	1.5%
Middle East	6.84	7.35	7.78	7.69	8.16	8.98	9.49	1.0%
Africa	3.23	3.34	3.30	3.37	3.57	3.80	4.09	0.8%
Brazil	2.52	2.65	2.84	2.94	3.15	3.47	3.80	1.5%
Other Central and South America	3.07	3.19	3.49	3.66	3.81	4.05	4.09	1.0%
Total non-OECD consumption	38.41	40.65	45.82	49.83	54.32	58.62	61.26	1.7%
Total liquids consumption	84.35	87.05	91.76	96.33	101.51	106.35	109.50	0.9%
OPEC production[9]	33.34	34.58	37.30	39.23	41.91	44.05	45.89	1.1%
Non-OPEC production[9]	51.01	52.47	54.46	57.10	59.60	62.30	63.61	0.8%
Net Eurasia exports	10.25	10.53	11.11	12.60	13.94	14.85	15.54	1.6%
OPEC market share (percent)	39.5	39.7	40.7	40.7	41.3	41.4	41.9	--

[1]Weighted average price delivered to U.S. refiners.
[2]Includes production of crude oil (including lease condensate and shale oil/tight oil), natural gas plant liquids, other hydrogen and hydrocarbons for refinery feedstocks, and refinery gains.
[3]OPEC = Organization of Petroleum Exporting Countries - Algeria, Angola, Ecuador, Iran, Iraq, Kuwait, Libya, Nigeria, Qatar, Saudi Arabia, the United Arab Emirates, and Venezuela.
[4]OECD Europe = Organization for Economic Cooperation and Development - Austria, Belgium, Czech Republic, Denmark, Finland, France, Germany, Greece, Hungary, Iceland, Ireland, Italy, Luxembourg, the Netherlands, Norway, Poland, Portugal, Slovakia, Slovenia, Spain, Sweden, Switzerland, Turkey, and the United Kingdom.
[5]Other Europe and Eurasia = Albania, Armenia, Azerbaijan, Belarus, Bosnia and Herzegovina, Bulgaria, Croatia, Estonia, Georgia, Kazakhstan, Kyrgyzstan, Latvia, Lithuania, Macedonia, Malta, Moldova, Montenegro, Romania, Serbia, Tajikistan, Turkmenistan, Ukraine, and Uzbekistan.
[6]Other Asia = Afghanistan, Bangladesh, Bhutan, Brunei, Cambodia (Kampuchea), Fiji, French Polynesia, Guam, Hong Kong, Indonesia, Kiribati, Laos, Malaysia, Macau, Maldives, Mongolia, Myanmar (Burma), Nauru, Nepal, New Caledonia, Niue, North Korea, Pakistan, Papua New Guinea, Philippines, Samoa, Singapore, Solomon Islands, Sri Lanka, Taiwan, Thailand, Tonga, Vanuatu, and Vietnam.
[7]Includes liquids produced from energy crops, natural gas, coal, extra-heavy oil, bitumen (oil sands), and kerogen (oil shale, not to be confused with shale oil/tight oil). Includes both OPEC and non-OPEC producers in the regional breakdown.
[8]Includes both OPEC and non-OPEC consumers in the regional breakdown.
[9]Includes both petroleum and other liquids production.
- - = Not applicable.
Note: Totals may not equal sum of components due to independent rounding. Data for 2009 and 2010 are model results and may differ slightly from official EIA data reports.
Sources: 2009 and 2010 low sulfur light crude oil price: U.S. Energy Information Administration (EIA), Form EIA-856, "Monthly Foreign Crude Oil Acquisition Report." 2009 and 2010 imported crude oil price: EIA, Annual Energy Review 2010, DOE/EIA-0384(2010) (Washington, DC, October 2011). 2009 quantities derived from: EIA, International Energy Statistics database as of November 2009. **2010 quantities and projections:** EIA, AEO2012 National Energy Modeling System run REF2012.D020112C and EIA, Generate World Oil Balance Model.

This page intentionally left blank

Appendix B
Economic growth case comparisons

Table B1. Total energy supply, disposition, and price summary
(quadrillion Btu per year, unless otherwise noted)

Supply, disposition, and prices	2010	2015 Low economic growth	2015 Reference	2015 High economic growth	2025 Low economic growth	2025 Reference	2025 High economic growth	2035 Low economic growth	2035 Reference	2035 High economic growth
Production										
Crude oil and lease condensate	11.59	13.23	13.23	13.25	13.53	13.77	13.79	12.86	12.89	13.12
Natural gas plant liquids	2.78	3.33	3.33	3.33	3.91	3.93	3.93	3.93	3.94	3.95
Dry natural gas	22.10	24.02	24.22	24.28	26.17	26.91	27.64	27.48	28.60	30.05
Coal[1]	22.06	19.71	20.24	20.79	20.27	22.25	23.65	21.91	24.14	25.33
Nuclear / uranium[2]	8.44	8.68	8.68	8.68	9.60	9.60	9.60	9.14	9.28	10.13
Hydropower	2.51	2.89	2.90	2.90	2.95	2.99	3.02	3.00	3.04	3.10
Biomass[3]	4.05	4.41	4.45	4.49	6.04	6.26	6.30	8.37	9.07	9.58
Other renewable energy[4]	1.34	2.08	1.99	2.18	2.21	2.22	2.42	2.44	2.81	3.64
Other[5]	0.64	0.60	0.60	0.60	0.68	0.69	0.71	0.83	0.91	0.93
Total	75.50	78.96	79.64	80.50	85.36	88.61	91.06	89.95	94.67	99.83
Imports										
Crude oil	20.14	18.34	18.87	19.43	15.20	16.23	17.55	15.30	16.90	18.50
Liquid fuels and other petroleum[6]	5.02	4.19	4.32	4.45	3.72	4.08	4.40	3.63	4.14	4.75
Natural gas[7]	3.81	3.67	3.73	3.76	2.61	2.75	2.89	2.74	2.84	2.86
Other imports[8]	0.52	0.34	0.44	0.47	0.97	1.07	0.95	0.73	0.81	0.96
Total	29.49	26.54	27.37	28.11	22.50	24.14	25.79	22.40	24.69	27.07
Exports										
Liquid fuels and other petroleum[9]	4.81	4.90	5.00	5.08	4.32	4.46	4.57	4.68	4.95	5.11
Natural gas[10]	1.15	1.93	1.93	1.92	3.55	3.51	3.48	4.29	4.17	4.07
Coal	2.10	2.73	2.73	2.73	2.78	2.82	2.82	3.09	3.13	3.18
Total	8.06	9.57	9.66	9.74	10.66	10.79	10.87	12.06	12.25	12.37
Discrepancy[11]	-1.23	-0.03	-0.08	-0.09	-0.01	-0.03	-0.06	0.25	0.18	0.15
Consumption										
Liquid fuels and other petroleum[12]	37.25	36.09	36.72	37.38	34.78	36.58	38.19	35.17	37.70	40.23
Natural gas	24.71	25.73	26.00	26.09	25.21	26.14	27.04	25.93	27.26	28.83
Coal[13]	20.76	17.17	17.80	18.36	18.23	20.02	21.30	19.16	21.15	22.43
Nuclear / uranium[2]	8.44	8.68	8.68	8.68	9.60	9.60	9.60	9.14	9.28	10.13
Hydropower	2.51	2.89	2.90	2.90	2.95	2.99	3.02	3.00	3.04	3.10
Biomass[14]	2.88	3.01	3.04	3.06	3.95	4.17	4.21	4.96	5.44	5.78
Other renewable energy[4]	1.34	2.08	1.99	2.18	2.21	2.22	2.42	2.44	2.81	3.64
Other[15]	0.29	0.30	0.30	0.30	0.28	0.28	0.28	0.24	0.24	0.25
Total	98.16	95.96	97.43	98.96	97.20	101.99	106.05	100.04	106.93	114.38
Prices (2010 dollars per unit)										
Petroleum (dollars per barrel)										
Low sulfur light crude oil[16]	79.39	116.06	116.91	117.83	130.58	132.56	134.77	142.51	144.98	147.82
Imported crude oil[16]	75.87	113.12	113.97	114.90	118.61	121.21	124.15	130.33	132.95	136.68
Natural gas (dollars per million Btu)										
at Henry hub	4.39	4.06	4.29	4.36	5.10	5.63	6.17	6.60	7.37	7.58
at the wellhead[17]	4.06	3.64	3.84	3.91	4.54	5.00	5.46	5.83	6.48	6.66
Natural gas (dollars per thousand cubic feet)										
at the wellhead[17]	4.16	3.73	3.94	4.00	4.65	5.12	5.59	5.97	6.64	6.82
Coal (dollars per ton)										
at the minemouth[18]	35.61	42.70	42.08	41.92	44.24	44.05	44.48	50.92	50.52	51.36
Coal (dollars per million Btu)										
at the minemouth[18]	1.76	2.11	2.08	2.08	2.24	2.23	2.25	2.57	2.56	2.60
Average end-use[19]	2.38	2.55	2.56	2.57	2.68	2.70	2.73	2.90	2.94	3.03
Average electricity (cents per kilowatthour)	9.8	9.9	9.7	9.6	9.7	9.7	9.9	9.8	10.1	10.5

Table B1. Total energy supply, disposition, and price summary (continued)
(quadrillion Btu per year, unless otherwise noted)

Supply, disposition, and prices	2010	Projections 2015			Projections 2025			Projections 2035		
		Low economic growth	Reference	High economic growth	Low economic growth	Reference	High economic growth	Low economic growth	Reference	High economic growth
Prices (nominal dollars per unit)										
Petroleum (dollars per barrel)										
Low sulfur light crude oil[16]	79.39	127 20	125.97	125.10	197 32	170.09	163.70	313.58	229 55	212.97
Imported crude oil[16]	75.87	123 98	122.81	121.98	179 23	155.52	150.79	286.76	210 51	196.92
Natural gas (dollars per million Btu)										
at Henry hub	4.39	4.45	4 62	4.63	7.70	7.23	7 50	14.52	11 67	10.92
at the wellhead[17]	4.06	3.99	4.14	4.15	6 86	6.42	6 63	12.82	10 26	9.59
Natural gas (dollars per thousand cubic feet)										
at the wellhead[17]	4.16	4.09	4 24	4.25	7 02	6.57	6.79	13.13	10 51	9.82
Coal (dollars per ton)										
at the minemouth[18]	35.61	46.80	45.34	44.50	66 85	56.52	54 03	112.04	80 00	74.00
Coal (dollars per million Btu)										
at the minemouth[18]	1.76	2.31	2.24	2.21	3 39	2.86	2.73	5.64	4 05	3.74
Average end-use[19]	2.38	2.79	2.76	2.73	4 05	3.47	3 32	6.37	4 66	4.36
Average electricity (cents per kilowatthour)	9.8	10.9	10.4	10.2	14.7	12.5	12 0	21.6	16 0	15.1

[1] Includes waste coal.
[2] These values represent the energy obtained from uranium when it is used in light water reactors. The total energy content of uranium is much larger, but alternative processes are required to take advantage of it.
[3] Includes grid-connected electricity from wood and wood waste; biomass, such as corn, used for liquid fuels production; and non-electric energy demand from wood. Refer to Table A17 for details.
[4] Includes grid-connected electricity from landfill gas; biogenic municipal waste; wind; photovoltaic and solar thermal sources; and non-electric energy from renewable sources, such as active and passive solar systems. Excludes electricity imports using renewable sources and nonmarketed renewable energy. See Table A17 for selected nonmarketed residential and commercial renewable energy data.
[5] Includes non-biogenic municipal waste, liquid hydrogen, methanol, and some domestic inputs to refineries.
[6] Includes imports of finished petroleum products, unfinished oils, alcohols, ethers, blending components, and renewable fuels such as ethanol.
[7] Includes imports of liquefied natural gas that is later re-exported.
[8] Includes coal, coal coke (net), and electricity (net). Excludes imports of fuel used in nuclear power plants.
[9] Includes crude oil, petroleum products, ethanol, and biodiesel.
[10] Includes re-exported liquefied natural gas and natural gas used for liquefaction at export terminals.
[11] Balancing item. Includes unaccounted for supply, losses, gains, and net storage withdrawals.
[12] Includes petroleum-derived fuels and non-petroleum derived fuels, such as ethanol and biodiesel, and coal-based synthetic liquids. Petroleum coke, which is a solid, is included. Also included are natural gas plant liquids and crude oil consumed as a fuel. Refer to Table A17 for detailed renewable liquid fuels consumption.
[13] Excludes coal converted to coal-based synthetic liquids and natural gas.
[14] Excludes grid-connected electricity from wood and wood waste, non-electric energy from wood, and biofuels heat and coproducts used in the production of liquid fuels, but excludes the energy content of the liquid fuels.
[15] Includes non-biogenic municipal waste, liquid hydrogen, and net electricity imports.
[16] Weighted average price delivered to U.S. refiners.
[17] Represents lower 48 onshore and offshore supplies.
[18] Includes reported prices for both open market and captive mines.
[19] Prices weighted by consumption; weighted average excludes residential and commercial prices, and export free-alongside-ship (f.a.s.) prices.
Btu = British thermal unit.
Note: Totals may not equal sum of components due to independent rounding. Data for 2010 are model results and may differ slightly from official EIA data reports.
Sources: 2010 natural gas supply values and natural gas wellhead price: U.S. Energy Information Administration (EIA), *Natural Gas Monthly*, DOE/EIA-0130(2011/07) (Washington, DC, July 2011). 2010 coal minemouth and delivered coal prices: EIA, *Annual Coal Report 2010*, DOE/EIA-0584(2010) (Washington, DC, November 2011). 2010 petroleum supply values: EIA, *Petroleum Supply Annual 2010*, DOE/EIA-0340(2010)/1 (Washington, DC, July 2011). 2010 low sulfur light crude oil price: EIA, Form EIA-856, "Monthly Foreign Crude Oil Acquisition Report." Other 2010 coal values: *Quarterly Coal Report, October-December 2010*, DOE/EIA-0121(2010/4Q) (Washington, DC, May 2011). Other 2010 values: EIA, *Annual Energy Review 2010*, DOE/EIA-0384(2010) (Washington, DC, October 2011). **Projections:** EIA, AEO2012 National Energy Modeling System runs LM2012.D022412A, REF2012.D020112C, and HM2012.D022412A.

Table B2. Energy consumption by sector and source
(quadrillion Btu per year, unless otherwise noted)

Sector and source	2010	Projections								
		2015			2025			2035		
		Low economic growth	Reference	High economic growth	Low economic growth	Reference	High economic growth	Low economic growth	Reference	High economic growth
Energy consumption										
Residential										
Liquefied petroleum gases	0.56	0.51	0.51	0.51	0.49	0.50	0.52	0.48	0.51	0.54
Kerosene	0.03	0.02	0.02	0.02	0.02	0.02	0.02	0.02	0.02	0.02
Distillate fuel oil	0.63	0.55	0.55	0.55	0.43	0.43	0.43	0.35	0.35	0.35
Liquid fuels and other petroleum subtotal	1.22	1.08	1.08	1.08	0.94	0.95	0.97	0.85	0.87	0.91
Natural gas	5.06	4.96	4.97	5.00	4.77	4.88	5.04	4.50	4.76	5.08
Coal	0.01	0.01	0.01	0.01	0.01	0.01	0.01	0.01	0.01	0.01
Renewable energy[1]	0.42	0.42	0.43	0.43	0.42	0.43	0.45	0.41	0.43	0.47
Electricity	4.95	4.68	4.75	4.82	4.97	5.23	5.58	5.35	5.86	6.57
Delivered energy	**11.66**	**11.15**	**11.24**	**11.34**	**11.11**	**11.51**	**12.05**	**11.12**	**11.93**	**13.04**
Electricity related losses	10.39	9.43	9.58	9.75	10.03	10.52	11.17	10.47	11.35	12.72
Total	**22.05**	**20.59**	**20.81**	**21.09**	**21.13**	**22.02**	**23.22**	**21.59**	**23.28**	**25.76**
Commercial										
Liquefied petroleum gases	0.14	0.14	0.14	0.14	0.15	0.15	0.15	0.15	0.16	0.16
Motor gasoline[2]	0.05	0.05	0.05	0.05	0.05	0.05	0.05	0.06	0.06	0.06
Kerosene	0.00	0.00	0.00	0.00	0.00	0.00	0.00	0.01	0.01	0.01
Distillate fuel oil	0.43	0.35	0.35	0.35	0.33	0.33	0.33	0.32	0.32	0.32
Residual fuel oil	0.08	0.08	0.08	0.08	0.08	0.08	0.08	0.08	0.08	0.08
Liquid fuels and other petroleum subtotal	0.72	0.62	0.62	0.62	0.62	0.62	0.62	0.62	0.62	0.63
Natural gas	3.28	3.43	3.41	3.42	3.56	3.53	3.51	3.70	3.69	3.71
Coal	0.06	0.06	0.06	0.06	0.06	0.06	0.06	0.06	0.06	0.06
Renewable energy[3]	0.11	0.11	0.11	0.11	0.11	0.11	0.11	0.11	0.11	0.11
Electricity	4.54	4.57	4.59	4.61	5.11	5.16	5.22	5.70	5.89	5.89
Delivered energy	**8.70**	**8.79**	**8.80**	**8.81**	**9.46**	**9.48**	**9.53**	**10.19**	**10.28**	**10.39**
Electricity related losses	9.52	9.21	9.27	9.32	10.30	10.38	10.44	11.15	11.23	11.40
Total	**18.22**	**18.00**	**18.06**	**18.13**	**19.76**	**19.86**	**19.97**	**21.34**	**21.50**	**21.79**
Industrial[4]										
Liquefied petroleum gases	2.00	1.80	1.83	1.83	2.06	2.17	2.18	2.01	2.15	2.20
Motor gasoline[2]	0.25	0.27	0.28	0.29	0.27	0.30	0.33	0.26	0.30	0.33
Distillate fuel oil	1.16	1.16	1.25	1.33	1.04	1.19	1.33	1.01	1.18	1.35
Residual fuel oil	0.12	0.09	0.09	0.09	0.08	0.08	0.09	0.08	0.08	0.09
Petrochemical feedstocks	0.94	1.00	1.01	1.01	1.22	1.29	1.29	1.21	1.30	1.33
Other petroleum[5]	3.59	3.29	3.44	3.60	2.81	3.11	3.45	2.80	3.19	3.60
Liquid fuels and other petroleum subtotal	8.05	7.61	7.89	8.15	7.48	8.13	8.68	7.36	8.21	8.89
Natural gas	6.76	7.04	7.19	7.34	6.81	7.32	7.62	6.49	7.18	7.84
Natural-gas-to-liquids heat and power	0.00	0.00	0.00	0.00	0.00	0.00	0.00	0.00	0.00	0.00
Lease and plant fuel[6]	1.37	1.42	1.43	1.43	1.54	1.57	1.60	1.57	1.63	1.71
Natural gas subtotal	8.14	8.46	8.62	8.77	8.35	8.89	9.22	8.06	8.81	9.55
Metallurgical coal	0.55	0.55	0.57	0.59	0.41	0.49	0.54	0.34	0.43	0.53
Other industrial coal	1.01	1.01	1.03	1.05	1.02	1.08	1.12	1.01	1.08	1.14
Coal-to-liquids heat and power	0.00	0.00	0.00	0.00	0.11	0.36	0.37	0.31	0.60	0.61
Net coal coke imports	-0.01	-0.01	-0.01	-0.00	-0.03	-0.03	-0.03	-0.05	-0.06	-0.07
Coal subtotal	1.56	1.55	1.59	1.63	1.52	1.90	2.00	1.60	2.06	2.21
Biofuels heat and coproducts	0.84	0.80	0.81	0.82	1.26	1.27	1.27	2.39	2.57	2.69
Renewable energy[7]	1.50	1.59	1.61	1.63	1.67	1.82	1.91	1.74	1.95	2.10
Electricity	3.28	3.34	3.44	3.53	3.22	3.52	3.75	3.01	3.33	3.67
Delivered energy	**23.37**	**23.35**	**23.96**	**24.53**	**23.49**	**25.53**	**26.83**	**24.17**	**26.94**	**29.11**
Electricity related losses	6.89	6.73	6.94	7.15	6.50	7.09	7.50	5.89	6.46	7.10
Total	**30.26**	**30.08**	**30.90**	**31.68**	**29.99**	**32.61**	**34.33**	**30.06**	**33.39**	**36.21**

Table B2. Energy consumption by sector and source (continued)
(quadrillion Btu per year, unless otherwise noted)

Sector and source	2010	Projections								
		2015			2025			2035		
		Low economic growth	Reference	High economic growth	Low economic growth	Reference	High economic growth	Low economic growth	Reference	High economic growth
Transportation										
Liquefied petroleum gases	0.04	0.04	0.04	0.04	0.04	0.04	0.05	0.04	0.05	0.06
E85[8]	0.00	0.01	0.01	0.01	0.40	0.30	0.21	1.14	1.22	1.22
Motor gasoline[2]	16.91	16.00	16.13	16.29	14.26	14.90	15.49	13.43	14.53	15.38
Jet fuel[9]	3.07	3.01	3.03	3.04	3.15	3.19	3.24	3.25	3.33	3.42
Distillate fuel oil[10]	5.77	6.35	6.55	6.77	6.50	7.03	7.51	7.06	7.44	8.27
Residual fuel oil	0.90	0.91	0.91	0.91	0.92	0.93	0.93	0.93	0.94	0.95
Other petroleum[11]	0.17	0.17	0.17	0.17	0.17	0.17	0.17	0.17	0.17	0.18
Liquid fuels and other petroleum subtotal	26.88	26.48	26.83	27.22	25.43	26.57	27.60	26.03	27.67	29.47
Pipeline fuel natural gas	0.65	0.68	0.68	0.69	0.65	0.67	0.69	0.66	0.69	0.74
Compressed / liquefied natural gas	0.04	0.06	0.06	0.06	0.11	0.11	0.12	0.16	0.16	0.17
Liquid hydrogen	0.00	0.00	0.00	0.00	0.00	0.00	0.00	0.00	0.00	0.00
Electricity	0.02	0.03	0.03	0.03	0.04	0.04	0.04	0.07	0.07	0.08
Delivered energy	**27.59**	**27.24**	**27.60**	**28.00**	**26.24**	**27.40**	**28.45**	**26.92**	**28.60**	**30.46**
Electricity related losses	0.05	0.05	0.05	0.05	0.08	0.08	0.09	0.13	0.14	0.15
Total	**27.63**	**27.30**	**27.65**	**28.05**	**26.32**	**27.49**	**28.54**	**27.05**	**28.75**	**30.62**
Delivered energy consumption for all sectors										
Liquefied petroleum gases	2.75	2.49	2.51	2.52	2.75	2.86	2.89	2.69	2.86	2.95
E85[8]	0.00	0.01	0.01	0.01	0.40	0.30	0.21	1.14	1.22	1.22
Motor gasoline[2]	17.21	16.32	16.46	16.63	14.58	15.25	15.87	13.75	14.88	15.77
Jet fuel[9]	3.07	3.01	3.03	3.04	3.15	3.19	3.24	3.25	3.33	3.42
Kerosene	0.04	0.03	0.03	0.03	0.03	0.03	0.03	0.03	0.03	0.03
Distillate fuel oil	7.99	8.41	8.69	9.00	8.30	8.99	9.61	8.74	9.29	10.29
Residual fuel oil	1.11	1.07	1.08	1.08	1.08	1.09	1.10	1.09	1.11	1.12
Petrochemical feedstocks	0.94	1.00	1.01	1.01	1.22	1.29	1.29	1.21	1.30	1.33
Other petroleum[12]	3.76	3.45	3.61	3.76	2.97	3.27	3.62	2.97	3.36	3.77
Liquid fuels and other petroleum subtotal	36.87	35.80	36.43	37.07	34.48	36.28	37.87	34.86	37.38	39.90
Natural gas	15.15	15.49	15.64	15.83	15.25	15.85	16.29	14.85	15.79	16.80
Natural-gas-to-liquids heat and power	0.00	0.00	0.00	0.00	0.00	0.00	0.00	0.00	0.00	0.00
Lease and plant fuel[6]	1.37	1.42	1.43	1.43	1.54	1.57	1.60	1.57	1.63	1.71
Pipeline natural gas	0.65	0.68	0.68	0.69	0.65	0.67	0.69	0.66	0.69	0.74
Natural gas subtotal	17.17	17.58	17.75	17.94	17.44	18.09	18.58	17.08	18.11	19.26
Metallurgical coal	0.55	0.55	0.57	0.59	0.41	0.49	0.54	0.34	0.43	0.53
Other coal	1.08	1.07	1.09	1.11	1.08	1.14	1.18	1.07	1.15	1.21
Coal-to-liquids heat and power	0.00	0.00	0.00	0.00	0.11	0.36	0.37	0.31	0.60	0.61
Net coal coke imports	-0.01	-0.01	-0.01	-0.00	-0.03	-0.03	-0.03	-0.05	-0.06	-0.07
Coal subtotal	1.62	1.62	1.65	1.70	1.58	1.96	2.06	1.67	2.12	2.28
Biofuels heat and coproducts	0.84	0.80	0.81	0.82	1.26	1.27	1.27	2.39	2.57	2.69
Renewable energy[13]	2.03	2.12	2.15	2.17	2.20	2.36	2.47	2.25	2.50	2.68
Liquid hydrogen	0.00	0.00	0.00	0.00	0.00	0.00	0.00	0.00	0.00	0.00
Electricity	12.79	12.61	12.81	12.98	13.34	13.96	14.60	14.13	15.06	16.20
Delivered energy	**71.32**	**70.54**	**71.59**	**72.69**	**70.30**	**73.92**	**76.86**	**72.39**	**77.75**	**83.01**
Electricity related losses	26.84	25.42	25.84	26.27	26.91	28.07	29.20	27.65	29.18	31.37
Total	**98.16**	**95.96**	**97.43**	**98.96**	**97.20**	**101.99**	**106.05**	**100.04**	**106.93**	**114.38**
Electric power[14]										
Distillate fuel oil	0.08	0.08	0.08	0.09	0.09	0.09	0.09	0.09	0.09	0.09
Residual fuel oil	0.30	0.21	0.21	0.22	0.21	0.22	0.23	0.22	0.23	0.24
Liquid fuels and other petroleum subtotal	0.38	0.29	0.29	0.30	0.30	0.31	0.32	0.31	0.32	0.34
Natural gas	7.54	8.15	8.25	8.15	7.77	8.04	8.46	8.84	9.16	9.58
Steam coal	19.13	15.56	16.15	16.67	16.65	18.06	19.24	17.50	19.03	20.15
Nuclear / uranium[15]	8.44	8.68	8.68	8.68	9.60	9.60	9.60	9.14	9.28	10.13
Renewable energy[16]	3.85	5.05	4.96	5.15	5.66	5.75	5.91	5.75	6.22	7.14
Electricity imports	0.09	0.10	0.10	0.10	0.08	0.08	0.08	0.04	0.04	0.04
Total[17]	**39.63**	**38.03**	**38.64**	**39.25**	**40.25**	**42.03**	**43.80**	**41.78**	**44.24**	**47.57**

Table B2. Energy consumption by sector and source (continued)
(quadrillion Btu per year, unless otherwise noted)

Sector and source	2010	Projections 2015			Projections 2025			Projections 2035		
		Low economic growth	Reference	High economic growth	Low economic growth	Reference	High economic growth	Low economic growth	Reference	High economic growth
Total energy consumption										
Liquefied petroleum gases	2.75	2.49	2.51	2.52	2.75	2.86	2.89	2.69	2.86	2.95
E85[8]	0.00	0.01	0.01	0.01	0.40	0.30	0.21	1.14	1.22	1.22
Motor gasoline[2]	17.21	16.32	16.46	16.63	14.58	15.25	15.87	13.75	14.88	15.77
Jet fuel[9]	3.07	3.01	3.03	3.04	3.15	3.19	3.24	3.25	3.33	3.42
Kerosene	0.04	0.03	0.03	0.03	0.03	0.03	0.03	0.03	0.03	0.03
Distillate fuel oil	8.07	8.50	8.78	9.08	8.39	9.07	9.70	8.83	9.38	10.38
Residual fuel oil	1.41	1.28	1.29	1.30	1.29	1.31	1.33	1.31	1.34	1.36
Petrochemical feedstocks	0.94	1.00	1.01	1.01	1.22	1.29	1.29	1.21	1.30	1.33
Other petroleum[12]	3.76	3.45	3.61	3.76	2.97	3.27	3.62	2.97	3.36	3.77
Liquid fuels and other petroleum subtotal	37.25	36.09	36.72	37.38	34.78	36.58	38.19	35.17	37.70	40.23
Natural gas	22.69	23.64	23.89	23.97	23.02	23.89	24.74	23.70	24.94	26.38
Natural-gas-to-liquids heat and power	0.00	0.00	0.00	0.00	0.00	0.00	0.00	0.00	0.00	0.00
Lease and plant fuel[6]	1.37	1.42	1.43	1.43	1.54	1.57	1.60	1.57	1.63	1.71
Pipeline natural gas	0.65	0.68	0.68	0.69	0.65	0.67	0.69	0.66	0.69	0.74
Natural gas subtotal	24.71	25.73	26.00	26.09	25.21	26.14	27.04	25.93	27.26	28.83
Metallurgical coal	0.55	0.55	0.57	0.59	0.41	0.49	0.54	0.34	0.43	0.53
Other coal	20.21	16.63	17.24	17.78	17.73	19.20	20.42	18.57	20.18	21.36
Coal-to-liquids heat and power	0.00	0.00	0.00	0.00	0.11	0.36	0.37	0.31	0.60	0.61
Net coal coke imports	-0.01	-0.01	-0.01	-0.01	-0.03	-0.03	-0.03	-0.05	-0.06	-0.07
Coal subtotal	20.76	17.17	17.80	18.36	18.23	20.02	21.30	19.16	21.15	22.43
Nuclear / uranium[15]	8.44	8.68	8.68	8.68	9.60	9.60	9.60	9.14	9.28	10.13
Biofuels heat and coproducts	0.84	0.80	0.81	0.82	1.26	1.27	1.27	2.39	2.57	2.69
Renewable energy[16]	5.88	7.18	7.11	7.33	7.85	8.11	8.38	8.00	8.71	9.82
Liquid hydrogen	0.00	0.00	0.00	0.00	0.00	0.00	0.00	0.00	0.00	0.00
Electricity imports	0.09	0.10	0.10	0.10	0.08	0.08	0.08	0.04	0.04	0.04
Total	**98.16**	**95.96**	**97.43**	**98.96**	**97.20**	**101.99**	**106.05**	**100.04**	**106.93**	**114.38**
Energy use and related statistics										
Delivered energy use	71.32	70.54	71.59	72.69	70.30	73.92	76.86	72.39	77.75	83.01
Total energy use	98.16	95.96	97.43	98.96	97.20	101.99	106.05	100.04	106.93	114.38
Ethanol consumed in motor gasoline and E85	1.11	1.21	1.22	1.23	1.55	1.55	1.54	1.99	2.15	2.23
Population (millions)	310.83	325.23	326.16	327.19	354.23	358.06	362.48	382.76	390.09	398.74
Gross domestic product (billion 2005 dollars)	13088	14401	14803	15235	17676	19185	20538	21630	24539	27084
Carbon dioxide emissions (million metric tons)	5633.6	5298.2	5407.2	5503.9	5226.8	5552.5	5823.7	5355.8	5757.9	6117.5

[1]Includes wood used for residential heating. See Table A4 and/or Table A17 for estimates of nonmarketed renewable energy consumption for geothermal heat pumps, solar thermal water heating, and electricity generation from wind and solar photovoltaic sources.
[2]Includes ethanol (blends of 15 percent or less) and ethers blended into gasoline.
[3]Excludes ethanol. Includes commercial sector consumption of wood and wood waste, landfill gas, municipal waste, and other biomass for combined heat and power. See Table A5 and/or Table A17 for estimates of nonmarketed renewable energy consumption for solar thermal water heating and electricity generation from wind and solar photovoltaic sources.
[4]Includes energy for combined heat and power plants, except those whose primary business is to sell electricity, or electricity and heat, to the public.
[5]Includes petroleum coke, asphalt, road oil, lubricants, still gas, and miscellaneous petroleum products.
[6]Represents natural gas used in well, field, and lease operations, and in natural gas processing plant machinery.
[7]Includes consumption of energy produced from hydroelectric, wood and wood waste, municipal waste, and other biomass sources. Excludes ethanol blends (15 percent or less) in motor gasoline.
[8]E85 refers to a blend of 85 percent ethanol (renewable) and 15 percent motor gasoline (nonrenewable). To address cold starting issues, the percentage of ethanol varies seasonally. The annual average ethanol content of 74 percent is used for this forecast.
[9]Includes only kerosene type.
[10]Diesel fuel for on- and off- road use.
[11]Includes aviation gasoline and lubricants.
[12]Includes unfinished oils, natural gasoline, motor gasoline blending components, aviation gasoline, lubricants, still gas, asphalt, road oil, petroleum coke, and miscellaneous petroleum products.
[13]Includes electricity generated for sale to the grid and for own use from renewable sources, and non-electric energy from renewable sources. Excludes ethanol and nonmarketed renewable energy consumption for geothermal heat pumps, buildings photovoltaic systems, and solar thermal water heaters.
[14]Includes consumption of energy by electricity-only and combined heat and power plants whose primary business is to sell electricity, or electricity and heat, to the public. Includes small power producers and exempt wholesale generators.
[15]These values represent the energy obtained from uranium when it is used in light water reactors. The total energy content of uranium is much larger, but alternative processes are required to take advantage of it.
[16]Includes conventional hydroelectric, geothermal, wood and wood waste, biogenic municipal waste, other biomass, wind, photovoltaic, and solar thermal sources. Excludes net electricity imports.
[17]Includes non-biogenic municipal waste not included above.
[18]Includes conventional hydroelectric, geothermal, wood and wood waste, biogenic municipal waste, other biomass, wind, photovoltaic, and solar thermal sources. Excludes ethanol, net electricity imports, and nonmarketed renewable energy consumption for geothermal heat pumps, buildings photovoltaic systems, and solar thermal water heaters.
Btu = British thermal unit.
Note: Totals may not equal sum of components due to independent rounding. Data for 2010 are model results and may differ slightly from official EIA data reports.
Sources: 2010 consumption based on: U.S. Energy Information Administration (EIA), *Annual Energy Review 2010*, DOE/EIA-0384(2010) (Washington, DC, October 2011). 2010 population and gross domestic product: IHS Global Insight Industry and Employment models, August 2011. 2010 carbon dioxide emissions: EIA, *Monthly Energy Review*, October 2011 DOE/EIA-0035(2011/10) (Washington, DC, October 2011). **Projections:** EIA, AEO2012 National Energy Modeling System runs LM2012.D022412A, REF2012.D020112C, and HM2012.D022412A.

Table B3. Energy prices by sector and source
(2010 dollars per million Btu, unless otherwise noted)

Sector and source	2010	Projections								
		2015			2025			2035		
		Low economic growth	Reference	High economic growth	Low economic growth	Reference	High economic growth	Low economic growth	Reference	High economic growth
Residential										
Liquefied petroleum gases	27.02	30.48	30.70	30.86	31.69	32.27	32.91	33.94	34.64	35.27
Distillate fuel oil	21.21	27.00	27.26	27.52	29.17	30.15	30.64	32.01	32.73	33.99
Natural gas	11.08	10.10	10.31	10.39	11.46	12.03	12.61	13.16	13.98	14.38
Electricity	33.69	35.59	34.59	34.31	34.30	34.08	34.20	34.14	34.58	35.27
Commercial										
Liquefied petroleum gases	23.52	27.21	27.42	27.57	28.39	28.97	29.59	30.62	31.30	31.89
Distillate fuel oil	20.77	23.72	23.98	24.23	25.89	26.86	27.30	28.58	29.18	30.43
Residual fuel oil	11.07	16.02	16.18	16.35	17.82	18.24	18.62	18.61	18.90	19.61
Natural gas	9.10	8.40	8.60	8.67	9.51	10.02	10.52	10.92	11.64	11.91
Electricity	29.73	29.65	29.03	28.97	28.81	29.00	29.51	28.42	29.48	30.79
Industrial[1]										
Liquefied petroleum gases	21.80	27.12	27.43	27.66	28.44	29.24	30.12	31.26	32.18	32.98
Distillate fuel oil	21.32	23.95	24.20	24.45	26.23	27.22	27.61	28.93	29.53	30.79
Residual fuel oil	10.92	18.95	19.21	19.45	20.54	21.23	21.59	21.12	21.65	22.44
Natural gas[2]	5.51	4.68	4.88	4.94	5.58	6.04	6.51	6.89	7.54	7.74
Metallurgical coal	5.84	7.30	7.22	7.20	8.24	8.11	8.08	9.24	9.11	9.11
Other industrial coal	2.71	3.27	3.27	3.27	3.38	3.38	3.39	3.61	3.64	3.69
Coal to liquids	--	1.27	1.26	1.26	2.27	2.08	2.14	2.34	2.38	2.42
Electricity	19.63	19.06	18.91	18.94	19.21	19.60	20.15	19.63	20.78	22.00
Transportation										
Liquefied petroleum gases[3]	26.88	31.71	31.93	32.09	32.80	33.38	34.04	35.02	35.74	36.31
E85[4]	25.21	28.85	29.03	29.26	27.92	28.81	31.30	31.02	31.96	33.04
Motor gasoline[5]	22.70	29.09	29.26	29.49	30.92	32.10	32.42	32.33	33.61	34.78
Jet fuel[6]	16.22	23.48	23.74	24.02	25.61	26.45	26.99	28.41	29.13	30.25
Diesel fuel (distillate fuel oil)[7]	21.87	27.28	27.56	27.83	29.18	30.42	30.85	31.53	32.40	33.80
Residual fuel oil	10.42	17.96	18.32	18.61	19.74	20.62	20.82	20.50	20.95	21.94
Natural gas[8]	13.20	12.17	12.40	12.51	12.51	13.29	13.86	13.42	14.51	14.87
Electricity	32.99	30.67	30.50	30.54	31.37	31.53	32.45	32.36	33.82	35.11
Electric power[9]										
Distillate fuel oil	18.73	22.50	22.77	23.04	24.44	25.35	25.88	27.17	27.80	29.02
Residual fuel oil	11.89	22.67	23.00	23.03	24.55	25.40	25.41	25.25	25.72	26.49
Natural gas	5.14	4.36	4.55	4.61	5.15	5.60	6.10	6.55	7.21	7.40
Steam coal	2.26	2.33	2.35	2.37	2.50	2.54	2.56	2.75	2.80	2.87
Average price to all users[10]										
Liquefied petroleum gases	17.28	22.78	22.99	23.18	23.62	24.19	24.91	25.96	26.63	27.37
E85[4]	25.21	28.85	29.03	29.26	27.92	28.81	31.30	31.02	31.96	33.04
Motor gasoline[5]	22.59	29.09	29.26	29.49	30.91	32.10	32.42	32.33	33.61	34.78
Jet fuel	16.22	23.48	23.74	24.02	25.61	26.45	26.99	28.41	29.13	30.25
Distillate fuel oil	21.65	26.61	26.87	27.14	28.65	29.81	30.23	31.09	31.91	33.27
Residual fuel oil	10.82	18.67	19.01	19.27	20.46	21.31	21.53	21.22	21.68	22.64
Natural gas	7.16	6.27	6.45	6.52	7.29	7.74	8.22	8.63	9.30	9.53
Metallurgical coal	5.84	7.30	7.22	7.20	8.24	8.11	8.08	9.24	9.11	9.11
Other coal	2.29	2.40	2.41	2.43	2.56	2.59	2.62	2.80	2.85	2.92
Coal to liquids	--	1.27	1.26	1.26	2.27	2.08	2.14	2.34	2.38	2.42
Electricity	28.68	29.05	28.38	28.23	28.55	28.54	28.90	28.73	29.56	30.64
Non-renewable energy expenditures by sector (billion 2010 dollars)										
Residential	251.69	247.63	246.72	248.83	253.92	266.75	285.47	270.07	298.72	336.43
Commercial	179.08	179.38	177.92	178.42	197.28	201.89	208.21	220.10	231.98	244.34
Industrial	198.98	214.83	223.88	231.79	232.07	261.92	285.16	242.72	282.31	317.58
Transportation	573.78	731.18	746.84	764.56	736.46	803.52	848.96	777.83	856.65	950.17
Total non-renewable expenditures	1203.54	1373.02	1395.36	1423.60	1419.73	1534.08	1627.80	1510.72	1669.66	1848.51
Transportation renewable expenditures	0.08	0.24	0.25	0.26	11.22	8.74	6.44	35.33	38.86	40.34
Total expenditures	**1203.62**	**1373.26**	**1395.61**	**1423.86**	**1430.95**	**1542.81**	**1634.24**	**1546.05**	**1708.52**	**1888.85**

Table B3. Energy prices by sector and source (continued)
(nominal dollars per million Btu, unless otherwise noted)

Sector and source	2010	Projections								
		2015			2025			2035		
		Low economic growth	Reference	High economic growth	Low economic growth	Reference	High economic growth	Low economic growth	Reference	High economic growth
Residential										
Liquefied petroleum gases	27.02	33.41	33.08	32.76	47.89	41.41	39.98	74.69	54.86	50.81
Distillate fuel oil	21.21	29.60	29.38	29.22	44.08	38.68	37.22	70.42	51.82	48.97
Natural gas	11.08	11.07	11.11	11.03	17.31	15.43	15.32	28.95	22.14	20.72
Electricity	33.69	39.01	37.27	36.43	51.84	43.72	41.53	75.12	54.76	50.81
Commercial										
Liquefied petroleum gases	23.52	29.82	29.54	29.27	42.91	37.17	35.94	67.37	49.56	45.95
Distillate fuel oil	20.77	26.00	25.83	25.73	39.13	34.47	33.15	62.88	46.20	43.85
Residual fuel oil	11.07	17.55	17.43	17.36	26.93	23.41	22.61	40.96	29.93	28.25
Natural gas	9.10	9.21	9.27	9.21	14.37	12.86	12.78	24.03	18.43	17.16
Electricity	29.73	32.49	31.28	30.75	43.53	37.21	35.84	62.54	46.67	44.37
Industrial[1]										
Liquefied petroleum gases	21.80	29.72	29.56	29.37	42.98	37.51	36.59	68.79	50.95	47.52
Distillate fuel oil	21.32	26.25	26.08	25.96	39.64	34.93	33.54	63.67	46.76	44.36
Residual fuel oil	10.92	20.77	20.70	20.64	31.03	27.24	26.22	46.48	34.28	32.33
Natural gas[2]	5.51	5.13	5.26	5.25	8.43	7.75	7.91	15.15	11.93	11.15
Metallurgical coal	5.84	8.00	7.78	7.64	12.45	10.40	9.81	20.34	14.42	13.13
Other industrial coal	2.71	3.59	3.52	3.47	5.11	4.34	4.12	7.95	5.77	5.32
Coal to liquids	--	1.39	1.36	1.34	3.42	2.67	2.60	5.15	3.78	3.49
Electricity	19.63	20.89	20.38	20.11	29.03	25.15	24.47	43.20	32.90	31.70
Transportation										
Liquefied petroleum gases[3]	26.88	34.76	34.41	34.07	49.57	42.83	41.35	77.05	56.59	52.31
E85[4]	25.21	31.62	31.28	31.06	42.19	36.97	38.02	68.26	50.61	47.60
Motor gasoline[5]	22.70	31.88	31.53	31.31	46.72	41.19	39.38	71.14	53.22	50.11
Jet fuel[6]	16.22	25.74	25.58	25.50	38.70	33.94	32.78	62.51	46.12	43.58
Diesel fuel (distillate fuel oil)[7]	21.87	29.90	29.69	29.55	44.10	39.03	37.47	69.37	51.29	48.70
Residual fuel oil	10.42	19.69	19.74	19.76	29.83	26.45	25.28	45.11	33.18	31.60
Natural gas[8]	13.20	13.34	13.36	13.29	18.91	17.05	16.84	29.54	22.97	21.42
Electricity	32.99	33.62	32.86	32.42	47.41	40.46	39.41	71.19	53.55	50.59
Electric power[9]										
Distillate fuel oil	18.73	24.66	24.53	24.46	36.93	32.52	31.43	59.79	44.02	41.80
Residual fuel oil	11.89	24.85	24.78	24.45	37.10	32.59	30.87	55.56	40.73	38.16
Natural gas	5.14	4.78	4.90	4.90	7.78	7.19	7.41	14.41	11.42	10.66
Steam coal	2.26	2.56	2.53	2.51	3.78	3.25	3.12	6.05	4.43	4.13

Table B3. Energy prices by sector and source (continued)
(nominal dollars per million Btu, unless otherwise noted)

Sector and source	2010	Projections								
		2015			2025			2035		
		Low economic growth	Reference	High economic growth	Low economic growth	Reference	High economic growth	Low economic growth	Reference	High economic growth
Average price to all users[10]										
Liquefied petroleum gases	17.28	24.97	24.78	24.61	35.69	31.04	30.26	57.13	42.17	39.44
E85[4]	25.21	31.62	31.28	31.06	42.19	36.97	38.02	68.26	50.61	47.60
Motor gasoline[5]	22.59	31.88	31.53	31.31	46.72	41.19	39.38	71.14	53.22	50.11
Jet fuel	16.22	25.74	25.58	25.50	38.70	33.94	32.78	62.51	46.12	43.58
Distillate fuel oil	21.65	29.16	28.96	28.81	43.29	38.24	36.72	68.42	50.52	47.93
Residual fuel oil	10.82	20.46	20.48	20.46	30.92	27.34	26.15	46.69	34.33	32.61
Natural gas	7.16	6.87	6.95	6.92	11.02	9.93	9.98	18.98	14.73	13.73
Metallurgical coal	5.84	8.00	7.78	7.64	12.45	10.40	9.81	20.34	14.42	13.13
Other coal	2.29	2.63	2.60	2.58	3.87	3.32	3.18	6.17	4.51	4.20
Coal to liquids	--	1.39	1.36	1.34	3.42	2.67	2.60	5.15	3.78	3.49
Electricity	28.68	31.84	30.58	29.97	43.14	36.62	35.11	63.22	46.80	44.14
Non-renewable energy expenditures by sector (billion nominal dollars)										
Residential	251.69	271.41	265.85	264.18	383.71	342.26	346.74	594.24	472.99	484.70
Commercial	179.08	196.61	191.71	189.42	298.11	259.04	252.89	484.30	367.31	352.03
Industrial	198.98	235.47	241.24	246.08	350.69	336.06	346.35	534.08	447.01	457.54
Transportation	573.78	801.41	804.75	811.72	1112.90	1030.98	1031.15	1711.49	1356.41	1368.93
Total non-renewable expenditures	1203.54	1504.89	1503.55	1511.41	2145.42	1968.35	1977.13	3324.10	2643.72	2663.20
Transportation renewable expenditures	0.08	0.27	0.27	0.27	16.95	11.21	7.82	77.73	61.53	58.11
Total expenditures	**1203.62**	**1505.16**	**1503.82**	**1511.69**	**2162.37**	**1979.56**	**1984.95**	**3401.83**	**2705.26**	**2721.31**

[1]Includes energy for combined heat and power plants, except those whose primary business is to sell electricity, or electricity and heat, to the public.
[2]Excludes use for lease and plant fuel.
[3]Includes Federal and State taxes while excluding county and local taxes.
[4]E85 refers to a blend of 85 percent ethanol (renewable) and 15 percent motor gasoline (nonrenewable). To address cold starting issues, the percentage of ethanol varies seasonally. The annual average ethanol content of 74 percent is used for this forecast.
[5]Sales weighted-average price for all grades. Includes Federal, State and local taxes.
[6]Kerosene-type jet fuel. Includes Federal and State taxes while excluding county and local taxes.
[7]Diesel fuel for on-road use. Includes Federal and State taxes while excluding county and local taxes.
[8]Natural gas used as a vehicle fuel. Includes estimated motor vehicle fuel taxes and estimated dispensing costs or charges.
[9]Includes electricity-only and combined heat and power plants whose primary business is to sell electricity, or electricity and heat, to the public.
[10]Weighted averages of end-use fuel prices are derived from the prices shown in each sector and the corresponding sectoral consumption.
Btu = British thermal unit.
- - = Not applicable.
Note: Data for 2010 are model results and may differ slightly from official EIA data reports.
Sources: 2010 prices for motor gasoline, distillate fuel oil, and jet fuel are based on prices in the U.S. Energy Information Administration (EIA), *Petroleum Marketing Annual 2009*, DOE/EIA-0487(2009) (Washington, DC, August 2010). 2010 residential and commercial natural gas delivered prices: EIA, *Natural Gas Monthly*, DOE/EIA-0130(2011/07) (Washington, DC, July 2011). 2010 industrial natural gas delivered prices are estimated based on: EIA, *Manufacturing Energy Consumption Survey* and industrial and wellhead prices from the *Natural Gas Annual 2009*, DOE/EIA-0131(2009) (Washington, DC, December 2010) and the *Natural Gas Monthly*, DOE/EIA-0130(2011/07) (Washington, DC, July 2011). 2010 transportation sector natural gas delivered prices are model results. 2010 electric power sector distillate and residual fuel oil prices: EIA, *Monthly Energy Review*, DOE/EIA-0035(2011/09) (Washington, DC, September 2010). 2010 electric power sector natural gas prices: EIA, *Electric Power Monthly*, DOE/EIA-0226, April 2010 and April 2011, Table 4.2, and EIA, *State Energy Data Report 2009*, DOE/EIA-0214(2009) (Washington, DC, June 2011). 2010 coal prices based on: EIA, *Quarterly Coal Report, October-December 2010*, DOE/EIA-0121(2010/4Q) (Washington, DC, May 2011) and EIA, AEO2012 National Energy Modeling System run REF2012.D020112C. 2010 electricity prices: EIA, *Annual Energy Review 2010*, DOE/EIA-0384(2010) (Washington, DC, October 2011). 2010 E85 prices derived from monthly prices in the Clean Cities Alternative Fuel Price Report. Projections: EIA, AEO2012 National Energy Modeling System runs LM2012.D022412A, REF2012.D020112C, and HM2012.D022412A.

Table B4. Macroeconomic indicators
(billion 2005 chain-weighted dollars, unless otherwise noted)

Indicators	2010	Projections								
		2015			2025			2035		
		Low economic growth	Reference	High economic growth	Low economic growth	Reference	High economic growth	Low economic growth	Reference	High economic growth
Real gross domestic product	**13088**	**14401**	**14803**	**15235**	**17676**	**19185**	**20538**	**21630**	**24539**	**27084**
Components of real gross domestic product										
Real consumption	9221	10007	10218	10510	11874	12697	13606	14594	16220	17889
Real investment	1715	2234	2457	2675	2956	3472	3982	3929	4836	5651
Real government spending	2557	2322	2355	2389	2420	2525	2601	2619	2818	2944
Real exports	1663	2243	2289	2322	3828	4235	4558	5846	6953	7979
Real imports	2085	2370	2463	2596	3258	3516	3909	5020	5690	6596
Energy intensity										
(thousand Btu per 2005 dollar of GDP)										
Delivered energy	5.45	4.90	4.84	4.77	3.98	3.85	3.74	3.35	3.17	3.06
Total energy	7.50	6.66	6.58	6.50	5.50	5.32	5.16	4.63	4.36	4.22
Price indices										
GDP chain-type price index (2005=1.000)	1.110	1.217	1.196	1.178	1.677	1.424	1.348	2.442	1.758	1.599
Consumer price index (1982-4=1)										
All-urban	2.18	2.47	2.42	2.36	3.53	2.95	2.78	5.38	3.72	3.36
Energy commodities and services	2.12	2.67	2.62	2.59	3.82	3.36	3.20	5.83	4.37	4.07
Wholesale price index (1982=1.00)										
All commodities	1.85	2.15	2.10	2.02	2.96	2.39	2.25	4.46	2.81	2.47
Fuel and power	1.86	2.31	2.29	2.27	3.41	3.01	2.92	5.44	4.12	3.85
Metals and metal products	2.08	2.45	2.43	2.45	2.85	2.57	2.53	3.39	2.64	2.56
Industrial commodities excluding energy	1.83	2.08	2.04	2.02	2.63	2.22	2.12	3.47	2.43	2.24
Interest rates (percent, nominal)										
Federal funds rate	0.17	3.31	3.26	2.50	5.75	4.29	3.58	7.56	4.30	3.59
10-year treasury note	3.21	6.62	4.67	4.09	8.03	5.06	4.49	8.22	5.18	4.47
AA utility bond rate	5.24	9.31	6.74	5.73	11.61	7.17	6.18	12.74	7.56	6.12
Value of shipments (billion 2005 dollars)										
Service sectors	20602	22047	22469	22970	26671	28029	29342	31392	33430	35331
Total industrial	5838	6407	6730	7072	7109	7973	8737	7606	8692	9954
Non-manufacturing	1578	1702	1873	2065	1885	2228	2554	2024	2407	2823
Manufacturing	4260	4705	4857	5008	5224	5745	6183	5583	6285	7131
Energy-intensive	1595	1633	1664	1692	1781	1901	1971	1854	2034	2155
Non-energy-intensive	2664	3072	3194	3316	3443	3844	4212	3729	4251	4976
Total shipments	**26440**	**28454**	**29199**	**30042**	**33780**	**36002**	**38079**	**38998**	**42122**	**45285**
Population and employment (millions)										
Population with armed forces overseas	310.8	325.2	326.2	327.2	354.2	358.1	362.5	382.8	390.1	398.7
Population, aged 16 and over	244.3	256.0	256.5	257.2	279.9	282.6	285.8	304.2	309.6	316.0
Population, over age 65	40.4	46.7	47.1	47.1	63.4	64.2	64.4	76.9	77.7	78.3
Employment, nonfarm	129.8	138.3	139.4	142.7	150.4	154.2	160.5	158.9	166.8	173.4
Employment, manufacturing	11.5	11.8	12.1	12.3	11.0	11.4	11.9	9.1	9.2	9.9
Key labor indicators										
Labor force (millions)	153.9	157.6	158.0	158.7	167.1	168.6	170.9	178.0	181.7	186.3
Non-farm labor productivity (1992=1.00)	1.10	1.14	1.16	1.18	1.33	1.42	1.47	1.55	1.75	1.85
Unemployment rate (percent)	9.63	8.11	7.51	7.10	6.04	5.54	5.05	6.15	5.54	5.09
Key indicators for energy demand										
Real disposable personal income	10062	10890	11035	11224	13862	14286	14978	17350	18217	19407
Housing starts (millions)	0.63	1.40	1.75	2.22	1.40	1.96	2.78	1.19	1.89	2.95
Commercial floorspace (billion square feet)	81.1	84.0	84.1	84.3	92.7	93.9	95.2	100.5	103.0	105.5
Unit sales of light-duty vehicles (millions)	11.55	15.34	16.16	16.69	16.20	17.79	18.85	15.31	18.64	20.55

GDP = Gross domestic product.
Btu = British thermal unit.
Sources: 2010: IHS Global Insight, Global Insight Industry and Employment models, August 2011. **Projections:** U.S. Energy Information Administration, AEO2012 National Energy Modeling System runs LM2012.D022412A, REF2012.D020112C, and HM2012.D022412A.

This page intentionally left blank

Appendix C
Price case comparisons

Table C1. Total energy supply, disposition, and price summary
(quadrillion Btu per year, unless otherwise noted)

Supply, disposition, and prices	2010	Projections 2015			2025			2035		
		Low oil price	Reference	High oil price	Low oil price	Reference	High oil price	Low oil price	Reference	High oil price
Production										
Crude oil and lease condensate	11.59	12.66	13.23	13.79	11.57	13.77	15.60	10.29	12.89	14.37
Natural gas plant liquids	2.78	3.15	3.33	3.34	3.84	3.93	4.01	3.80	3.94	4.00
Dry natural gas	22.10	24.02	24.22	24.44	26.20	26.91	27.65	27.80	28.60	29.38
Coal[1]	22.06	20.76	20.24	19.80	22.39	22.25	23.45	23.59	24.14	27.73
Nuclear / uranium[2]	8.44	8.68	8.68	8.68	9.60	9.60	9.60	9.42	9.28	9.26
Hydropower	2.51	2.90	2.90	2.90	2.99	2.99	2.98	3.05	3.04	3.04
Biomass[3]	4.05	4.52	4.45	4.67	6.14	6.26	7.14	7.92	9.07	11.33
Other renewable energy[4]	1.34	1.94	1.99	2.02	2.18	2.22	2.19	2.87	2.81	2.66
Other[5]	0.64	0.54	0.60	0.82	0.55	0.69	0.77	0.68	0.91	0.90
Total	75.50	79.18	79.64	80.46	85.46	88.61	93.38	89.43	94.67	102.65
Imports										
Crude oil	20.14	21.26	18.87	17.01	21.30	16.23	12.08	23.88	16.90	11.22
Liquid fuels and other petroleum[6]	5.02	4.97	4.32	3.89	5.08	4.08	3.43	5.40	4.14	3.26
Natural gas[7]	3.81	3.87	3.73	3.69	3.16	2.75	2.55	3.28	2.84	2.57
Other imports[8]	0.52	0.47	0.44	0.40	0.83	1.07	0.81	0.87	0.81	0.76
Total	29.49	30.58	27.37	24.98	30.37	24.14	18.88	33.42	24.69	17.82
Exports										
Liquid fuels and other petroleum[9]	4.81	5.16	5.00	4.95	4.51	4.46	4.58	4.89	4.95	5.02
Natural gas[10]	1.15	1.93	1.93	1.93	3.51	3.51	3.52	4.17	4.17	4.18
Coal	2.10	2.73	2.73	2.73	2.82	2.82	2.67	3.22	3.13	3.13
Total	8.06	9.82	9.66	9.62	10.84	10.79	10.76	12.28	12.25	12.33
Discrepancy[11]	-1.23	0.04	-0.08	0.01	0.09	-0.03	-0.01	0.23	0.18	0.27
Consumption										
Liquid fuels and other petroleum[12]	37.25	38.73	36.72	35.31	39.70	36.58	35.03	41.86	37.70	35.86
Natural gas	24.71	25.93	26.00	26.18	25.80	26.14	26.57	26.86	27.26	27.67
Coal[13]	20.76	18.35	17.80	17.30	20.17	20.02	20.39	21.05	21.15	22.69
Nuclear / uranium[2]	8.44	8.68	8.68	8.68	9.60	9.60	9.60	9.42	9.28	9.26
Hydropower	2.51	2.90	2.90	2.90	2.99	2.99	2.98	3.05	3.04	3.04
Biomass[14]	2.88	3.06	3.04	3.13	4.19	4.17	4.48	4.98	5.44	6.45
Other renewable energy[4]	1.34	1.94	1.99	2.02	2.18	2.22	2.19	2.87	2.81	2.66
Other[15]	0.29	0.30	0.30	0.30	0.28	0.28	0.28	0.24	0.24	0.24
Total	98.16	99.89	97.43	95.82	104.90	101.99	101.52	110.34	106.93	107.87
Prices (2010 dollars per unit)										
Petroleum (dollars per barrel)										
Low sulfur light crude oil[16]	79.39	58.36	116.91	182.10	59.41	132.56	193.48	62.38	144.98	200.36
Imported crude oil[16]	75.87	55.41	113.97	179.16	48.84	121.21	180.29	53.10	132.95	187.04
Natural gas (dollars per million Btu)										
at Henry hub	4.39	4.21	4.29	4.26	5.61	5.63	5.60	7.36	7.37	7.17
at the wellhead[17]	4.06	3.78	3.84	3.81	4.98	5.00	4.97	6.47	6.48	6.31
Natural gas (dollars per thousand cubic feet)										
at the wellhead[17]	4.16	3.87	3.94	3.91	5.10	5.12	5.09	6.63	6.64	6.46
Coal (dollars per ton)										
at the minemouth[18]	35.61	39.93	42.08	44.26	41.50	44.05	45.62	47.24	50.52	51.12
Coal (dollars per million Btu)										
at the minemouth[18]	1.76	1.98	2.08	2.18	2.10	2.23	2.31	2.40	2.56	2.62
Average end-use[19]	2.38	2.42	2.56	2.68	2.51	2.70	2.81	2.73	2.94	3.07
Average electricity (cents per kilowatthour)	9.8	9.5	9.7	9.9	9.5	9.7	9.9	10.0	10.1	10.2

Table C1. Total energy supply, disposition, and price summary (continued)
(quadrillion Btu per year, unless otherwise noted)

Supply, disposition, and prices	2010	Projections 2015			Projections 2025			Projections 2035		
		Low oil price	Reference	High oil price	Low oil price	Reference	High oil price	Low oil price	Reference	High oil price
Prices (nominal dollars per unit)										
Petroleum (dollars per barrel)										
Low sulfur light crude oil[16 5]	79.39	62.81	125.97	195.67	77.32	170.09	245.37	98.91	229.55	314.93
Imported crude oil[16]	75.87	59.64	122.81	192.52	63.56	155.52	228.64	84.19	210.51	294.00
Natural gas (dollars per million Btu)										
at Henry hub	4.39	4.54	4.62	4.57	7.30	7.23	7.10	11.67	11.67	11.26
at the wellhead[17]	4.06	4.07	4.14	4.10	6.48	6.42	6.30	10.26	10.26	9.91
Natural gas (dollars per thousand cubic feet)										
at the wellhead[17]	4.16	4.16	4.24	4.20	6.64	6.57	6.46	10.51	10.51	10.15
Coal (dollars per ton)										
at the minemouth[18]	35.61	42.97	45.34	47.56	54.01	56.52	57.86	74.91	80.00	80.35
Coal (dollars per million Btu)										
at the minemouth[18]	1.76	2.13	2.24	2.34	2.74	2.86	2.93	3.81	4.05	4.12
Average end-use[19]	2.38	2.61	2.76	2.88	3.27	3.47	3.56	4.33	4.66	4.83
Average electricity (cents per kilowatthour)	9.8	10.2	10.4	10.6	12.4	12.5	12.6	15.9	16.0	16.0

[1] Includes waste coal.
[2] These values represent the energy obtained from uranium when it is used in light water reactors. The total energy content of uranium is much larger, but alternative processes are required to take advantage of it.
[3] Includes grid-connected electricity from wood and wood waste; biomass, such as corn, used for liquid fuels production; and non-electric energy demand from wood. Refer to Table A17 for details.
[4] Includes grid-connected electricity from landfill gas; biogenic municipal waste; wind; photovoltaic and solar thermal sources; and non-electric energy from renewable sources, such as active and passive solar systems. Excludes electricity imports using renewable sources and nonmarketed renewable energy. See Table A17 for selected nonmarketed residential and commercial renewable energy data.
[5] Includes non-biogenic municipal waste, liquid hydrogen, methanol, and some domestic inputs to refineries.
[6] Includes imports of finished petroleum products, unfinished oils, alcohols, ethers, blending components, and renewable fuels such as ethanol.
[7] Includes imports of liquefied natural gas that is later re-exported.
[8] Includes coal, coal coke (net), and electricity (net). Excludes imports of fuel used in nuclear power plants.
[9] Includes crude oil, petroleum products, ethanol, and biodiesel.
[10] Includes re-exported liquefied natural gas and natural gas used for liquefaction at export terminals.
[11] Balancing item. Includes unaccounted for supply, losses, gains, and net storage withdrawals.
[12] Includes petroleum-derived fuels and non-petroleum derived fuels, such as ethanol and biodiesel, and coal-based synthetic liquids. Petroleum coke, which is a solid, is included. Also included are natural gas plant liquids and crude oil consumed as a fuel. Refer to Table A17 for detailed renewable liquid fuels consumption.
[13] Excludes coal converted to coal-based synthetic liquids and natural gas.
[14] Includes grid-connected electricity from wood and wood waste, non-electric energy from wood, and biofuels heat and coproducts used in the production of liquid fuels, but excludes the energy content of the liquid fuels.
[15] Includes non-biogenic municipal waste, liquid hydrogen, and net electricity imports.
[16] Weighted average price delivered to U.S. refiners.
[17] Represents lower 48 onshore and offshore supplies.
[18] Includes reported prices for both open market and captive mines.
[19] Prices weighted by consumption; weighted average excludes residential and commercial prices, and export free-alongside-ship (f.a.s.) prices.
Btu = British thermal unit.
Note: Totals may not equal sum of components due to independent rounding. Data for 2010 are model results and may differ slightly from official EIA data reports.
Sources: 2010 natural gas supply values and natural gas wellhead price: U.S. Energy Information Administration (EIA), *Natural Gas Monthly*, DOE/EIA-0130(2011/07) (Washington, DC, July 2011). 2010 coal minemouth and delivered coal prices: EIA, *Annual Energy Review 2010*, DOE/EIA-0384(2010) (Washington, DC, October 2011). 2010 petroleum supply values: EIA, *Petroleum Supply Annual 2010*, DOE/EIA-0340(2010)/1 (Washington, DC, July 2011). 2010 low sulfur light crude oil price: EIA, Form EIA-856, "Monthly Foreign Crude oil Acquisition Report." Other 2010 coal values: *Quarterly Coal Report, October-December 2010*, DOE/EIA-0121(2010/4Q) (Washington, DC, May 2011). Other 2010 values: EIA, *Annual Energy Review 2010*, DOE/EIA-0384(2010) (Washington, DC, October 2011). **Projections:** EIA, AEO2012 National Energy Modeling System runs LP2012.D022112A, REF2012.D020112C, and HP2012.D022112A.

Table C2. Energy consumption by sector and source
(quadrillion Btu per year, unless otherwise noted)

Sector and source	2010	2015			2025			2035		
		Low oil price	Reference	High oil price	Low oil price	Reference	High oil price	Low oil price	Reference	High oil price
Energy consumption										
Residential										
Liquefied petroleum gases	0.56	0.54	0.51	0.49	0.55	0.50	0.48	0.55	0.51	0.48
Kerosene	0.03	0.03	0.02	0.02	0.03	0.02	0.02	0.02	0.02	0.02
Distillate fuel oil	0.63	0.61	0.55	0.51	0.49	0.43	0.40	0.41	0.35	0.33
Liquid fuels and other petroleum subtotal	1.22	1.17	1.08	1.02	1.07	0.95	0.90	0.99	0.87	0.82
Natural gas	5.06	4.98	4.97	4.98	4.88	4.88	4.90	4.74	4.76	4.78
Coal	0.01	0.01	0.01	0.01	0.01	0.01	0.01	0.01	0.01	0.01
Renewable energy[1]	0.42	0.37	0.43	0.48	0.36	0.43	0.48	0.35	0.43	0.47
Electricity	4.95	4.78	4.75	4.71	5.27	5.23	5.20	5.90	5.86	5.83
Delivered energy	**11.66**	**11.31**	**11.24**	**11.19**	**11.58**	**11.51**	**11.48**	**11.98**	**11.93**	**11.91**
Electricity related losses	10.39	9.68	9.58	9.47	10.66	10.52	10.34	11.58	11.35	11.02
Total	**22.05**	**20.99**	**20.81**	**20.66**	**22.24**	**22.02**	**21.82**	**23.56**	**23.28**	**22.93**
Commercial										
Liquefied petroleum gases	0.14	0.16	0.14	0.12	0.18	0.15	0.13	0.19	0.16	0.14
Motor gasoline[2]	0.05	0.06	0.05	0.04	0.06	0.05	0.05	0.07	0.06	0.06
Kerosene	0.00	0.01	0.00	0.00	0.01	0.00	0.00	0.01	0.01	0.01
Distillate fuel oil	0.43	0.41	0.35	0.32	0.41	0.33	0.30	0.41	0.32	0.30
Residual fuel oil	0.08	0.13	0.08	0.06	0.14	0.08	0.06	0.14	0.08	0.07
Liquid fuels and other petroleum subtotal	0.72	0.76	0.62	0.55	0.79	0.62	0.56	0.81	0.62	0.57
Natural gas	3.28	3.42	3.41	3.42	3.51	3.53	3.55	3.64	3.69	3.72
Coal	0.06	0.06	0.06	0.06	0.06	0.06	0.06	0.06	0.06	0.06
Renewable energy[3]	0.11	0.11	0.11	0.11	0.11	0.11	0.11	0.11	0.11	0.11
Electricity	4.54	4.61	4.59	4.57	5.19	5.16	5.14	5.81	5.80	5.77
Delivered energy	**8.70**	**8.96**	**8.80**	**8.70**	**9.66**	**9.48**	**9.41**	**10.43**	**10.28**	**10.23**
Electricity related losses	9.52	9.34	9.27	9.18	10.50	10.38	10.21	11.41	11.23	10.90
Total	**18.22**	**18.30**	**18.06**	**17.89**	**20.16**	**19.86**	**19.62**	**21.84**	**21.50**	**21.13**
Industrial[4]										
Liquefied petroleum gases	2.00	1.86	1.83	1.80	2.22	2.17	2.13	2.23	2.15	2.11
Motor gasoline[2]	0.25	0.28	0.28	0.28	0.31	0.30	0.30	0.32	0.30	0.29
Distillate fuel oil	1.16	1.28	1.25	1.24	1.25	1.19	1.17	1.29	1.18	1.16
Residual fuel oil	0.12	0.12	0.09	0.09	0.13	0.08	0.07	0.14	0.08	0.07
Petrochemical feedstocks	0.94	1.01	1.01	1.01	1.30	1.29	1.28	1.32	1.30	1.29
Other petroleum[5]	3.59	3.82	3.44	3.23	3.82	3.11	2.89	4.10	3.19	2.83
Liquid fuels and other petroleum subtotal	8.05	8.39	7.89	7.65	9.03	8.13	7.83	9.40	8.21	7.76
Natural gas	6.76	7.17	7.19	7.21	7.19	7.32	7.38	7.18	7.18	7.29
Natural-gas-to-liquids heat and power	0.00	0.00	0.00	0.00	0.00	0.00	0.07	0.00	0.00	0.07
Lease and plant fuel[6]	1.37	1.42	1.43	1.44	1.53	1.57	1.63	1.54	1.63	1.71
Natural gas subtotal	8.14	8.59	8.62	8.65	8.72	8.89	9.09	8.71	8.81	9.07
Metallurgical coal	0.55	0.58	0.57	0.56	0.48	0.49	0.49	0.44	0.43	0.43
Other industrial coal	1.01	1.03	1.03	1.02	1.04	1.08	1.08	1.05	1.08	1.09
Coal-to-liquids heat and power	0.00	0.00	0.00	0.00	0.10	0.36	1.12	0.10	0.60	2.74
Net coal coke imports	-0.01	-0.00	-0.01	-0.01	-0.03	-0.03	-0.03	-0.06	-0.06	-0.06
Coal subtotal	1.56	1.60	1.59	1.58	1.60	1.90	2.67	1.54	2.06	4.21
Biofuels heat and coproducts	0.84	0.85	0.81	0.86	1.19	1.27	1.73	1.99	2.57	3.63
Renewable energy[7]	1.50	1.63	1.61	1.63	1.90	1.82	1.75	2.10	1.95	1.87
Electricity	3.28	3.52	3.44	3.40	3.57	3.52	3.51	3.40	3.33	3.32
Delivered energy	**23.37**	**24.57**	**23.96**	**23.76**	**26.02**	**25.53**	**26.58**	**27.14**	**26.94**	**29.85**
Electricity related losses	6.89	7.11	6.94	6.84	7.21	7.09	6.98	6.68	6.46	6.27
Total	**30.26**	**31.69**	**30.90**	**30.60**	**33.24**	**32.61**	**33.56**	**33.82**	**33.39**	**36.12**

Table C2. Energy consumption by sector and source (continued)
(quadrillion Btu per year, unless otherwise noted)

Sector and source	2010	Projections								
		2015			2025			2035		
		Low oil price	Reference	High oil price	Low oil price	Reference	High oil price	Low oil price	Reference	High oil price
Transportation										
Liquefied petroleum gases	0.04	0.04	0.04	0.05	0.04	0.04	0.05	0.05	0.05	0.05
E85[8]	0.00	0.01	0.01	0.37	0.02	0.30	1.49	0.20	1.22	2.63
Motor gasoline[2]	16.91	17.23	16.13	14.85	17.02	14.90	12.48	17.96	14.53	11.70
Jet fuel[9]	3.07	3.04	3.03	3.01	3.20	3.19	3.18	3.34	3.33	3.32
Distillate fuel oil[10]	5.77	6.71	6.55	6.45	7.08	7.03	7.14	7.58	7.44	7.57
Residual fuel oil	0.90	0.91	0.91	0.91	0.92	0.93	0.93	0.94	0.94	0.94
Other petroleum[11]	0.17	0.17	0.17	0.17	0.17	0.17	0.17	0.17	0.17	0.17
Liquid fuels and other petroleum subtotal	26.88	28.11	26.83	25.81	28.45	26.57	25.44	30.24	27.67	26.40
Pipeline fuel natural gas	0.65	0.68	0.68	0.69	0.66	0.67	0.69	0.67	0.69	0.69
Compressed / liquefied natural gas	0.04	0.05	0.06	0.08	0.06	0.11	0.21	0.07	0.16	0.30
Liquid hydrogen	0.00	0.00	0.00	0.00	0.00	0.00	0.00	0.00	0.00	0.00
Electricity	0.02	0.03	0.03	0.03	0.03	0.04	0.06	0.05	0.07	0.11
Delivered energy	**27.59**	**28.86**	**27.60**	**26.61**	**29.20**	**27.40**	**26.40**	**31.03**	**28.60**	**27.49**
Electricity related losses	0.05	0.05	0.05	0.06	0.07	0.08	0.12	0.10	0.14	0.20
Total	**27.63**	**28.92**	**27.65**	**26.67**	**29.27**	**27.49**	**26.52**	**31.12**	**28.75**	**27.69**
Delivered energy consumption for all sectors										
Liquefied petroleum gases	2.75	2.60	2.51	2.46	2.98	2.86	2.79	3.02	2.86	2.79
E85[8]	0.00	0.01	0.01	0.37	0.02	0.30	1.49	0.20	1.22	2.63
Motor gasoline[2]	17.21	17.57	16.46	15.17	17.39	15.25	12.82	18.35	14.88	12.05
Jet fuel[9]	3.07	3.04	3.03	3.01	3.20	3.19	3.18	3.34	3.33	3.32
Kerosene	0.04	0.04	0.03	0.03	0.04	0.03	0.03	0.04	0.03	0.03
Distillate fuel oil	7.99	9.01	8.69	8.52	9.24	8.99	9.02	9.69	9.29	9.36
Residual fuel oil	1.11	1.16	1.08	1.06	1.19	1.09	1.06	1.21	1.11	1.08
Petrochemical feedstocks	0.94	1.01	1.01	1.01	1.30	1.29	1.28	1.32	1.30	1.29
Other petroleum[12]	3.76	3.98	3.61	3.39	3.98	3.27	3.05	4.27	3.36	3.00
Liquid fuels and other petroleum subtotal	36.87	38.42	36.43	35.02	39.35	36.28	34.73	41.44	37.38	35.55
Natural gas	15.15	15.62	15.64	15.68	15.63	15.85	16.04	15.62	15.79	16.08
Natural-gas-to-liquids heat and power	0.00	0.00	0.00	0.00	0.00	0.00	0.07	0.00	0.00	0.07
Lease and plant fuel[6]	1.37	1.42	1.43	1.44	1.53	1.57	1.63	1.54	1.63	1.71
Pipeline natural gas	0.65	0.68	0.68	0.69	0.66	0.67	0.69	0.67	0.69	0.69
Natural gas subtotal	17.17	17.72	17.75	17.81	17.82	18.09	18.43	17.83	18.11	18.55
Metallurgical coal	0.55	0.58	0.57	0.56	0.48	0.49	0.49	0.44	0.43	0.43
Other coal	1.08	1.09	1.09	1.08	1.11	1.14	1.15	1.11	1.15	1.16
Coal-to-liquids heat and power	0.00	0.00	0.00	0.00	0.10	0.36	1.12	0.10	0.60	2.74
Net coal coke imports	-0.01	-0.00	-0.01	-0.01	-0.03	-0.03	-0.03	-0.06	-0.06	-0.06
Coal subtotal	1.62	1.67	1.65	1.64	1.67	1.96	2.74	1.60	2.12	4.28
Biofuels heat and coproducts	0.84	0.85	0.81	0.86	1.19	1.27	1.73	1.99	2.57	3.63
Renewable energy[13]	2.03	2.10	2.15	2.22	2.37	2.36	2.34	2.56	2.50	2.45
Liquid hydrogen	0.00	0.00	0.00	0.00	0.00	0.00	0.00	0.00	0.00	0.00
Electricity	12.79	12.94	12.81	12.71	14.07	13.96	13.91	15.16	15.06	15.02
Delivered energy	**71.32**	**73.71**	**71.59**	**70.26**	**76.47**	**73.92**	**73.87**	**80.58**	**77.75**	**79.48**
Electricity related losses	26.84	26.19	25.84	25.55	28.44	28.07	27.65	29.76	29.18	28.39
Total	**98.16**	**99.89**	**97.43**	**95.82**	**104.90**	**101.99**	**101.52**	**110.34**	**106.93**	**107.87**
Electric power[14]										
Distillate fuel oil	0.08	0.09	0.08	0.08	0.09	0.09	0.09	0.09	0.09	0.09
Residual fuel oil	0.30	0.22	0.21	0.21	0.27	0.22	0.22	0.33	0.23	0.23
Liquid fuels and other petroleum subtotal	0.38	0.30	0.29	0.29	0.36	0.31	0.31	0.42	0.32	0.32
Natural gas	7.54	8.22	8.25	8.37	7.97	8.04	8.14	9.03	9.16	9.12
Steam coal	19.13	16.68	16.15	15.66	18.50	18.06	17.65	19.45	19.03	18.41
Nuclear / uranium[15]	8.44	8.68	8.68	8.68	9.60	9.60	9.60	9.42	9.28	9.26
Renewable energy[16]	3.85	4.94	4.96	4.96	5.80	5.75	5.59	6.34	6.22	6.07
Electricity imports	0.09	0.10	0.10	0.10	0.08	0.08	0.08	0.04	0.04	0.04
Total[17]	**39.63**	**39.13**	**38.64**	**38.26**	**42.50**	**42.03**	**41.56**	**44.91**	**44.24**	**43.41**

Table C2. Energy consumption by sector and source (continued)
(quadrillion Btu per year, unless otherwise noted)

Sector and source	2010	Projections								
		2015			2025			2035		
		Low oil price	Reference	High oil price	Low oil price	Reference	High oil price	Low oil price	Reference	High oil price
Total energy consumption										
Liquefied petroleum gases	2.75	2.60	2.51	2.46	2.98	2.86	2.79	3.02	2.86	2.79
E85[8]	0.00	0.01	0.01	0.37	0.02	0.30	1.49	0.20	1.22	2.63
Motor gasoline[2]	17.21	17.57	16.46	15.17	17.39	15.25	12.82	18.35	14.88	12.05
Jet fuel[9]	3.07	3.04	3.03	3.01	3.20	3.19	3.18	3.34	3.33	3.32
Kerosene	0.04	0.04	0.03	0.03	0.04	0.03	0.03	0.04	0.03	0.03
Distillate fuel oil	8.07	9.10	8.78	8.60	9.33	9.07	9.10	9.78	9.38	9.45
Residual fuel oil	1.41	1.38	1.29	1.27	1.46	1.31	1.28	1.55	1.34	1.31
Petrochemical feedstocks	0.94	1.01	1.01	1.01	1.30	1.29	1.28	1.32	1.30	1.29
Other petroleum[12]	3.76	3.98	3.61	3.39	3.98	3.27	3.05	4.27	3.36	3.00
Liquid fuels and other petroleum subtotal	37.25	38.73	36.72	35.31	39.70	36.58	35.03	41.86	37.70	35.86
Natural gas	22.69	23.84	23.89	24.05	23.60	23.89	24.17	24.65	24.94	25.20
Natural-gas-to-liquids heat and power	0.00	0.00	0.00	0.00	0.00	0.00	0.07	0.00	0.00	0.07
Lease and plant fuel[6]	1.37	1.42	1.43	1.44	1.53	1.57	1.63	1.54	1.63	1.71
Pipeline natural gas	0.65	0.68	0.68	0.69	0.66	0.67	0.69	0.67	0.69	0.69
Natural gas subtotal	24.71	25.93	26.00	26.18	25.80	26.14	26.57	26.86	27.26	27.67
Metallurgical coal	0.55	0.58	0.57	0.56	0.48	0.49	0.49	0.44	0.43	0.43
Other coal	20.21	17.77	17.24	16.74	19.61	19.20	18.80	20.56	20.18	19.57
Coal-to-liquids heat and power	0.00	0.00	0.00	0.00	0.10	0.36	1.12	0.10	0.60	2.74
Net coal coke imports	-0.01	-0.00	-0.01	-0.01	-0.03	-0.03	-0.03	-0.06	-0.06	-0.06
Coal subtotal	20.76	18.35	17.80	17.30	20.17	20.02	20.39	21.05	21.15	22.69
Nuclear / uranium[15]	8.44	8.68	8.68	8.68	9.60	9.60	9.60	9.42	9.28	9.26
Biofuels heat and coproducts	0.84	0.85	0.81	0.86	1.19	1.27	1.73	1.99	2.57	3.63
Renewable energy[18]	5.88	7.05	7.11	7.18	8.16	8.11	7.93	8.91	8.71	8.52
Liquid hydrogen	0.00	0.00	0.00	0.00	0.00	0.00	0.00	0.00	0.00	0.00
Electricity imports	0.09	0.10	0.10	0.10	0.08	0.08	0.08	0.04	0.04	0.04
Total	**98.16**	**99.89**	**97.43**	**95.82**	**104.90**	**101.99**	**101.52**	**110.34**	**106.93**	**107.87**
Energy use and related statistics										
Delivered energy use	71.32	73.71	71.59	70.26	76.47	73.92	73.87	80.58	77.75	79.48
Total energy use	98.16	99.89	97.43	95.82	104.90	101.99	101.52	110.34	106.93	107.87
Ethanol consumed in motor gasoline and E85	1.11	1.30	1.22	1.36	1.56	1.55	2.14	1.77	2.15	2.80
Population (millions)	310.83	326.16	326.16	326.16	358.06	358.06	358.06	390.09	390.09	390.09
Gross domestic product (billion 2005 dollars)	13088	14990	14803	14666	19146	19185	19380	24596	24539	24703
Carbon dioxide emissions (million metric tons)	5633.6	5592.8	5407.2	5251.2	5770.9	5552.5	5450.8	6049.1	5757.9	5737.1

[1] Includes wood used for residential heating. See Table A4 and/or Table A17 for estimates of nonmarketed renewable energy consumption for geothermal heat pumps, solar thermal water heating, and electricity generation from wind and solar photovoltaic sources.
[2] Includes ethanol (blends of 15 percent or less) and ethers blended into gasoline.
[3] Excludes ethanol. Includes commercial sector consumption of wood and wood waste, landfill gas, municipal waste, and other biomass for combined heat and power. See Table A5 and/or Table A17 for estimates of nonmarketed renewable energy consumption for solar thermal water heating and electricity generation from wind and solar photovoltaic sources.
[4] Includes energy for combined heat and power plants, except those whose primary business is to sell electricity, or electricity and heat, to the public.
[5] Includes petroleum coke, asphalt, road oil, lubricants, still gas, and miscellaneous petroleum products.
[6] Represents natural gas used in well, field, and lease operations, and in natural gas processing plant machinery.
[7] Includes consumption of energy produced from hydroelectric, wood and wood waste, municipal waste, and other biomass sources. Excludes ethanol blends (15 percent or less) in motor gasoline.
[8] E85 refers to a blend of 85 percent ethanol (renewable) and 15 percent motor gasoline (nonrenewable). To address cold starting issues, the percentage of ethanol varies seasonally. The annual average ethanol content of 74 percent is used for this forecast.
[9] Includes only kerosene type.
[10] Diesel fuel for on- and off- road use.
[11] Includes aviation gasoline and lubricants.
[12] Includes unfinished oils, natural gasoline, motor gasoline blending components, aviation gasoline, lubricants, still gas, asphalt, road oil, petroleum coke, and miscellaneous petroleum products.
[13] Includes electricity generated for sale to the grid and for own use from renewable sources, and non-electric energy from renewable sources. Excludes ethanol and nonmarketed renewable energy consumption for geothermal heat pumps, buildings photovoltaic systems, and solar thermal water heaters.
[14] Includes consumption of energy by electricity-only and combined heat and power plants whose primary business is to sell electricity, or electricity and heat, to the public. Includes small power producers and exempt wholesale generators.
[15] These values represent the energy obtained from uranium when it is used in light water reactors. The total energy content of uranium is much larger, but alternative processes are required to take advantage of it.
[16] Includes conventional hydroelectric, geothermal, wood and wood waste, biogenic municipal waste, other biomass, wind, photovoltaic, and solar thermal sources. Excludes net electricity imports.
[17] Includes non-biogenic municipal waste not included above.
[18] Includes conventional hydroelectric, geothermal, wood and wood waste, biogenic municipal waste, other biomass, wind, photovoltaic, and solar thermal sources. Excludes ethanol, net electricity imports, and nonmarketed renewable energy consumption for geothermal heat pumps, buildings photovoltaic systems, and solar thermal water heaters.
Btu = British thermal unit.
Note: Totals may not equal sum of components due to independent rounding. Data for 2010 are model results and may differ slightly from official EIA data reports.
Sources: 2010 consumption based on: U.S. Energy Information Administration (EIA), *Annual Energy Review 2010*, DOE/EIA-0384(2010) (Washington, DC, October 2011). 2010 population and gross domestic product: IHS Global Insight Industry and Employment models, August 2011. 2010 carbon dioxide emissions: EIA, *Monthly Energy Review, October 2011* DOE/EIA-0035(2011/10) (Washington, DC, October 2011). Projections: EIA, AEO2012 National Energy Modeling System runs LP2012.D021112A, REF2012.D020112C, and HP2012.D022112A.

Table C3. Energy prices by sector and source
(2010 dollars per million Btu, unless otherwise noted)

Sector and source	2010	Projections 2015			Projections 2025			Projections 2035		
		Low oil price	Reference	High oil price	Low oil price	Reference	High oil price	Low oil price	Reference	High oil price
Residential										
Liquefied petroleum gases	27.02	22.54	30.70	39.69	22.18	32.27	40.42	23.49	34.64	42.03
Distillate fuel oil	21.21	16.55	27.26	38.29	17.27	30.15	39.23	18.46	32.73	40.00
Natural gas	11.08	10.22	10.31	10.30	11.96	12.03	12.02	13.97	13.98	13.86
Electricity	33.69	34.06	34.59	35.24	33.37	34.08	34.73	34.31	34.58	35.00
Commercial										
Liquefied petroleum gases	23.52	19.33	27.42	36.38	19.00	28.97	37.09	20.30	31.30	38.66
Distillate fuel oil	20.77	13.91	23.98	34.68	14.39	26.86	35.89	15.51	29.18	36.36
Residual fuel oil	11.07	5.99	16.18	27.80	6.25	18.24	28.32	6.90	18.90	28.11
Natural gas	9.10	8.52	8.60	8.59	9.98	10.02	10.01	11.66	11.64	11.49
Electricity	29.73	28.52	29.03	29.65	28.32	29.00	29.71	29.30	29.48	29.84
Industrial[1]										
Liquefied petroleum gases	21.80	16.98	27.43	38.87	16.33	29.24	39.62	17.95	32.18	41.60
Distillate fuel oil	21.32	14.50	24.20	34.82	14.95	27.22	36.32	16.19	29.53	36.60
Residual fuel oil	10.92	9.51	19.21	30.20	9.60	21.23	30.43	9.97	21.65	30.61
Natural gas[2]	5.51	4.78	4.88	4.88	5.99	6.04	6.01	7.52	7.54	7.38
Metallurgical coal	5.84	7.04	7.22	7.35	7.86	8.11	8.24	8.85	9.11	9.23
Other industrial coal	2.71	3.11	3.27	3.38	3.18	3.38	3.52	3.38	3.64	3.86
Coal to liquids	--	1.17	1.26	1.32	2.02	2.08	2.26	2.26	2.38	2.64
Electricity	19.63	18.58	18.91	19.26	19.11	19.60	19.96	20.61	20.78	20.97
Transportation										
Liquefied petroleum gases[3]	26.88	23.86	31.93	40.71	23.47	33.38	41.43	24.77	35.74	43.04
E85[4]	25.21	18.16	29.03	38.11	17.18	28.81	41.93	16.59	31.96	39.01
Motor gasoline[6]	22.70	18.53	29.26	41.14	18.20	32.10	43.26	18.49	33.61	42.09
Jet fuel[6]	16.22	12.62	23.74	35.26	12.80	26.45	35.89	13.96	29.13	36.89
Diesel fuel (distillate fuel oil)[7]	21.87	17.99	27.56	38.22	18.14	30.42	39.66	19.15	32.40	39.63
Residual fuel oil	10.42	8.64	18.32	29.02	8.67	20.62	29.37	8.76	20.95	29.86
Natural gas[8]	13.20	12.28	12.40	12.45	13.05	13.29	13.41	14.26	14.51	14.47
Electricity	32.99	30.37	30.50	30.24	30.91	31.53	33.04	33.26	33.82	34.36
Electric power[9]										
Distillate fuel oil	18.73	12.06	22.77	33.56	12.54	25.35	34.16	13.56	27.80	35.05
Residual fuel oil	11.89	13.08	23.00	33.74	12.12	25.40	34.30	11.20	25.72	34.59
Natural gas	5.14	4.46	4.55	4.54	5.58	5.60	5.59	7.18	7.21	7.04
Steam coal	2.26	2.22	2.35	2.47	2.34	2.54	2.68	2.56	2.80	3.00
Average price to all users[10]										
Liquefied petroleum gases	17.28	14.64	22.99	32.23	13.90	24.19	32.57	15.28	26.63	34.20
E85[4]	25.21	18.16	29.03	38.11	17.18	28.81	41.93	16.59	31.96	39.01
Motor gasoline[6]	22.59	18.53	29.26	41.14	18.19	32.10	43.26	18.49	33.61	42.09
Jet fuel	16.22	12.62	23.74	35.26	12.80	26.45	35.89	13.96	29.13	36.89
Distillate fuel oil	21.65	17.16	26.87	37.56	17.45	29.81	39.04	18.54	31.91	39.12
Residual fuel oil	10.82	9.17	19.01	29.82	9.16	21.31	30.21	9.22	21.68	30.63
Natural gas	7.16	6.36	6.45	6.43	7.70	7.74	7.74	9.26	9.30	9.18
Metallurgical coal	5.84	7.04	7.22	7.35	7.86	8.11	8.24	8.85	9.11	9.23
Other coal	2.29	2.28	2.41	2.53	2.39	2.59	2.73	2.61	2.85	3.06
Coal to liquids	--	1.17	1.26	1.32	2.02	2.08	2.26	2.26	2.38	2.64
Electricity	28.68	27.87	28.38	28.94	27.88	28.54	29.14	29.31	29.56	29.92
Non-renewable energy expenditures by sector (billion 2010 dollars)										
Residential	251.69	236.40	246.72	256.77	255.31	266.75	275.38	289.49	298.72	304.24
Commercial	179.08	171.63	177.92	184.03	193.67	201.89	208.38	225.40	231.98	235.90
Industrial	198.98	175.07	223.88	279.09	194.55	261.92	313.03	212.90	282.31	323.54
Transportation	573.78	489.96	746.84	998.67	491.22	803.52	976.23	537.61	856.65	958.30
Total non-renewable expenditures	1203.52	1073.06	1395.36	1718.56	1134.76	1534.08	1773.02	1265.39	1669.66	1821.97
Transportation renewable expenditures	0.08	0.18	0.25	14.01	0.39	8.74	62.29	3.32	38.86	102.69
Total expenditures	**1203.62**	**1073.25**	**1395.61**	**1732.58**	**1135.15**	**1542.81**	**1835.31**	**1268.71**	**1708.52**	**1924.66**

Table C3. Energy prices by sector and source (continued)
(nominal dollars per million Btu, unless otherwise noted)

Sector and source	2010	Projections								
		2015			2025			2035		
		Low oil price	Reference	High oil price	Low oil price	Reference	High oil price	Low oil price	Reference	High oil price
Residential										
Liquefied petroleum gases	27.02	24.26	33.08	42.65	28.87	41.41	51.27	37.25	54.86	66.07
Distillate fuel oil	21.21	17.81	29.38	41.14	22.48	38.68	49.75	29.27	51.82	62.87
Natural gas	11.08	11.00	11.11	11.06	15.57	15.43	15.25	22.15	22.14	21.78
Electricity	33.69	36.66	37.27	37.86	43.43	43.72	44.05	54.40	54.76	55.02
Commercial										
Liquefied petroleum gases	23.52	20.80	29.54	39.09	24.73	37.17	47.04	32.18	49.56	60.77
Distillate fuel oil	20.77	14.97	25.83	37.27	18.73	34.47	45.51	24.59	46.20	57.15
Residual fuel oil	11.07	6.44	17.43	29.87	8.13	23.41	35.92	10.94	29.93	44.18
Natural gas	9.10	9.17	9.27	9.23	12.99	12.86	12.69	18.48	18.43	18.06
Electricity	29.73	30.70	31.28	31.86	36.86	37.21	37.68	46.46	46.67	46.91
Industrial[1]										
Liquefied petroleum gases	21.80	18.28	29.56	41.77	21.25	37.51	50.25	28.46	50.95	65.39
Distillate fuel oil	21.32	15.61	26.08	37.41	19.46	34.93	46.06	25.67	46.76	57.53
Residual fuel oil	10.92	10.23	20.70	32.45	12.49	27.24	38.59	15.80	34.28	48.11
Natural gas[2]	5.51	5.14	5.26	5.24	7.80	7.75	7.63	11.92	11.93	11.60
Metallurgical coal	5.84	7.57	7.78	7.90	10.23	10.40	10.45	14.04	14.42	14.51
Other industrial coal	2.71	3.35	3.52	3.63	4.13	4.34	4.46	5.36	5.77	6.06
Coal to liquids	--	1.26	1.36	1.42	2.63	2.67	2.86	3.58	3.78	4.14
Electricity	19.63	19.99	20.38	20.69	24.87	25.15	25.31	32.68	32.90	32.96
Transportation										
Liquefied petroleum gases[3]	26.88	25.68	34.41	43.74	30.54	42.83	52.54	39.27	56.59	67.66
E85[4]	25.21	19.55	31.28	40.95	22.36	36.97	53.17	26.31	50.61	61.31
Motor gasoline[5]	22.70	19.94	31.53	44.21	23.68	41.19	54.86	29.32	53.22	66.16
Jet fuel[6]	16.22	13.59	25.58	37.89	16.66	33.94	45.51	22.13	46.12	57.99
Diesel fuel (distillate fuel oil)[7]	21.87	19.36	29.69	41.07	23.61	39.03	50.30	30.37	51.29	62.29
Residual fuel oil	10.42	9.30	19.74	31.18	11.28	26.45	37.25	13.89	33.18	46.93
Natural gas[8]	13.20	13.22	13.36	13.38	16.98	17.05	17.00	22.61	22.97	22.75
Electricity	32.99	32.69	32.86	32.50	40.22	40.46	41.90	52.74	53.55	54.01
Electric power[9]										
Distillate fuel oil	18.73	12.98	24.53	36.06	16.32	32.52	43.32	21.50	44.02	55.10
Residual fuel oil	11.89	14.07	24.78	36.26	15.77	32.59	43.50	17.77	40.73	54.38
Natural gas	5.14	4.80	4.90	4.88	7.27	7.19	7.09	11.38	11.42	11.06
Steam coal	2.26	2.39	2.53	2.65	3.04	3.25	3.40	4.06	4.43	4.72

Table C3. Energy prices by sector and source (continued)
(nominal dollars per million Btu, unless otherwise noted)

Sector and source	2010	Projections 2015			2025			2035		
		Low oil price	Reference	High oil price	Low oil price	Reference	High oil price	Low oil price	Reference	High oil price
Average price to all users[10]										
Liquefied petroleum gases	17.28	15.75	24.78	34.64	18.08	31.04	41.30	24.23	42.17	53.76
E85[4]	25.21	19.55	31.28	40.95	22.36	36.97	53.17	26.31	50.61	61.31
Motor gasoline[5]	22.59	19.94	31.53	44.21	23.68	41.19	54.86	29.31	53.22	66.16
Jet fuel	16.22	13.59	25.58	37.89	16.66	33.94	45.51	22.13	46.12	57.99
Distillate fuel oil	21.65	18.47	28.96	40.36	22.71	38.24	49.51	29.39	50.52	61.50
Residual fuel oil	10.82	9.87	20.48	32.04	11.92	27.34	38.32	14.63	34.33	48.14
Natural gas	7.16	6.84	6.95	6.91	10.02	9.93	9.82	14.69	14.73	14.42
Metallurgical coal	5.84	7.57	7.78	7.90	10.23	10.40	10.45	14.04	14.42	14.51
Other coal	2.29	2.45	2.60	2.72	3.11	3.32	3.47	4.14	4.51	4.81
Coal to liquids	- -	1.26	1.36	1.42	2.63	2.67	2.86	3.58	3.78	4.14
Electricity	28.68	30.00	30.58	31.10	36.28	36.62	36.96	46.48	46.80	47.03
Non-renewable energy expenditures by sector (billion nominal dollars)										
Residential	251.69	254.44	265.85	275.92	332.26	342.26	349.24	459.02	472.99	478.21
Commercial	179.08	184.73	191.71	197.75	252.04	259.04	264.27	357.40	367.31	370.80
Industrial	198.98	188.43	241.24	299.90	253.19	336.06	396.99	337.58	447.01	508.54
Transportation	573.78	527.35	804.75	1073.14	639.27	1030.98	1238.06	852.44	1356.41	1506.27
Total non-renewable expenditures	1203.54	1154.96	1503.55	1846.71	1476.75	1968.35	2248.56	2006.43	2643.72	2863.82
Transportation renewable expenditures	0.08	0.20	0.27	15.06	0.51	11.21	78.99	5.26	61.53	161.41
Total expenditures	**1203.62**	**1155.16**	**1503.82**	**1861.77**	**1477.26**	**1979.56**	**2327.55**	**2011.69**	**2705.26**	**3025.22**

[1]Includes energy for combined heat and power plants, except those whose primary business is to sell electricity, or electricity and heat, to the public.
[2]Excludes use for lease and plant fuel.
[3]Includes Federal and State taxes while excluding county and local taxes.
[4]E85 refers to a blend of 85 percent ethanol (renewable) and 15 percent motor gasoline (nonrenewable). To address cold starting issues, the percentage of ethanol varies seasonally. The annual average ethanol content of 74 percent is used for this forecast.
[5]Sales weighted-average price for all grades. Includes Federal, State and local taxes.
[6]Kerosene-type jet fuel. Includes Federal and State taxes while excluding county and local taxes.
[7]Diesel fuel for on-road use. Includes Federal and State taxes while excluding county and local taxes.
[8]Natural gas used as a vehicle fuel. Includes estimated motor vehicle fuel taxes and estimated dispensing costs or charges.
[9]Includes electricity-only and combined heat and power plants whose primary business is to sell electricity, or electricity and heat, to the public.
[10]Weighted averages of end-use fuel prices are derived from the prices shown in each sector and the corresponding sectoral consumption.
Btu = British thermal unit.
- - = Not applicable.
Note: Data for 2010 are model results and may differ slightly from official EIA data reports.
Sources: 2010 prices for motor gasoline, distillate fuel oil, and jet fuel are based on prices in the U.S. Energy Information Administration (EIA), *Petroleum Marketing Annual 2009*, DOE/EIA-0487(2009) (Washington, DC, August 2010). 2010 residential and commercial natural gas delivered prices: EIA, *Natural Gas Monthly*, DOE/EIA-0130(2011/07) (Washington, DC, July 2011). 2010 industrial natural gas delivered prices are estimated based on: EIA, *Manufacturing Energy Consumption Survey* and industrial and wellhead prices from the *Natural Gas Annual 2009*, DOE/EIA-0131(2009) (Washington, DC, December 2010) and the *Natural Gas Monthly*, DOE/EIA-0130(2011/07) (Washington, DC, July 2011). 2010 transportation sector natural gas delivered prices are model results. 2010 electric power sector distillate and residual fuel oil prices: EIA, *Monthly Energy Review*, DOE/EIA-0035(2011/09) (Washington, DC, September 2010). 2010 electric power sector natural gas prices: EIA, *Electric Power Monthly*, DOE/EIA-0226, April 2010 and April 2011, Table 4.2, and EIA, *State Energy Data Report 2009*, DOE/EIA-0214(2009) (Washington, DC, June 2011). 2010 coal prices based on: EIA, *Quarterly Coal Report, October-December 2010*, DOE/EIA-0121(2010/4Q) (Washington, DC, May 2011) and EIA, AEO2012 National Energy Modeling System run REF2012.D020112C. 2010 electricity prices: EIA, *Annual Energy Review 2010*, DOE/EIA-0384(2010) (Washington, DC, October 2011). 2010 E85 prices derived from monthly prices in the Clean Cities Alternative Fuel Price Report. **Projections:** EIA, AEO2012 National Energy Modeling System runs LP2012.D022112A, REF2012.D020112C, and HP2012.D022112A.

Table C4. Liquid fuels supply and disposition
(million barrels per day, unless otherwise noted)

Supply and disposition	2010	Projections							
		2015			2025			2035		
		Low oil price	Reference	High oil price	Low oil price	Reference	High oil price	Low oil price	Reference	High oil price
Crude oil										
Domestic crude production[1]	5.47	5.88	6.15	6.41	5 38	6.40	7 25	4.79	5.99	6.68
Alaska	0 60	0.46	0.46	0.46	0 34	0.40	0.68	0.00	0.27	0.36
Lower 48 states	4 87	5.42	5 69	5.95	5 04	6.00	6 57	4.79	5.72	6.32
Net imports	9.17	9.63	8 52	7.64	9 58	7.24	5.32	10.74	7.52	4.91
Gross imports	9 21	9.66	8 56	7.67	9 61	7.27	5.36	10.77	7.55	4.95
Exports	0 04	0.03	0 03	0.03	0 03	0.03	0.04	0.03	0.03	0.04
Other crude supply[2]	0 08	0.00	0 00	0.00	0 00	0.00	0 00	0.00	0.00	0.00
Total crude supply	**14.72**	**15.52**	**14.67**	**14.05**	**14.96**	**13.64**	**12.56**	**15.53**	**13.51**	**11.59**
Other petroleum supply	3.50	3.33	3.25	2.98	4.21	3.80	3.29	4.13	3.52	2.81
Natural gas plant liquids	2 07	2.40	2 56	2.56	2 94	3.01	3 07	2.91	3.01	3.06
Net product imports	0 39	-0.01	-0 25	-0.50	0 33	-0.12	-0 62	0.31	-0.34	-0.94
Gross refined product imports[3]	1 23	0.97	0 78	0.61	1 06	0.79	0 51	1.14	0.82	0.55
Unfinished oil imports	0 61	0.74	0 64	0.56	0 67	0.51	0 38	0.74	0.50	0.26
Blending component imports	0.74	0.69	0 66	0.63	0.71	0.65	0 61	0.73	0.66	0.61
Exports	2.19	2.41	2 32	2.30	2.12	2.07	2.13	2.31	2.31	2.36
Refinery processing gain[4]	1 07	0.94	0 95	0.92	0 95	0.91	0 84	0.91	0.85	0.69
Product stock withdrawal	-0 03	0.00	0 00	0.00	0 00	0.00	0.00	0.00	0.00	0.00
Other non-petroleum supply	1.00	1.24	1.22	1.46	1.61	1.86	2.84	2.18	2.96	4.87
Supply from renewable sources	0 87	1.11	1 05	1.20	1.42	1.48	2 01	1.92	2.37	3.24
Ethanol	0 85	1.00	0 94	1.05	1 20	1.19	1.64	1.36	1.65	2.15
Domestic production	0 88	0.99	0 94	0.99	1.18	1.17	1.47	1.35	1.59	1.96
Net imports	-0 02	0.01	0 00	0.06	0 02	0.02	0.17	0.01	0.06	0.19
Biodiesel	0 01	0.08	0 09	0.12	0.12	0.12	0.13	0.13	0.13	0.14
Domestic production	0 02	0.08	0 09	0.11	0.12	0.12	0.13	0.13	0.13	0.14
Net imports	-0 01	0.00	0 00	0.00	0 00	0.00	0.00	0.00	-0.00	-0.00
Other biomass-derived liquids[5]	0 00	0.02	0 03	0.03	0.10	0.16	0 24	0.44	0.59	0.95
Liquids from gas	0 00	0.00	0 00	0.00	0 00	0.00	0 06	0.00	0.00	0.06
Liquids from coal	0 00	0.00	0 00	0.00	0 05	0.17	0 52	0.05	0.28	1.27
Other[6]	0.13	0.14	0.17	0.26	0.15	0.21	0.24	0.20	0.31	0.30
Total primary supply[7]	**19.22**	**20.09**	**19.14**	**18.49**	**20.79**	**19.29**	**18.69**	**21.84**	**19.99**	**19.27**
Liquid fuels consumption										
by fuel										
Liquefied petroleum gases	2 27	2.00	1 94	1.90	2 30	2.21	2.15	2.32	2.21	2.15
E85[8]	0 00	0.01	0 01	0.25	0 02	0.21	1.02	0.14	0.83	1.80
Motor gasoline[9]	8 99	9.48	8 88	8.19	9.45	8.29	6 97	9.97	8.09	6.55
Jet fuel[10]	1.43	1.47	1.46	1.45	1 55	1.54	1.54	1.61	1.61	1.60
Distillate fuel oil[11]	3 80	4.34	4.19	4.10	4.45	4.33	4 34	4.67	4.48	4.51
Diesel	3 32	3.82	3.71	3.66	3 99	3.92	3.96	4.24	4.11	4.16
Residual fuel oil	0 54	0.60	0 56	0.55	0 63	0.57	0 56	0.67	0.58	0.57
Other[12]	2.14	2.23	2 06	1.97	2 38	2.06	1.95	2.51	2.10	1.94
by sector										
Residential and commercial	1.12	1.12	1 00	0.92	1 09	0.94	0 87	1.07	0.91	0.84
Industrial[13]	4 31	4.41	4.17	4.05	4 83	4.41	4.26	5.00	4.44	4.22
Transportation	13 82	14.47	13 80	13.31	14 69	13.71	13 26	15.64	14.41	13.90
Electric power[14]	0.17	0.14	0.13	0.13	0.16	0.14	0.14	0.19	0.14	0.14
Total	**19.17**	**20.14**	**19.10**	**18.41**	**20.77**	**19.20**	**18.53**	**21.90**	**19.90**	**19.12**
Discrepancy[15]	0.05	-0.05	0.05	0.08	0.01	0.10	0.16	-0.06	0.09	0.15

Table C4. Liquid fuels supply and disposition (continued)
(million barrels per day, unless otherwise noted)

Supply and disposition	2010	Projections								
		2015			2025			2035		
		Low oil price	Reference	High oil price	Low oil price	Reference	High oil price	Low oil price	Reference	High oil price
Domestic refinery distillation capacity[16]	17.6	17.6	17.5	17.1	16.8	15.5	14.6	17.1	15.2	13.8
Capacity utilization rate (percent)[17]	86.0	90.3	85.9	84.0	91.0	90.1	88.0	93.0	90.8	85.7
Net import share of product supplied (percent)	49.6	47.9	43.2	38.9	47.8	37.0	26.0	50.7	36.2	21.6
Net expenditures for imported crude oil and petroleum products (billion 2010 dollars)	243.07	207.99	373.00	523.15	189.41	344.58	384.81	226.36	389.97	363.97

[1]Includes lease condensate.
[2]Strategic petroleum reserve stock additions plus unaccounted for crude oil and crude stock withdrawals minus crude product supplied.
[3]Includes other hydrocarbons and alcohols.
[4]The volumetric amount by which total output is greater than input due to the processing of crude oil into products which, in total, have a lower specific gravity than the crude oil processed.
[5]Includes pyrolysis oils, biomass-derived Fischer-Tropsch liquids, and renewable feedstocks used for the on-site production of diesel and gasoline.
[6]Includes domestic sources of other blending components, other hydrocarbons, and ethers.
[7]Total crude supply plus other petroleum supply plus other non-petroleum supply.
[8]E85 refers to a blend of 85 percent ethanol (renewable) and 15 percent motor gasoline (nonrenewable). To address cold starting issues, the percentage of ethanol varies seasonally. The annual average ethanol content of 74 percent is used for this forecast.
[9]Includes ethanol and ethers blended into gasoline.
[10]Includes only kerosene type.
[11]Includes distillate fuel oil and kerosene from petroleum and biomass feedstocks.
[12]Includes aviation gasoline, petrochemical feedstocks, lubricants, waxes, asphalt, road oil, still gas, special naphthas, petroleum coke, crude oil product supplied, methanol, and miscellaneous petroleum products.
[13]Includes consumption for combined heat and power, which produces electricity and other useful thermal energy.
[14]Includes consumption of energy by electricity-only and combined heat and power plants whose primary business is to sell electricity, or electricity and heat, to the public. Includes small power producers and exempt wholesale generators.
[15]Balancing item. Includes unaccounted for supply, losses, and gains.
[16]End-of-year operable capacity.
[17]Rate is calculated by dividing the gross annual input to atmospheric crude oil distillation units by their operable refining capacity in barrels per calendar day.
Note: Totals may not equal sum of components due to independent rounding. Data for 2010 are model results and may differ slightly from official EIA data reports.
Sources: 2010 product supplied based on: U.S. Energy Information Administration (EIA), *Annual Energy Review 2010*, DOE/EIA-0384(2010) (Washington, DC, October 2011). Other 2010 data: EIA, *Petroleum Supply Annual 2010*, DOE/EIA-0340(2010)/1 (Washington, DC, July 2011). **Projections:** EIA, AEO2012 National Energy Modeling System runs LP2012.D022112A, REF2012.D020112C, and HP2012.D022112A.

Table C5. Petroleum product prices
(2010 dollars per gallon, unless otherwise noted)

Sector and fuel	2010	2015 Low oil price	2015 Reference	2015 High oil price	2025 Low oil price	2025 Reference	2025 High oil price	2035 Low oil price	2035 Reference	2035 High oil price
Crude oil prices (2010 dollars per barrel)										
Low sulfur light	79.39	58.36	116.91	182.10	59.41	132.56	193.48	62.38	144.98	200.36
Imported crude oil[1]	75.87	55.41	113.97	179.16	48.84	121.21	180.29	53.10	132.95	187.04
Delivered sector product prices										
Residential										
Liquefied petroleum gases	2.288	1.909	2.600	3.361	1.878	2.733	3.423	1.989	2.934	3.560
Distillate fuel oil	2.941	2.295	3.781	5.310	2.395	4.181	5.441	2.560	4.539	5.547
Commercial										
Distillate fuel oil	2.866	1.917	3.303	4.778	1.982	3.699	4.942	2.136	4.019	5.008
Residual fuel oil	1.657	0.896	2.421	4.161	0.935	2.731	4.240	1.033	2.830	4.207
Residual fuel oil (2010 dollars per barrel)	69.58	37.63	101.70	174.76	39.28	114.70	178.07	43.37	118.85	176.71
Industrial[2]										
Liquefied petroleum gases	1.846	1.438	2.323	3.292	1.383	2.476	3.355	1.520	2.725	3.523
Distillate fuel oil	2.932	1.991	3.322	4.780	2.053	3.737	4.986	2.223	4.054	5.025
Residual fuel oil	1.634	1.423	2.876	4.521	1.436	3.178	4.554	1.492	3.241	4.582
Residual fuel oil (2010 dollars per barrel)	68.62	59.77	120.80	189.87	60.33	133.47	191.28	62.65	136.12	192.45
Transportation										
Liquefied petroleum gases	2.276	2.021	2.704	3.447	1.987	2.827	3.508	2.097	3.026	3.645
Ethanol (E85)[3]	2.402	1.731	2.766	3.631	1.638	2.746	3.996	1.581	3.046	3.717
Ethanol wholesale price	1.712	2.356	2.228	2.622	2.215	2.333	2.741	1.985	2.159	2.571
Motor gasoline[4]	2.756	2.240	3.538	4.974	2.185	3.855	5.196	2.219	4.034	5.053
Jet fuel[5]	2.190	1.704	3.205	4.760	1.728	3.571	4.845	1.884	3.932	4.981
Diesel fuel (distillate fuel oil)[6]	2.998	2.465	3.776	5.237	2.486	4.168	5.435	2.624	4.439	5.430
Residual fuel oil	1.560	1.294	2.742	4.344	1.298	3.086	4.397	1.311	3.136	4.469
Residual fuel oil (2010 dollars per barrel)	65.53	54.33	115.15	182.43	54.50	129.62	184.67	55.06	131.73	187.70
Electric power[7]										
Distillate fuel oil	2.598	1.673	3.157	4.655	1.739	3.515	4.737	1.880	3.856	4.861
Residual fuel oil	1.780	1.957	3.443	5.051	1.814	3.802	5.135	1.677	3.850	5.178
Residual fuel oil (2010 dollars per barrel)	74.77	82.21	144.60	212.13	76.19	159.70	215.65	70.44	161.71	217.49
Refined petroleum product prices[8]										
Liquefied petroleum gases	1.464	1.239	1.947	2.729	1.177	2.049	2.758	1.294	2.255	2.896
Motor gasoline[4]	2.743	2.240	3.538	4.974	2.185	3.855	5.196	2.219	4.034	5.053
Jet fuel[5]	2.190	1.704	3.205	4.760	1.728	3.571	4.845	1.884	3.932	4.981
Distillate fuel oil	2.975	2.355	3.687	5.153	2.394	4.089	5.355	2.543	4.376	5.366
Residual fuel oil	1.619	1.372	2.845	4.464	1.371	3.189	4.523	1.381	3.246	4.585
Residual fuel oil (2010 dollars per barrel)	68.00	57.63	119.50	187.48	57.57	133.95	189.96	57.99	136.32	192.56
Average	**2.528**	**2.059**	**3.316**	**4.691**	**2.015**	**3.600**	**4.808**	**2.101**	**3.830**	**4.785**

Table C5. Petroleum product prices (continued)
(nominal dollars per gallon, unless otherwise noted)

Sector and fuel	2010	Projections								
		2015			2025			2035		
		Low oil price	Reference	High oil price	Low oil price	Reference	High oil price	Low oil price	Reference	High oil price
Crude oil prices (nominal dollars per barrel)										
Low sulfur light	79.39	62.81	125.97	195.67	77.32	170.09	245.37	98.91	229.55	314.93
Imported crude oil[1]	75.87	59.64	122.81	192.52	63.56	155.52	228.64	84.19	210.51	294.00
Delivered sector product prices										
Residential										
Liquefied petroleum gases	2.288	2.054	2.801	3.612	2.445	3.507	4.341	3.154	4.645	5.595
Distillate fuel oil	2.941	2.470	4.074	5.706	3.117	5.365	6.901	4.060	7.188	8.719
Commercial										
Distillate fuel oil	2.866	2.063	3.559	5.135	2.580	4.747	6.268	3.387	6.364	7.872
Residual fuel oil	1.657	0.964	2.609	4.471	1.217	3.504	5.377	1.637	4.481	6.613
Industrial[2]										
Liquefied petroleum gases	1.846	1.548	2.503	3.537	1.800	3.177	4.255	2.410	4.315	5.537
Distillate fuel oil	2.932	2.143	3.580	5.136	2.671	4.795	6.323	3.524	6.419	7.898
Residual fuel oil	1.634	1.532	3.099	4.858	1.869	4.077	5.776	2.365	5.132	7.202
Transportation										
Liquefied petroleum gases	2.276	2.175	2.914	3.704	2.586	3.627	4.449	3.326	4.792	5.729
Ethanol (E85)[3]	2.402	1.863	2.981	3.902	2.131	3.523	5.067	2.507	4.823	5.843
Ethanol wholesale price	1.712	2.535	2.400	2.818	2.883	2.994	3.477	3.147	3.419	4.041
Motor gasoline[4]	2.756	2.411	3.812	5.345	2.843	4.946	6.589	3.519	6.388	7.943
Jet fuel[5]	2.190	1.834	3.454	5.115	2.249	4.582	6.144	2.988	6.226	7.829
Diesel fuel (distillate fuel oil)[6]	2.998	2.653	4.069	5.628	3.235	5.348	6.893	4.161	7.029	8.535
Residual fuel oil	1.560	1.392	2.954	4.668	1.689	3.960	5.576	2.079	4.966	7.025
Electric power[7]										
Distillate fuel oil	2.598	1.801	3.402	5.002	2.263	4.510	6.008	2.982	6.105	7.641
Residual fuel oil	1.780	2.107	3.710	5.427	2.361	4.879	6.512	2.659	6.096	8.140
Refined petroleum product prices[8]										
Liquefied petroleum gases	1.464	1.334	2.098	2.933	1.531	2.629	3.498	2.052	3.571	4.552
Motor gasoline[4]	2.743	2.411	3.812	5.345	2.843	4.946	6.589	3.519	6.387	7.942
Jet fuel[5]	2.190	1.834	3.454	5.115	2.249	4.582	6.144	2.988	6.226	7.829
Distillate fuel oil	2.975	2.534	3.973	5.537	3.115	5.246	6.791	4.032	6.930	8.434
Residual fuel oil (nominal dollars per barrel)	68.00	62.03	128.77	201.46	74.93	171.87	240.90	91.95	215.84	302.67
Average	2.528	2.216	3.573	5.041	2.623	4.620	6.097	3.331	6.064	7.520

[1] Weighted average price delivered to U.S. refiners.
[2] Includes energy for combined heat and power plants, except those whose primary business is to sell electricity, or electricity and heat, to the public.
[3] E85 refers to a blend of 85 percent ethanol (renewable) and 15 percent motor gasoline (nonrenewable). To address cold starting issues, the percentage of ethanol varies seasonally. The annual average ethanol content of 74 percent is used for this forecast.
[4] Sales weighted-average price for all grades. Includes Federal, State and local taxes.
[5] Includes only kerosene type.
[6] Diesel fuel for on-road use. Includes Federal and State taxes while excluding county and local taxes.
[7] Includes electricity-only and combined heat and power plants whose primary business is to sell electricity, or electricity and heat, to the public. Includes small power producers and exempt wholesale generators.
[8] Weighted averages of end-use fuel prices are derived from the prices in each sector and the corresponding sectoral consumption.
Note: Data for 2010 are model results and may differ slightly from official EIA data reports.
Sources: 2010 low sulfur light crude oil price: U.S. Energy Information Administration (EIA), Form EIA-856, "Monthly Foreign Crude oil Acquisition Report." 2010 imported crude oil price: EIA, *Annual Energy Review 2010*, DOE/EIA-0384(2010) (Washington, DC, October 2011). 2010 prices for motor gasoline, distillate fuel oil, and jet fuel are based on: EIA, *Petroleum Marketing Annual 2009*, DOE/EIA-0487(2009) (Washington, DC, August 2010). 2010 residential, commercial, industrial, and transportation sector petroleum product prices are derived from: EIA, Form EIA-782A, "Refiners'/Gas Plant Operators' Monthly Petroleum Product Sales Report." 2010 electric power prices based on: *Monthly Energy Review*, DOE/EIA-0035(2011/09) (Washington, DC, September 2011). 2010 E85 prices derived from monthly prices in the Clean Cities Alternative Fuel Price Report. 2010 wholesale ethanol prices derived from Bloomberg U.S. average rack price. **Projections:** EIA, AEO2012 National Energy Modeling System runs LP2012.D022112A, REF2012.D020112C, and HP2012.D022112A.

Table C6. International liquids supply and disposition summary
(million barrels per day, unless otherwise noted)

Supply and disposition	2010	Projections								
		2015			2025			2035		
		Low oil price	Reference	High oil price	Low oil price	Reference	High oil price	Low oil price	Reference	High oil price
Crude oil prices (2010 dollars per barrel)										
Low sulfur light	79.39	58.36	116.91	182.10	59.41	132.56	193.48	62.38	144.98	200.36
Imported crude oil[1]	75.87	55.41	113.97	179.16	48.84	121.21	180.29	53.10	132.95	187.04
Crude oil prices (nominal dollars per barrel)[1]										
Low sulfur light	79.39	62.81	125.97	195.67	77.32	170.09	245.37	98.91	229.55	314.93
Imported crude oil[1]	75.87	59.64	122.81	192.52	63.56	155.52	228.64	84.19	210.51	294.00
Petroleum liquids production[2]										
OPEC[3]										
Middle East	23.43	29.09	25.46	23.39	33.98	29.77	28.26	35.70	33.94	32.96
North Africa	3.89	4.01	3.62	3.48	3.66	3.37	3.41	3.12	3.27	3.28
West Africa	4.45	5.57	5.09	4.86	5.92	5.40	5.47	5.74	5.26	5.27
South America	2.29	2.37	2.13	2.05	2.06	1.92	1.94	1.63	1.72	1.72
Total OPEC petroleum production	34.05	41.03	36.30	33.78	45.62	40.46	39.09	46.18	44.19	43.24
Non-OPEC										
OECD										
United States (50 states)	8.79	9.36	9.82	10.15	9.42	10.53	11.40	8.81	10.15	10.72
Canada	1.91	1.79	1.79	1.82	1.77	1.82	1.85	1.75	1.78	1.87
Mexico	2.98	2.65	2.65	2.59	1.46	1.58	1.50	1.27	1.68	1.67
OECD Europe[4]	4.36	3.72	3.70	3.63	3.03	3.15	3.01	2.79	2.83	2.82
Japan	0.13	0.15	0.14	0.14	0.15	0.15	0.15	0.15	0.16	0.16
Australia and New Zealand	0.62	0.55	0.55	0.54	0.52	0.54	0.52	0.52	0.53	0.53
Total OECD petroleum production	18.80	18.22	18.65	18.88	16.34	17.78	18.42	15.29	17.14	17.76
Non-OECD										
Russia	10.14	9.74	10.04	9.79	9.73	11.06	10.38	8.96	12.16	12.02
Other Europe and Eurasia[5]	3.22	3.68	3.67	3.58	4.02	4.37	4.11	3.27	4.54	4.49
China	4.27	4.32	4.29	4.21	4.55	4.79	4.52	4.66	4.70	4.67
Other Asia[6]	3.77	3.80	3.79	3.73	3.23	3.38	3.22	2.97	3.00	2.99
Middle East	1.58	1.43	1.43	1.40	1.12	1.18	1.11	0.97	0.97	0.97
Africa	2.41	2.41	2.40	2.36	2.55	2.68	2.54	2.67	2.68	2.67
Brazil	2.19	2.73	2.72	2.66	3.47	3.87	3.64	3.32	4.45	4.40
Other Central and South America	2.01	2.30	2.29	2.26	2.36	2.47	2.35	2.64	2.65	2.63
Total non-OECD petroleum	29.59	30.40	30.63	29.99	31.02	33.80	31.86	29.47	35.15	34.83
Total petroleum liquids production	82.44	89.66	85.58	82.65	92.98	92.04	89.37	90.93	96.47	95.83
Other liquids production[7]										
United States (50 states)	0.90	1.10	1.05	1.14	1.45	1.62	2.42	1.96	2.59	4.38
Other North America	1.93	2.55	2.51	2.90	4.09	3.75	4.78	5.53	5.16	6.53
OECD Europe[3]	0.22	0.28	0.23	0.27	0.37	0.26	0.30	0.45	0.28	0.32
Middle East	0.01	0.13	0.17	0.14	0.23	0.24	0.21	0.22	0.24	0.22
Africa	0.21	0.27	0.28	0.28	0.42	0.38	0.39	0.53	0.40	0.41
Central and South America	1.20	2.15	1.78	2.06	4.07	2.61	2.97	5.75	3.17	3.51
Other	0.13	0.21	0.16	0.24	0.81	0.61	1.15	1.75	1.18	1.69
Total other liquids production	4.61	6.70	6.18	7.01	11.43	9.47	12.22	16.19	13.02	17.07
Total production	87.05	96.36	91.76	89.67	104.42	101.51	101.59	107.13	109.50	112.90

Table C6. International liquids supply and disposition summary (continued)
(million barrels per day, unless otherwise noted)

Supply and disposition	2010	Projections								
		2015			2025			2035		
		Low oil price	Reference	High oil price	Low oil price	Reference	High oil price	Low oil price	Reference	High oil price
Liquids consumption[8]										
OECD										
United States (50 states)	19.17	20.14	19.10	18.41	20.77	19.20	18 53	21.90	19.90	19.12
United States territories	0.28	0 32	0.31	0.30	0 32	0.34	0 34	0.31	0.36	0 38
Canada	2.21	2 27	2.15	2.09	2.46	2.25	2 22	2.56	2.35	2.40
Mexico	2.34	2 50	2.38	2.30	2.78	2.50	2 32	3.20	2.68	2.43
OECD Europe[3]	14.58	14 86	14.14	13.69	15 97	14.65	13 85	16.10	14.74	13 93
Japan	4.45	4 80	4.51	4.35	5.14	4.62	4 33	4.92	4.42	4.14
South Korea	2.24	2 39	2.25	2.18	2.73	2.46	2 31	2.93	2.56	2 39
Australia and New Zealand	1.13	1.16	1.11	1.07	1 25	1.17	1 09	1.30	1.23	1.13
Total OECD consumption	46.40	48.43	45.95	44.38	51.42	47.19	44.97	53.23	48.24	45.90
Non-OECD										
Russia	2.93	3.14	3.02	2.96	2 88	2.91	2 93	2.71	2.97	3.12
Other Europe and Eurasia[5]	2.08	2 37	2.30	2.26	2 35	2.45	2.44	2.32	2.63	2 69
China	9.19	12 64	12.10	12.06	15 65	16.03	17 21	16.35	18.50	20 87
India	3.18	3 88	3.70	3.64	5 22	5.40	5.78	4.93	5.80	6 54
Other Asia	6.73	7 56	7.28	7.19	8.44	8.85	9.15	8.48	9.89	10.78
Middle East	7.35	8 26	7.78	7.72	8 35	8.16	8 51	9.03	9.49	10.46
Africa	3.34	3.44	3.30	3.24	3.43	3.57	3 57	3.47	4.09	4 21
Brazil	2.65	3 00	2.84	2.78	3 01	3.15	3 22	3.13	3.80	4.13
Other Central and South America	3.19	3 63	3.49	3.42	3 67	3.81	3 82	3.49	4.09	4 21
Total non-OECD consumption	40.65	47.92	45.82	45.29	52.99	54.32	56.62	53.90	61.26	67.00
Total liquids consumption	87.05	96.36	91.76	89.67	104.42	101.51	101.59	107.13	109.50	112.90
OPEC production[9]	34.58	42.18	37.30	34.88	47 89	41.91	40 63	49.42	45.89	45 01
Non-OPEC production[9]	52.47	54.18	54.46	54.79	56 52	59.60	60 97	57.71	63.61	67 89
Net Eurasia exports	10.53	10 64	11.11	10.81	12 00	13.94	12.75	10.52	15.54	15.10
OPEC market share (percent)	39.7	43.8	40.7	38 9	45.9	41 3	40.0	46.1	41 9	39.9

[1]Weighted average price delivered to U.S. refiners.
[2]Includes production of crude oil (including lease condensate and shale oil/tight oil), natural gas plant liquids, other hydrogen and hydrocarbons for refinery feedstocks, and refinery gains.
[3]OPEC = Organization of Petroleum Exporting Countries - Algeria, Angola, Ecuador, Iran, Iraq, Kuwait, Libya, Nigeria, Qatar, Saudi Arabia, the United Arab Emirates, and Venezuela.
[4]OECD Europe = Organization for Economic Cooperation and Development - Austria, Belgium, Czech Republic, Denmark, Finland, France, Germany, Greece, Hungary, Iceland, Ireland, Italy, Luxembourg, the Netherlands, Norway, Poland, Portugal, Slovakia, Spain, Sweden, Switzerland, Turkey, and the United Kingdom.
[5]Other Europe and Eurasia = Albania, Armenia, Azerbaijan, Belarus, Bosnia and Herzegovina, Bulgaria, Croatia, Estonia, Georgia, Kazakhstan, Kyrgyzstan, Latvia, Lithuania, Macedonia, Malta, Moldova, Montenegro, Romania, Serbia, Slovenia, Tajikistan, Turkmenistan, Ukraine, and Uzbekistan.
[6]Other Asia = Afghanistan, Bangladesh, Bhutan, Brunei, Cambodia (Kampuchea), Fiji, French Polynesia, Guam, Hong Kong, Indonesia, Kiribati, Laos, Malaysia, Macau, Maldives, Mongolia, Myanmar (Burma), Nauru, Nepal, New Caledonia, Niue, North Korea, Pakistan, Papua New Guinea, Philippines, Samoa, Singapore, Solomon Islands, Sri Lanka, Taiwan, Thailand, Tonga, Vanuatu, and Vietnam.
[7]Includes liquids produced from energy crops, natural gas, coal, extra-heavy oil, bitumen (oil sands), and kerogen (oil shale, not to be confused with shale oil/tight oil). Includes both OPEC and non-OPEC producers in the regional breakdown.
[8]Includes both OPEC and non-OPEC consumers in the regional breakdown.
[9]Includes both petroleum and other liquids production.
Note: Totals may not equal sum of components due to independent rounding. Data for 2010 are model results and may differ slightly from official EIA data reports.
Sources: 2010 low sulfur light crude oil price: U.S. Energy Information Administration (EIA), Form EIA-856, "Monthly Foreign Crude oil Acquisition Report." 2010 imported crude oil price: EIA, *Annual Energy Review 2010*, DOE/EIA-0384(2010) (Washington, DC, October 2011). **2010 quantities and projections:** EIA, AEO2012 National Energy Modeling System runs LP2012.D022112A, REF2012.D020112C, and HP2012.D022112A and EIA, Generate World Oil Balance Model.

This page intentionally left blank

Appendix D
Results from side cases

Table D1. Key results for residential and commercial sector technology cases

Energy consumption	2010	2015				2025			
		Integrated 2011 Demand Technology	Reference	Integrated High Demand Technology	Integrated Best Available Demand Technology	Integrated 2011 Demand Technology	Reference	Integrated High Demand Technology	Integrated Best Available Demand Technology
Residential									
Energy consumption (quadrillion Btu)									
Liquefied petroleum gases	0.56	0.52	0.51	0 51	0.50	0.52	0.50	0.48	0.48
Kerosene	0.03	0.03	0.02	0 02	0.02	0.02	0.02	0 02	0 02
Distillate fuel oil	0.63	0.56	0.55	0 54	0.53	0.46	0.43	0.41	0 39
Liquid fuels and other petroleum subtotal	1.22	1.10	1.08	1 07	1 05	1.00	0.95	0.91	0 88
Natural gas	5.06	5.03	4.97	4 83	4.63	5.12	4.88	4 51	4 00
Coal	0.01	0.01	0.01	0 01	0.01	0.01	0.01	0 01	0 01
Renewable energy[1]	0.42	0.43	0.43	0.42	0.41	0.47	0.43	0.41	0 37
Electricity	4.95	4.83	4.75	4 53	4.28	5.48	5.23	4.74	4.10
Delivered energy	11.66	11.40	11.24	10.85	10.38	12.08	11.51	10.57	9.36
Electricity related losses	10.39	9.75	9.58	9 09	8.52	10.98	10.52	9 53	8.17
Total	22.05	21.15	20.81	19.95	18.90	23.07	22.02	20.10	17.53
Delivered energy intensity (million Btu per household)	102.1	96.0	94.6	91.4	87.4	91.1	86.8	79.7	70.6
Nonmarketed renewables consumption (quadrillion Btu)	0.02	0.08	0.08	0.08	0.09	0.10	0.10	0.11	0.13
Commercial									
Energy consumption (quadrillion Btu)									
Liquefied petroleum gases	0.14	0.14	0.14	0.14	0.14	0.15	0.15	0.15	0.15
Motor gasoline[2]	0.05	0.05	0.05	0 05	0.05	0.05	0.05	0 05	0 05
Kerosene	0.00	0.00	0.00	0 00	0.00	0.00	0.00	0 00	0 00
Distillate fuel oil	0.43	0.35	0.35	0 35	0.35	0.33	0.33	0 32	0 32
Residual fuel oil	0.08	0.08	0.08	0 08	0.08	0.08	0.08	0 08	0 08
Liquid fuels and other petroleum subtotal	0.72	0.62	0.62	0 62	0.62	0.62	0.62	0.61	0 61
Natural gas	3.28	3.42	3.41	3 39	3.41	3.53	3.53	3.48	3 56
Coal	0.06	0.06	0.06	0 06	0.06	0.06	0.06	0 06	0 06
Renewable energy[3]	0.11	0.11	0.11	0.11	0.11	0.11	0.11	0.11	0.11
Electricity	4.54	4.64	4.59	4.42	4.26	5.39	5.16	4 62	4.17
Delivered energy	8.70	8.85	8.80	8.60	8.46	9.71	9.48	8.87	8.50
Electricity related losses	9.52	9.38	9.27	8 88	8.48	10.79	10.38	9 29	8 30
Total	18.22	18.24	18.06	17.48	16.94	20.50	19.86	18.16	16.80
Delivered energy intensity (thousand Btu per square foot)	107.3	105.3	104.6	102.2	100.6	103.4	101.0	94.5	90.5
Commercial sector generation									
Net summer generation capacity (megawatts)									
Natural gas	711	843	865	900	914	1455	1955	2605	3066
Solar photovoltaic	1197	1251	1253	1254	1262	1490	1578	1753	2235
Wind	83	90	91	94	106	106	132	138	225
Electricity generation (billion kilowatthours)									
Natural gas	5.17	6.13	6.29	6 54	6.64	10.58	14.22	18 95	22 30
Solar photovoltaic	1.87	1.96	1.96	1 96	1.97	2.34	2.51	2 80	3 58
Wind	0.10	0.12	0.12	0.12	0.14	0.14	0.18	0.19	0 31
Nonmarketed renewables consumption (quadrillion Btu)	0.03	0.04	0.04	0.05	0.05	0.04	0.04	0.07	0.08

[1]Includes wood used for residential heating. See Table A4 and/or Table A17 for estimates of nonmarketed renewable energy consumption for geothermal heat pumps, solar thermal hot water heating, and solar photovoltaic electricity generation.
[2]Includes ethanol (blends of 15 percent or less) and ethers blended into gasoline.
[3]Includes commercial sector consumption of wood and wood waste, landfill gas, municipal solid waste, and other biomass for combined heat and power.
Btu = British thermal unit.
Note: Totals may not equal sum of components due to independent rounding. Data for 2010 are model results and may differ slightly from official EIA data reports.
Source: U.S. Energy Information Administration, AEO2012 National Energy Modeling System, runs FROZTECH.D030812A, REF2012.D020112C, HIGHTECH.D032812A, and BESTTECH.D032812A.

Results from side cases

2035				Annual Growth 2010-2035 (percent)			
Integrated 2011 Demand Technology	Reference	Integrated High Demand Technology	Integrated Best Available Demand Technology	Integrated 2011 Demand Technology	Reference	Integrated High Demand Technology	Integrated Best Available Demand Technology
0.53	0.51	0.48	0.47	-0.2%	-0.4%	-0.6%	-0.7%
0.02	0.02	0.02	0.02	-1.2%	-1.7%	-2.1%	-2.4%
0.40	0.35	0.32	0.29	-1.8%	-2.3%	-2.7%	-3.1%
0.95	0.87	0.82	0.78	-1.0%	-1.3%	-1.6%	-1.8%
5.23	4.76	4.28	3.67	0.1%	-0.2%	-0.7%	-1.3%
0.01	0.01	0.00	0.00	-0.5%	-1.1%	-1.5%	-1.8%
0.50	0.43	0.39	0.34	0.6%	0.1%	-0.3%	-0.9%
6.23	5.86	5.26	4.45	0.9%	0.7%	0.2%	-0.4%
12.91	**11.93**	**10.75**	**9.24**	**0.4%**	**0.1%**	**-0.3%**	**-0.9%**
12.14	11.35	10.31	8.65	0.6%	0.4%	-0.0%	-0.7%
25.05	**23.28**	**21.06**	**17.89**	**0.5%**	**0.2%**	**-0.2%**	**-0.8%**
88.7	81.9	73.8	63.4	-0.6%	-0.9%	-1.3%	-1.9%
0.10	0.11	0.14	0.19	6.4%	6.9%	7.7%	9.2%
0.15	0.16	0.16	0.16	0.3%	0.3%	0.4%	0.4%
0.06	0.06	0.06	0.06	0.4%	0.4%	0.4%	0.4%
0.01	0.01	0.01	0.01	0.7%	0.7%	0.7%	0.7%
0.32	0.32	0.30	0.30	-1.2%	-1.2%	-1.4%	-1.5%
0.08	0.08	0.08	0.08	-0.1%	-0.0%	-0.0%	-0.0%
0.62	0.62	0.61	0.60	-0.6%	-0.5%	-0.7%	-0.7%
3.63	3.69	3.64	3.74	0.4%	0.5%	0.4%	0.5%
0.06	0.06	0.06	0.06	-0.0%	-0.0%	-0.0%	-0.0%
0.11	0.11	0.11	0.11	0.0%	0.0%	0.0%	0.0%
6.07	5.80	4.87	4.33	1.2%	1.0%	0.3%	-0.2%
10.49	**10.28**	**9.28**	**8.84**	**0.8%**	**0.7%**	**0.3%**	**0.1%**
11.82	11.23	9.54	8.41	0.9%	0.7%	0.0%	-0.5%
22.32	**21.50**	**18.82**	**17.25**	**0.8%**	**0.7%**	**0.1%**	**-0.2%**
101.9	99.8	90.1	85.8	-0.2%	-0.3%	-0.7%	-0.9%
2514	4795	6609	7235	5.2%	7.9%	9.3%	9.7%
1832	2311	3177	5546	1.7%	2.7%	4.0%	6.3%
178	270	269	375	3.1%	4.8%	4.8%	6.2%
18.29	34.88	48.08	52.63	5.2%	7.9%	9.3%	9.7%
2.88	3.74	5.17	9.02	1.7%	2.8%	4.2%	6.5%
0.24	0.38	0.38	0.53	3.5%	5.3%	5.3%	6.7%
0.04	0.05	0.11	0.12	1.0%	1.7%	4.8%	5.1%

Table D2. Key results for integrated technology cases

Consumption and emissions	2010	2015			2025			2035		
		Integrated 2011 Technology	Reference	Integrated High Technology	Integrated 2011 Technology	Reference	Integrated High Technology	Integrated 2011 Technology	Reference	Integrated High Technology
Energy consumption by sector (quadrillion Btu)										
Residential	11.66	11.39	11.24	10.87	12.08	11.51	10.60	12.90	11.93	10.80
Commercial	8.70	8.85	8.80	8.62	9.70	9.48	8.90	10.48	10.28	9.33
Industrial[1]	23.37	23.99	23.96	24.03	25.24	25.53	25.88	25.68	26.94	27.69
Transportation	27.59	27.61	27.60	27.48	27.45	27.40	26.80	28.57	28.60	27.64
Electric power[2]	39.63	39.09	38.64	37.46	43.38	42.03	39.08	46.11	44.24	40.45
Total	98.16	98.00	97.43	96.02	103.43	101.99	98.25	108.09	106.93	102.23
Energy consumption by fuel (quadrillion Btu)										
Liquid fuels and other petroleum[3]	37.25	36.77	36.72	36.54	36.67	36.58	35.84	37.67	37.70	36.52
Natural gas	24.71	26.02	26.00	25.69	26.77	26.14	25.13	28.64	27.26	25.23
Coal	20.76	18.14	17.80	16.64	20.73	20.02	17.87	21.89	21.15	18.45
Nuclear / uranium	8.44	8.68	8.68	8.68	9.60	9.60	9.34	9.14	9.28	9.55
Renewable energy[4]	6.72	8.10	7.92	8.17	9.38	9.38	9.80	10.48	11.29	12.24
Other[5]	0.29	0.30	0.30	0.30	0.28	0.28	0.27	0.26	0.24	0.24
Total	98.16	98.00	97.43	96.02	103.43	101.99	98.25	108.09	106.93	102.23
Energy intensity (thousand Btu per 2005 dollar of GDP)	7.50	6.62	6.58	6.49	5.39	5.32	5.12	4.41	4.36	4.17
Carbon dioxide emissions by sector (million metric tons)										
Residential	353	343	338	331	341	324	302	342	312	284
Commercial	229	231	231	230	237	237	233	242	246	242
Industrial[1]	909	964	963	962	993	992	983	1015	1011	995
Transportation	1872	1865	1864	1856	1829	1820	1772	1883	1859	1787
Electric power[6]	2271	2040	2011	1884	2268	2179	1942	2446	2330	1992
Total	5634	5443	5407	5263	5668	5552	5232	5928	5758	5300
Carbon dioxide emissions by fuel (million metric tons)										
Petroleum	2349	2332	2329	2315	2275	2261	2201	2327	2300	2208
Natural gas	1283	1368	1367	1350	1407	1374	1320	1508	1435	1327
Coal	1990	1731	1699	1586	1974	1906	1700	2081	2012	1753
Other[7]	12	12	12	12	12	12	12	12	12	12
Total	5634	5443	5407	5263	5668	5552	5232	5928	5758	5300
Carbon dioxide emissions (tons per person)	18.1	16.7	16.6	16.1	15.8	15.5	14.6	15.2	14.8	13.6

[1]Includes energy for combined heat and power plants, except those whose primary business is to sell electricity, or electricity and heat, to the public.
[2]Includes electricity-only and combined heat and power plants whose primary business is to sell electricity, or electricity and heat, to the public.
[3]Includes petroleum-derived fuels and non-petroleum derived fuels, such as ethanol and biodiesel, and coal-based synthetic liquids. Petroleum coke, which is a solid, is included. Also included are natural gas plant liquids, crude oil consumed as a fuel, and liquid hydrogen.
[4]Includes grid-connected electricity from conventional hydroelectric; wood and wood waste; landfill gas; biogenic municipal solid waste; other biomass; wind; photovoltaic and solar thermal sources; and non-electric energy from renewable sources, such as active and passive solar systems, and wood; and both the ethanol and gasoline components of E85, but not the ethanol component of blends less than 85 percent. Excludes electricity imports using renewable sources and nonmarketed renewable energy.
[5]Includes non-biogenic municipal waste, liquid hydrogen, and net electricity imports.
[6]Includes electricity-only and combined heat and power plants whose primary business is to sell electricity, or electricity and heat, to the public.
[7]Includes emissions from geothermal power and nonbiogenic emissions from municipal solid waste.
Btu = British thermal unit.
GDP = Gross domestic product.
Note: Includes end-use, fossil electricity, and renewable technology assumptions. Totals may not equal sum of components due to independent rounding. Data for 2010 are model results and may differ slightly from official EIA data reports.
Source: U.S. Energy Information Administration, AEO2012 National Energy Modeling System runs LTRKITEN.D031312A, REF2012.D020112C, and HTRKITEN.D032812A.

Table D3. Key results for transportation sector light-duty vehicle efficiency cases

Consumption and indicators	2010	2015 Reference	2015 CAFE Standards	2025 Reference	2025 CAFE Standards	2035 Reference	2035 CAFE Standards
Level of travel							
(billion vehicle miles traveled)							
Light-duty vehicles less than 8,501 pounds	2662	2710	2710	3111	3129	3583	3650
Commercial light trucks[1]	64	70	70	83	83	92	93
Freight trucks greater than 10,000 pounds	234	273	273	317	318	345	346
(billion seat miles available)							
Air	999	1028	1028	1120	1120	1208	1208
(billion ton miles traveled)							
Rail	1559	1503	1505	1782	1789	1871	1878
Domestic shipping	522	549	549	604	604	627	625
Energy efficiency indicators							
(miles per gallon)							
Tested new light-duty vehicle[2]	28.3	31.5	31.5	36.8	48.1	37.9	49.0
New car[2]	33.3	36.4	36.4	41.2	55.6	42.8	56.9
New light truck[2]	24.3	26.7	26.7	31.0	39.6	31.5	39.8
Light-duty stock[3]	20.4	21.5	21.5	25.6	27.5	28.2	34.5
New commercial light truck[1]	15.7	16.7	16.7	18.9	22.5	19.1	23.3
Stock commercial light truck[1]	14.4	15.2	15.2	18.0	19.0	19.0	22.5
Freight truck	6.7	6.8	6.8	7.7	7.7	8.1	8.1
(seat miles per gallon)							
Aircraft	62.3	62.8	62.8	65.2	65.2	69.3	69.3
(ton miles per thousand Btu)							
Rail	3.4	3.5	3.5	3.5	3.5	3.5	3.5
Domestic shipping	2.4	2.4	2.4	2.5	2.5	2.5	2.5
Energy use (quadrillion Btu)							
by mode							
Light-duty vehicles	16.06	15.39	15.39	14.73	13.78	15.46	12.84
Commercial light trucks[1]	0.55	0.58	0.58	0.58	0.55	0.61	0.52
Bus transportation	0.25	0.26	0.26	0.29	0.29	0.31	0.31
Freight trucks	4.82	5.51	5.51	5.66	5.67	5.84	5.87
Rail, passenger	0.05	0.05	0.05	0.06	0.06	0.06	0.06
Rail, freight	0.45	0.43	0.44	0.51	0.51	0.53	0.53
Shipping, domestic	0.22	0.23	0.23	0.25	0.25	0.25	0.25
Shipping, international	0.86	0.87	0.87	0.88	0.88	0.89	0.89
Recreational boats	0.25	0.26	0.26	0.27	0.27	0.29	0.29
Air	2.52	2.55	2.55	2.71	2.71	2.79	2.79
Military use	0.77	0.66	0.66	0.66	0.66	0.74	0.74
Lubricants	0.14	0.13	0.13	0.14	0.14	0.14	0.14
Pipeline fuel	0.65	0.68	0.68	0.67	0.67	0.69	0.68
Total	27.59	27.60	27.60	27.40	26.44	28.60	25.92
by fuel							
Liquefied petroleum gases	0.04	0.04	0.04	0.04	0.04	0.05	0.04
E85[4]	0.00	0.01	0.01	0.30	0.44	1.22	1.37
Motor gasoline[5]	16.91	16.13	16.13	14.90	13.81	14.53	11.82
Jet fuel[6]	3.07	3.03	3.03	3.19	3.19	3.33	3.33
Distillate fuel oil[7]	5.77	6.55	6.55	7.03	7.02	7.44	7.31
Residual fuel oil	0.90	0.91	0.91	0.93	0.93	0.94	0.94
Other petroleum[8]	0.17	0.17	0.17	0.17	0.17	0.17	0.17
Liquid fuels and other petroleum	26.88	26.83	26.83	26.57	25.60	27.67	24.99
Pipeline fuel natural gas	0.65	0.68	0.68	0.67	0.67	0.69	0.68
Compressed/liquefied natural gas	0.04	0.06	0.06	0.11	0.11	0.16	0.15
Liquid hydrogen	0.00	0.00	0.00	0.00	0.00	0.00	0.00
Electricity	0.02	0.03	0.03	0.04	0.05	0.07	0.09
Delivered energy	27.59	27.60	27.60	27.40	26.44	28.60	25.92
Electricity related losses	0.05	0.05	0.05	0.08	0.10	0.14	0.18
Total	27.63	27.65	27.65	27.49	26.54	28.75	26.11

[1]Commercial trucks 8,500 to 10,000 pounds.
[2]Environmental Protection Agency rated miles per gallon.
[3]Combined car and light truck "on-the-road" estimate.
[4]E85 refers to a blend of 85 percent ethanol (renewable) and 15 percent motor gasoline (nonrenewable). To address cold starting issues, the percentage of ethanol varies seasonally. The annual average ethanol content of 74 percent is used for this forecast.
[5]Includes ethanol (blends of 15 percent or less) and ethers blended into gasoline.
[6]Includes only kerosene type.
[7]Diesel fuel for on- and off- road use.
[8]Includes aviation gasoline and lubricants.
CAFE = Corporate average fuel economy.
Btu = British thermal unit.
Note: Totals may not equal sum of components due to independent rounding. Data for 2010 are model results and may differ slightly from official EIA data reports.
Source: U.S. Energy Information Administration, AEO2012 National Energy Modeling System runs REF2012.D020112C and CAFEY.D032112A.

Table D4. Key results for HD NGV Potential case

Sales, consumption, and efficiency	2010	2015 Heavy Duty Vehicle Reference	2015 Heavy Duty Natural Gas Vehicle Potential	2025 Heavy Duty Vehicle Reference	2025 Heavy Duty Natural Gas Vehicle Potential	2035 Heavy Duty Vehicle Reference	2035 Heavy Duty Natural Gas Vehicle Potential
Truck sales by size class (millions)	0.36	0.56	0.56	0.65	0.65	0.80	0.81
Medium	0.21	0.29	0.29	0.33	0.33	0.40	0.40
Diesel	0.13	0.20	0.20	0.24	0.20	0.28	0.21
Motor gasoline	0.07	0.08	0.08	0.08	0.07	0.10	0.08
Liquefied petroleum gases	0.00	0.00	0.00	0.00	0.00	0.01	0.01
Natural gas	0.00	0.00	0.01	0.01	0.06	0.02	0.11
Heavy	0.15	0.27	0.27	0.32	0.32	0.40	0.40
Diesel	0.15	0.26	0.25	0.30	0.22	0.37	0.23
Motor gasoline	0.00	0.01	0.01	0.01	0.01	0.02	0.01
Liquefied petroleum gases	0.00	0.00	0.00	0.00	0.00	0.00	0.00
Natural gas	0.00	0.00	0.01	0.00	0.08	0.01	0.16
Consumption by size class (quadrillion Btu)	4.82	5.50	5.51	5.66	5.68	5.85	5.93
Medium	0.83	1.03	1.03	1.12	1.12	1.15	1.16
Diesel	0.56	0.72	0.71	0.79	0.72	0.83	0.65
Motor gasoline	0.26	0.30	0.30	0.28	0.27	0.26	0.21
Liquefied petroleum gases	0.01	0.01	0.01	0.01	0.01	0.02	0.02
Natural gas	0.01	0.01	0.02	0.03	0.12	0.05	0.28
Heavy	3.99	4.47	4.48	4.55	4.56	4.71	4.77
Diesel	3.87	4.36	4.32	4.44	3.82	4.57	3.11
Motor gasoline	0.11	0.09	0.09	0.08	0.07	0.08	0.06
Liquefied petroleum gases	0.01	0.01	0.01	0.01	0.01	0.01	0.01
Natural gas	0.00	0.01	0.06	0.02	0.66	0.05	1.59
New truck fuel efficiency by size class (gasoline equivalent miles per gallon)	6.63	7.41	7.38	8.11	7.88	8.22	7.82
Medium	11.92	13.42	13.34	15.06	14.32	15.43	14.12
Diesel	13.50	14.49	14.49	16.29	16.29	16.37	16.35
Motor gasoline	10.13	10.49	10.49	11.87	11.87	13.07	13.07
Liquefied petroleum gases	9.95	10.56	10.56	12.11	12.11	13.39	13.39
Natural gas	9.17	9.99	9.99	11.07	11.07	11.07	11.07
Heavy	5.79	6.82	6.80	7.46	7.29	7.58	7.29
Diesel	5.79	6.85	6.85	7.50	7.49	7.63	7.59
Motor gasoline	5.50	5.35	5.35	5.45	5.45	5.46	5.46
Liquefied petroleum gases	5.15	5.58	5.58	5.75	5.75	5.75	5.75
Natural gas	5.56	6.04	6.35	6.40	6.87	6.42	6.95
Stock fuel efficiency by size class (gasoline equivalent miles per gallon)	6.66	6.83	6.82	7.72	7.61	8.12	7.81
Medium	11.48	12.06	12.05	13.90	13.60	14.99	14.04
Diesel	13.87	13.89	13.89	15.54	15.49	16.27	16.23
Motor gasoline	9.23	9.66	9.66	10.82	10.79	12.35	12.30
Liquefied petroleum gases	8.67	9.59	9.59	11.31	11.31	12.87	12.86
Natural gas	8.69	9.32	9.49	10.85	10.95	11.05	11.06
Heavy	6.05	6.16	6.16	7.05	6.97	7.44	7.22
Diesel	6.07	6.19	6.18	7.09	7.04	7.50	7.44
Motor gasoline	5.36	5.34	5.34	5.38	5.38	5.44	5.44
Liquefied petroleum gases	5.43	5.43	5.43	5.62	5.62	5.71	5.71
Natural gas	5.51	5.75	6.06	6.31	6.79	6.41	6.92

[1]Includes lease condensate.
[2]Includes natural gas plant liquids, refinery processing gain, other crude oil supply, and stock withdrawals.
[3]Includes liquids, such as ethanol and biodiesel, derived from biomass, natural gas, and coal. Includes net imports of ethanol and biodiesel.
- - = Not applicable.
Btu = British thermal unit.
Note: Totals may not equal sum of components due to independent rounding. Data for 2010 are model results and may differ slightly from official EIA data reports.
Sources: 2010 data based on: Oak Ridge National Laboratory, *Transportation Energy Data Book: Edition 28 and Annual* (Oak Ridge, TN, 2009); U.S. Department of Commerce, Bureau of the Census, "Vehicle Inventory and Use Survey," EC02TV (Washington, DC, December 2004); Federal Highway Administration, *Highway Statistics 2007* (Washington, DC, October 2008); U.S. Energy Information Administration (EIA), *Annual Energy Review 2010*, DOE/EIA-0384(2010) (Washington, DC, October 2011); and EIA, AEO2012 National Energy Modeling System run RFNGV12.D050412A. **Projections:** EIA, AEO2012 National Energy Modeling System runs RFNGV12.D050412A and NOSUBNGV12.D050412A.

Table D5. Energy consumption and carbon dioxide emissions for extended policy cases

Consumption and emissions	2010	2015			2025			2035		
		Reference	No Sunset	Extended Policies	Reference	No Sunset	Extended Policies	Reference	No Sunset	Extended Policies
Energy consumption by sector (quadrillion Btu)										
Residential	11.66	11.24	11.21	11.22	11.51	11.34	11.03	11.93	11.58	10.92
Commercial	8.70	8.80	8.79	8.78	9.48	9.49	9.20	10.28	10.31	9.79
Industrial[1]	23.37	23.96	23.95	23.96	25.53	25.73	25.42	26.94	26.99	26.60
Transportation	27.59	27.60	27.59	27.59	27.40	27.43	26.41	28.60	28.57	25.42
Electric power[2]	39.63	38.64	38.60	38.53	42.03	41.63	40.45	44.24	43.95	42.24
Total	98.16	97.43	97.35	97.30	101.99	101.78	99.11	106.93	106.64	100.79
Energy consumption by fuel (quadrillion Btu)										
Liquid fuels and other petroleum[3]	37.25	36.72	36.72	36.71	36.58	36.57	35.44	37.70	37.62	34.20
Natural gas	24.71	26.00	25.98	26.00	26.14	25.93	25.52	27.26	26.37	25.42
Coal	20.76	17.80	17.84	17.82	20.02	19.96	19.27	21.15	20.59	19.82
Nuclear / uranium	8.44	8.68	8.68	8.68	9.60	9.60	9.50	9.28	9.16	9.05
Renewable energy[4]	6.72	7.92	7.82	7.79	9.38	9.45	9.10	11.29	12.66	12.05
Other[5]	0.29	0.30	0.30	0.30	0.28	0.27	0.27	0.24	0.24	0.24
Total	98.16	97.43	97.35	97.30	101.99	101.78	99.11	106.93	106.64	100.79
Energy intensity (thousand Btu per 2005 dollar of GDP)	7.50	6.58	6.58	6.58	5.32	5.30	5.16	4.36	4.35	4.11
Carbon dioxide emissions by sector (million metric tons)										
Residential	353	338	337	338	324	322	319	312	307	293
Commercial	229	231	231	231	237	238	232	246	248	236
Industrial[1]	909	963	962	963	992	993	983	1011	1016	991
Transportation	1872	1864	1864	1863	1820	1813	1749	1859	1853	1642
Electric power[6]	2271	2011	2015	2012	2179	2161	2084	2330	2221	2133
Total	5634	5407	5409	5407	5552	5526	5367	5758	5645	5295
Carbon dioxide emissions by fuel (million metric tons)										
Petroleum	2349	2329	2329	2328	2261	2251	2180	2300	2289	2061
Natural gas	1283	1367	1366	1367	1374	1363	1341	1435	1387	1337
Coal	1990	1699	1702	1700	1906	1901	1835	2012	1957	1885
Other[7]	12	12	12	12	12	12	12	12	12	12
Total	5634	5407	5409	5407	5552	5526	5367	5758	5645	5295
Carbon dioxide emissions (tons per person)	18.1	16.6	16.6	16.6	15.5	15.4	15.0	14.8	14.5	13.6

[1]Includes energy for combined heat and power plants, except those whose primary business is to sell electricity, or electricity and heat, to the public.
[2]Includes electricity-only and combined heat and power plants whose primary business is to sell electricity, or electricity and heat, to the public.
[3]Includes petroleum-derived fuels and non-petroleum derived fuels, such as ethanol and biodiesel, and coal-based synthetic liquids. Petroleum coke, which is a solid, is included. Also included are natural gas plant liquids, crude oil consumed as a fuel, and liquid hydrogen.
[4]Includes grid-connected electricity from conventional hydroelectric; wood and wood waste; landfill gas; biogenic municipal solid waste; other biomass; wind; photovoltaic and solar thermal sources; and non-electric energy from renewable sources, such as active and passive solar systems, and wood; and both the ethanol and gasoline components of E85, but not the ethanol component of blends less than 85 percent. Excludes electricity imports using renewable sources and nonmarketed renewable energy.
[5]Includes non-biogenic municipal waste and net electricity imports.
[6]Includes electricity-only and combined heat and power plants whose primary business is to sell electricity, or electricity and heat, to the public.
[7]Includes emissions from geothermal power and nonbiogenic emissions from municipal solid waste.
Btu = British thermal unit.
GDP = Gross domestic product.
Note: Includes end-use, fossil electricity, and renewable technology assumptions. Totals may not equal sum of components due to independent rounding. Data for 2010 are model results and may differ slightly from official EIA data reports.
Source: U.S. Energy Information Administration, AEO2012 National Energy Modeling System runs REF2012.D020112C, NOSUNSET.D032112A, and EXTENDED.D050612B.

Results from side cases

Table D6. Electricity generation and generating capacity in extended policy cases
(gigawatts, unless otherwise noted)

Net summer capacity, generation, consumption, and emissions	2010	2015			2025			2035		
		Reference	No Sunset	Extended Policies	Reference	No Sunset	Extended Policies	Reference	No Sunset	Extended Policies
Capacity	1036.1	1042.0	1020.7	1011.3	1091.1	1088.5	1059.4	1190.0	1232.9	1167.6
Electric power sector[1]	1006.5	998.7	977.3	967.6	1033.3	1004.8	976.6	1112.5	1098.0	1032.8
Pulverized coal	312.8	280.7	271.7	264.2	272.8	265.8	257.0	273.6	265.7	256.9
Coal gasification combined-cycle	0.5	0.9	0.9	0.9	1.8	1.8	1.7	1.7	1.7	1.5
Conventional natural gas combined-cycle	198.0	212.4	212.4	212.5	213.5	213.0	212.4	218.8	215.7	213.6
Advanced natural gas combined-cycle	0.0	1.2	1.0	1.3	10.3	4.7	2.4	53.4	20.5	8.4
Conventional combustion turbine	137.6	136.3	133.5	133.0	132.3	129.7	127.8	130.3	129.2	126.8
Advanced combustion turbine	0.0	5.2	3.7	4.0	23.2	11.7	6.8	41.5	24.9	10.2
Fuel cells	0.0	0.0	0.0	0.0	0.0	0.0	0.0	0.0	0.0	0.0
Nuclear / uranium	101.2	103.6	103.6	103.6	114.7	114.7	113.6	110.9	109.3	108.1
Oil and natural gas steam	108.1	90.7	85.2	84.2	89.6	83.3	81.4	87.9	83.1	80.6
Renewable sources	126.1	145.3	143.0	141.6	152.1	157.5	151.2	170.2	224.4	203.8
Pumped storage	22.2	22.2	22.2	22.2	22.2	22.2	22.2	22.2	22.2	22.2
Distributed generation	0.0	0.2	0.1	0.1	0.8	0.5	0.3	2.1	1.3	0.5
Combined heat and power[2]	29.6	43.3	43.4	43.7	57.8	83.7	82.8	77.5	134.9	134.9
Fossil fuels / other	22.0	25.7	25.7	26.0	34.4	35.7	35.8	47.0	49.9	49.6
Renewable fuels	7.6	17.6	17.7	17.7	23.4	48.0	47.0	30.6	85.0	85.3
Cumulative additions	0.0	69.8	65.8	65.3	126.7	140.0	124.8	235.0	290.9	240.4
Electric power sector[1]	0.0	56.1	52.0	51.2	98.5	85.9	71.6	187.1	185.6	135.2
Pulverized coal	0.0	8.7	8.7	8.7	8.7	8.7	8.7	9.4	8.7	8.7
Coal gasification combined-cycle	0.0	0.6	0.6	0.6	1.5	1.5	1.5	1.5	1.5	1.5
Conventional natural gas combined-cycle	0.0	14.5	14.5	14.5	15.8	15.3	14.7	21.1	18.0	15.9
Advanced natural gas combined-cycle	0.0	1.2	1.0	1.3	10.3	4.7	2.4	53.4	20.5	8.4
Conventional combustion turbine	0.0	5.0	5.0	5.0	5.0	5.0	5.0	5.0	5.0	5.0
Advanced combustion turbine	0.0	5.2	3.7	4.0	23.2	11.7	6.8	41.5	24.9	10.2
Nuclear / uranium	0.0	1.1	1.1	1.1	6.8	6.8	6.8	8.5	6.9	6.8
Renewable sources	0.0	19.6	17.3	15.9	26.4	31.8	25.5	44.5	98.7	78.1
Distributed generation	0.0	0.2	0.1	0.1	0.8	0.5	0.3	2.1	1.3	0.5
Combined heat and power[2]	0.0	13.7	13.8	14.1	28.2	54.1	53.2	47.9	105.3	105.3
Fossil fuels / other	0.0	3.7	3.8	4.1	12.4	13.7	13.9	25.0	27.9	27.6
Renewable fuels	0.0	10.0	10.0	10.0	15.8	40.3	39.3	22.9	77.4	77.7
Cumulative retirements	0.0	65.2	82.5	91.4	78.9	94.9	108.8	88.4	101.3	116.2
Generation by fuel (billion kilowatthours)	4126	4152	4147	4142	4556	4559	4427	4992	5004	4813
Electric power sector[1]	3971	3956	3950	3944	4279	4229	4106	4586	4498	4310
Coal	1831	1562	1565	1563	1741	1736	1673	1834	1781	1711
Petroleum	34	26	26	26	27	27	26	28	28	27
Natural gas	898	1028	1030	1030	1006	971	938	1196	1030	976
Nuclear / uranium	807	830	830	830	917	917	909	887	875	865
Renewable sources	395	508	498	493	584	574	557	634	780	728
Pumped storage	2	2	2	2	2	2	2	2	2	2
Distributed generation	0	0	0	0	2	1	1	4	2	1
Combined heat and power[2]	155	197	197	198	277	330	321	406	506	502
Fossil fuels / other	122	142	142	144	198	206	206	281	298	294
Renewable fuels	34	55	55	55	78	124	115	125	208	208
Average electricity price (cents per kilowatthour)	9.8	9.7	9.8	9.8	9.7	9.6	9.6	10.1	9.9	9.6

[1]Includes electricity-only and combined heat and power plants whose primary business is to sell electricity, or electricity and heat, to the public. Includes small power producers and exempt wholesale generators.
[2]Includes combined heat and power plants and electricity-only plants in the commercial and industrial sectors. Includes small on-site generating systems in the residential, commercial, and industrial sectors used primarily for own-use generation, but which may also sell some power to the grid. Excludes off-grid photovoltaics and other generators not connected to the distribution or transmission systems.
Note: Totals may not equal sum of components due to independent rounding. Data for 2010 are model results and may differ slightly from official EIA data reports.
Source: U.S. Energy Information Administration, AEO2012 National Energy Modeling System runs REF2012.D020112C, NOSUNSET.D032112A, and EXTENDED.D050612B.

Table D7. Key results for advanced nuclear plant life cases
(gigawatts, unless otherwise noted)

Net summer capacity, generation, emissions, and fuel prices	2010	2015			2025			2035		
		Low Nuclear	Reference	High Nuclear	Low Nuclear	Reference	High Nuclear	Low Nuclear	Reference	High Nuclear
Capacity										
Coal steam	313.4	280.7	281.6	281.3	273.4	274.7	275.3	276.2	275.2	275.4
Oil and natural gas steam	108.1	88.2	90.7	91.0	87.0	89.6	89.4	84.5	87.9	86.9
Combined cycle	198.0	212.6	213.6	213.8	224.1	223.8	219.0	279.8	272.2	257.3
Combustion turbine / diesel	137.6	138.1	141.5	141.3	150.8	155.5	155.4	168.1	171.8	172.6
Nuclear / uranium	101.2	103.1	103.6	103.6	108.2	114.7	121.4	77.9	110.9	122.7
Pumped storage	22.2	22.2	22.2	22.2	22.2	22.2	22.2	22.2	22.2	22.2
Fuel cells	0.0	0.0	0.0	0.0	0.0	0.0	0.0	0.0	0.0	0.0
Renewable sources	126.1	145.4	145.3	145.0	153.2	152.1	151.4	175.7	170.2	167.4
Distributed generation (natural gas)	0.0	0.1	0.2	0.2	0.7	0.8	0.8	1.7	2.1	2.1
Combined heat and power[1]	29.6	43.4	43.3	43.3	57.8	57.8	58.0	78.6	77.5	77.4
Total	**1036.1**	**1033.8**	**1042.0**	**1041.6**	**1077.4**	**1091.1**	**1093.0**	**1164.8**	**1190.0**	**1183.9**
Cumulative additions										
Coal steam	0.0	9.3	9.3	9.3	10.2	10.2	10.2	13.2	10.9	10.4
Oil and natural gas steam	0.0	0.0	0.0	0.0	0.0	0.0	0.0	0.0	0.0	0.0
Combined cycle	0.0	14.7	15.7	15.9	26.4	26.1	21.3	82.1	74.5	59.6
Combustion turbine / diesel	0.0	8.6	10.2	10.2	25.7	28.2	28.0	44.7	46.5	46.0
Nuclear / uranium	0.0	1.1	1.1	1.1	6.8	6.8	13.5	6.8	8.5	14.8
Pumped storage	0.0	0.0	0.0	0.0	0.0	0.0	0.0	0.0	0.0	0.0
Fuel cells	0.0	0.0	0.0	0.0	0.0	0.0	0.0	0.0	0.0	0.0
Renewable sources	0.0	19.7	19.6	19.3	27.5	26.4	25.7	50.0	44.5	41.7
Distributed generation	0.0	0.1	0.2	0.2	0.7	0.8	0.8	1.7	2.1	2.1
Combined heat and power[1]	0.0	13.8	13.7	13.7	28.2	28.2	28.4	49.0	47.9	47.7
Total	**0.0**	**67.2**	**69.8**	**69.7**	**125.5**	**126.7**	**127.9**	**247.5**	**235.0**	**222.4**
Cumulative retirements	**0.0**	**70.4**	**65.2**	**65.4**	**85.0**	**78.9**	**78.3**	**119.6**	**88.4**	**81.9**
Generation by fuel (billion kilowatthours)										
Coal	1831	1570	1562	1565	1760	1741	1727	1853	1834	1822
Petroleum	34	26	26	26	27	27	27	28	28	28
Natural gas	898	1022	1028	1026	1029	1006	972	1361	1196	1136
Nuclear / uranium	807	826	830	830	866	917	970	625	887	979
Pumped storage	2	2	2	2	2	2	2	2	2	2
Renewable sources	395	508	508	507	585	584	585	653	634	632
Distributed generation	0	0	0	0	2	2	2	3	4	4
Combined heat and power[1]	155	197	197	197	277	277	278	412	406	404
Total	**4124**	**4151**	**4152**	**4152**	**4547**	**4556**	**4562**	**4936**	**4992**	**5006**
Carbon dioxide emissions by the electric power sector (million metric tons)[2]										
Petroleum	33	23	23	23	24	24	24	24	25	25
Natural gas	399	436	438	437	445	427	415	545	485	467
Coal	1828	1547	1539	1543	1737	1717	1703	1823	1809	1798
Other[3]	12	12	12	12	12	12	12	12	12	12
Total	**2271**	**2017**	**2011**	**2014**	**2207**	**2179**	**2154**	**2404**	**2330**	**2301**
Prices to the electric power sector[2] (2010 dollars per million Btu)										
Petroleum	13.32	22.93	22.93	22.94	25.38	25.38	25.38	26.53	26.31	26.13
Natural gas	5.14	4.52	4.55	4.54	5.70	5.60	5.46	8.03	7.21	7.00
Coal	2.26	2.36	2.35	2.35	2.54	2.54	2.53	2.81	2.80	2.78

[1]Includes combined heat and power plants and electricity-only plants in commercial and industrial sectors. Includes small on-site generating systems in the residential, commercial, and industrial sectors used primarily for own-use generation, but which may also sell some power to the grid. Excludes off-grid photovoltaics and other generators not connected to the distribution or transmission systems.
[2]Includes electricity-only and combined heat and power plants whose primary business is to sell electricity, or electricity and heat, to the public.
[3]Includes emissions from geothermal power and nonbiogenic emissions from municipal solid waste.
Btu = British thermal unit.
Note: Totals may not equal sum of components due to independent rounding. Data for 2010 are model results and may differ slightly from official EIA data reports.
Source: U.S. Energy Information Administration, AEO2012 National Energy Modeling System runs LOWNUC12.D022312A, REF2012.D020112C, and HINUC12.D022312A.

Table D8. Key results for Low Renewable Technology Cost case

Capacity, generation, and emissions	2010	2015		2025		2035	
		Reference	Low Renewable Technology Cost	Reference	Low Renewable Technology Cost	Reference	Low Renewable Technology Cost
Net summer capacity (gigawatts)							
Electric power sector[1]							
Conventional hydropower	78.03	78.55	78.76	80.14	81.34	81.25	84.36
Geothermal[2]	2.37	2.86	2.58	4.45	4.37	6.30	6.82
Municipal waste[3]	3.30	3.36	3.36	3.36	3.36	3.36	3.36
Wood and other biomass[4]	2.45	2.72	2.72	2.72	2.82	2.89	4.31
Solar thermal	0.47	1.36	1.36	1.36	1.36	1.36	1.36
Solar photovoltaic	0.38	2.02	2.05	2.30	5.12	8.18	34.27
Wind	39.05	54.46	61.41	57.77	65.59	66.85	105.87
Total	**126.06**	**145.34**	**152.25**	**152.10**	**163.96**	**170.19**	**240.35**
End-use sector[5]							
Conventional hydropower	0.33	0.33	0.33	0.33	0.33	0.33	0.33
Geothermal	0.00	0.00	0.00	0.00	0.00	0.00	0.00
Municipal waste[6]	0.35	0.35	0.35	0.35	0.35	0.35	0.35
Wood and other biomass	4.56	5.73	5.89	8.44	10.52	13.81	17.21
Solar photovoltaic	2.05	8.98	9.19	11.69	14.29	13.33	23.29
Wind	0.36	2.25	3.18	2.60	4.06	2.74	5.26
Total	**7.65**	**17.64**	**18.95**	**23.41**	**29.55**	**30.57**	**46.43**
Generation (billion kilowatthours)							
Electric power sector[1]							
Coal	1831	1562	1547	1741	1731	1834	1780
Petroleum	34	26	26	27	27	28	28
Natural gas	898	1028	1018	1006	974	1196	1037
Total fossil	**2764**	**2616**	**2591**	**2774**	**2732**	**3058**	**2846**
Conventional hydropower	255.32	295.43	296.17	305.00	310.24	310.08	321.78
Geothermal	15.67	18.68	16.42	31.53	30.91	46.54	50.89
Municipal waste[7]	16.56	14.66	14.66	14.67	14.67	14.67	14.67
Wood and other biomass[4]	11.51	21.28	24.10	63.90	68.89	49.28	78.41
Dedicated plants	10.15	10.13	12.58	13.30	12.84	10.37	23.13
Cofiring	1.36	11.15	11.52	50.60	56.05	38.92	55.28
Solar thermal	0.82	2.86	2.86	2.86	2.86	2.86	2.86
Solar photovoltaic	0.46	3.61	3.68	4.37	11.91	20.19	84.04
Wind	94.49	150.97	174.49	161.49	188.46	190.67	310.55
Total renewable	**394.82**	**507.49**	**532.38**	**583.81**	**627.94**	**634.30**	**863.20**
End-use sector[5]							
Total fossil	**106**	**123**	**123**	**180**	**177**	**262**	**260**
Conventional hydropower[8]	1.76	1.75	1.75	1.75	1.75	1.75	1.75
Geothermal	0.00	0.00	0.00	0.00	0.00	0.00	0.00
Municipal waste[6]	2.02	2.79	2.79	2.79	2.79	2.79	2.79
Wood and other biomass	26.10	33.30	34.27	52.34	67.01	96.17	118.46
Solar photovoltaic	3.21	13.88	14.20	18.22	22.41	20.91	37.06
Wind	0.47	2.88	3.92	3.36	5.09	3.56	6.78
Total renewable	**33.56**	**54.59**	**56.92**	**78.45**	**99.05**	**125.17**	**166.82**
Carbon dioxide emissions by the electric power sector (million metric tons)[1]							
Coal	1828	1539	1525	1717	1706	1809	1754
Petroleum	33	23	23	24	24	25	25
Natural gas	399	438	434	427	416	485	435
Other[9]	12	12	12	12	12	12	12
Total	**2271**	**2011**	**1993**	**2179**	**2157**	**2330**	**2225**

[1]Includes electricity-only and combined heat and power plants whose primary business is to sell electricity, or electricity and heat, to the public.
[2]Includes hydrothermal resources only (hot water and steam).
[3]Includes all municipal waste, landfill gas, and municipal sewage sludge. Incremental growth is assumed to be for landfill gas facilities. All municipal waste is included, although a portion of the municipal waste stream contains petroleum-derived plastics and other non-renewable sources.
[4]Includes projections for energy crops after 2010.
[5]Includes combined heat and power plants and electricity-only plants in the commercial and industrial sectors; and small on-site generating systems in the residential, commercial, and industrial sectors used primarily for own-use generation, but which may also sell some power to the grid. Excludes off-grid photovoltaics and other generators not connected to the distribution or transmission systems.
[6]Includes municipal waste, landfill gas, and municipal sewage sludge. All municipal waste is included, although a portion of the municipal waste stream contains petroleum-derived plastics and other non-renewable sources.
[7]Includes biogenic municipal waste, landfill gas, and municipal sewage sludge. Incremental growth is assumed to be for landfill gas facilities.
[8]Represents own-use industrial hydroelectric power.
[9]Includes emissions from geothermal power and nonbiogenic emissions from municipal solid waste.
Note: Totals may not equal sum of components due to independent rounding. Data for 2010 are model results and may differ slightly from official EIA data reports.
Source: U.S. Energy Information Administration, AEO2012 National Energy Modeling System runs REF2012.D020112C, and LORENCST12.D041312A.

Table D9. Key results for environmental cases

Net summer capacity, generation, emissions, and fuel prices	2010	2035 Reference	2035 Reference 05	2035 High EUR	2035 Low Gas Price 05	2035 Greenhouse Gas $15	2035 Greenhouse Gas $25
Capacity (gigawatts)							
Coal steam	313.4	275.2	261.6	268.3	254.2	124.3	39.1
Oil and natural gas steam	108.1	87.9	86.5	88.1	90.7	81.9	72.3
Combined cycle	198.0	272.2	276.2	273.1	285.6	298.0	312.7
Combustion turbine / diesel	137.6	171.8	173.9	181.5	178.4	154.7	142.9
Nuclear / uranium	101.2	110.9	111.1	109.3	109.3	160.5	225.0
Pumped storage	22.2	22.2	22.2	22.2	22.2	22.2	22.2
Renewable sources	126.1	170.2	174.2	159.4	165.3	227.6	257.6
Distributed generation (natural gas)	0.0	2.1	2.0	5.2	5.6	0.3	0.2
Combined heat and power[1]	29.6	77.5	78.3	80.8	81.2	96.7	105.2
Total	**1036.1**	**1190.0**	**1186.0**	**1187.8**	**1192.5**	**1166.0**	**1177.3**
Cumulative additions (gigawatts)							
Coal steam	0.0	10.9	11.1	10.2	10.6	10.2	10.3
Combined cycle	0.0	74.5	78.4	75.4	87.9	100.3	115.0
Combustion turbine / diesel	0.0	46.5	43.4	52.1	48.0	38.9	24.7
Nuclear / uranium	0.0	8.5	8.7	6.9	6.9	58.1	122.7
Renewable sources	0.0	44.5	48.5	33.7	39.6	101.9	131.9
Distributed generation	0.0	2.1	2.0	5.2	5.6	0.3	0.2
Combined heat and power[1]	0.0	47.9	48.7	51.2	51.6	67.0	75.6
Total	**0.0**	**235.0**	**240.8**	**234.6**	**250.2**	**376.8**	**480.4**
Cumulative retirements (gigawatts)	**0.0**	**88.4**	**98.3**	**90.2**	**101.1**	**254.1**	**346.6**
Generation by fuel (billion kilowatthours)							
Coal	1831	1834	1752	1748	1664	699	102
Petroleum	34	28	27	29	28	24	21
Natural gas	898	1196	1253	1347	1404	1351	1306
Nuclear / uranium	807	887	889	875	875	1268	1782
Pumped storage	5	2	2	2	2	2	2
Renewable sources	395	634	642	601	618	888	876
Distributed generation	0	4	4	16	16	0	0
Combined heat and power[1]	155	406	410	426	428	512	545
Total	**4126**	**4992**	**4979**	**5044**	**5034**	**4743**	**4634**
Emissions by the electric power sector[2]							
Carbon dioxide (million metric tons)	2271	2330	2263	2310	2238	1228	555
Sulfur dioxide (million short tons)	5.11	1.71	1.68	1.54	1.57	0.61	0.15
Nitrogen oxides (million short tons)	2.06	1.96	1.93	1.93	1.93	0.85	0.42
Mercury (short tons)	34.70	7.86	7.57	7.49	7.15	3.40	0.91
Retrofits (gigawatts)							
Scrubber	0.00	47.57	19.91	52.97	18.31	30.07	25.69
Nitrogen oxide controls							
Combustion	0.00	7.97	6.08	4.16	1.51	2.38	2.38
Selective catalytic reduction post-combustion	0.00	19.17	10.29	13.44	6.10	7.67	5.91
Selective non-catalytic reduction post-combustion	0.00	0.71	0.71	0.71	0.71	0.70	2.50
Prices to the electric power sector[2] (2010 dollars per million Btu)							
Natural gas	5.14	7.21	7.35	6.03	6.14	9.37	11.10
Coal	2.26	2.80	2.77	2.73	2.70	6.64	9.45

[1]Includes combined heat and power plants and electricity-only plants in commercial and industrial sectors. Includes small on-site generating systems in the residential, commercial, and industrial sectors used primarily for own-use generation, but which may also sell some power to the grid. Excludes off-grid photovoltaics and other generators not connected to the distribution or transmission systems.
[2]Includes electricity-only and combined heat and power plants whose primary business to sell electricity, or electricity and heat, to the public.
EUR = Estimated ultimate recovery.
Btu = British thermal unit.
Note: Totals may not equal sum of components due to independent rounding. Data for 2010 are model results and may differ slightly from official EIA data reports.
Source: U.S. Energy Information Administration, AEO2012 National Energy Modeling System runs REF2012.D020112C, REF12_R05.D030712A, HEUR12.D022212A, HEUR12_R05.D022312A, CO2FEE15.D031312A, and CO2FEE25.D031312A.

Table D10. Natural gas supply and disposition, oil and gas resource cases
(trillion cubic feet per year, unless otherwise noted)

Supply, disposition, and prices	2010	2015 Low EUR	2015 Reference	2015 High EUR	2015 High TRR	2025 Low EUR	2025 Reference	2025 High EUR	2025 High TRR	2035 Low EUR	2035 Reference	2035 High EUR	2035 High TRR
Natural gas prices													
(2010 dollars per million Btu)													
Henry Hub spot price	4.39	4.58	4.29	3.94	3.10	6.93	5.63	4.77	3.45	8.26	7.37	5.99	4.25
Average lower 48 wellhead	4.06	4.10	3.84	3.54	2.80	6.11	5.00	4.26	3.11	7.24	6.48	5.31	3.81
(2010 dollars per thousand cubic feet)													
Average lower 48 wellhead	4.16	4.19	3.94	3.62	2.87	6.25	5.12	4.36	3.19	7.41	6.64	5.43	3.90
Dry gas production[2]	21.58	22.80	23.65	24.38	26.54	24.25	26.28	27.81	30.85	26.11	27.93	30.07	34.15
Lower 48 onshore	18.66	20.62	21.48	22.20	24.37	21.48	23.64	25.24	28.60	21.19	24.97	27.19	31.66
Associated-dissolved	1.40	1.47	1.52	1.58	1.70	1.31	1.41	1.50	1.60	0.90	1.00	1.13	1.29
Non-associated	17.26	19.15	19.96	20.62	22.68	20.17	22.23	23.74	27.00	20.28	23.97	26.07	30.37
Tight gas	5.68	6.13	6.08	6.01	5.88	6.40	6.17	6.02	5.86	6.30	6.14	5.93	5.76
Shale gas	4.99	7.35	8.24	8.99	11.24	8.88	11.26	12.98	16.44	9.74	13.63	16.01	20.53
Coalbed methane	1.99	1.85	1.83	1.80	1.74	1.84	1.77	1.73	1.69	1.80	1.76	1.70	1.66
Other	4.59	3.81	3.82	3.82	3.82	3.04	3.03	3.02	3.02	2.44	2.44	2.43	2.42
Lower 48 offshore	2.56	1.89	1.88	1.88	1.87	2.51	2.38	2.31	1.99	3.12	2.72	2.64	2.27
Associated-dissolved	0.71	0.55	0.55	0.55	0.55	0.71	0.67	0.67	0.59	0.84	0.73	0.71	0.60
Non-associated	1.85	1.34	1.33	1.33	1.32	1.81	1.71	1.65	1.40	2.28	2.00	1.93	1.67
Alaska	0.36	0.29	0.29	0.29	0.29	0.25	0.25	0.25	0.25	1.80	0.23	0.23	0.22
Supplemental natural gas[3]	0.07	0.06	0.06	0.06	0.06	0.06	0.06	0.06	0.06	0.06	0.06	0.06	0.06
Net imports	2.58	1.77	1.73	1.65	1.42	-0.39	-0.79	-1.06	-1.62	-1.16	-1.36	-1.73	-2.35
Pipeline[4]	2.21	1.61	1.56	1.49	1.27	0.22	-0.13	-0.40	-0.95	-0.50	-0.70	-1.07	-1.69
Liquefied natural gas	0.37	0.17	0.16	0.16	0.15	-0.61	-0.66	-0.66	-0.66	-0.66	-0.66	-0.66	-0.66
Total supply	24.22	24.64	25.45	26.09	28.02	23.92	25.55	26.81	29.30	25.01	26.63	28.40	31.86
Consumption by sector													
Residential	4.94	4.83	4.85	4.88	4.94	4.69	4.76	4.82	4.92	4.59	4.64	4.72	4.84
Commercial	3.20	3.30	3.33	3.37	3.47	3.32	3.44	3.54	3.71	3.50	3.60	3.75	3.97
Industrial[5]	6.60	6.99	7.01	7.07	7.20	6.96	7.14	7.26	7.51	6.85	7.00	7.24	7.61
Electric power[6]	7.38	7.40	8.08	8.56	10.07	6.74	7.87	8.78	10.54	7.67	8.96	10.13	12.62
Transportation[7]	0.04	0.06	0.06	0.06	0.06	0.11	0.11	0.12	0.12	0.15	0.16	0.17	0.18
Pipeline fuel	0.63	0.66	0.67	0.67	0.69	0.64	0.66	0.67	0.69	0.72	0.67	0.69	0.74
Lease and plant fuel[8]	1.34	1.35	1.39	1.43	1.55	1.44	1.53	1.60	1.78	1.54	1.60	1.70	1.91
Total	24.13	24.59	25.39	26.04	27.97	23.90	25.53	26.79	29.28	25.01	26.63	28.40	31.87
Discrepancy[9]	0.10	0.05	0.05	0.05	0.05	0.02	0.02	0.02	0.02	-0.00	-0.00	-0.01	-0.01
Lower 48 end of year reserves	260.50	265.85	274.79	283.88	298.90	280.90	299.77	318.24	347.21	291.70	311.58	333.43	371.70

[1] Represents lower 48 onshore and offshore supplies.
[2] Marketed production (wet) minus extraction losses.
[3] Synthetic natural gas, propane air, coke oven gas, refinery gas, biomass gas, air injected for Btu stabilization, and manufactured gas commingled and distributed with natural gas.
[4] Includes any natural gas regasified in the Bahamas and transported via pipeline to Florida.
[5] Includes energy for combined heat and power plants, except those whose primary business is to sell electricity, or electricity and heat, to the public.
[6] Includes consumption of energy by electricity-only and combined heat and power plants whose primary business is to sell electricity, or electricity and heat, to the public. Includes small power producers and exempt wholesale generators.
[7] Natural gas used as a vehicle fuel.
[8] Represents natural gas used in field gathering and processing plant machinery.
[9] Balancing item. Natural gas lost as a result of converting flow data measured at varying temperatures and pressures to a standard temperature and pressure and the merger of different data reporting systems which vary in scope, format, definition, and respondent type. In addition, 2010 values include net storage injections.
EUR = Estimated ultimate recovery.
TRR = Technically recoverable resources.
Note: Totals may not equal sum of components due to independent rounding. Data for 2010 are model results and may differ slightly from official EIA data reports.
Sources: 2010 supply values; lease, plant, and pipeline fuel consumption; and wellhead price: U.S. Energy Information Administration (EIA), *Natural Gas Monthly*, DOE/EIA-0130(2011/07) (Washington, DC, July 2011). Other 2010 consumption based on: EIA, *Annual Energy Review 2010*, DOE/EIA-0384(2010) (Washington, DC, October 2011).
Projections: EIA, AEO2012 National Energy Modeling System runs LEUR12.D022212A, REF2012.D020112C, HEUR12.D022212A., and HTRR12.D050412A

Table D11. Liquid fuels supply and disposition, oil and gas resource cases
(million barrels per day, unless otherwise noted)

Supply, disposition, and prices	2010	2015				2025				2035			
		Low EUR	Reference	High EUR	High TRR	Low EUR	Reference	High EUR	High TRR	Low EUR	Reference	High EUR	High TRR
Prices													
(2010 dollars per barrel)													
Low sulfur light crude oil[1]	79.39	117.84	116.91	116.11	113.74	134.54	132.56	130.60	127.97	146.78	144.98	143.27	139.78
Imported crude oil[1]	75.87	114.90	113.97	113.17	110.80	123.99	121.21	118.63	115.77	135.38	132.95	131.20	127.55
Crude oil supply													
Domestic production[2]	5.47	5.91	6.15	6.38	7.09	5.82	6.40	6.95	7.69	5.49	5.99	6.62	7.76
Alaska	0.60	0.46	0.46	0.46	0.46	0.40	0.40	0.40	0.34	0.27	0.27	0.27	0.38
Lower 48 onshore	3.21	3.85	4.09	4.32	5.04	3.77	4.43	5.00	5.98	3.22	3.99	4.67	5.97
Lower 48 offshore	1.67	1.60	1.60	1.60	1.59	1.65	1.57	1.54	1.36	2.00	1.74	1.69	1.41
Net imports	9.17	8.80	8.52	8.28	7.57	7.87	7.24	6.68	5.89	8.12	7.52	6.90	5.65
Other crude oil supply	0.08	0.00	0.00	0.00	0.00	0.00	0.00	0.00	0.00	0.00	0.00	0.00	0.00
Total crude oil supply	14.72	14.71	14.67	14.65	14.66	13.69	13.64	13.63	13.58	13.61	13.51	13.52	13.40
Other petroleum supply	3.50	3.17	3.25	3.33	3.40	3.66	3.80	3.94	4.13	3.40	3.52	3.73	4.02
Natural gas plant liquids	2.07	2.43	2.56	2.68	2.97	2.67	3.01	3.27	3.91	2.66	3.01	3.33	4.04
Net product imports[3]	0.39	-0.20	-0.25	-0.30	-0.54	0.08	-0.12	-0.24	-0.69	-0.12	-0.34	-0.43	-0.89
Refinery processing gain[4]	1.07	0.94	0.95	0.94	0.97	0.90	0.91	0.91	0.91	0.86	0.85	0.83	0.86
Product stock withdrawal	-0.03	0.00	0.00	0.00	0.00	0.00	0.00	0.00	0.00	0.00	0.00	0.00	0.00
Other non-petroleum supply	1.00	1.22	1.22	1.22	1.22	1.87	1.86	1.86	1.85	2.91	2.96	2.87	2.81
From renewable sources[5]	0.87	1.05	1.05	1.05	1.05	1.48	1.48	1.48	1.49	2.33	2.37	2.32	2.27
From non-renewable sources[6]	0.13	0.17	0.17	0.17	0.16	0.38	0.38	0.37	0.36	0.58	0.58	0.55	0.53
Total primary supply[7]	19.22	19.10	19.14	19.20	19.27	19.21	19.29	19.42	19.56	19.91	19.99	20.11	20.23
Refined petroleum products supplied													
Residential and commercial	1.12	1.00	1.00	1.00	1.00	0.93	0.94	0.94	0.95	0.90	0.91	0.91	0.92
Industrial[8]	4.31	4.17	4.17	4.19	4.19	4.38	4.41	4.44	4.46	4.41	4.44	4.46	4.47
Transportation	13.82	13.78	13.80	13.82	13.88	13.66	13.71	13.79	13.88	14.37	14.41	14.49	14.57
Electric power[9]	0.17	0.13	0.13	0.13	0.13	0.14	0.14	0.14	0.14	0.14	0.14	0.15	0.14
Total	19.17	19.07	19.10	19.14	19.21	19.11	19.20	19.31	19.44	19.83	19.90	20.01	20.10
Discrepancy[10]	0.05	0.03	0.05	0.06	0.07	0.10	0.10	0.11	0.12	0.09	0.09	0.11	0.12
Lower 48 end of year reserves (billion barrels)[2]	18.33	19.39	20.55	21.66	23.49	21.36	23.64	25.77	27.83	22.68	24.23	26.27	29.06

[1]Weighted average price delivered to U.S. refiners.
[2]Includes lease condensate.
[3]Includes net imports of finished petroleum products, unfinished oils, other hydrocarbons, alcohols, ethers, and blending components.
[4]The volumetric amount by which total output is greater than input due to the processing of crude oil into products which, in total, have a lower specific gravity than the crude oil processed.
[5]Includes ethanol (including imports), biodiesel (including imports), pyrolysis oils, biomass-derived Fischer-Tropsch liquids, and renewable feedstocks for the production of green diesel and gasoline.
[6]Includes alcohols, ethers, domestic sources of blending components, other hydrocarbons, natural gas converted to liquid fuel, and coal converted to liquid fuel.
[7]Total crude supply plus natural gas plant liquids, other inputs, refinery processing gain, and net product imports.
[8]Includes consumption for combined heat and power, which produces electricity and other useful thermal energy.
[9]Includes consumption of energy by electricity-only and combined heat and power plants whose primary business is to sell electricity, or electricity and heat, to the public. Includes small power producers and exempt wholesale generators.
[10]Balancing item. Includes unaccounted for supply, losses and gains.
EUR = Estimated ultimate recovery.
TRR = Technically recoverable resources.
Note: Totals may not equal sum of components due to independent rounding. Data for 2010 are model results and may differ slightly from official EIA data reports.
Sources: 2010 product supplied data and imported crude oil price based on: U.S. Energy Information Administration (EIA), *Annual Energy Review 2010*, DOE/EIA-0384(2010) (Washington, DC, October 2011). 2010 imported low sulfur light crude oil price: EIA, Form EIA-856, "Monthly Foreign Crude Oil Acquisition Report." Other 2010 data: EIA, *Petroleum Supply Annual 2010*, DOE/EIA-0340(2010)/1 (Washington, DC, July 2011). Projections: EIA, AEO2012 National Energy Modeling System runs LEUR12.D022212A, REF2012.D020112C, HEUR12.D022212A, and HTRR.D050412A.

Table D12. Volumetric and mass representations of liquid fuels production cases
(volume in million barrels per day, mass in billion tons, unless otherwise noted)

Supply and disposition	2000 Volume	2000 Mass	2011 PMM Volume	2011 LFMM Volume	2011 LFMM Mass	2035 PMM Volume	2035 LFMM Volume	2035 LFMM Mass
Primary feedstocks[1]								
Crude oil[2]	15.36	0.83	15.37	14.87	0.83	14.05	13.73	0.78
Natural gas[3]	0.00	0.00	0.00	0.00	0.00	0.00	2.95	0.03
Natural gas plant liquids[4]	1.91	0.07	2.16	1.21	0.09	3.01	0.30	0.11
Coal[5]	0.00	0.00	0.00	0.00	0.00	0.28	0.27	0.09
Biomass[6]	0.10	0.01	0.92	13.99	0.14	2.37	14.64	0.31
Total primary feedstocks	17.37	0.91	18.45	--	1.06	19.71	--	1.32
Refined products[1]								
Residual fuel oil	0.91	0.04	0.47	0.52	0.03	0.58	0.58	0.03
Middle distillates[7]	2.55	0.26	3.21	5.90	0.30	3.73	6.69	0.34
Biodiesel[8]	0.00	0.00	0.05	0.02	0.00	0.13	0.01	0.00
Gasoline blendstocks[9]	8.37	0.37	7.84	8.57	0.41	6.94	7.73	0.37
Ethanol[10]	0.10	0.00	0.86	0.95	0.05	1.65	1.61	0.08
Chemicals[11]	2.62	0.10	2.11	2.17	0.05	2.10	3.20	0.08
Solid products[12]	--	0.05	--	--	0.07	--	--	0.08
Fuel consumption and other[13]	--	0.10	--	0.00	0.15	0.00	0.00	0.34
Total refined products	14.55	0.91	14.54	18.13	1.06	15.13	19.82	1.32
End use products								
Residual fuel oil	0.91	0.04	0.47	0.50	0.03	0.58	0.57	0.03
Heating oil[14]	1.17	0.03	0.62	0.53	0.03	0.37	0.37	0.02
Diesel fuel[15]	2.55	0.16	3.27	3.40	0.17	4.11	4.19	0.21
Jet fuel	1.73	0.08	1.44	1.51	0.08	1.61	1.67	0.08
Motor Gasoline[16]	8.47	0.38	8.76	9.29	0.44	8.09	8.32	0.40
E85[17]	0.00	0.00	0.00	0.00	0.00	0.83	0.84	0.04
Liquefied petroleum gases	2.43	0.02	2.26	0.46	0.01	2.21	0.74	0.01
Chemical feedstocks[18]	0.40	0.07	0.33	1.70	0.06	0.57	2.47	0.06
Agricultural products[19]	--	0.00	--	--	0.05	--	--	0.06
Biomass heat and power[20]	--	0.00	--	--	0.00	--	--	0.02
Other[21]	1.91	0.04	1.89	0.34	0.02	1.79	0.36	0.02
Total end use products	19.57	0.82	19.04	17.73	0.89	20.16	19.53	0.95

[1]Includes domestic production and net imports.
[2]Includes unfinished oils and lease condensate.
[3]Natural gas that remains after the liquefiable hydrocarbon portion has been removed from the gas stream at lease and/or plant separation facilities. Volume in billion cubic feet per day.
[4]Liquids in the natural gas production stream that stay in gaseous form at the surface and are separated at a gas processing plant. Once extracted, these liquids are separated into distinct products, or "fractions", such as propane, butane, and ethane.
[5]Coal input to the coal-to-liquids process. Volume in million barrels per day fuel oil equivalent.
[6]Biological material from living, or recently living organisms such as grain crops, sugars, cellulosic biomass, or renewable oils. Volume in million barrels per day fuel oil equivalent.
[7]Includes all fuels that meet ASTM D396 and D975 (#4 and lighter) and D1655/D6615, including those derived from fossil and renewable feedstock.
[8]Methyl ester based fuel produced from fatty acids in renewable oils.
[9]Includes all blendstocks that meet ASTM D4814, including those derived from fossil and renewable feedstock.
[10]Includes denaturant.
[11]Includes liquefied petroleum gases and petrochemical feestocks.
[12]Includes petroleum coke, distillers grains, sulfur, and asphalt sales.
[13]Includes fuels burned for internal use, heat and power sales, solid waste, and process emissions.
[14]A distillate fuel oil for use in atomizing type burners for domestic heating or for use in medium capacity commercial-industrial burner units.
[15]For on-road use.
[16]Includes ethanol and ethers blended into motor gasoline.
[17]E85 refers to a blend of 85 percent ethanol (renewable) and 15 percent motor gasoline (nonrenewable). To address cold starting issues, the percentage of ethanol varies seasonally. The annual average ethanol content of 74 percent is used for this forecast.
[18]Includes petrochemical feedstocks and chemicals from Fischer-Tropsch processes, such as coal-to-liquids, biomass-to-liquids, and natural gas-to-liquids.
[19]Non-liquid co-products for use in the agricultural sector. Includes dried distiller grains.
[20]Heat and power generated from the burning of residual biomass.
[21]Includes petroleum coke, asphalt, road oil, and still gas.
-- = Not applicable.
PMM = Petroleum market module.
LFMM = Liquid fuels market module.
Note: PMM and LFMM projections do not exactly match due to differences in accounting for additional materials and updated refinery stream representations. Totals may not equal sum of components due to independent rounding. Data for 2000 are model results and may differ slightly from official EIA data reports.
Sources: 2000 product supplied data and imported crude oil price based on: U.S. Energy Information Administration (EIA), *Annual Energy Review 2010*, DOE/EIA-0384(2010) (Washington, DC, October 2011). 2000 crude oil production: EIA, *Petroleum Supply Annual 2001*, DOE/EIA-0340(2001)/1 (Washington, DC, June 2002). Other 2000 data: EIA, *Petroleum Supply Annual 2000*, DOE/EIA-0340(2000)/1 (Washington, DC, June 2001). **Projections:** EIA, AEO2012 National Energy Modeling System runs REF2012.D020112C, and REF_LFMM.D050312A.

Table D13. Key results for No GHG Concern case
(million short tons per year, unless otherwise noted)

Supply, disposition, and prices	2010	2015 Reference	2015 No GHG Concern	2025 Reference	2025 No GHG Concern	2035 Reference	2035 No GHG Concern
Production[1]	1084	993	1016	1118	1169	1212	1339
Appalachia	336	300	301	271	263	291	301
Interior	156	151	156	163	173	198	216
West	592	542	558	684	733	722	822
Waste coal supplied[2]	14	15	18	16	16	19	24
Net imports[3]	-64	-95	-97	-71	-57	-94	-88
Total supply[4]	1034	914	936	1064	1128	1138	1276
Consumption by sector							
Residential and commercial	3	3	3	3	3	3	3
Coke plants	21	22	22	19	19	17	17
Other industrial[5]	52	50	50	52	52	53	53
Coal-to-liquids heat and power	0	0	0	19	47	34	90
Coal-to-liquids liquids production	0	0	0	18	44	32	85
Electric power[6]	975	839	861	952	962	998	1028
Total coal use	1051	914	936	1063	1127	1137	1276
Average minemouth price[7]							
(2010 dollars per short ton)	35.61	42.08	41.83	44.05	43.14	50.52	49.88
(2010 dollars per million Btu)	1.76	2.08	2.07	2.23	2.21	2.56	2.54
Delivered prices[8]							
(2010 dollars per short ton)							
Coke plants	153.59	189.11	188.05	212.18	212.06	238.32	237.86
Other industrial[5]	59.28	70.14	70.04	72.77	73.23	78.53	79.88
Coal to liquids	- -	18.65	18.62	39.03	36.06	41.54	43.46
Electric power[6]							
(2010 dollars per short ton)	44.27	45.17	44.94	48.13	48.40	53.31	55.05
(2010 dollars per million Btu)	2.26	2.35	2.34	2.54	2.55	2.80	2.87
Average	47.17	49.95	49.60	51.90	51.28	56.48	56.89
Exports[9]	120.41	140.89	140.22	163.43	163.15	177.66	176.61
Cumulative electricity generating capacity additions (gigawatts)[10]							
Coal	0.0	9.1	9.1	13.5	18.4	16.6	39.9
Conventional	0.0	8.7	8.7	8.7	9.1	9.4	21.8
Advanced without sequestration	0.0	0.6	0.6	0.6	0.7	0.6	2.0
Advanced with sequestration	0.0	0.0	0.0	0.9	0.9	0.9	0.9
End-use generators[11]	0.0	-0.1	-0.1	3.4	7.8	5.6	15.2
Petroleum	0.0	0.1	0.1	0.1	0.1	0.1	0.1
Natural gas	0.0	29.1	28.0	63.3	61.4	141.6	128.9
Nuclear / uranium	0.0	1.1	1.1	6.8	6.8	8.5	7.4
Renewables[12]	0.0	29.6	29.3	42.2	41.3	67.4	58.2
Other	0.0	0.8	0.8	0.8	0.8	0.8	0.8
Total	0.0	69.8	68.4	126.7	128.8	235.0	235.3
Liquids from coal (million barrels per day)	0.00	0.00	0.00	0.17	0.38	0.28	0.73

[1]Includes anthracite, bituminous coal, subbituminous coal, and lignite.
[2]Includes waste coal consumed by the electric power and industrial sectors. Waste coal supplied is counted as a supply-side item to balance the same amount of waste coal included in the consumption data.
[3]Excludes imports to Puerto Rico and the U.S. Virgin Islands.
[4]Production plus waste coal supplied plus net imports.
[5]Includes consumption for combined heat and power plants, except those plants whose primary business is to sell electricity, or electricity and heat, to the public. Excludes all coal use in the coal-to-liquids process.
[6]Includes all electricity-only and combined heat and power plants whose primary business is to sell electricity, or electricity and heat, to the public.
[7]Includes reported prices for both open market and captive mines.
[8]Prices weighted by consumption tonnage; weighted average excludes residential and commercial prices, and export free-alongside-ship (f.a.s.) prices.
[9]F.a.s. price at U.S. port of exit.
[10]Cumulative additions after December 31, 2010. Includes all additions of electricity only and combined heat and power plants projected for the electric power, industrial, and commercial sectors.
[11]Includes combined heat and power plants and electricity-only plants in the commercial and industrial sectors; and small on-site generating systems in the residential, commercial, and industrial sectors used primarily for own-use generation, but which may also sell some power to the grid.
[12]Includes conventional hydroelectric, geothermal, wood, wood waste, municipal waste, landfill gas, other biomass, solar, and wind power. Facilities co-firing biomass and coal are classified as coal.
- - = Not applicable.
Btu = British thermal unit.
GHG = Greenhouse gas.
Note: Totals may not equal sum of components due to independent rounding. Data for 2010 are model results and may differ slightly from official EIA data reports.
Sources: 2010 data based on: U.S. Energy Information Administration (EIA), *Annual Coal Report 2010*, DOE/EIA-0584(2010) (Washington, DC, November 2011); EIA, *Quarterly Coal Report, October-December 2010*, DOE/EIA-0121(2010/4Q) (Washington, DC, May 2011); and EIA, AEO2012 National Energy Modeling System run REF2012.D020112C.
Projections: EIA, AEO2012 National Energy Modeling System runs REF2012.D020112C and NOGHGCONCERN.D031212A.

Results from side cases

Table D14. Key results for coal cost cases
(million short tons per year, unless otherwise noted)

Supply, disposition, and prices	2010	2020			2035			Annual growth 2010-2035 (percent)		
		Low Coal Cost	Reference	High Coal Cost	Low Coal Cost	Reference	High Coal Cost	Low Coal Cost	Reference	High Coal Cost
Production[1]	**1084**	**1096**	**1034**	**962**	**1336**	**1212**	**946**	**0.8%**	**0.4%**	**-0.5%**
Appalachia	336	281	262	253	309	291	261	-0.3%	-0.6%	-1.0%
Interior	156	168	159	159	194	198	202	0.9%	1.0%	1.0%
West	592	647	613	550	833	722	483	1.4%	0.8%	-0.8%
Waste coal supplied[2]	14	13	15	18	14	19	40	0.2%	1.4%	4.4%
Net imports[3]	-64	-78	-67	-73	-87	-94	-59	1.2%	1.5%	-0.3%
Total supply[4]	**1034**	**1031**	**982**	**907**	**1263**	**1138**	**927**	**0.8%**	**0.4%**	**-0.4%**
Consumption by sector										
Residential and commercial	3	3	3	3	3	3	3	-0.2%	-0.3%	-0.4%
Coke plants	21	19	18	18	17	17	16	-0.8%	-1.0%	-1.1%
Other industrial[5]	52	51	51	50	53	53	52	0.1%	0.0%	-0.0%
Coal-to-liquids heat and power	0	15	13	12	57	34	29	--	--	--
Coal-to-liquids liquids production	0	14	12	11	54	32	27	--	--	--
Electric power[6]	975	929	885	812	1079	998	800	0.4%	0.1%	-0.8%
Total coal use	**1051**	**1031**	**982**	**907**	**1263**	**1137**	**926**	**0.7%**	**0.3%**	**-0.5%**
Average minemouth price[7]										
(2010 dollars per short ton)	35.61	32.70	40.96	52.91	25.80	50.52	106.78	-1.3%	1.4%	4.5%
(2010 dollars per million Btu)	1.76	1.64	2.06	2.65	1.31	2.56	5.24	-1.2%	1.5%	4.5%
Delivered prices[8]										
(2010 dollars per short ton)										
Coke plants	153.59	165.27	198.45	239.32	136.73	238.32	413.77	-0.5%	1.8%	4.0%
Other industrial[5]	59.28	60.23	70.89	84.14	50.11	78.53	127.31	-0.7%	1.1%	3.1%
Coal to liquids	--	34.43	40.67	49.20	25.22	41.54	68.76	--	--	--
Electric power[6]										
(2010 dollars per short ton)	44.27	39.19	45.98	55.09	34.16	53.31	94.16	-1.0%	0.7%	3.1%
(2010 dollars per million Btu)	2.26	2.04	2.41	2.89	1.77	2.80	4.79	-1.0%	0.9%	3.0%
Average	**47.17**	**42.38**	**49.99**	**60.26**	**35.44**	**56.48**	**100.09**	**-1.1%**	**0.7%**	**3.1%**
Exports[9]	120.41	121.34	155.03	187.16	96.75	177.66	338.54	-0.9%	1.6%	4.2%
Cumulative electricity generating capacity additions (gigawatts)[10]										
Coal	0.0	12.9	12.5	12.2	30.7	16.6	14.5	--	--	--
Conventional	0.0	8.7	8.7	8.7	19.8	9.4	8.7	--	--	--
Advanced without sequestration	0.0	0.6	0.6	0.6	1.0	0.6	0.6	--	--	--
Advanced with sequestration	0.0	0.9	0.9	0.9	0.9	0.9	0.9	--	--	--
End-use generators[11]	0.0	2.7	2.3	2.1	9.0	5.6	4.3	--	--	--
Petroleum	0.0	0.1	0.1	0.1	0.1	0.1	0.1	--	--	--
Natural gas	0.0	36.6	39.7	43.1	128.1	141.6	131.7	--	--	--
Nuclear / uranium	0.0	6.8	6.8	6.8	7.3	8.5	7.7	--	--	--
Renewables[12]	0.0	34.2	34.5	41.0	67.9	67.4	65.9	--	--	--
Other	0.0	0.8	0.8	0.8	0.8	0.8	0.8	--	--	--
Total	**0.0**	**91.3**	**94.3**	**104.0**	**234.9**	**235.0**	**220.6**	--	--	--
Liquids from coal (million barrels per day)	0.00	0.14	0.12	0.11	0.45	0.28	0.21	--	--	--

Table D14. Key results for coal cost cases (continued)
(million short tons per year, unless otherwise noted)

Supply, disposition, and prices	2010	2020			2035			Annual growth 2010-2035 (percent)		
		Low Coal Cost	Reference	High Coal Cost	Low Coal Cost	Reference	High Coal Cost	Low Coal Cost	Reference	High Coal Cost
Cost indices (constant dollar index, 2010=1.000)										
Transportation rate multipliers										
Eastern railroads	1.000	0.970	1.067	1.170	0.780	1.044	1.300	-1.0%	0.2%	1.1%
Western railroads	1.000	0.870	0.963	1.050	0.750	0.999	1.250	-1.1%	-0.0%	0.9%
Mine equipment costs										
Underground	1.000	0.914	1.000	1.094	0.786	1.000	1.270	-1.0%	0.0%	1.0%
Surface	1.000	0.914	1.000	1.094	0.786	1.000	1.270	-1.0%	0.0%	1.0%
Other mine supply costs										
East of the Mississippi: all mines	1.000	0.914	1.000	1.094	0.786	1.000	1.270	-1.0%	0.0%	1.0%
West of the Mississippi: underground	1.000	0.914	1.000	1.094	0.786	1.000	1.270	-1.0%	0.0%	1.0%
West of the Mississippi: surface	1.000	0.914	1.000	1.094	0.786	1.000	1.270	-1.0%	0.0%	1.0%
Coal mining labor productivity (short tons per miner per hour)	5.55	6.29	4.92	3.67	8.06	3.88	1.68	1.5%	-1.4%	-4.7%
Average coal miner wage (2010 dollars per year)	77,466	84,135	92,285	100,436	78,164	99,537	124,954	0.0%	1.0%	1.9%

[1] Includes anthracite, bituminous coal, subbituminous coal, and lignite.
[2] Includes waste coal consumed by the electric power and industrial sectors. Waste coal supplied is counted as a supply-side item to balance the same amount of waste coal included in the consumption data.
[3] Excludes imports to Puerto Rico and the U.S. Virgin Islands.
[4] Production plus waste coal supplied plus net imports.
[5] Includes consumption for combined heat and power plants, except those plants whose primary business is to sell electricity, or electricity and heat, to the public. Excludes all coal use in the coal to liquids process.
[6] Includes all electricity-only and combined heat and power plants whose primary business is to sell electricity, or electricity and heat, to the public.
[7] Includes reported prices for both open market and captive mines.
[8] Prices weighted by consumption tonnage; weighted average excludes residential and commercial prices, and export free-alongside-ship (f.a.s.) prices.
[9] F.a.s. price at U.S. port of exit.
[10] Cumulative additions after December 31, 2010. Includes all additions of electricity only and combined heat and power plants projected for the electric power, industrial, and commercial sectors.
[11] Includes combined heat and power plants and electricity-only plants in the commercial and industrial sectors; and small on-site generating systems in the residential, commercial, and industrial sectors used primarily for own-use generation, but which may also sell some power to the grid.
[12] Includes conventional hydroelectric, geothermal, wood, wood waste, municipal waste, landfill gas, other biomass, solar, and wind power. Facilities co-firing biomass and coal are classified as coal.
- - = Not applicable.
Btu = British thermal unit.
Note: Totals may not equal sum of components due to independent rounding. Data for 2010 are model results and may differ slightly from official EIA data reports.
Sources: 2010 data based on: U.S. Energy Information Administration (EIA), *Annual Coal Report 2010*, DOE/EIA-0584(2010) (Washington, DC, November 2011); EIA, *Quarterly Coal Report, October-December 2010*, DOE/EIA-0121(2010/4Q) (Washington, DC, May 2011); U.S. Department of Labor, Bureau of Labor Statistics, Average Hourly Earnings of Production Workers: Coal Mining, Series ID: ceu1021210008; and EIA, AEO2012 National Energy Modeling System run REF2012.D020112C. **Projections:** EIA, AEO2012 National Energy Modeling System runs LCCST12.D031312A, REF2012.D020112C, and HCCST12.D031312A.

This page intentionally left blank

Appendix E
NEMS overview and brief description of cases

The National Energy Modeling System

Projections in the *Annual Energy Outlook 2012 (AEO2012)* are generated using the National Energy Modeling System (NEMS) [142], developed and maintained by the Office of Energy Analysis of the U.S. Energy Information Administration (EIA). In addition to its use in developing the *Annual Energy Outlook (AEO)* projections, NEMS is also used to complete analytical studies for the U.S. Congress, the Executive Office of the President, other offices within the U.S. Department of Energy (DOE), and other Federal agencies. NEMS is also used by other nongovernment groups, such as the Electric Power Research Institute, Duke University, Georgia Institute of Technology, and OnLocation, Inc. In addition, the *AEO* projections are used by analysts and planners in other government agencies and nongovernment organizations.

The projections in NEMS are developed with the use of a market-based approach, subject to regulations and standards. For each fuel and consuming sector, NEMS balances energy supply and demand, accounting for economic competition among the various energy fuels and sources. The time horizon of NEMS extends to 2035. To represent regional differences in energy markets, the component modules of NEMS function at the regional level: the nine Census divisions for the end-use demand modules; production regions specific to oil, natural gas, and coal supply and distribution; 22 regions and subregions of the North American Electric Reliability Corporation for electricity; and the five Petroleum Administration for Defense Districts (PADDs) for refineries.

NEMS is organized and implemented as a modular system. The modules represent each of the fuel supply markets, conversion sectors, and end-use consumption sectors of the energy system. The modular design also permits the use of the methodology and level of detail most appropriate for each energy sector. NEMS executes each of the component modules to solve for prices of energy delivered to end users and the quantities consumed, by product, region, and sector. The delivered fuel prices encompass all the activities necessary to produce, import, and transport fuels to end users. The information flows also include other data on such areas as economic activity, domestic production, and international petroleum supply. NEMS calls each supply, conversion, and end-use demand module in sequence until the delivered prices of energy and the quantities demanded have converged within tolerance, thus achieving an economic equilibrium of supply and demand in the consuming sectors. A solution is reached annually through the projection horizon. Other variables, such as petroleum product imports, crude oil imports, and several macroeconomic indicators, also are evaluated for convergence.

Each NEMS component represents the impacts and costs of legislation and environmental regulations that affect that sector. NEMS accounts for all combustion-related carbon dioxide (CO_2) emissions, as well as emissions of sulfur dioxide, nitrogen oxides, and mercury from the electricity generation sector.

The version of NEMS used for *AEO2012* generally represents current legislation and environmental regulations, including recent government actions, for which implementing regulations were available as of December 31, 2011, such as: the Mercury and Air Toxics Standards (MATS) [143] issued by the U.S. Environmental Protection Agency (EPA) in December 2011; the Cross-State Air Pollution Rule (CSAPR) [144] as finalized by the EPA in July 2011; the new fuel efficiency standards for medium- and heavy-duty vehicles (HDVs) published by the EPA and the National Highway Traffic Safety Administration (NHTSA) in September 2011 [145]; California's cap-and-trade program authorized by Assembly Bill (AB) 32, the Global Warming Solutions Act of 2006 [146]; the EPA policy memo regarding compliance of surface coal mining operations in Appalachia [147], issued on July 21, 2011; and the American Recovery and Reinvestment Act of 2009 (ARRA2009) [148], which was enacted in mid-February 2009.

The potential impacts of proposed Federal and State legislation, regulations, or standards—or of sections of legislation that have been enacted but require funds or implementing regulations that have not been provided or specified—are not reflected in NEMS. However, many pending provisions are examined in alternative cases included in *AEO2012* or in other analyses completed by EIA.

In general, the historical data presented with the *AEO2012* projections are based on EIA's *Annual Energy Review 2010*, published in October 2011 [149]; however, data were taken from multiple sources. In some cases, only partial or preliminary data were available for 2010. Historical numbers are presented for comparison only and may be estimates. Source documents should be consulted for the official data values. Footnotes to the *AEO2012* appendix tables indicate the definitions and sources of historical data.

Where possible, the *AEO2012* projections for 2011 and 2012 incorporate short-term projections from EIA's December 2011 *Short-Term Energy Outlook (STEO)*. For short-term energy projections, readers are referred to monthly updates of the *STEO* [150].

Component modules

The component modules of NEMS represent the individual supply, demand, and conversion sectors of domestic energy markets and also include international and macroeconomic modules. In general, the modules interact through values representing prices or expenditures for energy delivered to the consuming sectors and the quantities of end-use energy consumption.

Macroeconomic Activity Module

The Macroeconomic Activity Module (MAM) provides a set of macroeconomic drivers to the energy modules and receives energy-related indicators from the NEMS energy components as part of the macroeconomic feedback mechanism within NEMS.

Key macroeconomic variables used in the energy modules include gross domestic product (GDP), disposable income, value of industrial shipments, new housing starts, sales of new light-duty vehicles (LDVs), interest rates, and employment. Key energy indicators fed back to the MAM include aggregate energy prices and costs. The MAM uses the following models from IHS Global Insight: Macroeconomic Model of the U.S. Economy, National Industry Model, and National Employment Model. In addition, EIA has constructed a Regional Economic and Industry Model to project regional economic drivers, and a Commercial Floorspace Model to project 13 floorspace types in 9 Census divisions. The accounting framework for industrial value of shipments uses the North American Industry Classification System (NAICS).

International Energy Module

The International Energy Module (IEM) uses assumptions of economic growth and expectations of future U.S. and world petroleum and other liquids production and consumption, by year, to project the interaction of U.S. and international petroleum and other liquids markets. The IEM computes world oil prices, provides a world crude-like liquids supply curve, generates a worldwide oil supply/demand balance for each year of the projection period, and computes initial estimates of crude oil and light and heavy petroleum product imports to the United States by PADD regions. The supply-curve calculations are based on historical market data and a world oil supply/demand balance, which is developed from reduced-form models of international petroleum and other liquids supply and demand, current investment trends in exploration and development, and long-term resource economics by country and territory. The oil production estimates include both conventional and other liquids supply recovery technologies.

In interacting with the rest of NEMS, the IEM changes the oil price—which is defined as the price of light, low-sulfur crude oil delivered to Cushing, Oklahoma (PADD 2)—in response to changes in expected production and consumption of crude oil and other liquids in the United States.

Residential and Commercial Demand Modules

The Residential Demand Module projects energy consumption in the residential sector by Census division, housing type, and end use, based on delivered energy prices, the menu of equipment available, the availability of renewable sources of energy, and changes in the housing stock. The Commercial Demand Module projects energy consumption in the commercial sector by Census division, building type, and category of end use, based on delivered prices of energy, availability of renewable sources of energy, and changes in commercial floorspace.

Both modules estimate the equipment stock for the major end-use services, incorporating assessments of advanced technologies, representations of renewable energy technologies, and the effects of both building shell and appliance standards. The modules also include projections of distributed generation. The Commercial Demand Module also incorporates combined heat and power (CHP) technology. Both modules incorporate changes to "normal" heating and cooling degree-days by Census division, based on a 10-year average and on State-level population projections. The Residential Demand Module projects an increase in the average square footage of both new construction and existing structures, based on trends in new construction and remodeling.

Industrial Demand Module

The Industrial Demand Module (IDM) projects the consumption of energy for heat and power, as well as the consumption of feedstocks and raw materials in each of 21 industry groups, subject to the delivered prices of energy and macroeconomic estimates of employment and the value of shipments for each industry. As noted in the description of the MAM, the representation of industrial activity in NEMS is based on the NAICS. The industries are classified into three groups—energy-intensive manufacturing, non-energy-intensive manufacturing, and nonmanufacturing. Of the eight energy-intensive manufacturing industries, seven are modeled in the IDM, including energy-consuming components for boiler/steam/cogeneration, buildings, and process/assembly use of energy. Energy demand for petroleum refining (the eighth energy-intensive manufacturing industry) is modeled in the Petroleum Market Module (PMM), as described below, but the projected consumption is reported under the industrial totals.

There are several updates and upgrades in the representations of select industries. The base year for the bulk chemical industry has been updated to 2006 in keeping with updates to EIA's 2006 Manufacturing Energy Consumption Survey [151]. *AEO2012* also includes an upgraded representation for the cement and lime industries and agriculture. Instead of assuming that technological development for a particular process occurs on a predetermined (exogenous) path based on engineering judgment, these upgrades allow IDM technological change to be modeled endogenously, while using more detailed process representation. The upgrade allows for technological change, and therefore energy intensity, to respond to economic, regulatory, and other conditions. For subsequent *AEO*s, other industries represented in the IDM projections will be similarly upgraded.

A generalized representation of CHP is included. A revised methodology for CHP systems, implemented for *AEO2012*, simulates the utilization of installed CHP systems based on historical utilization rates and is driven by end-use electricity demand. To evaluate the economic benefits of additional CHP capacity, the model also includes an updated appraisal incorporating historical rather than assumed capacity factors and regional acceptance rates for new CHP facilities. The evaluation of CHP systems still uses a discount rate, which is equal to the projected 10-year Treasury bill rate plus a risk premium.

Transportation Demand Module

The Transportation Demand Module projects consumption of energy in the transportation sector—including petroleum products, electricity, methanol, ethanol, compressed natural gas (CNG), and hydrogen—by transportation mode, subject to delivered energy prices and macroeconomic variables such as disposable personal income, GDP, population, interest rates, and industrial shipments. The Transportation Demand Module includes legislation and regulations, such as the Energy Policy Act of 2005 (EPACT2005), the Energy Improvement and Extension Act of 2008 (EIEA2008), and the ARRA2009, which contain tax credits for the purchase of alternatively fueled vehicles. Fleet vehicles are also modeled, allowing for analysis of legislative proposals specific to those markets. Representations of LDV Corporate Average Fuel Economy (CAFE) and greenhouse gas (GHG) emissions standards, HDV fuel consumption and GHG emissions standards, and biofuels consumption in the module reflect standards enacted by NHTSA and the EPA, as well as provisions in the Energy Independence and Security Act of 2007 (EISA2007).

The air transportation component of the Transportation Demand Module explicitly represents air travel in domestic and foreign markets and includes the industry practice of parking aircraft in both domestic and international markets to reduce operating costs, as well as the movement of aging aircraft from passenger to cargo markets. For passenger travel and air freight shipments, the module represents regional fuel use in regional, narrow-body, and wide-body aircraft. An infrastructure constraint, which is also modeled, can potentially limit overall growth in passenger and freight air travel to levels commensurate with industry-projected infrastructure expansion and capacity growth.

Electricity Market Module

There are three primary submodules of the Electricity Market Module—capacity planning, fuel dispatching, and finance and pricing. The capacity expansion submodule uses the stock of existing generation capacity, the cost and performance of future generation capacity, expected fuel prices, expected financial parameters, expected electricity demand, and expected environmental regulations to project the optimal mix of new generation capacity that should be added in future years. The fuel dispatching submodule uses the existing stock of generation equipment types, their operation and maintenance costs and performance, fuel prices to the electricity sector, electricity demand, and all applicable environmental regulations to determine the least-cost way to meet that demand. The submodule also determines transmission and pricing of electricity. The finance and pricing submodule uses capital costs, fuel costs, macroeconomic parameters, environmental regulations, and load shapes to estimate generation costs for each technology.

All specifically identified options promulgated by the EPA for compliance with the Clean Air Act Amendments of 1990 are explicitly represented in the capacity expansion and dispatch decisions. All financial incentives for power generation expansion and dispatch specifically identified in EPACT2005 have been implemented. Several States, primarily in the Northeast, have enacted air emission regulations for CO_2 that affect the electricity generation sector, and those regulations are represented in *AEO2012*. The *AEO2012* Reference case also imposes a limit on power sector CO_2 emissions for plants serving California, to represent the power sector impacts of California's AB 32. The *AEO2012* Reference case reflects the CSAPR as finalized by the EPA on July 6, 2011, requiring reductions in emissions from power plants that contribute to ozone and fine particle pollution in 28 States. Reductions in mercury emissions from coal- and oil-fired power plants also are reflected through the inclusion of the mercury and air toxics standards for power plants, finalized by the EPA on December 16, 2011.

Although currently there is no Federal legislation in place that restricts GHG emissions, regulators and the investment community have continued to push energy companies to invest in technologies that are less GHG-intensive. The trend is captured in the *AEO2012* Reference case through a 3-percentage-point increase in the cost of capital, when evaluating investments in new coal-fired power plants, new coal-to-liquids (CTL) plants without carbon capture and storage (CCS), and for pollution control retrofits.

Renewable Fuels Module

The Renewable Fuels Module (RFM) includes submodules representing renewable resource supply and technology input information for central-station, grid-connected electricity generation technologies, including conventional hydroelectricity, biomass (dedicated biomass plants and co-firing in existing coal plants), geothermal, landfill gas, solar thermal electricity, solar photovoltaics (PV), and both onshore and offshore wind energy. The RFM contains renewable resource supply estimates representing the regional opportunities for renewable energy development. Investment tax credits (ITCs) for renewable fuels are incorporated, as currently enacted, including a permanent 10-percent ITC for business investment in solar energy (thermal nonpower uses as well as power uses) and geothermal power (available only to those projects not accepting the production tax credit [PTC] for geothermal power). In addition, the module reflects the increase in the ITC to 30 percent for solar energy systems installed before January 1, 2017. The extension of the credit to individual homeowners under EIEA2008 is reflected in the Residential and Commercial Demand Modules.

PTCs for wind, geothermal, landfill gas, and some types of hydroelectric and biomass-fueled plants also are represented. They provide a credit of up to 2.2 cents per kilowatthour for electricity produced in the first 10 years of plant operation. For *AEO2012*, new wind plants coming on line before January 1, 2013, are eligible to receive the PTC; other eligible plants must be in service before January 1, 2014. As part of the ARRA2009, plants eligible for the PTC may instead elect to receive a 30-percent ITC or an equivalent direct grant. *AEO2012* also accounts for new renewable energy capacity resulting from State renewable portfolio standard programs, mandates, and goals, as described in *Assumptions to the Annual Energy Outlook 2012* [152].

Oil and Gas Supply Module

The Oil and Gas Supply Module represents domestic crude oil and natural gas supply within an integrated framework that captures the interrelationships among the various sources of supply—onshore, offshore, and Alaska—by all production techniques, including natural gas recovery from coalbeds and low-permeability formations of sandstone and shale. The framework analyzes cash flow and profitability to compute investment and drilling for each of the supply sources, based on the prices for crude oil and natural gas, the domestic recoverable resource base, and the state of technology. Oil and natural gas production activities are modeled for 12 supply regions, including 6 onshore, 3 offshore, and 3 Alaskan regions.

The Onshore Lower 48 Oil and Gas Supply Submodule evaluates the economics of future exploration and development projects for crude oil and natural gas at the play level. Crude oil resources include conventional resources as well as highly fractured continuous zones, such as the Austin chalk and Bakken shale formations. Production potential from advanced secondary recovery techniques (such as infill drilling, horizontal continuity, and horizontal profile) and enhanced oil recovery (such as CO_2 flooding, steam flooding, polymer flooding, and profile modification) are explicitly represented. Natural gas resources include high-permeability carbonate and sandstone, tight gas, shale gas, and coalbed methane.

Domestic crude oil production quantities are used as inputs to the PMM in NEMS for conversion and blending into refined petroleum products. Supply curves for natural gas are used as inputs to the Natural Gas Transmission and Distribution Module (NGTDM) for determining natural gas wellhead prices and domestic production.

Natural Gas Transmission and Distribution Module

The NGTDM represents the transmission, distribution, and pricing of natural gas, subject to end-use demand for natural gas and the availability of domestic natural gas and natural gas traded on the international market. The module tracks the flows of natural gas and determines the associated capacity expansion requirements in an aggregate pipeline network, connecting the domestic and foreign supply regions with 12 lower 48 U.S. demand regions. The 12 lower 48 regions align with the 9 Census divisions, with three subdivided, and Alaska handled separately. The flow of natural gas is determined for both a peak and off-peak period in the year, assuming a historically based seasonal distribution of natural gas demand. Key components of pipeline and distributor tariffs are included in separate pricing algorithms. An algorithm is included to project the addition of CNG retail fueling capability. The module also accounts for foreign sources of natural gas, including pipeline imports and exports to Canada and Mexico, as well as liquefied natural gas (LNG) imports and exports. For *AEO2012*, LNG exports and re-exports were set exogenously and assumed to reach and maintain a total level of 903 billion cubic feet per year by 2020.

Petroleum Market Module

The PMM projects prices of petroleum products, crude oil and product import activity, and domestic refinery operations, subject to demand for petroleum products, availability and price of imported petroleum, and domestic production of crude oil, natural gas liquids, and biofuels—ethanol, biodiesel, biomass-to-liquids (BTL), CTL, gas-to-liquids (GTL), and coal-and-biomass-to-liquids (CBTL). Costs, performance, and first dates of commercial availability for the advanced other liquids technologies [153] are reviewed and updated annually.

The module represents refining activities in the five PADDs, as well as a less detailed representation of refining activities in the rest of the world. It models the costs of automotive fuels, such as conventional and reformulated gasoline, and includes production of biofuels for blending in gasoline and diesel. Fuel ethanol and biodiesel are included in the PMM, because they are commonly blended into petroleum products. The module allows ethanol blending into gasoline at 10 percent or less by volume (E10), 15 percent by volume (E15) in States that lack explicit language capping ethanol volume or oxygen content, and up to 85 percent by volume (E85) for use in flex-fuel vehicles.

The PMM includes representation of the Renewable Fuels Standard (RFS) included in EISA2007, which mandates the use of 36 billion gallons of ethanol equivalent renewable fuel by 2022. Both domestic and imported ethanol count toward the RFS. Domestic ethanol production is modeled for three feedstock categories: corn, cellulosic plant materials, and advanced feedstock materials. Starch-based ethanol plants are numerous (more than 190 are now in operation, with a total maximum sustainable nameplate capacity of more than 14 billion gallons annually), and they are based on a well-known technology that converts starch and sugar into ethanol. Ethanol from cellulosic sources is a new technology with only a few small pilot plants in operation. Ethanol from advanced feedstocks—defined as plants that ferment and distill grains other than corn and reduce GHG emissions by at least 50 percent—is also a new technology modeled in the PMM.

Fuels produced by Fischer-Tropsch synthesis and through a pyrolysis process are also modeled in the PMM, based on their economics relative to competing feedstocks and products. The five processes modeled are CTL, CBTL, GTL, BTL, and pyrolysis.

Coal Market Module

The Coal Market Module (CMM) simulates mining, transportation, and pricing of coal, subject to end-use demand for coal differentiated by heat and sulfur content. U.S. coal production is represented in the CMM by 41 separate supply curves—differentiated by region, mine type, coal rank, and sulfur content. The coal supply curves respond to capacity utilization of mines, mining capacity, labor productivity, and factor input costs (mining equipment, mining labor, and fuel requirements). Projections of

U.S. coal distribution are determined by minimizing the cost of coal supplied, given coal demands by region and sector, environmental restrictions, and accounting for minemouth prices, transportation costs, and coal supply contracts. Over the projection horizon, coal transportation costs in the CMM vary in response to changes in the cost of rail investments.

The CMM produces projections of U.S. steam and metallurgical coal exports and imports in the context of world coal trade, determining the pattern of world coal trade flows that minimizes production and transportation costs while meeting a specified set of regional world coal import demands, subject to constraints on export capacities and trade flows. The international coal market component of the module computes trade in 3 types of coal for 17 export regions and 20 import regions. U.S. coal production and distribution are computed for 14 supply regions and 16 demand regions.

Annual Energy Outlook 2012 cases

Table E1 provides a summary of the cases produced as part of *AEO2012*. For each case, the table gives the name used in *AEO2012*, a brief description of the major assumptions underlying the projections, and a reference to the pages in the body of the report and in this appendix where the case is discussed. The text sections following Table E1 describe the various cases. The Reference case assumptions for each sector are described in *Assumptions to the Annual Energy Outlook 2012* [154]. Regional results and other details of the projections are available at website www.eia.gov/aeo/supplement.

Macroeconomic growth cases

In addition to the *AEO2012* Reference case, Low Economic Growth and High Economic Growth cases were developed to reflect the uncertainty in projections of economic growth. The alternative cases are intended to show the effects of alternative growth assumptions on energy market projections. The cases are described as follows:

- In the Reference *case*, population grows by 0.9 percent per year, nonfarm employment by 1.0 percent per year, and labor productivity by 1.9 percent per year from 2010 to 2035. Economic output as measured by real GDP increases by 2.5 percent per year from 2010 through 2035, and growth in real disposable income per capita averages 1.5 percent per year.

- The Low Economic Growth case assumes lower growth rates for population (0.8 percent per year) and labor productivity (1.5 percent per year), resulting in lower nonfarm employment (0.8 percent per year), higher prices and interest rates, and lower growth in industrial output. In the Low Economic Growth case, economic output as measured by real GDP increases by 2.0 percent per year from 2010 through 2035, and growth in real disposable income per capita averages 1.3 percent per year.

- The High Economic Growth case assumes higher growth rates for population (1.0 percent per year) and labor productivity (2.2 percent per year), resulting in higher nonfarm employment (1.2 percent per year). With higher productivity gains and employment growth, inflation and interest rates are lower than in the Reference case, and consequently economic output grows at a higher rate (3.0 percent per year) than in the Reference case (2.5 percent). Disposable income per capita grows by 1.6 percent per year, compared with 1.5 percent in the Reference case.

Oil price cases

The oil price in *AEO2012* is defined as the average price of light, low-sulfur crude oil delivered in Cushing, Oklahoma, and is similar to the price for light, sweet crude oil traded on the New York Mercantile Exchange, referred to as West Texas Intermediate (WTI). *AEO2012* also includes a projection of the U.S. annual average refiners' acquisition cost of imported crude oil, which is more representative of the average cost of all crude oils used by domestic refiners.

The historical record shows substantial variability in oil prices, and there is arguably even more uncertainty about future prices in the long term. *AEO2012* considers three oil price cases (Reference, Low Oil Price, and High Oil Price) to allow an assessment of alternative views on the future course of oil prices.

The Low and High Oil Price cases reflect a wide range of potential price paths, resulting from variation in demand by countries outside the Organization for Economic Cooperation and Development (OECD) for petroleum and other liquid fuels due to different levels of economic growth. The Low and High Oil Price cases also reflect different assumptions about decisions by members of the Organization of the Petroleum Exporting Countries (OPEC) regarding the preferred rate of oil production and about the future finding and development costs and accessibility of conventional oil resources outside the United States.

- In the Reference case, real oil prices rise from a $93 per barrel (2010 dollars) in 2011 to $145 per barrel in 2035. The Reference case represents EIA's current judgment regarding exploration and development costs and accessibility of oil resources. It also assumes that OPEC producers will choose to maintain their share of the market and will schedule investments in incremental production capacity so that OPEC's conventional oil production will represent about 40 percent of the world's total petroleum and other liquids production over the projection period.

- In the Low Oil Price case, crude oil prices are only $62 per barrel (2010 dollars) in 2035, compared with $145 per barrel in the Reference case. In the Low Oil Price case, the low price results from lower demand for petroleum and other liquid fuels in the non-OECD nations. Lower demand is derived from lower economic growth relative to the Reference case. In this case, GDP growth in the non-OECD countries is reduced by 1.5 percentage points relative to Reference case in each projection year, beginning in 2015. The OECD projections are affected only by the price impact. On the supply side, OPEC countries increase

Table E1. Summary of the *AEO2012* cases

Case name	Description	Reference in text	Reference in Appendix E
Reference	Baseline economic growth (2.5 percent per year from 2010 through 2035), oil price, and technology assumptions. Complete projection tables in Appendix A. Light, sweet crude oil prices rise to about $145 per barrel (2010 dollars) in 2035. Assumes RFS target to be met as soon as possible.	--	--
Low Economic Growth	Real GDP grows at an average annual rate of 2.0 percent from 2010 to 2035. Other energy market assumptions are the same as in the Reference case. Partial projection tables in Appendix B.	p. 72	p. 221
High Economic Growth	Real GDP grows at an average annual rate of 3.0 percent from 2010 to 2035. Other energy market assumptions are the same as in the Reference case. Partial projection tables in Appendix B.	p. 72	p. 221
Low Oil Price	Low prices result from a combination of low demand for petroleum and other liquid fuels in the non-OECD nations and higher global supply. Lower demand is measured by lower economic growth relative to the Reference case. In this case, GDP growth in the non-OECD is reduced by 1.5 percentage points in each projection year relative to Reference case assumptions, beginning in 2015. On the supply side, OPEC increases its market share to 46 percent, and the costs of other liquids production technologies are lower than in the Reference case. Light, sweet crude oil prices fall to $62 per barrel in 2035. Partial projection tables in Appendix C.	p. 74	p. 221
High Oil Price	High prices result from a combination of higher demand for petroleum and other liquid fuels in the non-OECD nations and lower global supply. Higher demand is measured by higher economic growth relative to the Reference case. In this case, GDP growth rates for China and India are raised by 1.0 percentage point relative to the Reference case in 2012 and decline to 0.3 percentage point above the Reference case in 2035. GDP growth rates for other non-OECD regions average about 0.5 percentage point above the Reference case. OPEC market share remains at about 40 percent throughout the projection, and non-OPEC petroleum production expands more slowly in the short to middle term relative to the Reference case. Light, sweet crude oil prices rise to $200 per barrel (2010 dollars) in 2035. Partial projection tables in Appendix C.	p. 74	p. 224
No Sunset	Begins with the Reference case and assumes extension of all existing energy policies and legislation that contain sunset provisions, except those requiring additional funding (e.g., loan guarantee programs) and those that involve extensive regulatory analysis, such as CAFE improvements and periodic updates of efficiency standards. Partial projection tables in Appendix D.	p. 18	p. 229
Extended Policies	Begins with the No Sunset case but excludes extension of tax credits for blenders and for other biofuels that were included in the No Sunset case. Assumes an increase in the capacity limitations on the ITC and extension of the program. The case includes additional rounds of efficiency standards for residential and commercial products, as well as new standards for products not yet covered, adds multiple rounds of national building codes by 2026, and increases LDV fuel economy standards in the transportation sector to 62 miles per gallon in 2035. Partial projection tables in Appendix D.	p. 18	p. 230
Transportation: CAFE Standards	Explores energy and market impacts assuming that LDV CAFE and GHG emissions standards proposed for model years 2017-2025 are enacted. Partial projection tables in Appendix D.	p. 29	p. 226
Transportation: High Technology Battery	Explores the impact of significant improvement in vehicle battery and non-battery system cost and performance on new LDV sales, energy consumption, and GHG emissions. Partial projection tables in Appendix D.	p. 31	p. 226
Transportation: HDV Reference	Incorporates revised CNG and LNG pricing assumptions and HDV market acceptance relative to the *AEO2012* Reference case. Partial projection tables in Appendix D.	p.40	p. 226
Transportation: HD NGV Potential	Using the HDV Reference case, explores energy and market issues associated with the assumed expansion of natural gas refueling infrastructure for the HDV market. Partial projection tables in Appendix D.	p. 39	p. 226

Table E1. Summary of the *AEO2012* cases (continued)

Case name	Description	Reference in text	Reference in Appendix E
Electricity: Low Nuclear	Assumes that all nuclear plants are limited to a 60-year life (31 gigawatts of retirements), uprates are limited to the 1 gigawatt that has been reported to EIA, and planned additions are the same as in the Reference case. Partial projection tables in Appendix D.	p. 51	p. 226
Electricity: High Nuclear	Assumes that all nuclear plants are life-extended beyond 60 years (except for one announced retirement), and uprates are the same as in the Reference case. New plants include those under construction and plants that have a scheduled U.S. Nuclear Regulatory Commission (NRC) or Atomic Safety and Licensing Board hearing and use a currently certified design (e.g., AP1000). Partial projection tables in Appendix D.	p. 52	p. 227
Electricity: Reference 05	Includes CSAPR and MATS as in the Reference case, with reduced 5-year environmental investment recovery. Partial projection tables in Appendix D.	p. 47	p. 227
Electricity: Low Gas Price 05	Includes CSAPR and MATS as in the Reference case, with reduced 5-year environmental investment recovery combined with the High Estimated Ultimate Recovery (EUR) case. Partial projection tables in Appendix D.	p. 47	p. 227
Renewable Fuels: Low Renewable Technology Cost	Costs for new nonhydropower renewable generating technologies start 20 percent lower in 2012 and decline to 40 percent lower than Reference case levels in 2035. Capital costs of renewable other liquid fuel technologies start 20 percent lower in 2012 and decline to approximately 40 percent lower than Reference case levels in 2035. Partial projection tables in Appendix D.	p. 208	p. 227
Petroleum: LFMM	Changes in the refining industry in the past and prospective future are discussed in the context of the development of the Liquid Fuels Market Module (LFMM) developed for NEMS. Provides overview of large-scale trends and highlights of specific issues that may require further analysis. Partial projection tables in Appendix D.	p. 43	p. 228
Oil and Gas: Low EUR	EUR per tight oil or shale gas well is 50 percent lower than in the Reference case.	p. 60	p. 227
Oil and Gas: High EUR	The EUR per tight oil and shale gas well is 50 percent higher than in the Reference case. Partial projection tables in Appendix D	p. 60	p. 227
Oil and Gas: High Technically Recoverable Resources (TRR)	The well spacing for all tight oil and shale gas plays is 8 wells per square mile (i.e., each well has an average drainage area of 80 acres), and the EUR for tight oil and shale gas wells is 50 percent higher than in the Reference case. Partial projection tables in Appendix D.	p. 60	p. 227
Coal: Low Coal Cost	Regional productivity growth rates for coal mining are approximately 2.8 percent per year higher than in the Reference case, and coal mining wages, mine equipment, and coal transportation rates in 2035 are between 21 and 25 percent lower than in the Reference case. Partial projection tables in Appendix D.	p. 101	p. 228
Coal: High Coal Cost	Regional productivity growth rates for coal mining are approximately 2.8 percent per year lower than in the Reference case, and coal mining wages, mine equipment, and coal transportation rates in 2035 are between 25 and 27 percent higher than in the Reference case. Partial projection tables in Appendix D.	p. 214	p. 228
Integrated 2011 Demand Technology	Referred to in text as "2011 Demand Technology." Assumes future equipment purchases in the residential and commercial sectors are based only on the range of equipment available in 2011. Energy efficiency of new industrial plant and equipment is held constant at the 2012 level over the projection period. Partial projection tables in Appendix D.	p. 27	p. 224
Integrated Best Available Demand Technology	Referred to in text as "Best Available Demand Technology." Assumes all future equipment purchases in the residential and commercial sectors are made from a menu of technologies that includes only the most efficient models available in a particular year for each fuel, regardless of cost. Partial projection tables in Appendix D.	p. 27	p. 225

Table E1. Summary of the *AEO2012* cases (continued)

Case name	Description	Reference in text	Reference in Appendix E
Integrated High Demand Technology	Referred to in text as "High Demand Technology." Assumes earlier availability, lower costs, and higher efficiencies for more advanced residential and commercial equipment. For new residential and commercial construction, building shell efficiencies are assumed to meet ENERGY STAR requirements after 2016. Industrial sector assumes earlier availability, lower costs, and higher efficiency for more advanced equipment and a more rapid rate of improvement in the recovery of biomass byproducts from industrial processes. In the transportation sector, the characteristics of conventional and alternative-fuel LDVs reflect more optimistic assumptions about incremental improvements in fuel economy and costs. Freight trucks are assumed to see more rapid improvement in fuel efficiency for engine and emissions control technologies. More optimistic assumptions for fuel efficiency improvements are also made for the air, rail, and shipping sectors. Partial projection tables in Appendix D.	p. 27	p. 225
Integrated 2011 Technology	Referred to in text as "2011 Technology." Combination of the Integrated 2011 Demand Technology case with the assumption that costs of new power plants do not improve from 2012 levels throughout the projection. Partial projection tables in Appendix D.	p. 202	p. 229
Integrated High Technology	Referred to in text as "High Technology." Combination of the Integrated High Demand Technology case and the Low Renewable Technology Cost case. Also assumes that costs for new nuclear and fossil-fired power plants are lower than Reference case levels, by 20 percent in 2012 and 40 percent in 2035. Partial projection tables in Appendix D.	p. 202	p. 229
No GHG Concern	No GHG emissions reduction policy is enacted, and market investment decisions are not altered in anticipation of such a policy. Partial projection tables in Appendix D.	p. 102	p. 229
GHG15	Applies a price for CO_2 emissions throughout the economy, starting at $15 per metric ton in 2013 and rising by 5 percent per year through 2035. The price is set to target the same reduction in CO_2 emissions as in the *Annual Energy Outlook 2011* (*AEO2011*) GHG Price Economywide case. Partial projection tables in Appendix D.	p. 46	p. 229
GHG25	Applies a price for CO_2 emissions throughout the economy, starting at $25 per metric ton in 2013 and rising by 5 percent per year through 2035. The price is set at the same dollar amount as in the *AEO2011* GHG Price Economywide case. Partial projection tables in Appendix D.	p. 46	p. 229

their conventional oil production to obtain a 46-percent share of total world petroleum and other liquids production, and oil resources outside the United States are more accessible and/or less costly to produce (as a result of technology advances, more attractive fiscal regimes, or both) than in the Reference case.

- In the High Oil Price case, oil prices reach about $200 per barrel (2010 dollars) in 2035. In the High Oil Price case, the high prices result from higher demand for petroleum and other liquid fuels in the non-OECD nations. Higher demand is measured by higher economic growth relative to the Reference case. In this case, GDP growth in the non-OECD region is raised by 0.1 to 1.0 percentage point relative to the Reference case in each projection year, starting in 2012. GDP growth rates for China and India are raised by 1.0 percentage points relative to the Reference case in 2012, declining to 0.3 percentage point above the Reference case in 2035. GDP growth rates for most other non-OECD regions average about 0.5 percentage point above the Reference case in each projection year. The OECD projections are affected only by the price impact. On the supply side, OPEC countries are assumed to reduce their market share somewhat, and oil resources outside the United States are assumed to be less accessible and/or more costly to produce than in the Reference case.

Buildings sector cases

In addition to the *AEO2012* Reference case, three technology-focused cases using the Demand Modules of NEMS were developed to examine the effects of changes in technology. Buildings sector assumptions for the Integrated 2011 Demand Technology case and the Integrated High Demand Technology case are also used in the appropriate Integrated Technology cases.

Residential sector assumptions for the technology-focused cases are as follows:

- For the Integrated 2011 Demand Technology case it is assumed that all future residential equipment purchases are based only on the range of equipment available in 2011. Existing building shell efficiencies are assumed to be fixed at 2011 levels (no further improvements). For new construction, building shell technology options are constrained to those available in 2011.

- For the Integrated High Demand Technology case it is assumed that residential advanced equipment is available earlier, at lower costs, and/or at higher efficiencies [155]. For new construction, building shell efficiencies are assumed to meet ENERGY STAR requirements after 2016. Consumers evaluate investments in energy efficiency at a 7-percent real discount rate.
- For the Integrated Best Available Demand Technology case it is assumed that all future residential equipment purchases are made from a menu of technologies that includes only the most efficient models available in a particular year for each fuel, regardless of cost. For new construction, building shell efficiencies are assumed to meet the criteria for the most efficient components after 2011.

Commercial sector assumptions for the technology-focused cases are as follows:

- For the Integrated 2011 Demand Technology case it is assumed that all future commercial equipment purchases are based only on the range of equipment available in 2011. Building shell efficiencies are assumed to be fixed at 2011 levels.
- For the Integrated High Demand Technology case it is assumed that commercial advanced equipment is available earlier, at lower costs, and/or with higher efficiencies than in the Reference case [156]. Energy efficiency investments are evaluated at a 7-percent real discount rate. Building shell efficiencies for new and existing buildings in 2035 assume a 25-percent improvement relative to the Reference case.
- For the Integrated Best Available Demand Technology case it is assumed that all future commercial equipment purchases are made from a menu of technologies that includes only the most efficient models available in a particular year for each fuel, regardless of cost. Building shell efficiencies for new and existing buildings in 2035 assume a 50-percent improvement relative to the Reference case.

The Residential and Commercial Demand Modules of NEMS were also used to complete the Low Renewable Technology Cost case, which is discussed in more detail below, in the renewable fuels cases section. In combination with assumptions for electricity generation from renewable fuels in the electric power sector and industrial sector, this sensitivity case analyzes the impacts of changes in generating technologies that use renewable fuels and in the availability of renewable energy sources. For the Residential and Commercial Demand Modules:

- The Low Renewable Technology Cost case assumes greater improvements in residential and commercial PV and wind systems than in the Reference case. The assumptions for capital cost estimates are 20 percent below Reference case assumptions in 2012 and decline to at least 40 percent lower than Reference case costs in 2035.

The No Sunset and Extended Policies cases described below in the cross-cutting integrated cases discussion also include assumptions in the Residential and Commercial Demand Modules of NEMS. The Extended Policies case builds on the No Sunset case and adds multiple rounds of appliance standards and building codes as described below.

- The No Sunset case assumes that selected policies with sunset provisions will be extended indefinitely rather than allowed to sunset as the law currently prescribes. For the residential sector, these extensions include: personal tax credits for selected end-use equipment, including furnaces, heat pumps, and central air conditioning; personal tax credits for PV installations, solar water heaters, small wind turbines, and geothermal heat pumps; and manufacturer tax credits for refrigerators, dishwashers, and clothes washers, passed on to consumers at 100 percent of the tax credit value. For the commercial sector, business ITCs for PV installations, solar water heaters, small wind turbines, geothermal heat pumps, and CHP are extended to the end of the projection. The business tax credit for solar technologies remains at the current 30-percent level without reverting to 10 percent as scheduled.
- The Extended Policies case includes updates to appliance standards, as prescribed by the timeline in DOE's multiyear plan, and introduces new standards for products currently not covered by DOE. Efficiency levels for the updated residential appliance standards are based on current ENERGY STAR guidelines. Residential end-use technologies subject to updated standards are not eligible for No Sunset incentives in addition to the standards. Efficiency levels for updated commercial equipment standards are based on the technology menu from the *AEO2012* Reference case and purchasing specifications for Federal agencies designated by the Federal Energy Management Program (FEMP). The case also adds national building codes to reach 30-percent improvement relative to the 2006 International Energy Conservation Code (IECC 2006) for residential households and to American Society of Heating, Refrigerating, and Air-Conditioning Engineers (ASHRAE) Standard 90.1-2004 for commercial buildings by 2020, with additional rounds of improved codes in 2023 and 2026.

Industrial sector cases

In addition to the *AEO2012* Reference case, two technology-focused cases using the IDM of NEMS were developed that examine the effects of less rapid and more rapid technology change and adoption. The energy intensity changes discussed in this section exclude the refining industry, which is modeled separately from the IDM in the PMM. Different assumptions for the IDM were also used as part of the Integrated Low Renewable Technology Cost case, No Sunset case, and Extended Policies case, but each is structured on a set of the initial industrial assumptions used for the Integrated 2011 Demand Technology case and Integrated High Demand Technology case. For the industrial sector, assumptions for those two technology-focused cases are as follows:

- For the Integrated 2011 Demand Technology case, the energy efficiency of new industrial plant and equipment is held constant at the 2012 level over the projection period. Changes in aggregate energy intensity may result both from changing equipment and production efficiency and from changing composition of output within an individual industry. Because all *AEO2012* side cases are integrated runs, potential feedback effects from energy market interactions are captured. Hence, the level and composition of overall industrial output varies from the Reference case, and any change in energy intensity in the two technology side cases is attributable to process and efficiency changes and increased use of CHP, as well as changes in the level and composition of overall industrial output.

- For the Integrated High Demand Technology case, the IDM assumes earlier availability, lower costs, and higher efficiency for more advanced equipment [157] and a more rapid rate of improvement in the recovery of biomass byproducts from industrial processes—i.e., 0.7 percent per year, as compared with 0.4 percent per year in the Reference case. The same assumption is incorporated in the Low Renewable Technology Cost case, which focuses on electricity generation. Although the choice of the 0.7-percent annual rate of improvement in byproduct recovery is an assumption in the High Demand Technology case, it is based on the expectation of higher recovery rates and substantially increased use of CHP in that case. Due to integration with other NEMS modules, potential feedback effects from energy market interactions are captured.

The industrial No Sunset and Extended Policies cases described below in the cross-cutting integrated cases discussion also include assumptions in the IDM of NEMS. The Extended Policies case builds on the No Sunset case and modifies select industrial assumptions, which are as follows:

- The No Sunset case and Extended Policies case include an assumption for CHP that extends the existing industrial CHP ITC through the end of the projection period. Additionally, the Extended Policies case includes an increase in the capacity limitations on the ITC by increasing the cap on CHP equipment from 15 megawatts to 25 megawatts and eliminating the system-wide cap of 50 megawatts. These assumptions are based on the current proposals in H.R. 2750 and H.R. 2784 of the 112th Congress.

Transportation sector cases

In addition to the *AEO2012* Reference case, the NEMS Transportation Demand Module was used to examine the effects of advanced technology costs and efficiency improvement on technology adoption and vehicle fuel economy as part of the Integrated High Demand Technology case [158]. For the Integrated High Demand Technology case, the characteristics of conventional and alternative-fuel LDVs reflect more optimistic assumptions about incremental improvements in fuel economy and costs. In the freight truck sector, the High Demand Technology case assumes more rapid incremental improvement in fuel efficiency and lower costs for engine and emissions control technologies. More optimistic assumptions for fuel efficiency improvements are also made for the air, rail, and shipping sectors.

Three additional integrated cases were developed to examine the potential energy impacts associated with the implementation of proposed model year 2017 to 2025 LDV CAFE standards, the impact of the successful development of advanced batteries, and the impact of the penetration of HDVs using LNG. The specific cases include:

- The CAFE Standards case examines the energy, GHG, and vehicle market impacts of increasing LDV fuel economy standards to reflect those proposed by the EPA and NHTSA for model years 2017-2025. Fuel economy standards are assumed to remain constant after model year 2025.
- The High Technology Battery case examines the energy, GHG emissions, and sales impacts on new LDVs associated with rapid improvement in battery cost and non-battery systems performance.
- The HDV Reference case incorporates revised pricing assumptions for CNG and LNG highway fuels and HDV market acceptance.
- The HD NGV Potential case examines the energy and GHG impacts associated with assumed significant increases in LNG refueling infrastructure to enable market adoption of natural gas use by HDVs in long-haul corridors relative to the HDV Reference case.

Electricity sector cases

In addition to the Reference case, several integrated cases with alternative electric power assumptions were developed to support discussions in the "Issues in focus" section of *AEO2012*. Two alternative cases were run for nuclear power plants, to address uncertainties about the operating lives of existing reactors, the potential for new nuclear capacity, and capacity uprates at existing plants. These scenarios are discussed in the "Issues in focus" article, "Nuclear power in *AEO2012*."

In addition, two alternative cases were run to analyze uncertainties related to the lifetimes of coal-fired power plants due to recent environmental regulations and potential GHG legislation in the future. Over the next few years, electricity generators will begin taking steps to comply with a number of new environmental regulations, primarily by adding environmental controls at existing coal-fired power plants. The additional cases examine the impacts of shorter economic recovery periods for the environmental controls, with the natural gas prices used in the *AEO2012* Reference case and lower natural gas prices.

Nuclear cases

- The Low Nuclear case assumes that all existing nuclear plants are retired after 60 years of operation. In the Reference case, existing plants are assumed to run as long as they continue to be economic, implicitly assuming that a second 20-year license renewal will be obtained for most plants that reach 60 years before 2035. The Low Nuclear case was run to analyze the impact of additional nuclear retirements, which could occur if the oldest plants do not receive a second license extension. In this case, 31 gigawatts of nuclear capacity is assumed to be retired by 2035. The Low Nuclear case assumes that no new nuclear capacity will be added throughout the projection, excluding capacity already planned or under construction. The case also assumes that only those capacity uprates reported to EIA will be completed (1 gigawatt). The Reference case assumes additional uprates based on NRC surveys and industry reports.

- The High Nuclear case assumes that all existing nuclear units will receive a second license renewal and operate beyond 60 years (excluding one announced retirement). In the Reference case, beyond the announced retirement of Oyster Creek, an additional 5.5 gigawatts of nuclear capacity is assumed to be retired through 2035, reflecting uncertainty about the impacts and/or costs of future aging. This case was run to provide a more optimistic outlook, with all licenses renewed and all plants continuing to operate economically beyond 60 years. The High Nuclear case also assumes that additional planned nuclear capacity is completed based on combined license applications issued by the NRC. The Reference case assumes that 6.8 gigawatts of planned capacity is added, compared with 13.5 gigawatts of planned capacity additions in the High Nuclear case.

Environmental Rules cases

- The Reference 05 case assumes that the economic recovery period for investments in new environmental controls in the electric power sector is reduced from 20 years to 5 years.
- The Low Gas Price 05 case uses more optimistic assumptions about future volumes of shale gas production, leading to lower natural gas prices, combined with the 5-year recovery period for new environmental controls in the electric power sector. The domestic shale gas resource assumption comes from the High EUR case.

Renewable fuels cases

In addition to the *AEO2012* Reference case, EIA developed a case with alternative assumptions about renewable fuels to examine the effects of more aggressive improvement in the cost of renewable technologies.

- In the Low Renewable Technology Cost case, the levelized costs of new nonhydropower renewable generating technologies are assumed to start at 20 percent below Reference case assumptions in 2012 and decline to 40 percent below the Reference case costs for the same resources in 2035. In general, lower costs are represented by reducing the capital costs of new plant construction. Biomass fuel supplies also are assumed to be 40 percent less expensive than for the same resource quantities used in the Reference case. Assumptions for other generating technologies are unchanged from those in the Reference case. In the Low Renewable Technology Cost case, the rate of improvement in recovery of biomass byproducts from industrial processes also is increased.
- In the No Sunset case and the Extended Policies case, expiring Federal tax credits targeting renewable electricity are assumed to be permanently extended. This applies to the PTC, which is a tax credit of 2.2 cents per kilowatthour available for the first 10 years of production by new generators using wind, geothermal, and certain biomass fuels, or a tax credit of 1.1 cents per kilowatthour available for the first 10 years of production by new generators using geothermal energy, certain hydroelectric technologies, and biomass fuels not eligible for the full credit of 2.2 cents per kilowatthour. This tax credit is scheduled to expire on December 31, 2012, for wind and 1 year later for other eligible technologies. The same schedule applies to the 30-percent ITC, which is available to new solar installations through December 31, 2016, and may also be claimed in lieu of the PTC for eligible technologies, expiring concurrently with the PTC expiration dates indicated above.

Oil and gas supply cases

The sensitivity of the *AEO2012* projections to changes in assumptions regarding technically recoverable tight oil and shale gas resources are examined in two cases:

- In the Low EUR case, the EUR per tight oil or shale gas well is assumed to be 50 percent lower than in the Reference case, increasing the per-unit cost of developing the resource. The total unproved TRR of tight oil is decreased to 17 billion barrels, and the shale gas resource is decreased to 241 trillion cubic feet, as compared with unproved resource estimates of 33 billion barrels of tight oil and 482 of shale gas in the Reference case as of January 1, 2010.
- In the High EUR case, the EUR per tight oil and shale gas well is assumed to be 50 percent higher than in the Reference case, decreasing the per-unit cost of developing the resource. The total unproved technically recoverable tight oil resource is increased to 50 billion barrels, and the shale gas resource is increased to 723 trillion cubic feet.
- In the High TRR case, the well spacing for all tight oil and shale gas plays is assumed to be 8 wells per square mile (i.e., each well has an average drainage area of 80 acres), and the EUR for tight oil and shale gas wells is assumed to be 50 percent higher than in the Reference case. The total unproved technically recoverable tight oil resource is increased to 89 billion barrels, and the shale gas resource is increased to 1,091 trillion cubic feet, more than twice the Reference case assumptions for tight oil and shale gas resources.

Petroleum market cases

Production of petroleum and other liquid fuels has evolved and changed significantly in recent years as a result of changes in the mix of feedstocks, production regions, technologies, regulation and policy, and international markets. To better reflect those changes, a new LFMM has been developed for use as part of NEMS. The intent is to use the LFMM in developing the *Annual Energy Outlook 2013 (AEO2013)*. The LFMM was designed as a data-driven tool using a generalized algebraic modeling system. The LFMM uses nine types of crude oil (compared to five types in the current model). The LFMM configuration uses nine refining regions instead of the traditional five PADDs—eight domestic regions and one maritime Canada and Caribbean region that captures imports of refined products into the northeastern United States.

Market conditions and regulations have resulted in the implementation of new technologies using nonpetroleum feedstocks such as grains, biomass, pyrolysis oils, coal, biomass, and natural gas. The EISA2007 RFS mandates the use 36 billion gallons of renewable fuels by 2022, and the LFMM allows analysis of different renewable fuel capacities required to meet the mandate. Because the LFMM is a data-driven model, new technologies can be added easily to help in analysis of the RFS mandate. In addition, the LFMM has extensive representation of the RFS and other policies that affect its implementation. The technologies associated with the RFS have high development costs, and capital recovery is uncertain. That uncertainty can be analyzed by varying the market penetration rates for the technologies under different assumptions. Further, to accommodate evolving international markets, LFMM uses different approaches while interfacing with NEMS PMM. The new interface is able to work with newer crude types, as well as changes in prices for crude oil and petroleum products.

For *AEO2012*, an LFMM case was developed to test the new model and compare results with those produced by the PMM—which is the current model used for *AEO2012*—for the Reference, Low Economic Growth, High Economic Growth, Low Oil Price, and High Oil Price cases produced using the current version of the NEMS. The intent is to highlight areas where the two models produce significantly different results and explore the basis of those differences so that EIA will be able to ensure that the LFMM is ready for use as part of *AEO2013*.

Coal market cases

Two alternative coal cost cases examine the impacts on U.S. coal supply, demand, distribution, and prices that result from alternative assumptions about mining productivity, labor costs, mine equipment costs, and coal transportation rates. The alternative productivity and cost assumptions are applied in every year from 2012 through 2035. For the coal cost cases, adjustments to the Reference case assumptions for coal mining productivity are based on variation in the average annual productivity growth of 2.8 percent observed since 2000. Transportation rates are lowered (in the Low Coal Cost case) or raised (in the High Coal Cost case) from Reference case levels to achieve a 25-percent change in rates relative to the Reference case in 2035. The Low and High Coal Cost cases represent fully integrated NEMS runs, with feedback from the macroeconomic activity, international, supply, conversion, and enduse demand modules.

- In the Low Coal Cost case, the average annual growth rates for coal mining productivity are higher than those in the Reference case and are applied at the supply curve level. As an example, the average annual productivity growth rate for Wyoming's Southern Powder River Basin supply curve is increased from -1.8 percent in the Reference case for the years 2012 through 2035 to 0.8 percent in the Low Coal Cost case. Coal mining wages, mine equipment costs, and other mine supply costs all are assumed to be about 21 percent lower in 2035 in real terms in the Low Coal Cost case than in the Reference case. Coal transportation rates, excluding the impact of fuel surcharges, are assumed to be 25 percent lower in 2035.

- In the High Coal Cost case, the average annual productivity growth rates for coal mining are lower than those in the Reference case and are applied as described in the Low Coal Cost case. Coal mining wages, mine equipment costs, and other mine supply costs in 2035 are assumed to be about 27 percent higher than in the Reference case, and coal transportation rates in 2035 are assumed to be 25 percent higher.

Additional details of the productivity, wage, mine equipment cost, and coal transportation rate assumptions for the Reference and alternative coal cost cases are provided in Appendix D.

Cross-cutting integrated cases

A series of cross-cutting integrated cases are used in *AEO2012* to analyze specific cases with broader sectoral impacts. For example, three integrated technology progress cases analyze the impacts of more rapid and slower technology improvement rates in the demand sector (partially described in the sector-specific sections above), and two other integrated technology cases examine the impacts of more rapid and slower technology improvement rates across both demand and supply/conversion sectors. In addition, two cases also were run with alternative assumptions about expectations of future regulation of GHG emissions.

Integrated technology cases

In the demand sectors (residential, commercial, industrial, and transportation), technology improvement typically means greater efficiency of energy use and/or reduced cost. In the energy supply/conversion sectors (electricity generation, natural gas and petroleum and other liquids supply, petroleum refining, etc.), technology improvement tends to mean greater availability of energy supplies and/or reduced cost of production (and ultimately prices). When alternative cases that examine the impacts of variation in the rate of technology improvement are completed, combining the demand and supply/conversion sectors, the impacts on energy markets are sometimes masked because of the offsetting nature of technology improvements in the two areas.

Two sets of alternative cases are used in *AEO2012* to examine the potential impacts of variation in the rate of technology improvement. The first set looks at impacts on the demand sector in isolation. The second set looks at the combined impacts of technology changes in both the demand and supply/conversion sectors. The three demand technology cases—Integrated 2011 Demand Technology, Integrated Best Available Demand Technology, and Integrated High Demand Technology—examine the impacts on the end-use demand sectors of variations in the rate of technology improvement, independent of the offsetting impacts of variations in technology improvement in the supply/conversion sectors.

EIA also completed two fully integrated technology cases that examine combined impacts on the demand and supply/conversion sectors. The Integrated 2011 Technology case combines the assumptions from the Integrated 2011 Demand Technology case with an assumption that the costs of new fossil, nuclear, and nonhydroelectric renewable power plants are fixed at 2012 levels and do not improve due to learning during the projection period. The Integrated High Technology case combines the assumptions from the Integrated High Demand Technology and the Low Renewable Technology Cost case with an assumption that the costs of new nuclear and fossil-fired power plants are lower than assumed in the Reference case, with costs 20 percent lower than Reference case levels in 2012 and 40 percent lower than Reference case levels in 2035.

Greenhouse gas cases

On May 13, 2010, the EPA promulgated standards for GHG emissions in the "Prevention of Significant Deterioration and Title V Greenhouse Gas Tailoring Rule" [159]. The rule sets up levels of CO_2-equivalent emissions at new and existing facilities that make major modifications that increase GHG emissions which trigger coverage of the facilities in the New Source Review and Title V permitting program. As a result of this and prior actions, regulators and the investment community are beginning to push energy companies to invest in less GHG-intensive technologies. To reflect the market reaction to potential future GHG regulation, a 3-percentage-point increase in the cost of capital is assumed for investments in new coal-fired power plants without CCS and new CTL plants without CCS in the Reference case and all other *AEO2012* cases except the No GHG Concern, GHG15, and GHG25 cases. Those assumptions affect cost evaluations for the construction of new capacity but not the actual operating costs when a new plant begins operation.

The three alternative GHG cases are used to provide a range of potential outcomes, from no concern about future GHG legislation to the imposition of a specific economywide carbon allowance price. *AEO2012* includes two economywide CO_2 price cases, the GHG15 and GHG25 cases, which examine the impacts of economywide carbon allowance prices. In the GHG15 case, the price is set at $15 per metric ton CO_2 in 2013. In the GHG25 case, the price is set at $25 per metric ton CO_2 in 2013. In both cases the price begins to rise in 2014 at 5 percent per year. The GHG cases are intended to measure the sensitivity of the *AEO2012* assumptions to different CO_2 prices that are consistent with previously proposed legislation. At the time the *AEO2012* was completed, no legislation including a GHG price was pending, but the EPA is developing technology-based CO_2 standards for new coal-fired power plants. In the two GHG cases for *AEO2012*, no assumptions are made with regard to offsets, bonus allowances for CCS, or specific allocation of allowances.

The No GHG Concern case was run without any adjustment for concern about potential GHG regulations (without the 3-percentage-point increase in the cost of capital). In the No GHG Concern case, the same cost of capital is used to evaluate all new capacity builds, regardless of type.

No Sunset case

In addition to the *AEO2012* Reference case, a No Sunset case was run assuming that selected policies with sunset provisions—such as the PTC, ITC, and tax credits for energy-efficient equipment in the buildings and industrial sectors—will be extended indefinitely rather than allowed to sunset as the law currently prescribes.

For the residential sector, the extensions include: (a) personal tax credits for selected end-use equipment, including furnaces, heat pumps, and central air conditioning; (b) personal tax credits for PV installations, solar water heaters, small wind turbines, and geothermal heat pumps; (c) manufacturer tax credits for refrigerators, dishwashers, and clothes washers, passed on to consumers at 100 percent of the tax credit value.

For the commercial sector, business ITCs for PV installations, solar water heaters, small wind turbines, geothermal heat pumps, and CHP are extended to the end of the projection. The business tax credit for solar technologies remains at the current 30-percent level without reverting to 10 percent as scheduled.

In the industrial sector, the existing ITC for industrial CHP, which currently ends in 2016, is extended to 2035.

For the refinery sector, blending credits are extended; the $1.00 per gallon biodiesel tax credit is extended; the $0.54 per gallon tariff on imported ethanol is extended; and the $1.01 per gallon PTC for cellulosic biofuels is extended.

For renewables, the PTC of 2.2 cents per kilowatthour for wind, geothermal, and certain biomass and the PTC of 1.1 cents per kilowatthour for hydroelectric and landfill gas resources, which currently are set to expire at the end of 2012 for wind and the end of 2013 for other eligible resources, are extended to 2035; and the 30-percent solar power ITC, which currently is scheduled to revert to 10 percent in 2016, is extended indefinitely.

Extended Policies case

In the Extended Policies case, assumptions for tax credit extensions are the same as in the No Sunset case described above with the exception of the PTC extension for cellulosic biofuels and the tax credits for residential equipment subject to updated Federal efficiency standards, which are dropped. Further, updates to Federal appliance efficiency standards are assumed to occur at regular intervals, and new standards for products not currently covered by DOE are assumed to be introduced. Finally, proposed rules by NHTSA and the EPA for national tailpipe CO_2-equivalent emissions and fuel economy standards for LDVs, including both passenger cars and light-duty trucks, are harmonized and incorporated in this case.

Updates to appliance standards are assumed to occur as prescribed by the timeline in DOE's multi-year plan, and new standards for products currently not covered by DOE are introduced by 2019. The efficiency levels chosen for the updated residential appliance standards are based on current ENERGY STAR guidelines. Residential end-use technologies subject to updated standards are not eligible for No Sunset incentives in addition to the standards. The efficiency levels chosen for updated commercial equipment standards are based on the technology menu from the *AEO2011* Reference case and either FEMP-designated purchasing specifications for Federal agencies or ENERGY STAR guidelines. National building codes are added to reach 30-percent improvement relative to IECC 2006 for residential households and ASHRAE 90.1-2004 for commercial buildings by 2020, with additional rounds of improvements in 2023 and 2026.

In the industrial sector, the ITC for industrial CHP is further extended to cover all system sizes rather than applying only to systems under 50 megawatts; and the CHP equipment cap is increased from 15 megawatts to 25 megawatts. These extensions are consistent with previously proposed legislation (S. 1639) or pending legislation (H.R. 2750 and 2784).

For transportation, the Extended Policies case assumes that the standards are further increased, so that the minimum fuel economy standard achieved by LDVs continues to increase through 2035.

Endnotes for Appendix E

Links current as of April 2012

142. U.S. Energy Information Administration, *The National Energy Modeling System: An Overview 2009*, DOE/EIA-0581(2009) (Washington, DC: October 2009), website www.eia.gov/oiaf/aeo/overview.
143. U.S. Environmental Protection Agency, "Mercury and Air Toxics Standards," website www.epa.gov/mats.
144. U.S. Environmental Protection Agency, "Cross-State Air Pollution Rule (CSAPR)," website epa.gov/airtransport. CSAPR was scheduled to begin on January 1, 2012; however, the U.S. Court of Appeals for the D.C. Circuit issued a stay delaying implementation while it addresses legal challenges to the rule that have been raised by several power companies and States. CSAPR is included in *AEO2012* despite the stay, because the Court of Appeals had not made a final ruling at the time *AEO2012* was published.
145. U.S. Environmental Protection Agency and National Highway Traffic Safety Administration, "Greenhouse Gas Emissions Standards and Fuel Efficiency Standards for Medium- and Heavy-Duty Engines and Vehicles; Final Rule," *Federal Register*, Vol. 76, No. 179 (September 15, 2011), pp. 57106-57513, website www.gpo.gov/fdsys/pkg/FR-2011-09-15/html/2011-20740.htm.
146. California Air Resources Board (ARB), "California Cap on Greenhouse Gas Emissions and Market-Based Compliance Mechanisms," Article 5 § 95800 to 96023, website www.arb.ca.gov/cc/capandtrade/capandtrade.htm.
147. U.S. Environmental Protection Agency, "July 21, 2011 Final Memorandum: Improving EPA Review of Appalachian Surface Coal Mining Operations Under the Clean Water Act, National Environmental Policy Act, and the Environmental Justice Executive Order," website water.epa.gov/lawsregs/guidance/wetlands/mining.cfm.
148. For the complete text of the American Recovery and Reinvestment Act of 2009, see website www.gpo.gov/fdsys/pkg/PLAW-111publ5/html/PLAW-111publ5.htm.
149. U.S. Energy Information Administration, *Annual Energy Review 2010*, DOE/EIA-0384(2010) (Washington, DC: October 2011), website www.eia.gov/aer.
150. U.S. Energy Information Administration, "Short-Term Energy Outlook," website www.eia.gov/forecasts/steo. Portions of the preliminary information were also used to initialize the NEMS Petroleum Market Module projection.
151. U.S. Energy Information Administration, "Manufacturing Energy Consumption Survey," website www.eia.doe.gov/emeu/mecs.
152. U.S. Energy Information Administration, *Assumptions to the Annual Energy Outlook 2012*, DOE/EIA-0554(2012) (Washington, DC: June 2012), website www.eia.gov/forecasts/aeo/assumptions.
153. Alternative other liquids technologies include all biofuels technologies plus CTL and GTL.
154. U.S. Energy Information Administration, *Assumptions to the Annual Energy Outlook 2012*, DOE/EIA-0554(2012) (Washington, DC: June 2012), website www.eia.gov/forecasts/aeo/assumptions.
155. High technology assumptions for the residential sector are based on U.S. Energy Information Administration, *EIA—Technology Forecast Updates—Residential and Commercial Building Technologies—Advanced Case* (Navigant Consulting, Inc. with SAIC, September 2011), and *EIA—Technology Forecast Updates—Residential and Commercial Building Technologies—Advanced Case: Residential and Commercial Lighting, Commercial Refrigeration, and Commercial Ventilation Technologies* (Navigant Consulting, Inc., September 2008).
156. High technology assumptions for the commercial sector are based on Energy Information Administration, *EIA—Technology Forecast Updates—Residential and Commercial Building Technologies—Advanced Case* (Navigant Consulting, Inc. with SAIC, September 2011), and *EIA—Technology Forecast Updates—Residential and Commercial Building Technologies—Advanced Case: Residential and Commercial Lighting, Commercial Refrigeration, and Commercial Ventilation Technologies* (Navigant Consulting, Inc., September 2008).
157. These assumptions are based in part on Energy Information Administration, *Industrial Technology and Data Analysis Supporting the NEMS Industrial Model* (FOCIS Associates, October 2005).
158. U.S. Energy Information Administration, *Documentation of Technologies Included in the NEMS Fuel Economy Model for Passenger Cars and Light Trucks* (Energy and Environmental Analysis, September 2003).
159. U.S. Environmental Protection Agency, "Final Rule: Prevention of Significant Deterioration and Title V Greenhouse Gas Tailoring Rule," website www.epa.gov/nsr/documents/20100413fs.pdf.

This page intentionally left blank

Appendix F
Regional Maps

Figure F1. United States Census Divisions

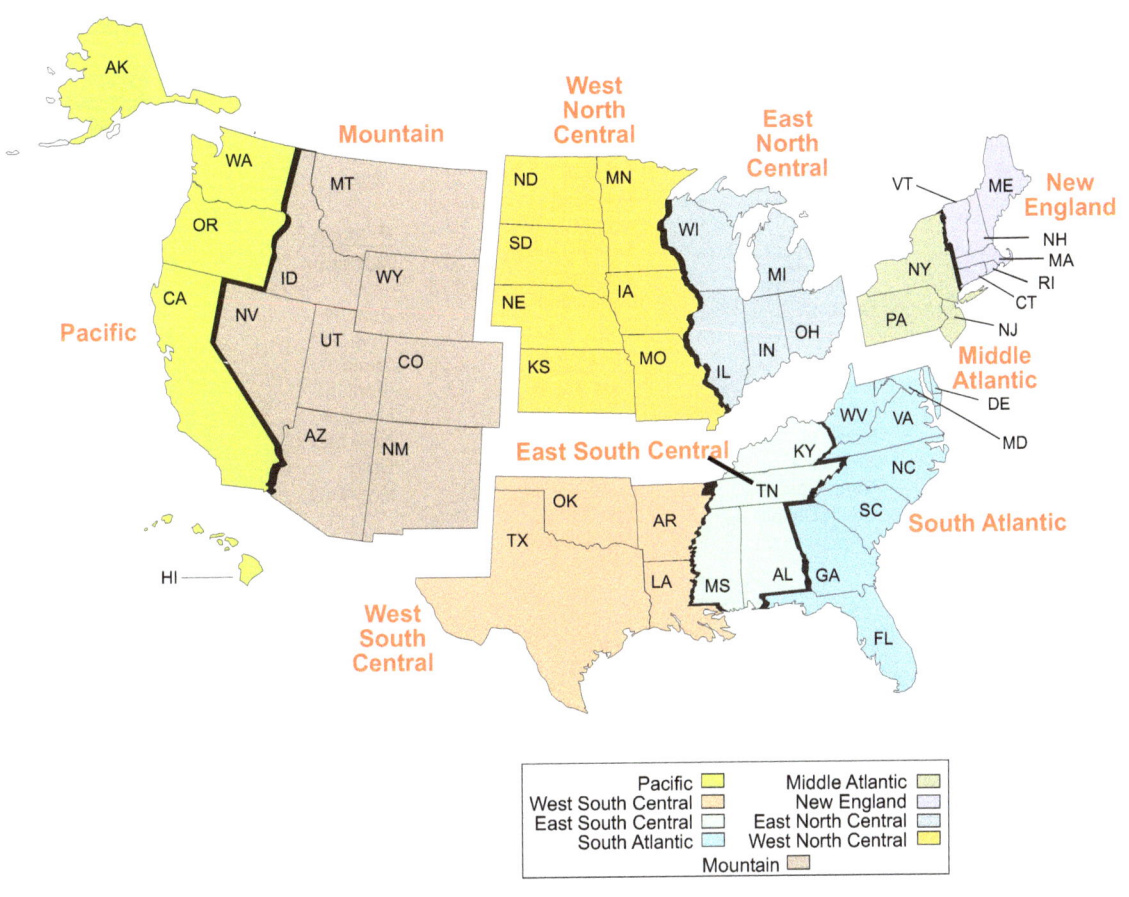

Regional maps

Figure F1. United States Census Divisions (continued)

Division 1
New England

Connecticut
Maine
Massachusetts
New Hampshire
Rhode Island
Vermont

Division 2
Middle Atlantic

New Jersey
New York
Pennsylvania

Division 3
East North Central

Illinois
Indiana
Michigan
Ohio
Wisconsin

Division 4
West North Central

Iowa
Kansas
Minnesota
Missouri
Nebraska
North Dakota
South Dakota

Division 5
South Atlantic

Delaware
District of
 Columbia
Florida
Georgia
Maryland
North Carolina
South Carolina
Virginia
West Virginia

Division 6
East South Central

Alabama
Kentucky
Mississippi
Tennessee

Division 7
West South Central

Arkansas
Louisiana
Oklahoma
Texas

Division 8
Mountain

Arizona
Colorado
Idaho
Montana
Nevada
New Mexico
Utah
Wyoming

Division 9
Pacific

Alaska
California
Hawaii
Oregon
Washington

Source: U.S. Energy Information Administration, Office of Energy Analysis.

Regional maps

Figure F2. Electricity market module regions

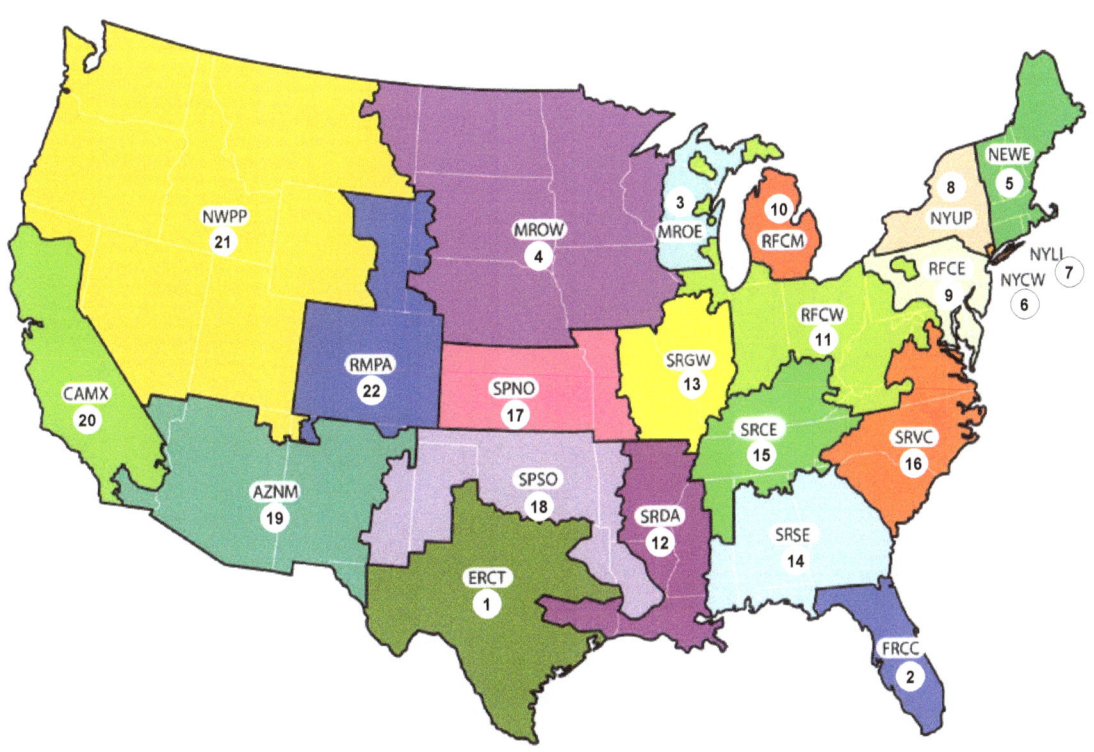

1.	ERCT	TRE All		12.	SRDA	SERC Delta
2.	FRCC	FRCC All		13.	SRGW	SERC Gateway
3.	MROE	MRO East		14.	SRSE	SERC Southeastern
4.	MROW	MRO West		15.	SRCE	SERC Central
5.	NEWE	NPCC New England		16.	SRVC	SERC VACAR
6.	NYCW	NPCC NYC/Westchester		17.	SPNO	SPP North
7.	NYLI	NPCC Long Island		18.	SPSO	SPP South
8.	NYUP	NPCC Upstate NY		19.	AZNM	WECC Southwest
9.	RFCE	RFC East		20.	CAMX	WECC California
10.	RFCM	RFC Michigan		21.	NWPP	WECC Northwest
11.	RFCW	RFC West		22.	RMPA	WECC Rockies

Source: U.S. Energy Information Administration, Office of Energy Analysis.

Regional maps

Figure F3. Petroleum Administration for Defense Districts

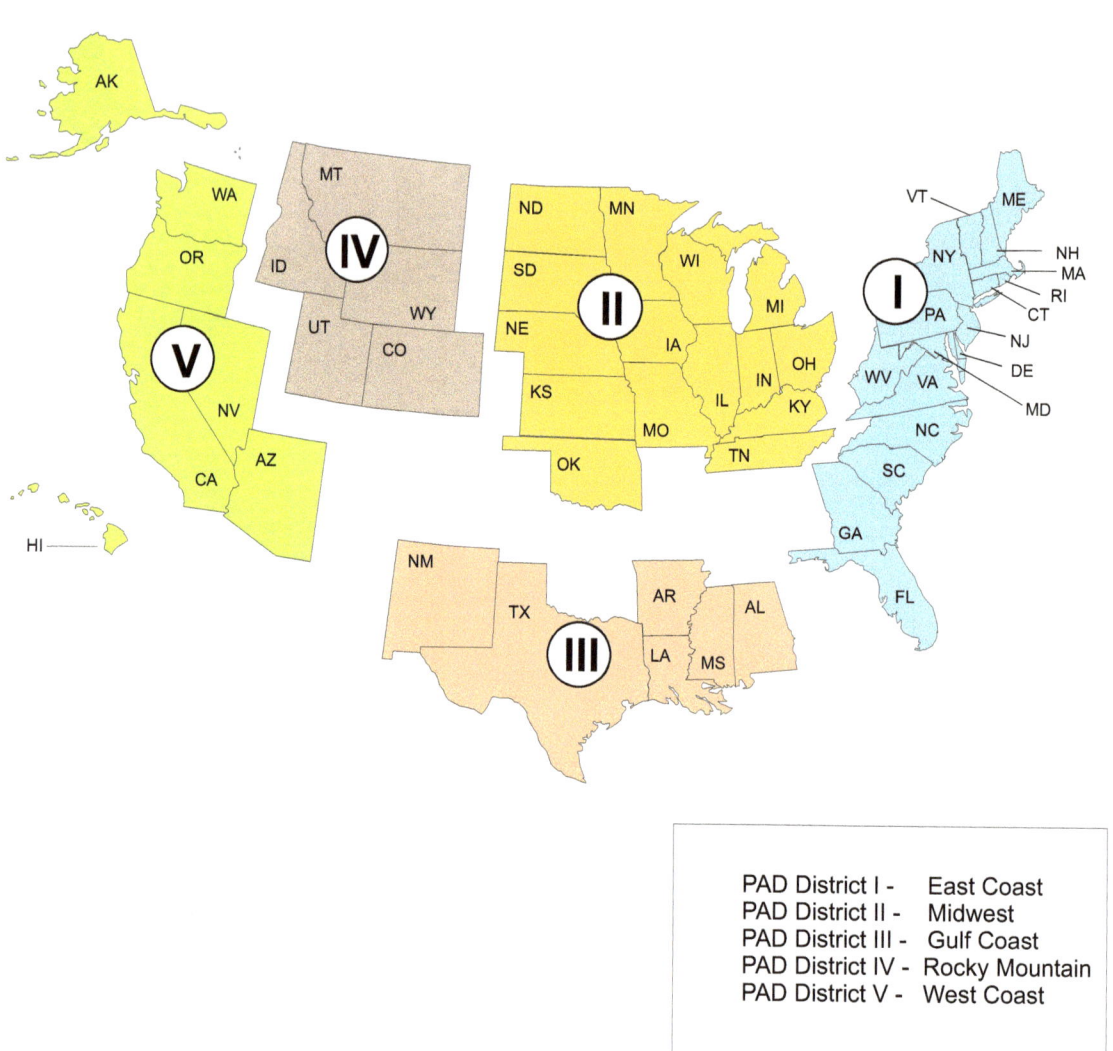

PAD District I - East Coast
PAD District II - Midwest
PAD District III - Gulf Coast
PAD District IV - Rocky Mountain
PAD District V - West Coast

Source: U.S. Energy Information Administration, Office of Energy Analysis.

Regional maps

Figure F4. Oil and gas supply model regions

Source: U.S. Energy Information Administration, Office of Energy Analysis.

Regional maps

Figure F5. Natural gas transmission and distribution model regions

Source: U.S. Energy Information Administration, Office of Energy Analysis.

Figure F6. Coal supply regions

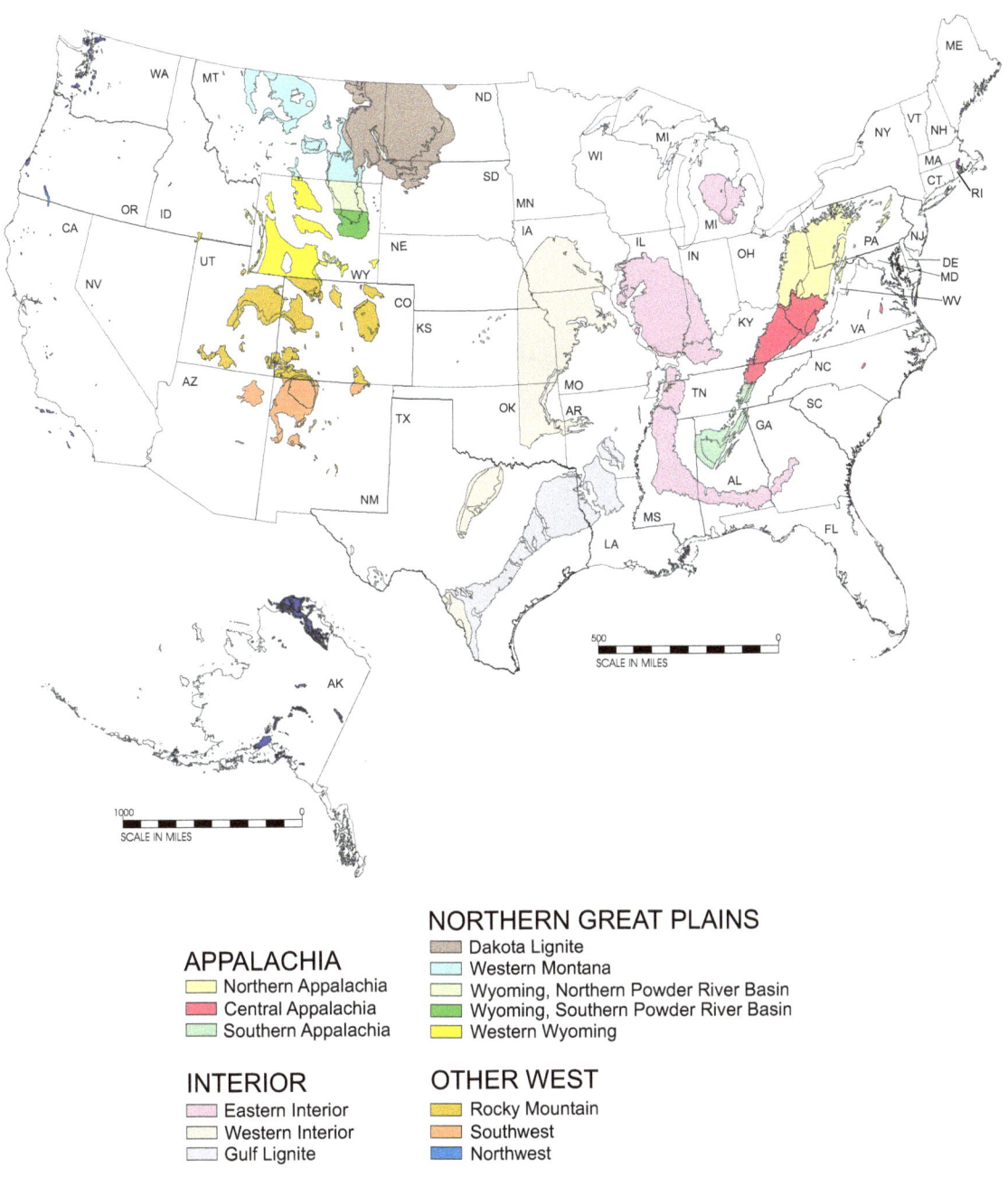

Source: U.S. Energy Information Administration, Office of Energy Analysis.

Regional maps

Figure F7. Coal demand regions

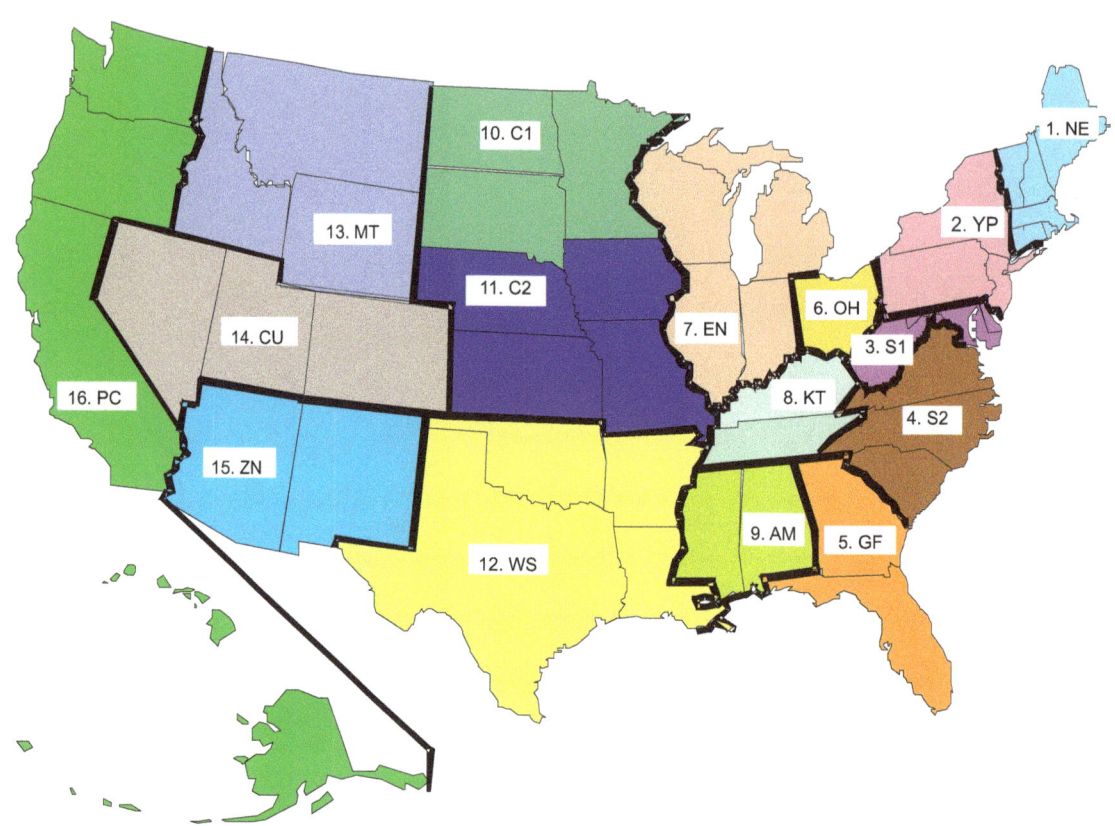

Region Code	Region Content
1. NE	CT,MA,ME,NH,RI,VT
2. YP	NY,PA,NJ
3. S1	WV,MD,DC,DE
4. S2	VA,NC,SC
5. GF	GA,FL
6. OH	OH
7. EN	IN,IL,MI,WI
8. KT	KY,TN

Region Code	Region Content
9. AM	AL,MS
10. C1	MN,ND,SD
11. C2	IA,NE,MO,KS
12. WS	TX,LA,OK,AR
13. MT	MT,WY,ID
14. CU	CO,UT,NV
15. ZN	AZ,NM
16. PC	AK,HI,WA,OR,CA

Source: U.S. Energy Information Administration, Office of Energy Analysis.

Appendix G
Conversion factors

Table G1. Heat rates

Fuel	Units	Approximate heat content
Coal[1]		
Production	million Btu per short ton	20.192
Consumption	million Btu per short ton	19.847
Coke plants	million Btu per short ton	26.297
Industrial	million Btu per short ton	20.433
Residential and commercial	million Btu per short ton	21.188
Electric power sector	million Btu per short ton	19.623
Imports	million Btu per short ton	24.719
Exports	million Btu per short ton	25.698
Coal coke	million Btu per short ton	24.800
Crude oil		
Production	million Btu per barrel	5.800
Imports[1]	million Btu per barrel	5.989
Petroleum products and other liquids		
Consumption[1]	million Btu per barrel	5.254
Motor gasoline[1]	million Btu per barrel	5.100
Jet fuel	million Btu per barrel	5.670
Distillate fuel oil[1]	million Btu per barrel	5.771
Diesel fuel[1]	million Btu per barrel	5.762
Residual fuel oil	million Btu per barrel	6.287
Liquefied petroleum gases[1]	million Btu per barrel	3.557
Kerosene	million Btu per barrel	5.670
Petrochemical feedstocks[1]	million Btu per barrel	5.510
Unfinished oils	million Btu per barrel	6.118
Imports[1]	million Btu per barrel	5.337
Exports[1]	million Btu per barrel	5.851
Ethanol	million Btu per barrel	3.561
Biodiesel	million Btu per barrel	5.359
Natural gas plant liquids		
Production[1]	million Btu per barrel	3.674
Natural gas[1]		
Production, dry	Btu per cubic foot	1,024
Consumption	Btu per cubic foot	1,024
End-use sectors	Btu per cubic foot	1,025
Electric power sector	Btu per cubic foot	1,022
Imports	Btu per cubic foot	1,025
Exports	Btu per cubic foot	1,009
Electricity consumption	Btu per kilowatthour	3,412

[1]Conversion factor varies from year to year. The value shown is for 2010.
Btu = British thermal unit.
Sources: U.S. Energy Information Administration (EIA), *Annual Energy Review 2010*, DOE/EIA-0384(2010) (Washington, DC, October 2011), and EIA, AEO2012 National Energy Modeling System run REF2012.D020112C.

THIS PAGE INTENTIONALLY LEFT BLANK

www.ingramcontent.com/pod-product-compliance
Ingram Content Group UK Ltd.
Pitfield, Milton Keynes, MK11 3LW, UK
UKHW051118200426
11947UKWH00043B/853